H. Kuzmany J. Fink M. Mehring S. Roth (Eds.)

Electronic Properties of Fullerenes

Proceedings of the International Winterschool on Electronic Properties of Novel Materials, Kirchberg, Tirol, March 6–13, 1993

With 238 Figures

Springer-Verlag
Berlin Heidelberg New York
London Paris Tokyo
Hong Kong Barcelona
Budapest

Professor Dr. Hans Kuzmany

Universität Wien, Institut für Festkörperphysik
Strudlhofgasse 4, A-1090 Wien, Austria

Dr. Jörg Fink

Kernforschungszentrum Karlsruhe, Institut für Nukleare Festkörperphysik
Postfach 3640, D-76021 Karlsruhe, Germany

Professor Dr. Michael Mehring

Universität Stuttgart, Physikalisches Institut II
Pfaffenwaldring 57, D-70550 Stuttgart, Germany

Dr. Siegmar Roth

Max-Planck-Institut für Festkörperforschung
Heisenbergstrasse 1, D-70569 Stuttgart, Germany

Series Editors:

Professor Dr., Dres. h. c. Manuel Cardona
Professor Dr., Dr. h. c. Peter Fulde
Professor Dr., Dr. h. c. Klaus von Klitzing
Professor Dr., Dres. h. c. Hans-Joachim Queisser

Max-Planck-Institut für Festkörperforschung, Heisenbergstrasse 1
D-70569 Stuttgart, Germany

Managing Editor:

Dr. Helmut K.V. Lotsch

Springer-Verlag, Tiergartenstrasse 17, D-69121 Heidelberg, Germany

ISBN-13: 978-3-642-85051-6 e-ISBN-13: 978-3-642-85049-3
DOI: 10.1007/978-3-642-85049-3

Library of Congress Cataloging-in-Publication Data. Electronic properties of fullerenes: proceedings of the international winterschool, Kirchberg, Tirol, March 6–13, 1993/[edited by] J. Fink ... [et al.]. p. cm. — (Springer seris in solid-state sciences: 117) "[Proceedings of] the International Winter School on the Electronic Properties of Fullerenes and Related Compounds" — Pref. Includes bibliographical references and index. ISBN 3-540-57415-8 (Berlin: acid-free). — ISBN 0-387-57415-8 (New York: acid-free) 1. Fullerenes—Electric Properties—Congresses. I. Fink, J. (Jörg), 1938– . II. International Winter School on the Electronic Properties of Fullerenes and Related Compounds (1993: Kirchberg am Wechsel, Austria) III. Series. QD181.C1E45 1993 546'.681—dc20 93-38404 CIP

Softcover reprint of the hardcover 1st edition 1993

Typesetting: Camera-ready copy from the authors/editors
54/3140 – 5 4 3 2 1 0 – Printed on acid-free paper

Preface

The *First International Winter School on the Electronic Properties of Fullerenes and Related Compounds* took place between March 6 –13, 1993, in Kirchberg (Tirol) Austria and, like its predecessors, was organized in cooperation with the "Bundesministerium für Wissenschaft und Forschung" in Austria, and with the "Bundesministerium für Forschung und Technologie" in Germany. Consistent with the ideology of the previous meetings, the goal was to provide an opportunity for experienced scientists from universities and industry to discuss their most recent results and for students and young scientists to become familiar with the present status of research and applications in this new but rapidly developing field.

Since 1991, research in the field of organic carbon materials has developed at a rapid pace due to the advent of the fullerenes and related materials. These forms of carbon are considered as a missing link between the previously discussed electroactive polymers and the oxidic superconductors. It was therefore challenging to select this topic for an international winter school in Kirchberg.

Although still in its infancy, research on the physics and chemistry of fullerenes and related compounds has already led to a wealth of results, which was reflected in the wide range of topics covered and the numerous discussions which emerged at the meeting. For C_{60} itself, preparation methods and crystal growth techniques continue to evolve, while the understanding of the electronic and structural properties of its solid state continues to pose challenges to experimental and theoretical physicists. The ever-expanding range of higher fullerenes and related materials, such as nanotubes and onions, poses a daunting but exciting task for researchers.

For synthetic chemists, fullerenes represent the basis of a whole new range of synthetic compounds. The prospect of a periodic table of endohedral fullerene complexes has been discussed, and exohedrally complexed metal-fullerenes have already attracted the attention of physicists. The first endohedral materials are now available.

The success of this winter school, the first to focus on fullerenes and related materials, resulted from the range of the topics of interest, as well as from the merging of the many disciplines represented by the participants. As pointed out in the final discussion, the fullerene is a unique material, not only due to its molecular structure and symmetry, but more so due to the diversity of phenomena observable in its solid state; structural phase transitions, glass transitions, metallic phases, superconducting phases, phototransformation and more, have already been observed. This diversity requires an interdisciplinary approach to the scientific research on these materials. We have met this challenge by providing an interdisciplinary discussion at this meeting.

This book summarizes the tutorial and research lectures as well as the poster displays presented at the winter school in Kirchberg. We acknowledge all the authors for their contributions and all those who took part in the discussions for their stimulating remarks, which were an essential ingredient to this exciting and informative event.

We are in particular grateful to the "Bundesministerium für Wissenschaft und Forschung" in Austria, the "Bundesministerium für Forschung und Technologie" in Germany and to the "US-Army Research, Development and Standardization Group", as well as to the many sponsors from industry, for their financial support. This support was more than just a great help; it was, in fact, indispensable in attaining the goal of the meeting.

Finally we are indebted, as ever, to the manager of the Hotel Sonnalp, Herr J.R. Jurgeit, and his staff for their continuous support and for their patience with the many special arrangements required during the meeting.

Vienna	*H. Kuzmany*
Karlsruhe	*J. Fink*
Stuttgart	*M. Mehring*
June 1993	*S. Roth*

Contents

Part IV Infrared and Raman Spectroscopy, Luminescence, Transient Optics, and NLO

Part VI Electrical Transport and Superconductivity

The Birth of C_{60}: Buckminsterfullerene

H. Kroto

School of Chemistry and Molecular Sciences, University of Sussex, Brighton BN1 9QJ, UK

Abstract. Almost exactly five years after C_{60}: Buckminsterfullerene (fullerene-60) was discovered serendipitously during a series of graphite laser vaporization experiments which were designed to simulate the chemistry in a red giant carbon star, the molecule has been isolated in macroscopic amounts. This breakthrough has triggered an explosion of research into its chemical and physical properties. The molecule has already exhibited a wide range of novel phenomena which promise exciting applications. Whether or not applications arise Buckminsterfullerene has a beauty and elegance that has excited the imaginations of scientists and laymen alike. It seems almost impossible to comprehend how the existence of the third well-characterised allotropic form of carbon could have evaded discovery until almost the end of the twentieth century. New fields of chemistry, physics and materials scince have been born and the articles contained in this volume cover some of the fascinating properties that have been uncovered and which pressage exciting implications for the future. This article surveys some of the key events which led to the birth of these new field and serves as an introduction to this volume.

1. Origins

The discovery of C_{60}: Buckminsterfullerene in 1985 and the extraction and characterisation of the molecule in 1990 are tributes to the value of interdisciplinary, fundamental science. The breakthroughs resulted from the dovetailing of advances in experimental techniques in the chemical physics of clusters with ideas which evolved from advances in our understanding of astrophysical processes involving inter/circum-stellar molecules and dust. The molecule had however been a twinkle in the eye of a few imaginative scientists some years beforehand. The earliest record of C_{60} in the literature is contained in an article by Eiji Osawa in KAGAKU (in Japanese) in 1970 [1]. Osawa conjectured that such a molecule would be stable and the following year he and Yoshida described it more fully in a book on aromatic molecules - again in Japanese [2]. A little later Bochvar and Gal'pern published a Hückel calculation on C_{60} [3] and in 1980 Davidson applied general group theoretical techniques to a range of highly symmetric molecules, one of which was C_{60} [4]. However, somewhat earlier in a remarkable article in 1966, David Jones (writing under the pseudonym of Daedalus in the NEW SCIENTIST) discussed the possibility of creating graphite balloons [5,6] similar to geodesic cages - these objects are essentially Giant

Springer Series in Solid-State Sciences, Volume 117
Electronic Properties of Fullerenes Eds.: H. Kuzmany, J. Fink, M. Mehring, and S. Roth

Fullerenes [7,8]. During the early 80's Orville Chapman at UCLA had probed some speculative synthetic procedures that promised possible routes to the molecule. The big surprise is however that the molecule forms spontaneously under our noses when 60 or so carbon atoms are allowed to aggregate during gas-phase nucleation. In general such conditions may occur relatively frequently, but they appear never previously to have been probed carefully enough. Often when such a situation has arisen many other processes tend to take place as well and so the fleeting, shadowy existence of this "Third Man" of the carbon family of allotropes has remained out of sight. The discovery had to wait until the right experiment could be carried out at the right time with the right instrument. In September 1985 this conjunction took place [9-11].

2. Discovery

Some long carbon chain molecules (the cyanopolyynes) which had originally been discovered in interstellar space during a Sussex/NRC(Ottawa) radioastronomy collaboration [12] were, by the early 80's, found to be streaming out of red giant carbon stars in particular the object IRC+10216, a mass losing star which is extremely bright in the infrared. At just this moment the laser vaporisation cluster beam technique developed by Rick Smalley and his co-workers at Rice University [13] was beginning to revolutionise the study of clusters by making small aggregates (with tens to hundreds of atoms) of refractory materials available for the first time. The clusters are produced in a supersonic expansion nozzle by a pulsed laser focused on a solid target under helium from a pulsed solenoid valve [13]. The skimmed beam is ionised by an excimer laser and the clusters detected by a time-of-flight mass spectrometer. When this technique was considered in the light of the likely circumstellar processes which produced carbon chains and particles it seemed to offer a way to simulate the nucleation conditions in a star such as IRC+10216 - if graphite were the target material in the apparatus. Graphite vaporisation in a Rice-type cluster beam apparatus had actually been carried out somewhat earlier by Rohlfing et al. [14] and Bloomfield et al [15] under conditions which had not highlighted the special nature of C_{60} - so the molecule was overlooked. A collaboration with Robert Curl and Rick Smalley at Rice University (Houston, Texas, USA) was initiated in 1985 and the experiments performed with graduate students Jim Heath, Sean O'Brien and Yuan Liu. The experiments immediately confirmed that long cyanopolyynes including those detected in space are formed in a plasma produced by a laser focused on a graphite target [16, 17]. During these experiments the serendipitous discovery was made that the C_{60} molecule was remarkably stable [9].

The result was rationalised on the basis of a closed fullerene cage structure consisting of 12 pentagons and 20 hexagons with the symmetry of a soccerball [9]. Geodesic and chemical factors inherent in such a structure seemed to account for the stability of the molecule and it was named Buckminsterfullerene after the designer/inventor of the geodesic domes. During the period 1985-90 unequivocal evidence for the exceptional stability of C_{60} was amassed which provided convincing evidence for the cage structure proposal [10,11,18].

Endohedral complexes were detected [19] and reactivities studied [20]. Other groups such as that of Busmann, Cambell, Hertel and co-workers who carried out detailed studies in Freiburg [21-23] concluded that the fragmentation behaviour was quite consistent with the fullerene proposal. In another systematic series of cluster beam measurements at the NRL in Washington, O'Keefe, Ross, Baronavski, McElvaney, Dunlap and their co-workers [24-28] obtained key electronic and dynamic information - including ionisation potentials etc. UPS spectra of the negative ion carbon clusters were measured at Rice [29-30]. All of these experimental studies added considerable weight to the fullerene structural proposal. A large number of theoretical papers also lent considerable weight to the proposal. Particularly important were those by Tom Schmalz, Doug Klein and their co-workers [31,32] who among other things, pointed out that as the pentagons in the Buckminsterfullerene (soccerball) isomer of C_{60} do not abut this isomer is likely to be more stable than all the other possible C_{60} isomers in which they cannot abut. The requirement that pentagons do not abut was subsequently recognised as having quantitative significance and the Pentagon Isolation Principle was formulated [33,34] which predicted the magic numbers observed experimentally - in particular the principle required that 60 and 70 be the first and second magic numbers [33,34]. Indeed using a generalised form of this rule, together with symmetry principles, a whole family of closed carbon cages was predicted to exist with 20+2n atoms (n = 0,2,3,4...) [33]. The study led to the conclusion that C_{28} should be very special and that tetravalent analogues such as $C_{28}H_4$ should be particularly stable [33,35]. Fowler and Woolrich discovered that fullerene cages, C_n, with n=60+6k (k=0,2,3,4..) should be electronically stable [36] - a fascinating analogue of Huckel's famous 4n+2 rule for aromatic compounds. Giant fullerenes with n = 240, 560, 960 etc. were studied and discovered to have beautiful quasi-icosahedral shapes [7].

A possible mechanism for the formation of C_{60} was proposed and included a suggestion that it might be a by-product of soot formation [20]. This prediction was verified when Gerhardt, Löffler and Homann in 1987 found that C_{60} was indeed readily detectable during the soot formation process [37].

3. Extraction

The original detection of C_{60} was made on minute amounts of the species in a beam using that most sensitive of detection devices - the mass spectrometer. During 1989-90 Wolfgang Krätschmer, Lowell Lamb, Kostas Fostiropoulos and Don Huffman (Max Planck Institut, Heidelberg, Germany and University of Arizona, Tucson, USA) put two camel humps together and got sixty [38,39]. They first presented a remarkable paper at an astronomy conference [39] in which they suggested that they might actually have detected the IR absorption signal of C_{60} in carbon dust deposited from by carbon arc struck under an inert gas. In August 1990 any remaining scepticism surrounding the assignment of these IR features, and of course the original fullerene-60 cage concept [9], was finally and emphatically laid to rest by their announcement that they had extracted macroscopic amounts of a soluble red material which contained

C_{60} and that the molecule was indeed a round pure carbon ball (38). Similar experiments were being carried at Sussex (with Jonathan Hare and Ala'a Abdul Sada) where it was also discovered independently that the arc-processed material showing the IR features gave rise to a prominent mass spectrometric line at 720 amu, consistent with C_{60} and that, furthermore, a benzene soluble red extract could be obtained from this material [40,35]. From a solution of the extract Krätschmer et al obtained crystals which were shown by X-ray analysis to consist of arrays of spherical molecules some 7Å in diameter - a result beautifully and unequivocally consistent with expectation for C_{60} [38]. At Sussex Roger Taylor showed that the soluble extract was chromatographically separable into two components, C_{60} and C_{70}. C_{60} gave an exquisitely delicate magenta solution. ^{13}C NMR analysis yielded a single resonance [40] which indicated that all the C atoms in the molecule are equivalent - a result perfectly commensurate with the proposed cage structure. Elegant further support for the fullerene proposal came from the ^{13}NMR spectrum of the C_{70} component (red in solution) which exhibited five lines, again in perfect agreement with expectation and most importantly proof of the existence of other members of the fullerene family. In a parallel investigation Meijer and Bethune [41,42] were able to identify C_{60} by mass spectrometry in the solid material deposited from a Smalley-type laser vaporisation nozzle. Subsequent studies by Johnson et al [43] confirmed the NMR measurement on C_{60} and Ajie et al [44] confirmed that the material could be chromatographically separated.

4. Epilogue

The fact that this new third form of carbon may be created spontaneously and has been under our noses since time immemorial is almost unbelievable. Since the fullerenes have been available in bulk quantities, the field has exploded and the chemical and physical properties of the new materials are now being investigated in minute detail. Already there are exciting results: the molecule undergoes a vast new range of novel reactions; it readily accepts and donates electrons suggesting battery applications; it forms high temperature superconducting compounds; complexes exhibit ferromagnetic behaviour. This volume has brought together reviews from groups which show how wide are the physico/chemical implications of the new material.

There is food for thought in the fact that this fascinating, new round world of synthetic chemistry and materials science was discovered as a consequence of the desire to understand the rôle of carbon in space and stars rather than the probing carbon's possible material applications. It was however dependent on the development of apparatus designed to probe clusters and develop an understanding of the ways in which atomic and molecular properties correlate with those of the bulk. Finally it is interesting to reflect on what new light the terrestrial experiments shed on the carbon content of space and stars. Although C_{60} appears to be unstable in the terrestrial environment it appears to be stable in beams where the conditions are similar to those in space. Thus it seems very likely that C_{60} is formed in those regions of space where polyynes and carbon dust are formed and there is reason to believe that it may be stable and a major constituent of the space between the stars [45,46]. It would certainly be a most fitting last chapter in the story of the discovery if it did turn out that C_{60} does indeed play a mysterious key role in space.

Acknowledgements. I am most grateful to the many co-workers and colleagues, particularly those named in the text who were most directly involved in the many aspects of the studies covered above. In addition to those mentioned in this short article I thank Michael Jura and Steve Wood. I particularly thank David Walton for advice.

References

1) Osawa, E. *Kagaku (Kyoto)* **1970**, *25*, 854-863 (in Japanese); *Chem. Abstr.* **1971**, *74*, 75698v.

2) Yoshida, Z.; Osawa, E. *Aromaticity*; Kagakudojin: Kyoto, 1971, 174-178, (in Japanese).

3) Bochvar, D. A.; Gal'pern, E. G. *Dokl. Akad. Nauk SSSR* **1973**, *209*, 610-612 (English translation *Proc. Acad. Sci. USSR* **1973**, *209*, 239-241).

4) Davidson, R. A. *Theor. Chim. Acta.* **1981**, *58*, 193-195.

5) Jones, D. E. H. *New. Sci.*, *32* (3 November 1966), p. 245.

6) Jones, D. E. H. *The Inventions of Daedalus*; Freeman: Oxford, 1982; pp. 118-119.

7) Kroto, H. W.; McKay, K. G. *Nature (London)*, **1988**, *331*, 328-331.

8) Kroto, H. W. *Chem. Brit.* **1990**, *26*, 40-45.

9) Kroto, H. W.; Heath, J. R.; O'Brien, S. C.; Curl, R. F.; Smalley, R. E. *Nature (London)* **1985**, *318*, 162-163.

10) Kroto, H. W. *Science* **1988**, *242*, 1139-1145.

11) Curl, R. F. and Smalley, R. E. *Science* **1988**, *242*, 1017-1022.

12) Kroto, H. W. *Chem. Soc. Rev.* **1982**, *11*, 435-491.

13) Dietz, T. G.; Duncan, M. A.; Powers, D. E.; Smalley, R. E. *J. Chem. Phys.* **1981**, *74*, 6511-6512.

14) Rohlfing, E. A.; Cox, D. M.; Kaldor, A. *J. Chem. Phys.* **1984**, *81*, 3322-3330.

15) Bloomfield, L. A.; Geusic, M. E.; Freeman, R. R.; Brown, W. L. *Chem. Phys. Lett.* **1985**, *121*, 33-37.

16) Heath, J. R.; Zhang, Q.; O'Brien, S. C.; Curl, R. F.; Kroto, H. W.; Smalley, R. E. *J. Am. Chem. Soc.* **1987**, *109*, 359-363.

17) Kroto, H. W.; Heath, J. R.; O'Brien, S. C; Curl, R. F.; Smalley, R. E. *Astrophys. J.* **1987**, *314*, 352-355.

18) Kroto, H. W.; Balm, S. P.; Allaf, A. W., *Chem Revs* **1991**, *91*, 1213

19) Heath, J. R.; O'Brien, S. C.; Zhang, Q.; Liu, Y.; Curl, R. F.; Kroto, H. W.; Smalley, R. E. *J. Am. Chem. Soc.* **1985**, *107*, 7779-7780.

20) Zhang, Q. L.; O'Brien, S. C.; Heath, J. R.; Liu, Y.; Curl, R. F.; Kroto, H. W.; Smalley, R. E. *J. Phys. Chem.* **1986**, *90*, 525-528.

21) Campbell, E. E. B.; Ulmer, G.; Hasselberger, B.; Busmann, H. -G.; Hertel, I. V. *J. Chem. Phys.* **1990**, *93*, 6900-6907.

22) Hasselberger, B; Busmann, H. -G.; Campbell, E. E. B. *Appl. Surf. Sci.* **1990**, *46*, 272-278.

23) Campbell, E. E. B.; Ulmer, G.; Busmann; H. -G.; Hertel, I. V. *Chem. Phys. Lett.* **1990**, *175*, 505-510.

24) O'Keefe, A.; Ross, M. M.; Baronavski, A. P. *Chem. Phys. Lett.* **1986**, *130*, 17-19.

25) McElvany, S. W.; Nelson, H. H.; Baronavski, A. P.; Watson, C. H.; Eyler, J. R. *Chem. Phys. Lett.* **1987**, *134*, 214-219.

26) McElvany, S. W.; Dunlap, B. I.; O'Keefe, J. *J. Chem. Phys.* **1987**, *86*, 715-725.

27) Zimmerman, J. A.; Eyler, J. R.; Bach, S. B. H.; McElvany, S. W. *J. Chem. Phys.*; **1991**, *94*, 3556-3562.

28) McElvany, S. W. *Int. J. Mass Spectrom. Ion Process* **1990**, *102*, 81-98.

29) Yang, S. H.; Pettiette, C. L.; Conceicao, J.; Cheshnovsky, O.; Smalley, R. E. *Chem. Phys. Lett.* **1987**, *139*, 233-238.

30) Cheshnovsky, O.; Yang, S. H.; Pettiette, C. L.; Craycraft, M. J.; Liu, Y.; Smalley, R. E. *Chem. Phys. Lett.* **1987**, *138*, 119-124.

31) Schmalz, T. G.; Seitz, W. A.; Klein, D. J.; Hite, G. E. *Chem. Phys. Lett.* **1986**, *130*, 203-207.

32) Klein, D. J.; Schmalz, T. G.; Hite, G. E.; Seitz, W. A. *J. Am. Chem. Soc.* **1986**, *108*, 1301-1302.

33) Kroto, H. W. *Nature (London)* **1987**, *329*, 529-531.

34) Schmalz, T. G.; Seitz, W. A.; Klein, D. J.; Hite, G. E. *J. Am. Chem. Soc.* **1988**, *110*, 1113-1127.

35) Kroto, H. W, *Angewandte Chemie* **1991**, *31*, 111-129

36) Fowler, P. W.; Woolrich, J. *Chem. Phys. Lett.*, **1986**, *127*, 78-83.

38) Gerhardt, Ph.; Loeffler, S.; Homann, K. *Chem. Phys. Lett.*, **1987**, *137*, 306-310.

37) Krätschmer, W.; Lamb, L. D.; Fostiropoulos, K.; Huffman, D. R. *Nature (London).*, **1990**, *347*, 354-358.

39) Krätschmer, W.; Fostiropoulos, K.; Huffman, D. R. *Dusty Objects in the Universe*, Bussoletti, E.; Vittone, A. A. eds.; Kluwer : Dordrecht, 1990 (Conference in 1989).

40) Taylor, R.; Hare, J. P.; Abdul-Sada, A. K.; Kroto, H. W. *J. Chem. Soc. Chem. Commun.* **1990**, 1423-1425.

41) Meijer, G.; Bethune, D. S. *Chem. Phys. Lett.* **1990**, *175*, 1-2.

42) Meijer, G.; Bethune, D. S. *J. Chem Phys.* **1990**, *93*, 7800-7802.

43) Johnson, R. D.; Meijer, G.; Bethune, D. S. *J. Am. Chem. Soc.* **1990**, *112*, 8983-8984.

44) Ajie, H.; Alvarez, M. M.; Anz, S. J.; Beck, R. D.; Diederich, F.; Fostiropoulos, K.; Huffman, D. R.; Kraetschmer, W.; Rubin, Y.; Schriver, K. E.; Sensharma, K.; Whetten, R. L. *J. Phys. Chem.* **1990**, *94,* 8630-8633.

45) Kroto, H. W.; Jura, M.; *Astron. Astrophys.,* in press.

46) Hare J. P.; Kroto, H. W. *Acc Chem Res,* **1992**, **25**, 106-112

Part I

Chemistry and Sample Preparation

Fullerene Intercalation Compounds

M.J. Rosseinsky[1,2], *D.W. Murphy*[2], *A.P. Ramirez*[2], *R.M. Fleming*[2], *and O. Zhou*[2]

[1]Inorganic Chemistry Laboratory, University of Oxford, South Parks Road, Oxford OX1 3QR, UK
[2]A.T.&T. Bell Laboratories, 600, Mountain Avenue, Murray Hill, NJ, 07974, USA

The past two decades have seen considerable research directed towards the synthesis and characterisation of molecular conductors. These systems have been thought of as characteristic 'low dimensional' conductors, due to the one dimensional chains or two dimensional planes formed by the flat organic components in their crystal structures. This low dimensionality makes many of these systems susceptible to electronically driven metal-to-insulator transitions, such as the Peierls or spin density wave. The synthesis of a quasi-spherical, redox active π-system (C_{60}) has resulted in the first class of three dimensional molecular conductors, A_3C_{60}. In this short article, we briefly address some recent developments in the physics and chemistry of these systems.

Solid solutions $A_{3-x}A'_xC_{60}$ have superconducting transition temperatures which increase monotonically with the fcc lattice parameter, a, from 8.0K at $Na_2Rb_{0.5}Cs_{0.5}C_{60}$ to 31.3K at Rb_2CsC_{60} [1],[2],[3]. Application of hydrostatic pressure to K_3C_{60} or Rb_3C_{60} reduces T_c, [4],[5] and the $T_c(a)$ relationship resulting from combined susceptibility and lattice parameter measurements under pressure is in good agreement with the "chemical pressure" for lattice parameters between 14.5 and 14.15Å [6]. The variation of T_c with lattice parameter is qualitatively consistent with a BCS-like dependence of T_c on the density of states at the Fermi energy, $N(E_f)$

$$T_c = E\exp(-1/\lambda) \quad \lambda = VN(E_f)\ .$$

Increasing the inter-C_{60} separation decreases the overlap between neighbouring t_{1u} wavefunctions, increasing $N(E_f)$ [1]. This does *not* identify the pair-mediating excitation E as a phonon: purely *electronic* pairing mechanisms also predict such a dependence on $N(E_f)$.

Na_2CsC_{60} has a T_c similar to that found for K_3C_{60} under 7kbar pressure, as expected for its lattice parameter of 14.132Å. However, the correlation between chemical

Springer Series in Solid-State Sciences, Volume 117
Electronic Properties of Fullerenes Eds.: H. Kuzmany, J. Fink, M. Mehring, and S. Roth

and hydrostatic pressure breaks down for Na_2RbC_{60}, which has $T_c \leq 3.5K$, considerably lower than expected for a = 14.091Å. High quality X-ray data clearly show that the C_{60}^{3-} anions in the anomalous Na_2RbC_{60} system do not adopt the merohedrally disordered orientation with their C_2 axes along the unit cell vectors in space group $Fm\bar{3}m$, found for K_3C_{60} by Stephens et al. [7]. Instead, the existence of primitive cubic reflections shows the anions adopt the $Pa\bar{3}$ structure [8] found for C_{60} itself below its orientational ordering phase transition [9],[10]. Refinement also confirms chemical ordering of Na^+ on the tetrahedral and Rb^+ on the octahedral sites. Differences in the details of the band structure resulting from the modified t_{1u} wavefunction overlap in the $Pa\bar{3}$ structure could reduce $N(E_f)$ sufficiently to account for the marked suppression of T_c in Na_2RbC_{60}, compared with that expected for an $Fm\bar{3}m$ phase of the same lattice parameter. Detailed normal state measurements on Na_2RbC_{60} would now be extremely useful to confirm this interpretation.

Our present knowledge of *normal state* vibrational and electronic properties is sufficient to make an assessment of the applicability of an electron-phonon coupling model to the A_3C_{60} systems. Raman [11] and inelastic neutron scattering [12] indicate the existence of symmetry-determined electron-phonon coupling of the t_{1u} electrons to the 8 H_g modes of the C_{60} molecule, of both low frequency radial and high frequency tangential character. Quantitative estimates of λ from the line broadening [13],[14] are in reasonable accord with theoretical calculations, but proponents of an electronic mechanism would argue that the existence of symmetry determined electron-phonon coupling does *not* prove that such coupling drives the superconductivity.

Measurement of $N(E_f)$ across the isostructural series is critical to the construction of a self-consistent model of the normal and superconducting states within a BCS-like framework, which requires that the normal state can be described as a Fermi liquid. To within 5%, the paramagnetic susceptibility of K_3C_{60} and Rb_3C_{60} is temperature independent, and yields a density of states, N_χ, of 14±1 states eV^{-1} $spin^{-1}$ $C_{60}{}^{-1}$ for K_3C_{60} and 19±0.6 states eV^{-1} $spin^{-1}$ $C_{60}{}^{-1}$ for Rb_3C_{60} [15]. This is qualitatively consistent with the dependence of T_c on lattice parameter. If we assume that N_χ is equal to N_b, the unenhanced density of states which determines T_c, application of the McMillan equation

$$T_c = \frac{\Omega_{ph}}{1.2} \exp\left(\frac{-1.04(1+\lambda_{ph})}{\lambda_{ph}-\mu^*(1+0.62\lambda_{ph})}\right)$$

$(\lambda_{ph} = N_b V)$,

with a conventional value of $\mu^* \simeq 0.1$-0.2, requires that Ω_{ph} is low ($\simeq$200K) and there is strong electron-phonon coupling ($\lambda_{ph} \simeq 2$). Although this interpretation is consistent with point-contact tunnelling measurements of the gap [16] $\left(\frac{2\Delta}{k_B T_c} = 5.3\right)$, the specific heat jump at $T_c(\Delta C)$ is too small to be consistent with such strong coupling [15].

The upper limit consistent with ΔC is $\lambda = 1$. For the normal state susceptibility to be consistent with this, electronic correlations must significantly enhance N_χ over N_b. Within the Fermi liquid picture, $N_\chi = \frac{N_b}{1+F_o{}^a}$, where $F_o{}^a = -IN_b$ is the spin antisymmetric part of the quasiparticle interaction, and I the Stoner parameter. All the data can then be rationalised within a weak-coupling picture for $\lambda_{ph} \simeq 0.5$ for K_3C_{60}, assuming V is a constant. The Stoner parameter I = 0.07 ($F_o{}^a = -0.42$, K_3C_{60}) and N_b (K_3C_{60}) = 6 states eV^{-1} $spin^{-1}$ $C_{60}{}^{-1}$ while band narrowing on expansion gives N_b (Rb_3C_{60}) = 8 states eV^{-1} spin $C_{60}{}^{-1}$. Recent EPR, SQUID [17] and ^{13}C nmr T_1T [18] measurements are consistent with this interpretation.

The narrow bandwidths and resulting low Fermi temperatures (T_F) suggest that possible temperature dependence of the susceptibility should be considered. A paramagnon theory [19] gives

$$\frac{\chi(T)}{\chi(T{=}O)} = 1 - \frac{3.2\pi^2}{24K^2}\left(\frac{T}{T_F}\right)^2$$

where $K = 1 - IN_b$. If the susceptibility is identified with an unenhanced (K = 1) density of states in a square band, the bandwidth $W = \frac{3}{2x14}$ eV (K_3C_{60}) and $T_F = \frac{3}{2x14}$ eV = 1250K (assuming half-filling). An 8% reduction of χ between 300K and 30K is then predicted. The correlated Fermi liquid, required for the weak coupling scenario outlined in the previous paragraph, has $T_F = \frac{3}{6x2}$ eV = 2900K, K = 0.6 and would display as 5% rise in susceptibility over this temperature range. It is very important to note that the lattice parameters of both K_3C_{60} and Rb_3C_{60} contract by approximately 0.1Å between 300K and 30K [20]. This should broaden the t_{1u} band, decreasing $N(E_f)$ on cooling and counteracting the expected rise in χ at low temperature. Given the small temperature dependence predicted theoretically and the opposing effect of thermal contraction, it seems that the absence of a strong temperature dependence of the susceptibility does not pose a major obstacle to the Fermi liquid model for the t_{1u} electrons in this system.

The large octahedral site makes face-centred cubic C_{60} a versatile intercalation host. The 'hard sphere' radius of this site is approximately 2.06Å, which allows the isolation of some unusual species. Na_{n+1}^{n+} (n = 3,5) clusters have been postulated in the sodalite cages of zeolites on the basis of EPR spectroscopy, but recent powder neutron diffraction measurements do not provide evidence for their existence [21]. Sodium cluster cations are found on the octahedral site of C_{60}, however. Na_6C_{60} has tetrahedral Na_4 clusters with a shortest Na-Na distance of 2.8Å [2], while a body centred Na_9 cube occurs in $Na_{11}C_{60}$ [22]. The extent of Na-Na covalency and Na-C charge transfer in systems containing these cluster species is an important issue, as it has implications for the position of the Fermi energy in superconducting Ca_5C_{60}, which also has cluster cations on the octahedral site [23].

The cation ordering in Na_2CsC_{60} is identical to that found for Na_2RbC_{60}. Intercalation of ammonia, driven by coordination to the small Na^+ cation, alters the cation distribution [24]. $(NH_3)_4Na_2CsC_{60}$ has one sodium and one cesium cation disordered over the tetrahedral sites. The remaining sodium cation now occupies the octahedral site as part of an $Na(NH_3)_4^+$ complex cation. Refinement of powder X-ray data indicates that the ammonia molecules occupy positions displaced along the <111> directions from the centre of the octahedral site at vertices of a cube, consistent with a tetrahedral geometry for the complex. Intercalation of the four ammonia molecules increases the lattice parameter from a = 14.132Å to a = 14.473Å. This produces an increase of T_c from 10.5K to 29.6K consistent with the T_c (a) relationship found for the ternary systems [1]. This system demonstrates that increasing the chemical complexity of species intercalated into the fullerene host enhances the possibilities for tuning the electronic properties of the resulting materials.

References

[1] R.M. Fleming et al., Nature 352, 787, 1991.
[2] M.J. Rosseinsky et al., Nature 358, 416, 1992.
[3] K. Tanigaki et al., Nature 356, 419, 1992.
[4] J.E. Schirber et al., Physica C 178, 137, 1991.
[5] G. Sparn et al., Science 252, 1829, 1991.
[6] O. Zhou et al., Science 255, 833, 1992.
[7] P.W. Stephens et al., Nature 351, 632, 1991.
[8] J.E. Schirber et al., preprint.
[9] P.A. Heiney et al., Phys. Rev. Lett. 66, 2911, 1991.
[10] W.I.F. David et al., Nature 353, 147, 1991.
[11] S.J. Duclos et al., Science 254, 1625, 1992.
[12] K. Prassides et al., Nature 354, 462, 1991.
[13] M.G. Mitch, S.J. Chase and J.S. Lannin, Phys. Rev. Lett. 68, 883, 1992.
[14] C. Christides et al., Europhysics Letters **19**, 629, 1992.
[15] A.P. Ramirez, M.J. Rosseinsky, D.W. Murphy and R.C. Haddon, Phys. Rev. Lett. 69, 1687, 1992.
[16] Z. Zhang et al., Nature 353, 333, 1991.
[17] A. Janossy et al., preprint.
[18] G. Quirion et al., Europhysics Letters **21**, 233, 1993.
[19] M.T. Beal-Monod, S.-K. Ma and D.R. Fredkin, Phys. Rev. Lett. 20, 929, 1968.
[20] R.M. Fleming, unpublished.
[21] P.A. Anderson, P.P. Edwards and A.R. Armstrong, ISIS Annual Report 1992, A38.
[22] T. Yildirim et al., Nature 380, 568, 1992.
[23] A.R. Kortan et al., Nature 355, 529, 1992.
[24] O. Zhou et al., Nature 362, 433, 1993.

Chemical and Structural Aspects of the Interaction Between C_{60} and Molecular Oxygen

H. Werner, M. Wohlers, D. Bublak, Th. Belz, W. Bensch, and R. Schlögl

Institut für Anorganische Chemie, Universität Frankfurt,
Niederurseler Hang, D-60439 Frankfurt, Germany

ABSTRACT: The spontaneous interaction between molecular oxygen and C_{60} was studied in the regimes of reversible and irreversible bonding. The molecular nature of oxygen intercalated at 300 K was shown by an isotope labelling experiment. Other gases such as nitrogen, argon, helium are intercalated with the same energetics as found for oxygen. The room temperature single crystal structure of C_{60} was determined showing the location of oxygen. A possible involvement of intercalated oxygen acting as blocking agent for the postulated free rotation and the parameters of spontaneous oxygen incorporation into fullerenes will be discussed.

1. INTRODUCTION

The reaction of oxygen with carbon in the form of graphite, carbon black or diamond is characterised by a complex sequence of interactions involving chemisorption, dissociation in several steps and product formation. All these steps are kinetically influenced by the chemical anisotropy of the substrates, they are topochemically and defect-controlled reactions[1]. Topochemical control of a heterogeneous reaction means that its rate is locally different depending on crystal orientation and local modifications of geometry and composition. Fullerenes, the third form of carbon offer the possibility to exclude the topochemical effects in the reaction mechanism as C_{60} is a van-der-Waals solid with only little energy stored in the stacking order defects[2].

Oxidation of C_{60} was studied with powder samples by IR analysis[3], by thermogravimetry[4] and with thin films in UHV by photoemission[5]. The intercalation of nitrogen and oxygen into C_{60} was detected by XPS[6] and a weak non-dissociative

Springer Series in Solid-State Sciences, Volume 117
Electronic Properties of Fullerenes Eds.: H. Kuzmany, J. Fink, M. Mehring, and S. Roth

interaction of the gases with carbon was concluded. High pressure exposures of C_{60} to oxygen and hydrogen with subsequent ^{13}C NMR investigation showed the octahedral voids in the fullerene lattice to be accessible for intercalation[7]. The authors were unable to find chemically bound atomic oxygen in their samples. We studied the initial reaction of C_{60} films with oxygen by X-ray absorption spectroscopy[8] and found evidence for the presence of molecular oxygen ($\pi*$ resonance at 532.5 eV close to the value of gaseous oxygen, molecular Rydberg states) with no detectable bonding interaction to the carbon valence electron system.

Formation of epoxides of C_{60} under mild conditions[9] and the detection of covalent C–O bonds[3] document, on the other hand, the possible attack of oxygen on the fullerene molecules. Such classical oxidations were also made responsible for the (photo)-degradation of solutions and solid C_{60} reported earlier[10]. In a recent detailed Raman study of photo-assisted structural transitions and oxygen diffusion in solid C_{60} films[11] it was noted that the presence of oxygen reduced the sensitivity of C_{60} against polymerisation without influencing the chemical bonding within single molecules to any significant extent. In a previous report[12] solid C_{60} was found, on the other hand, to be converted at about 473 K into an amorphous solid with similar structural parameters as oxidised graphitic carbon with an approximate composition of $C_{60}O_{12}$.

It was assumed that within the C_{60} crystals the highly symmetric buckyballs rotate freely at 300 K, a picture supported by NMR[13,14] and by X - ray diffraction studies[15]. This would preclude structure analysis at atomic resolution using diffraction methods. Therefore, in a pioneering study[16] rotation inhibiting OsO_4 units were attached onto the C_{60} units or recently a structure determination was carried out at very low temperature[17] where no molecular dynamics is possible. The purpose of the present study was to test if the molecular structure can be solved using conventional crystallographic methods with good quality crystals. Using the sublimation technique and pure C_{60} starting material we obtained about 0.1 mm long single crystals absolutely solvent-free[18] and suitable for X-ray analysis. A total of 25 crystals were investigated with rotation photographs showing frequently twinning of equally large individuals. Only crystals without this twinning were further investigated at 300 K.

The single crystal structure of the OsO_4 derivative of C_{60} yielded a football made up from flat pentagons and hexagons with a radius of 351.2 (3) pm and two types of C-C bonds of 138.8 (9) pm and 143.2 (5) pm length[16]. The synchrotron powder diffraction study[15] of pure C_{60} at 300 K was analysed with a model of freely rotating carbon atoms describing a sphere of charge with a radius of 352(1) pm. The authors further reported on a phase transition at 249 K from a centered F lattice into a simple cubic P lattice. A phase transition was also observed in NMR studies[13,14] which concluded in the high temperature form carbon atoms to be present in either rotationally narrowed mobile form or in fixed positions with a jump rotational motion. In both NMR studies the phase transition temperature was lower than 250 K and in one study[13] C_{60} was found to exhibit a strong EPR signal although the pure solid is an insulating diamagnet. These observations point to the presence of impurities in various samples influencing possibly the solid state dynamics.

One reason for studying the reaction of oxygen with carbons is to understand the chemistry of this fundamental gas-solid interface reaction for which fullerenes are perfect model compounds. The other reason is to understand the material properties of fullerenes, their chemical and structural integrity and their reactivity under ambient conditions. Such data are essential for planning and interpretation of all kinds of experiments. The fact that oxidation of fullerenes is a gas-solid interface reaction allows a drastic influence of sample properties and crystallinity on the kinetics of the reaction which leads to controversial views about the stability of fullerenes in air. The collection of experiments presented here is to illustrate the phenomenology of the oxygen-C_{60} interaction.

2. EXPERIMENTAL

The fullerene was prepared according to the Krätschmer method[19] in a modified commercial BUCKY II system using spectroscopic graphite rods (dia. 6 or 12 mm). The resulting fullerene black was soxhlet-extracted with toluene and C_{60} was isolated by column chromatography using alumina as stationary phase and hexane/toluene as solvent. Purity was ascertained by UV-VIS spectroscopy and by powder X-ray diffraction of the crystallized and dried samples. Organic solvents, still present in the C_{60} crystals were removed by heating the samples in high vauum to 670 K. One sample was used in this state for the experiments (sample A), another sample was sublimed in a quartz tube at 10^{-6} mbar and used after this final purification stage (sample B).

TPD experiments were carried out in a quartz cell attached to an UHV chamber equipped with a mass spectrometer and gas dosing facilities. The apparatus is described elsewhere[20]. Sample B was used as prepared, sample A was sublimed within the quartz cell at 850 K and $1x10^{-8}$ mbar pressure. The walls of the cell were then cleaned from C_{60} which was collected as a pile in the center of the tube. Ca. 100 mg of samples were used. The precisely linear heating ramps were 1 K s^{-1} if not otherwise stated. The intensities of the spectra are given in arbitrary units representing the ion current in Amperes. Exposures with 99.9995% oxygen were conducted for 30 min at atmospheric pressure and the sample temperatures stated. After 30 min pumping time the base pressure of the experiment of $1x10^{-8}$ mbar was reached. Samples were cooled to 300 K after exposure in oxygen. The $^{18}O_2$ source was at 470 mbar pressure and connected via glass joints to the samples. Dead volumes of normal oxygen caused the abundance of $^{18}O_2$ to be only ca. 65%.

Single crystal X-ray studies were performed on a CAD-4 instrument at 300 K (Cu ka radiation) and on a STOE AED II instrument in the temperature range of 100 K to 300 K (Mo kα radiation). The known[15,21] phase transition at 250±10K was found with all crystals at 256±3K. It is associated with a lowering of the crystal symmetry from F to P as concluded from the intensity changes with temperature of several reflections of the P system not allowed in F symmetry. Irrespective of the heating/cooling rate these changes occur reversibly and in a narrow temperature range. The densities of several well-characterised[22] crystals were determined by the flotation method and found to be 1.765 gcm^{-3} as an average over 7 crystals with values ranging from 1.74 to 1.77 gcm^{-3}. Analysis of single crystals for heavy elements by EDX showed no appreciable contamination with e.g. alkali or silicon. All handling with the material was done under Ar and structure determinations were carried out with crystals sealed under Ar in Lindemann capillaries.

Infrared spectra were recorded from powder samples with a BRUKER ISF 66 FT instrument using the pellet technique pressed with highly purified KBr in a glove box.

3. INTERCALATION EXPERIMENTS

Using sample A we first illustrate that oxygen is reversibly intercalated into C_{60}. Figure 1 shows a TPD experiment after exposure of a fresh specimen at 300 K. The first heating causes desorption in a two-peak structure centered at 370 K and 430 K. Subsequent heating in a second experiment causes no further desorption. The surface area of the samples was determined to be below 1 m^2g^{-1} and no porosity was seen in the shape of the nitrogen BET isotherm. This and the intensity of the desorption signal

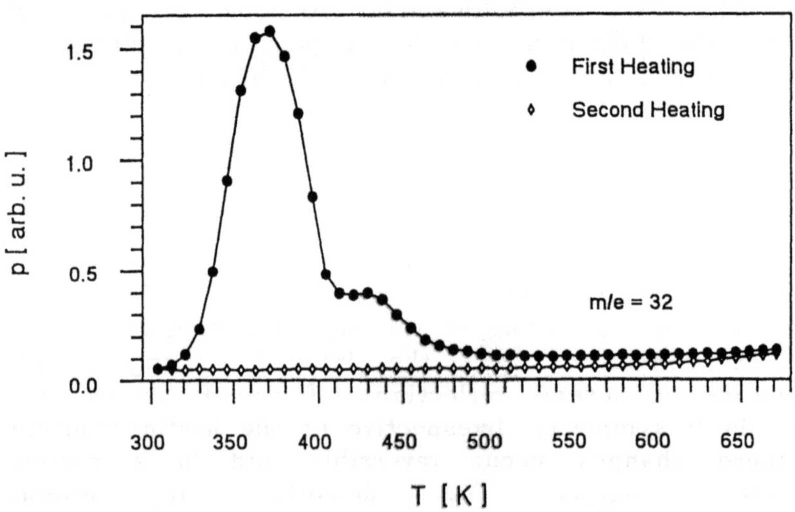

FIG.1: TPD of sample A after exposure to oxygen at 300 K.

exclude the location of the oxygen on the outer surface of the sample, the gas is not adsorbed but rather intercalated in the solid.

Reversibility of the absorption (intercalation) process requires the absence of oxidation of the host lattice detectable as CO/CO_2 desorption. This condition is fullfilled above 700 K is not possible as then significant sublimation sets in which prevents reproduction of the experiments. The mass fragments above 600 K are due to decomposition of oxygen adducts[12,23] of C_{60}.

The intensities of the m/e 28 and m/e 44 peaks indeed suggest the presence of a source of oxygen which is not the residual gas or the quartz cell as was tested with blank experiments. The resolution of the m/e 32 signal is worse as in Figure 2 as the heating ramp was not linear but stepped in this multi-mass experiment. The absence of oxidation after oxygen exposure prevails up to a critical temperature of 570 K above which formation set in of CO and CO_2.

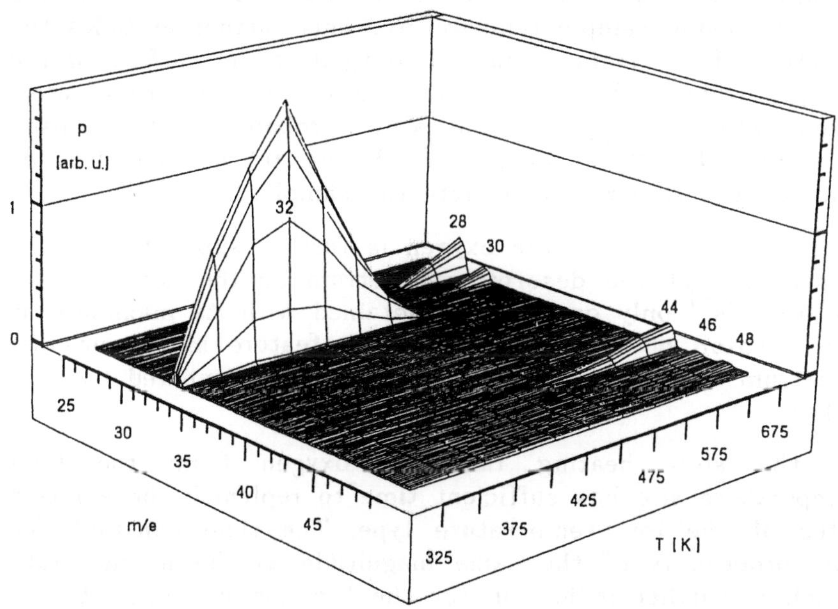

FIG. 2: Multi-mass TPD experiment with sample A showing the absence of any oxidation product desorption during release of intercalated oxygen.

The presence of two peaks in the desorption spectrum indicates the existence of two energetically different locations for oxygen. The alternative possibility of two chemically different kinds of oxygen is ruled out by the small difference in desorption energy. The high temperature desorption peaks grows relative to the low temperature feature with increasing sample temperature during exposure. At 470 K the two features are of equal intensity and the high temperature peak becomes asymmetric towards even higher temperatures. With increasing number of exposures the intensity under the high temperature peak also increased at a given exposure temperature. These findings indicate that the high temperature feature is a significantly thermally activated process whereas the low temperature process requires an activation energy close to 300 K (its intensity grew by a factor of 3.5 less than that of the high temperature peak with increasing sample temperature). The ageing effect indicates a morphological modification (e.g. defects from thermal stress during repeated thermal loads) of the sample reducing the activation barrier for the high temperature process. Both desorption features cannot be saturated i.e. their intensities grow with increasing exposure and increasing sample temperature. This further excludes the location of the main amount of oxygen at the surface of the sample which exhibits a *finite* number of adsorption sites. Even with a very low sticking probability the chosen exposure of $2x10^9$ Langmuir of the ca. 0.1 m^2 sample area should be sufficient to saturate the sample.

The bulk location of the oxygen is further supported by the dependence of the desorption spectrum on the heating rate. At 0.25 Ks^{-1} only one peak is obtained with its maximum at the position of the low temperature feature in the original spectrum. In Figure 3 the two spectra are compared to each other.

At the slow heating rate the oxygen from the high temperature site has sufficient time to replenish the emptied sites of the low temperature type. The time constant for this process is of the same magnitude as the heating rate which is further indication for the location of both sites in the bulk of the C_{60} crystals.

If both sites are within the bulk of the samples at least one of them should be susceptible to the defect

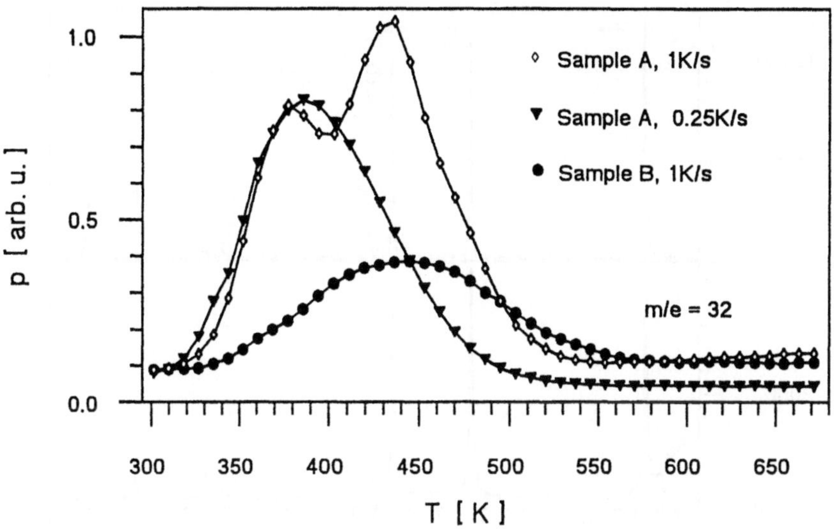

FIG. 3: TPD spectra after exposures to 450 K: trace 1: sample A at 1 Ks^{-1}, trace 2: sample A at 0.25 Ks^{-1}, trace 3: sample B at 1 Ks^{-1} Mass 32 was recorded.. Trace 1 was offset by 5 x 10^{-10} intensity units.

concentration in the crystals. This concentration is vastly different in samples A and B as can be seen from their powder diffraction patterns displayed in Figure 4. The extra intensity at the (111) reflection indicates the abundance of c-axis stacking order defects.

The oxygen desorption trace from sample B is shown also in Figure 3. With this information it is possible to assign the low temperature site to lattice defects and the high temperature sites to genuine voids in the regular C_{60} lattice. This assignment is also in line with the reported NMR data in ref.7 as their grinding procedure will probably have also increased the number of defects in the sample. The possibility of surface vs. bulk location put forward in their work can be ruled out on the basis of the present data.

The bulk localisation of the oxygen explains also the spurious oxygen source noted in the interpretation of Figure 2. TPD is in such cases a non-quantitative method as with varying temperatures no steady state of desorption is reached. This means that with an increasing number of experiments a reservoir of deep lying oxygen is slowly filled

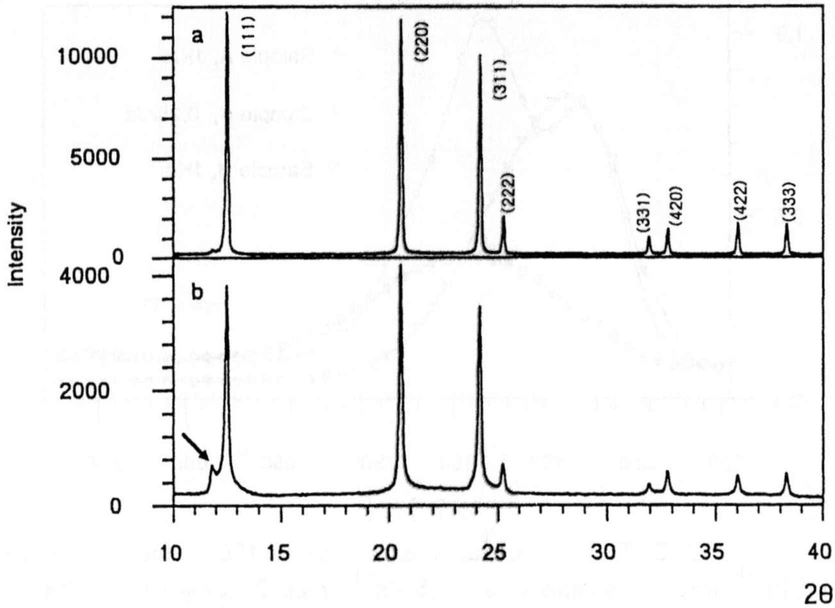

FIG. 4: Powder X-ray diffractograms (Co Kα radiation) of samples A and B illustrating their different crystal quality. Lattice parameters were determined to be 1416.7 pm for sample A and 1417.1 pm for sample B. The average FWHM of the reflections were 0.12° for sample A and 0.16° for sample B.

up from which the oxygen molecules cannot desorb quantitatively within one heating cycle. They will, however, be active at high temperatures and at isothermal conditions and oxidise their host lattice. In fresh materials this reservoir may be empty (from the preceeding clean-up) and thus no desorption is seen in the second heating cycle mentioned in Figure 1.

In order to elucidate the assumed molecular nature of the intercalated oxygen the following isotope labelling experiment was carried out. Sample A was emptied from oxygen by prolonged heating at 700 K in UHV. Subsequently it was transfered in a small glass vessel which was filled with a mixture of normal and $^{18}O_2$.

There it was loaded with the oxygen mixture by prolonged heating to 400 K. Residual gas analysis showed that after heating more $^{16}O_2$ was present than before heating. No

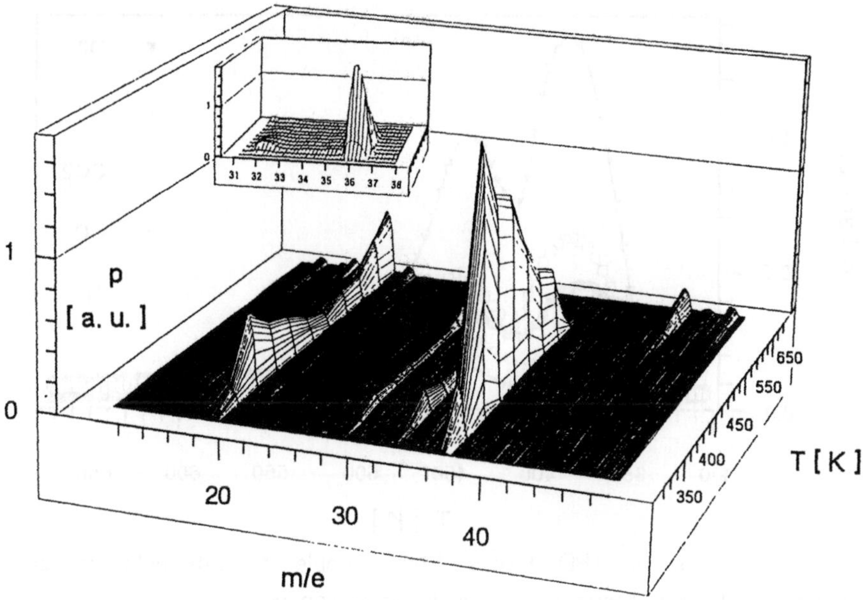

FIG. 5: Multi mass TPD of sample A after loading with $^{18}O_2$. The inset shows the absence of any significant cross-labelling (no peaks between m/e 32 and m/e 36).

detectable quantity of cross-labelled oxygen was found. After loading, the sample was transfered back to the TPD cell and the multi mass-TPD experiment displayed in Figure 5 was performed.

The experiment shows again no cross-labelling. This means that neither during loading nor during desorption the oxygen molecules are dissociated. At all instances the molecular O-O bonds remain intact.

In Figure 5 co-desorptions can be seen of normal oxygen, of traces of CO and of water together with the main $^{18}O_2$ desorption. These species are co-intercalated from a contaminated gas source (the $^{18}O_2$ reservoir) into the same sites. The intercalation of small molecules is thus not selective and occurs with very simliar energetics.

The desorption energies which may be deduced from the peak desorption temperatures are mere activation energies for molecular transport effects and do not represent species-selective binding energies onto the C_{60} molecules.

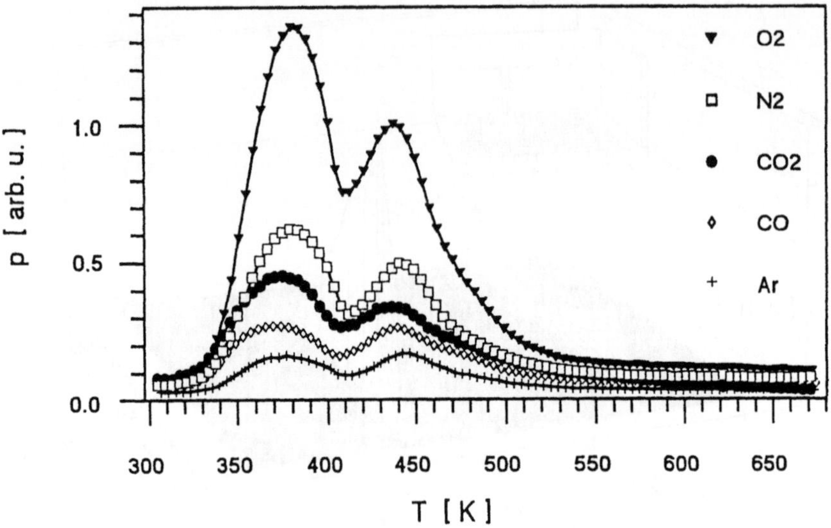

FIG. 6: TPD traces from sample A with different gases detected with their respective molecular ionisation peaks.

To further test this conclusion sample A was loaded with various gases including O_2, N_2, Ar, CO, CO_2 and H_2. All species yielded the same pattern seen in Figure 6. The intensity differences are due to cross section differences of the mass spectrometer and indicate further that for the smaller molecules the sample exposure temperature may have been with 400 K too high. The hypothesis of transport-limited desorption is supported by this experiment. The puzzling fact that H_2 shows the same pattern as all other gases may indicate, on the other hand, that the energetics seen in the TPD experiments of Fig. 6 are intrinsic to the solid host and not to the gaseous guest. In this case we would detect by the experiment the lattice dynamics of the van der Waals solid and not the chemical host-guest interaction.

4.SINGLE CRYSTAL STRUCTURE

The crystal structure at 300 K was determined using a data set collected with Cu kα radiation in the 2Θ range from 2° to 148° yielding 1532 observations. The density of the crystals as well as the volume of the unit cell require 4

C_{60} molecules per translation unit. The point symmetry of the molecule is *m35* of which *m3* is a subgroup satisfying the crystallographic translational requirements. It is plausible to adopt this symmetry for the crystal structure model of a face-centered cubic close packed arrangement of the C_{60} balls. The resulting space group in accordance with the point symmetry is *Fm3.* With this assumption the measured intensities reduce into 281 unique data with an internal R value of 2.55%. The lattice constant was at 300 K 1415.2(1) pm, at 200 K 1409.1(2) pm and at 100 K 1405.4(3) pm. The alternative space group *Fm3m*[24] requires a four-fold symmetry axis which would in a non-twinned crystal only be possible with a special packing of two pairs of molecules rotated by 90°. Both space groups were used in the least square refinement with *Fm3* leading to the better model.

The calculated density for pure C_{60} is 1.689 gcm^{-3} with 4 molecules in the unit cell. It is pointed out that the experimental density is significantly larger suggesting the presence of impurities besides the 4 molecules in the unit cell. These impurities are identified with several analytical methods as additional oxygen as shown in section 3.

In the early stages of refinement attempts were made to fit the observations with 3 unique C atoms with isotropic thermal parameters. Only one set of parameters and a restrained variation of the bond lenghts uniquely led to the undistorted truncated icosahedron with a diameter of 348 pm. This yielded R values between 30% and 40% depending on constraints and the number of observations used. In the following steps the main axes and orientations of the probability density elipsoids were fixed such as to simulate the spherical average surface of the buckyball. This drastically reduced the R value down to 14.3% using 270 observations and 24 parameters.

The highest residual electron density was located above 8 of the 24 hexagons at a distance to carbon of 148 pm. Keeping in mind the results of the TPD experiments, the discrepancy between observed and calculated density and noting that the (h00) reflections were non-zero in intensity, it was concluded that additional oxygen is attached to the buckyballs. An oxygen atom (justified by non-crystallographic observations) on a partially occupied

TABLE I Structural Parameters of C_{60} at 300 K

ATOM	X	Y	Z	Ueq
C1	0.2349 (5)	0. 0	0.0517 (7)	0.358 (4)
C2	0.2015 (5)	0.0845 (4)	0.0946(4)	0.482 (4)
C3	0.1722 (4)	0.1599 (2)	0.0469 (7)	0.430 (4)
O1	0.1379 (5)	0.1379 (5)	0.1379 (5)	0.150 (4)

special position was included in the refinement. With 8 atoms per unit cell ($C_{60}O_2$) a perfect match between the densities is achieved and the R value reduced finally to 11.25% (270 observations, 29 parameters) with the weighted R value coming down to 5.28%. The positions of the 4 unique atoms in the structure are given in Table 1

The resulting molecular shape of the modified C_{60} is shown in Figure 7. The oxygen atoms located 30 pm above the flat rings are part of oxygen molecules attached to the buckyballs. The other atom has too much orientational freedom to be located by X-ray diffraction. From the isotope labelling experiment it can be no doubt, however, that the nature of the additional oxygen is molecular.

Two sets of bond lengths of about 137 pm and 142 pm were obtained. As discussed by Bürgi et al.[25], these values are artificially shortened with no meaningful correction being possible in the present case. The model requires, however, 2 sets of bond lengths with a difference of ca. 5 pm in order to yield a buckyball as solution. It is noted that other models which do not expand into even-shaped buckyballs lead to better bond lenghts compared to the gas phase data[26] and to smaller R values which are, however, considered as chemically meaningless.

The C atoms exhibit large anisotropic displacement parameters (ADP) indicating large intermolecular disorder (projection of different rotamers). The overall effect of such high ADP on the calculated structure factors is their faster decay with increasing scattering angle compared to a

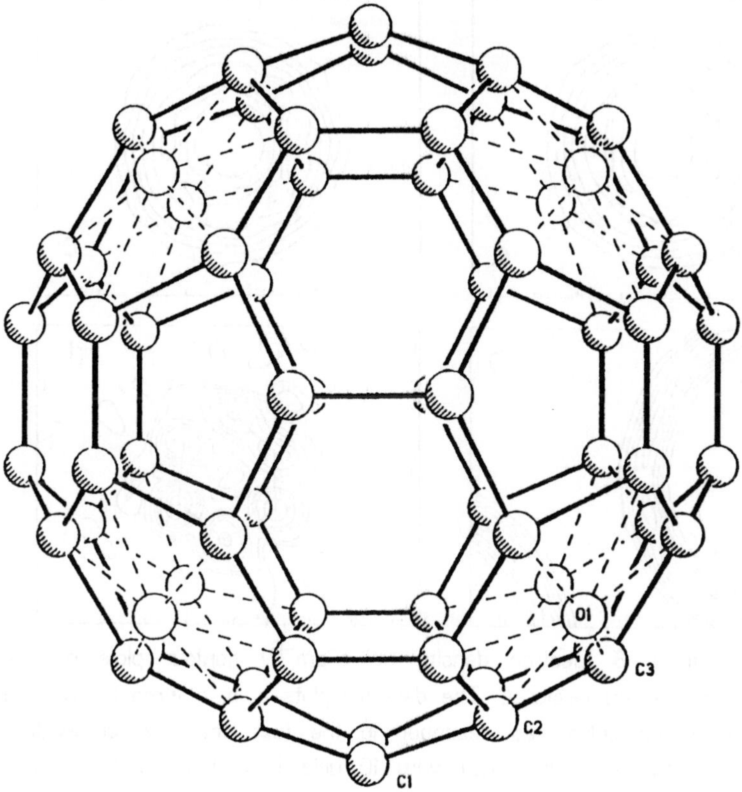

Figure 7 Plot of the $C_{60}O_2$ molecule as the average unit with the flat rings. The unique atoms (at the bottom) are labelled. Note the 8 partially occupied apical oxygen positions.

well ordered situation. Exactly this was observed with the progress of the experimental structure amplitudes. A significant improvement in the R values towards the internal R value may thus be achieved artificially by omitting suitable observations or physically meaningful by considering different rotamers in the model. This procedure was sucessfully adopted for the structure below the phase transition which was explained by superposition of two unequally populated rotamers[7]. Assigning the anomaly at 250 K to an order-disorder transition it may be seen as an increase in the number of rotamers from 2 at low temperature to a finite larger number at high temperature. Their minimal energy difference[7] renders it, however, inadequate to model these high-temperature rotamers in

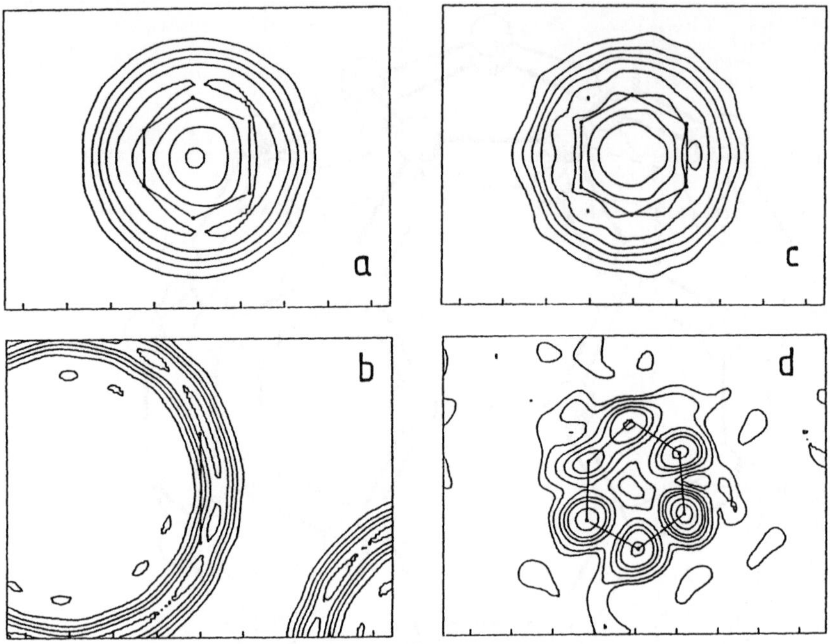

Figure 8 Electron density (in $e^{-}pm^{-6}$) contour plots of one hexagon at various temperatures. The data for plots a, b, c were taken from a Mo ka data set solved as described in the text, the low temperature structure was refined in R6 (Pa3) with 10 unique C atoms and a similar procedure as outlined in the text. No observations were suppressed. The resolution was set to 2 pm. Plot a) least squares plane through the hexagon, temperature 260 K, contours at 0.5, 1.0, 1.5, 2.0, 2.5, 2.8; Plot b) perpendicular section through the C_{60} unit (hexagon marked), temperature 258 K, contours at 0.5, 1.0, 1.5, 2.0, 2.5; Plot c) same view at 258 K and same contour heights as a); Plot d) same as a) and c), temperature 200 K, contours at 0.5, 1.0, 2.0, 3.0, 4.0, 6.0, 8.0. Note the anisotropic electron distribution at the highest contour levels. The interior of the sphere is free from electron density as seen by X-ray diffraction.

absence of experimental evidence for their relative orientation.

The finite number of rotamers differs from the picture of free rotation and leads to an averaged uneven electron distribution on the buckyball surface. This is demonstrated by electron density maps calculated within and perpendicular to the least squares plane of a hexagon. Results at various temperatures are shown in Figure 8. It is evident that the electron density in the high temperature form is smeared out but not uniformly distributed. Lowering the temperature clearly improves the localisation of the electron density at the ring corners due to a reduction of the number of projected rotational isomers. These observations must not be mistaken as improvement of the ordering within the carbon molecules. In addition, we found that the progress of the observed structure amplidudes with increasing scattering angle is anisotropic for different directions. Such a modulation is not in agreement with the expected scattering factor for a freely rotating buckyball.

It should be noted that the time averaging X-ray diffraction experiment can not distinguish between a hopping rotation or a static rotational disorder. Nevertheless, the measured "scattering law" is satisfactorily described with the model presented using one set of three unique C atoms.

5. INFRARED SPECTROSCOPY

IR exhibits enormous specifity in detecting carbon-oxygen bonds. As transmission experiment it is bulk-sensitive and requires no long-range ordering. It is a complementary method to desorption and X-ray diffraction. A short summary of the phenomena occuring upon exposure of C_{60} powder to air and heating it to 575 K is given in Figure 9.

The top trace shows clean sublimed C_{60} pressed as KBr pellet in a glove box and measured under protective gas. The 4 permitted IR bands of the buckyball molecule are clearly visible accompanied from a series of weak bands of non-permitted absorptions led by the peak at 1536 cm^{-1}. If this material is brought in contact with ambient air for 15 min and sealed afterwards in a KBr disk in a glove box the central spectrum of Figure 9 is obtained. The most intense band in the spectrum at 1383 cm^{-1} is due to the formation of an oxygen adduct, the same species which was detected in the single crystal structure and which gives rise to the double feature above 400 K in the TPD experiments. The broad line at 1640 cm^{-1} arises from molecular water as contaminant from the air.

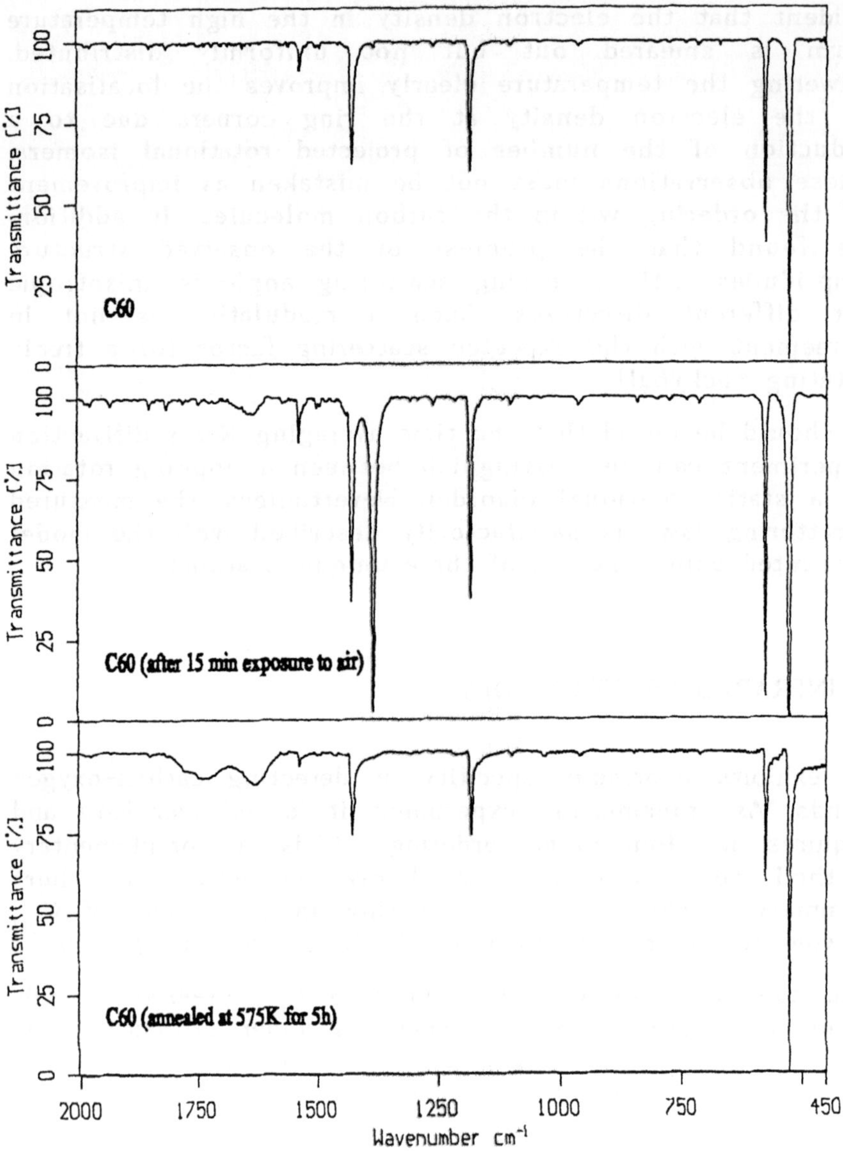

Fig. 9: FT IR spectra of KBr pellets of C_{60} after various treatments. To avoid post-contamination the pellets were pressed inside a glove box.

If such an oxygen contaminated sample is heated to 575 K, the onset temperature of CO/CO_2 production (see Figure 2), the loss of the additional peak is expected according to the TDS experiments. This is indeed the case as can be seen from the bottom spectrum in Figure 9. Two new features at 1777 cm^{-1} and 1743 cm^{-1} besides the water peak indicate the conversion of some C_{60} into carbonyl-containing partly oxidised molecules. A number of additional weak bands in the range of C–O bonds, and carbon-carbon multiple bonds point to a complicated structure of the partly destroyed fullerenes.

The small number of partly oxidised molecules relative to the still intact molecules indicates that the process of oxidative gasification occurs not isotropically but topotactically on the gas-solid interface. At the high activation temperature required for oxidation of C_{60} with molecular oxygen the clathrate with oxygen molecules is widely decomposed by de-intercalation of the molecules. This observation proves that a sample of contaminated fullerene can be cleaned by annealing in the bulk but the data show also that the surface of such a sample will be always damaged and will contain ill-defined (polymeric) carbonaceous species.

6. DISCUSSION

The low temperature intercalation reaction of small molecules was studied by TPD. An involvement of the outer surface of the C_{60} crystals as significant sorption site was excluded. Two bulk absorption sites were detected, one of which is intrinsic. The other site is associated with lattice defects, the presence of which reduces significantly the activation barrier for occupation.

Exposure of C_{60} to oxygen at ambient pressure (in the absence of irradiation) causes up to a critical temperature of 570 K no oxidation of the carbon material.

Intercalated oxygen is present in the absorbed state and during transport in the solid as molecule and is not dissociated up to temperatures of 373 K. We speculate that the O–O bonds are polarised to some extent by the intimate presence of the electron-rich C_{60} units.

This polarisation leads eventually to a bond formation fixing the oxygen molecule as a poorly stable peroxide to the fullerene which then easily dissociates forming an epoxide-like structure of yet unknown stoichiometry. This stage is the immediate precursor to irreversible damage of the carbon molecules associated with the removal of carbon atoms in the form of CO/CO_2 occuring at 570 K.

The involvement of defects in the uptake of oxygen into C_{60} allows to disintegrate the solid if the reaction with oxygen is carried out with defective (e.g. solvet-loaded) samples at rapidly changing temperatures as e.g. in DSC experiments. In this way the conflicting results about bulk oxidation/amorphisation[14] and structural integrity of thin films[13] may be explained. In addition, the kinetics of oxygen uptake will depend on the perfection of the solid C_{60} surface and may be considerably slowed down for perfect single crystals compared to the time scales of minutes found in the present powder experiments.

Pure crystals of C_{60} absorb in an unselective way a number of gases into their interstices. Contrary to the intercalation of solvent molecules, iodine or alkali metals, the present intercalation does not lead to structural changes of the crystals and cannot be readily detected by X-ray diffraction.

The selective adsorption of small molecules on graphitic carbon which is used in gas separation technology may be related to local defect geometries rather than to a homogeneous surface of sp^2 carbon. As the results of the present study are transport limited and not representative for chemisorption effects, further experiments with high surface area C_{60} are needed to clarify this point. In the present mode of gas absorption the possible analogy between activated carbon and C_{60} is not valid.

It is pointed out that oxygen cannot be intercalated into the two-dimensional voids of graphite under retention of the host electronic structure. Only under highly oxidising conditions at low temperatures (electrochemically or with very strong oxidising acids) an ill-defined graphite oxide may be obtained which contains covalent C-O bonds as is no longer graphitic.

The quantitative effect of this intercalation is estimated from the TPD peaks to be in the range of a few percent of all voids available. Sorption isotherms are currently measured to obtain more precise numbers.

The possible significance of the presence of these interstitial molecules on physical properties such as electrical conductivity was illustrated very recently[27] in a study which found dramatic changes in resistivity as function of oxygen sorption or desorption in a temperature range very similar to that of our observations. The polarisation of the oxygen molecules and a small charge transfer from C_{60} to oxygen following the large difference in electronegativity between C and O may account for the large conductivity effects in the semiconducting material.

Regardless of the actual mechanism of rotational hindering the present room temperature single crystal structure confirms the predicted[14] existence of distinct symmetric C_{60} molecules made from flat units with a variation in chemical bond lengths well within the known limits of conventional structures. A high degree of rotational disorder leads, however, to a structural overlay detectable by the time averaging method of X-ray diffraction but does not result in a featureless sphere[3] of equally distributed electron density. The single crystal intensity data of purified C_{60} at 300 K can sufficiently be interpreted with a conventional parameter refinement procedure adopting the well-documented point symmetry of the C_{60} molecule.

The present study finally highlights that even a molecular solid may be affected by small impurities such as oxygen very significantly. At present, no definitive experiment is available, however, to distinguish intrinsic reasons for a hindered rotation (e.g. non-spherical electron density ditribution in the molecule *per se*) from the difficult to avoid impurities as controlling factors for the molecular dynamics. As their effect has to be considered cooperative, an impurity level difficult to analyse may be sufficient to modify the dynamic behaviour of a C_{60} sample.

To answer the question from the introduction about the stabilty of C_{60} these experiments allow the following conclusions:

1. C_{60} *molecules* are stable against oxidation under ambient and dark conditions.

2. C_{60} *crystals* react with molecular oxygen and other permanent gases at ambient conditions to yield gas clathrates. The presence of solvent molecules or special

protective adduct forming agents will considerably reduce the extent of this reaction.

3. The kinetics of this process depends critically upon the crystallinity of the fullerene: saturation can be reached in minutes in powders and thin films and will take days in well-shaped single crystals.

4. Thermal treatment can remove most of the gases at temperatures up to 550K, oxidising conditions such as impure protective gases or poor vacuum will, however, increase the amount of irreversibly oxidised fullerene molecules above a threshold inevitably connected with the exposure of the C_{60} to oxygen. These broken carbon structures carry the EPR centers and will be finally converted to CO/CO_2 which may well be trapped again within the fullerene crystal.

ACKNOWLEDGEMENTS

The $^{18}O_2$ was kindly donated by Prof. Brauer (Frankfurt). Several helpful discussions with W. Krätschmer are gratefully acknowleged. Crystallographic problems were discussed with E. Egert, W. Batz H. Bartl and W. Paulus. We thank E. Egert for the access to the CAD 4 instrument. The project is financially supported by the Bundesminister für Forschung und Technologie. Financial support came also from the Fonds der Chemischen Industrie and from Hoechst AG (Frankfurt)

REFERENCES

1 F. Atamny, J. Blöcker, A. Dübotzky, H. Kurt, G. Loose, W. Mahdi, R. Schlögl, J. Mol. Phys., 1992, 76, 851

2 J.E. Fischer, P.A. Heiney, D.E. Luzzi, D.E. Cox, in: "Fullerenes", ed. G.S. Hammond, V.J.Kuck, ACS Sympsium Series, Vol. 481, 1992, p. 58

3 A.M. Vassallo, L.S.K. Pang, R.A. Cole-Clarke, M.A. Wilson, J. Am. Chem. Soc.,1991, 113, 7820

4 D.W. McKee, Carbon, 1991, 29,, 1057

5 G.H. Kroll, P.J. Benning, Y. Chen, T.R. Ohno, J.H. Weaver, L.P.F. Chibante, R.E. Smalley, Chem. Phys. Lett., 1991, 181, 112

6 V. Vijayakrishnan, A. Kumar Santra, T. Pradeep,R. Seshadri, R. Nagarajan, C.N.R. Rao, J. Chem. Soc., Chem. Commun, 1992, 198

7 R.A. Assink, J.E. Schirber, D.A. Loy, B. Morosin,G.A. Carlson, J. Mater. Res., 1992, 7, 2136

8 Th. Schedel-Niedrig, M. Keil, A.M. Bradshaw, H. Werner, R. Schlögl, W. Kirchner, Chem. Phys. Lett., submitted

9 K. Creegan, J.L. Robbins, W.K. Robbins, J.M. Millar, R.D. Sherwood, P.J. Tindall, D.M. Cox,, J. Am. Chem. Soc., 1992, 114, 1103

10 R. Taylor, J.P. Parsons, A.G. Avent, S.P. Rannard, T.J. Dennis, J.P. Hare, H.W. Kroto, D.R.M. Walton,Nature, 1991, 351, 277

11 P. Zhou, A.M. Wang, J.D. Robsertson, C. Eloi, M.S. Meier, S.L. Ren, X.X. Bi, P.C. Eklund, M.S. Dresselhaus, Appl. Phys. Lett., 1992, 60, 2871

12 H.S. Chen, A.R. Kortan, R.C. Haddon, M.L. Kaplan, C.H. Chen, A.M. Mujsce, H. Chou, D.A. Fleming, Appl. Phys. Lett., 1991, 59, 2956

13 C.S. Yannoni, R.D. Johnson, G. Meijer, D.S. Bethune, J.R. Salem. J. Phys. Chem., (1991) 95, 9,

14 R. Tycko, R.C. Haddon, G. Dabagh, S.H. Glarum, D.C. Douglass, A.M. Mujsce J. Phys. Chem., (1991) 95, 518,

15 P.A. Heiney, J.E. Fischer, A.R. Mc Ghie, W.J. Romanow, A.M. Dennenstein, J.P. McCauley jr., A.B. Smith III, Phys. Rev. Lett., (1991) 66, 2911,

16 J.M. Hawkins, A. Meyer, T.A. Lewis, S. Loren, F.J. Hollander, Science, (1991) 252, 312,

17 W.I.F. David, R.M. Isberson, J.C. Matthewman, K. Prassides, T.J.S. Dennis, J.P. Hare, H.W. Kroto, R. Taylor, D.R.M. Walton, Nature, (1991) 353, 147,

18 H. Werner, D. Bublak, U. Göbel, B. Henschke, W. Bensch, R. ,Schlögl, Angew. Chem., 1992, 31, 868

19 W. Krätschmer, L.D. Lamb, K. Fostiropoulos, D.R. Huffman, Nature, 1990, 347, 354

20 U. Göbel, M. Wesemann, W. Bensch, R. Schlögl, J. Anal. Chem., 1992, 334, 582

21 A. Dworkin, H. Szwarc, S. Leach, J.P. Hare, T.J. Dennis, H.W. Kroto, R. Taylor, D.R.M. Walton, C.R. Acad. Sci. Paris, (1991) 312, 979,

22 S. Liu, Y. Lu, M.M. Kappes, J. A. Ibers, Science, (1991) 254, 408,

23 J.M. Wood, B. Kahr, S.H. Hoke II, L. Dejarme, R.G. Cooks, D. Ben-Amotz, J. Am. Chem. Soc., 1991, 113, 5907

24 R. M. Fleming, B. Hessen, T. Siegrist, A.R. Kortan, P. Marsh, R. Tycko, G. Dabbagh, R.C. Haddon, in: ed. G.S. Hammond, V.J. Kuck, ACS Symposium Series, (1992) Vol. 481, p. 25,

25 H.B. Bürgi, E. Blanc, D. Schwarzenbach, S. Liu, Y. Lu, M.M. Kappes, J.A. Ibers, Angew. Chem., (1992) 104, 667,

26 K. Hedberg, L. Hedberg, D.S. Bethune, C.A. Brown, H.C. Dorn, R.D. Johnson, M. de Vries, Science, (1991) 254, 410,

27 T. Arai, Y. Murakami, H. Suematsu, K. Kikuchi, Y. Achiba, I. Ikemoto, Solid State Commun., 1992, 84, 827

A New Apparatus for Fullerene Production

L. Dunsch[1], *F. Ziegs*[1], *J. Fröhner*[1], *U. Kirbach*[1], *K. Klostermann*[2], *A. Bartl*[1], *and U. Feist*[1]

[1]Institut für Festkörper- und Werkstofforschung Dresden e.V., D-01069 Dresden, Germany

[2]Institut für Analytische Chemie der Technischen Universität Dresden, D-01069 Dresden, Germany

Abstract. The discovery by Krätschmer and co-workers to use a light-arc method for preparing macroscopic quantities of C_{60} and C_{70} has opened the availibity of fullerene physics and chemistry in preparative scale. Using the experiences at the Krätschmer apparatus we developed an equipment for fullerene production with a great variability of preparation conditions. At this equipment we can study the influence of preparation conditions on the yield and the preparation of higher and endohedral fullerenes.

1. Introduction

The preparation of fullerenes by arc burning of graphite rods under a helium atmosphere and reduced pressure [1] is the most promising way to get a variety of higher and endohedral fullerenes besides large amounts of C_{60} and C_{70}. To study the conditions of fullerene production and to vary the soot preparation for the application of metals to produce endohedral fullerenes an improved apparatus for arc burning was to be constructed. Thus the possibilities for an efficient production of fullerenes was established. Furthermore the separation of the product of arc burning by liquid extraction and chromatographic methods is to be improved to get better yields of the higher fullerenes. As a useful technique for characterisation mass spectroscopy with different methods of ionisation is the most promising spectroscopic method. The redox reactions at carbon electrodes are of interest to study the influence of metallic substrates on the electron transfer reaction.

2. Experimental

The apparatus for fullerene production by arc burning used in this study is a cylindrical steel vessel with 24.4 cm of inner diameter and a length of 30 cm. On opposite sides of the cylinder mantle two vacuum-tight passings (a) are mounted used for the rotating graphite rod holder (b) with electrical contacts (c). On the front side a window for observing the light arc is attached which can be cleaned contineously during vaporization [Fig. 1]. The steel chamber and the graphite rod holder are cooled by water. The vessel can pumped out to a pressure of 10^{-3} bar by a mechanical pump equipped with a cooling trap. Using a adjustable inlet value (d) the burning chamber can be filled with helium or other gases [Fig. 2]. Immediately

Springer Series in Solid-State Sciences, Volume 117
Electronic Properties of Fullerenes Eds.: H. Kuzmany, J. Fink, M. Mehring, and S. Roth

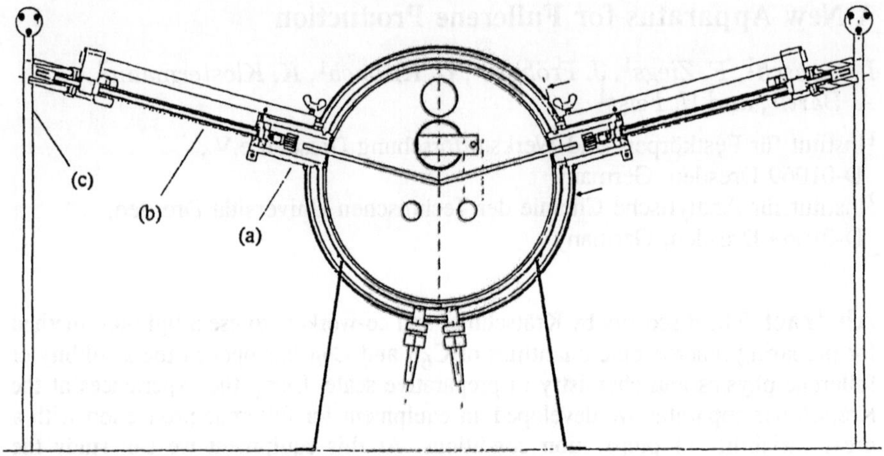

Fig. 1 Scheme of the new apparatus for fullerene production, front view (For explanation see experimental part).

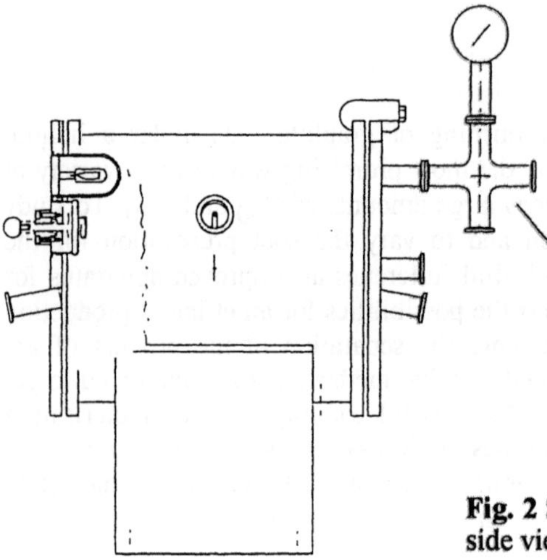

Fig. 2 Scheme of the fullerene apparatus, side view (see text).

before the burning process the graphite rods are cleaned by resistivity-heating of about 400°C in helium atmosphere. After this procedure the inert gas of the chamber is renewed.

The soot produced by arc burning was extracted with toluene resulting in a mixture of fullerenes followed by filtration over alumina column to remove suspended micro particles of soot.

At present the HPLC is the best method for the separation of fullerenes [2] because this method is a simple, reproducible and automatically working procedure. We have carried out the separation with a Hewlett-Packard 1050 system equipped with a 7.5 x 300 mm PL-gel column (10 µm, 500 Å) and a Foxy 200 fraction sampler which is programmed by a time-window procedure. Using a flow-rate of

1.5 ml/min of 1.2-dichlorethane as the solvent and repetitive periodical (run time 25 min) injections of 100 μl portions of fullerene solution the mass-separation of C_{60}, C_{70} and higher fullerene fraction was succesfully.

The best cutpoint in the HPLC profile of different fractions were found by mass spectroscopy which is the most important method to characterise the purity of the HPLC-fractions as well as the composition of the soot extract. The samples were measured with a mass spectrometer MAT 95 (FINNIGAN). This spectrometer allows the detection of fullerenes by electron impact, fast atom bombardment and chemical ionisation.

The different fullerenes were characterised at room temperature by cyclic voltammetry at glassy carbon, carbon fiber and platinum electrodes using the electrochemical system ECM 700 (ZWG Berlin) in a three electrode system with the calomel reference electrode. 10^{-3} molar fullerene solutions in methylene chloride with 0.1 molar tetrabutylammoniumtetrafluoroborate as supporting electrolyte were used.

3. Results

The apparatus for fullerene production by arc burning developed in this study was used for a contineous fullerene production without any technical problems after a first induction period of the new installation. Varying the current of the arc in the range of 85 to 240 A no significant influence of this parameter on the yield and the partition of the fullerenes could be found. In general it was shown that the C_{70} content of the soot produced with this apparatus was higher than in other procedures [3]. In the course of the experiments it was found that the apparatus has the following advantages:

- long term burning is possible because of water cooling of the apparatus wall and the electrode holders
- no vacuum oil interference appears as the cooling of the pumping gas
- stable arc burning due to the angle between the two graphite rods and the possibility of their rotation
- centric (to the arc) orientation of several inlets for the application of additional rods and differents sensors (temperature)
- variation of the arc modes by the electric supply
- contineous visual control of the arc using a contineous cleanable window
- simple removal of the soot

The efficiency of the HPLC separation procedure is demonstrated by the mass spectra of the different fractions [Fig. 3]. The C_{60} fraction has a purity of 99.6% while the C_{70} sample has a C_{60} impurity of less than 5%. Higher fullerenes were found in a decreasing amount at molecular weights higher than 84 [Fig. 3]

The effect of the type of ionisation on mass spectroscopic studies of fullerenes was studied.The result of the spectra is that the negative chemical ionisation of fullerenes with methane is the best ionisation method regarding the sensitivity, multiple ionisation and fragmentation. In case of the electron impact ionisation the fragmentation and multiple ionisation was very high. During the fast atom bombardment with caesium-ions the sensitivity of the detection of fullerenes

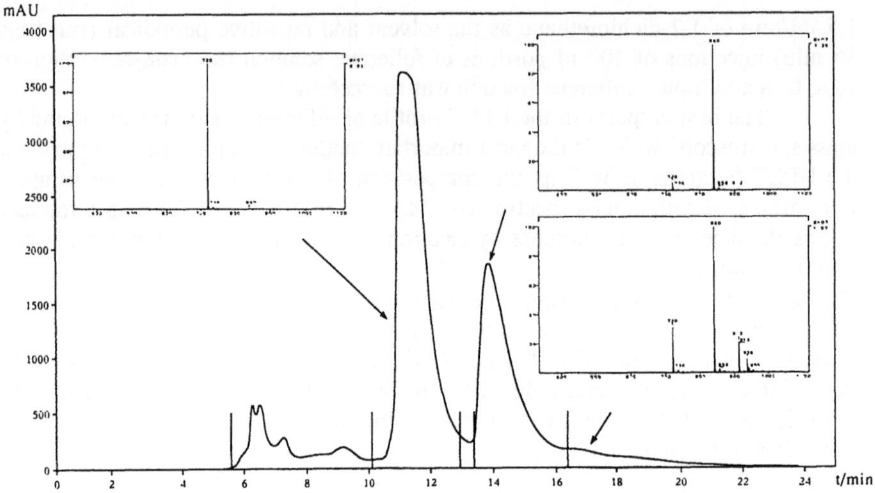

Fig. 3 HPLC-separation and mass spectra of the main fractions.

was about one thousandth of that of chemical ionisation. Because of the different evaporation temperatures of the fullerenes the average of all spectra during the evaporation of the sample was calculated. The result was a spectrum with the real composition of the sample.

Mass spectroscopy was furthermore used to get the purity of the HPLC-fractions and the composition of the soot extract. The soot extract contained as a main product C_{60} (70%), about 28% C_{70} and a minor content of higher fullerenes up to C_{110}, too. Using methane for chemical ionisation it is possible that the addition product can be found by mass spectroscopy (mass 736).

The electrochemical reduction at glassy carbon electrodes results in three redox couples at -550, -950 and -1410 mV vs. SCE. They are less reversible than those at metal electrodes (peak separation between 90 and 180 mV). Furthermore the peak shape of the reduction peaks led to the conclusion that the electrode reaction is not diffusion-controlled. The cyclic voltammogram of C_{70} is similiar to that of C_{60} which was already found by other groups [4]. The second redox peak of C_{60} is enlarged in the presence of oxygen what might be due to an epoxide formation. Higher fullerenes show an oxidation peak in the anodic potential region which was not found at C_{60} and C_{70} [5].

4. Conclusion

The new apparatus is an improved technical system allowing a stable arc burning of graphite rods under rotation. It offers the possibility of the variation of arc modes and contineous visual control of arc burning. The resulting soot contained 70 % of C_{60} and 28 % of C_{70}. Higher fullerenes up to C_{110} could be extracted with toluene. The separation of the fullerenes by HPLC is best done 1,2-dichloromethane as the eluent.

5. Literature

[1] W. Krätschmer, L. D. Lamb, K. Fostiropoulos and D.R. Huffman, Nature *347* (1990) 354; see also: R. Malhotra, D.C. Lorents, Y.K. Bae, C.H. Becker, D.S. Tse, L.E. Jusinski and E. D. Wachsman in: Fullerenes: Synthesis, Properties, and Chemistry of Large Carbon Clusters,(G.S. Hammond and V.J. Kuck, Eds) ACS Symp. Series 481, Washington, ACS 1992, p. 127 for a review see: W. Krätschmer and K. Fostiropoulos, Phys. in unserer Zeit *23* (1992) 105

[2] A. Gugel, M. Becker, D. Hammel, L. Mindach, J. Räder, T. Simon, M. Wagner and K. Müllen, Angew. Chem. *104* (1992) 666

[3] A. Mittelbach, W. Hönle, H. G. v. Schnering, J. Carlsen, R. Janiak and H. Quast, Angew. Chem. *104* (1992) 1681

[4] P.-M. Allemand, A.Koch, F. Wudl, Y. Rubin, F. Diederich, M.M. Alvarez, S.J. Anz and R.L. Whetten, J. Am. Chem. Soc. *113* (1993) 1050

[5] Q. Li, F. Wudl, C. Thilgen, R. L. Whetten, and F. Diederich, J. Am. Chem. Soc. *114* (1992) 3994

Implantation into Fullerites

K. Misof[1], *G. Vogl*[1], *P. Fratzl*[1], *R. Sielemann*[2], *B. Keck*[2], *and Y. Yoshida*[3]

[1]Institut für Festkörperphysik der Universität Wien, Strudlhofgasse 4, A-1090 Wien, Austria
[2]Hahn-Meitner-Institut Berlin GmbH, Bereich Schwerionenphysik, Glienicker Str. 100, D-14109 Berlin, Germany
[3]Shizuoka Institute of Science and Technology, Japan

Abstract. Fcc C_{60} and C_{70} were irradiated with high-energy heavy ions at low temperature. The landing positions of the iron ions were studied by means of in-beam Mössbauer spectroscopy. For both materials the Mössbauer data indicate that the iron atoms landed on interstitial sites between two fullerene molecules with two pentagons facing each other. In the C_{70} sample irradiation led to a fcc-to-rhombohedral phase transition.

1 Introduction

Solid C_{60} is a Van der Waals-bonded molecular crystal [1]. In a number of experiments phase transitions have been studied: Above 260K the C_{60} molecules form a fcc lattice with the soccer-ball shaped molecules completely freely rotating [2], whereas below 260K their rotation is partially hindered and the structure changes to simple cubic with four molecules per unit cell [3]. Below about 90 K the motion of the molecules is completely hindered [2]. Because of the rugby-ball shape of C_{70} the phase diagramme of this material may be more complex than that one of C_{60} [4,5]. Another interesting feature of C_{60} and C_{70} is the large inner space in the molecules. Some work was done to study the endohedral chemistry [6].
The aim of our work was to irradiate fullerenes C_{60} and C_{70} with high-energy ions in order to answer two questions: (a) What are the effects of the ion irradiation on the bulk material? (b) Is it possible to implant these ions endohedrally into the inner cage of the molecules ?
The ions were ^{57}Fe, suitable to study their positions by help of in-beam Mössbauer spectroscopy in analogy to our earlier studies on other materials [7].

2 Experiments and Results

C_{60} and C_{70} in powder form were purchased from Texas Fullerenes. Layers of 30 μm thickness were evaporated at about 450°C in a vacuum of $5 \cdot 10^{-5}$ torr onto Al backings. The structure of the samples was checked by means of X-ray diffraction measurements [8]. The C_{60} and C_{70} samples were then irradiated for 18 hours and 45 hours, respectively, at temperatures between 10K and 300K at the heavy-ion accelerator VICKSI of the Hahn-Meitner-Institut Berlin with iron ions (energy spectrum up to 50 MeV). According to calculations with the well-known TRIM programme the average penetration depth of 50 MeV iron ions into fullerenes is about 12μm. The accumulated density of iron ions in the sample was about 10^{12} particles/cm^2, therefore the concentration of iron atoms in the fullerenes was tiny (about 10^{-6} per one C_{70} molecule). After irradiation the samples were

Springer Series in Solid-State Sciences, Volume 117
Electronic Properties of Fullerenes Eds.: H. Kuzmany, J. Fink, M. Mehring, and S. Roth

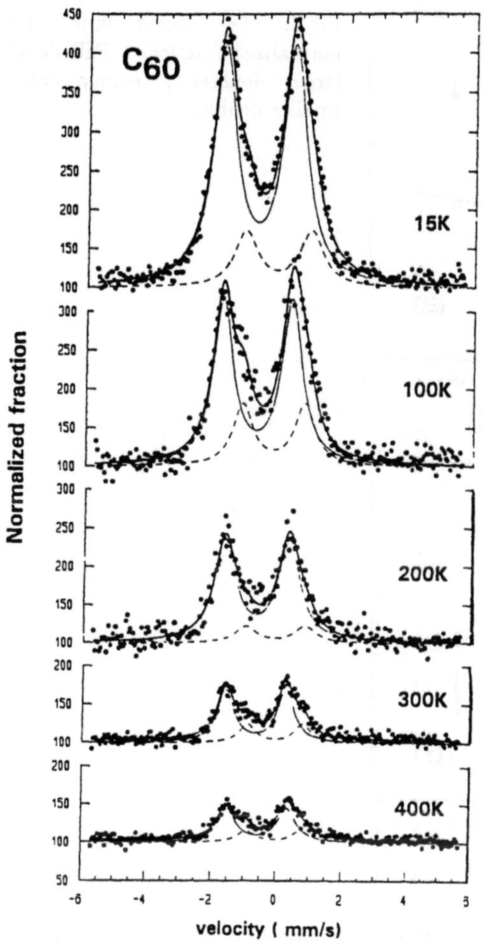

Fig.1: Mössbauer spectra of C_{60} at different temperatures. The solid line is a fit with two doublets. The fit indicates two different sites for the iron atoms.

warmed up to room temperature and again investigated by X-ray scattering. The Bragg peaks of the C_{60} sample remained unchanged, i.e. no phase transition was detected. The X-ray pattern of the C_{70} specimen, on the contrary, showed that in addition to fcc an additional rhombohedral phase was present [8].

The landing positions of the ^{57}Fe ions were studied by means of in-beam Mössbauer spectroscopy. Fig.1 shows the spectra of C_{60} at different temperatures. The Mössbauer data of C_{70} are similar to the spectra of C_{60}.

The data can be fitted with two doublets indicating two different well defined sites of the iron atoms. The quadrupole splittings of both components are very similar (QS1: 2.20 mm/s and QS2: 2.04 mm/s at 15 K), but the isomer shifts are different (IS1: -0.70mm/s and IS2: -0.05 mm/s at 15K). (To compare our data with absorber experiments you have to invert the sign of the isomer shift.) The temperature dependence of the isomer shift of the large component (Fig.2a) can be well described by the second-order Doppler shift.

The parameters for the larger doublet indicate a high-spin Fe(II) complex. The second component shows an isomer shift which indicates an Fe(0) or a low-spin Fe(III)

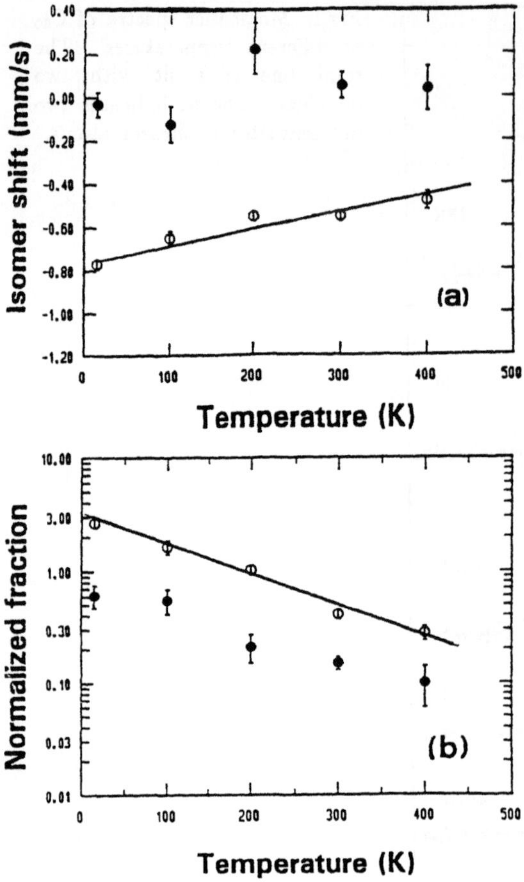

Fig.2: (a) isomer shift, (b) normalized fractions. Full dots, larger doublet ; empty dots, smaller doublet

compound. The fractions of the two doublets (proportional to Debye-Waller factor) show no anomalous behaviour (Fig.2b) and indicate a rather weak bonding for both sites (Debye-temperature is about 150K).

3 Discussion

The irradiation effect on the bulk material is discussed in detail elsewhere [8]. The data obtained by means of in-beam Mössbauer spectroscopy of C_{60} can be interpreted as follows: The large quadrupole splitting and isomer shift of the main component is similar to the Mössbauer data of the well known organometallic compound ferrocene [9]. This is a complex where an iron atom is situated between two pentagons. Therefore we interprete these data as follows: the iron atoms after having slowed down through a number of collisions landed on interstitial sites between two bucky-balls with two pentagons facing each other without damaging their near surrounding. This complex at higher temperatures hinders the free rotation of the bucky-balls bound to iron atoms , therefore we cannot observe an anomalous behaviour in the Debey-Waller factor at 260K. The second component is not understood yet. Its parameters are not as accurate as those of the larger component due to its smaller fraction.

We finally mention that we can exclude any iron dimer formation, since the Mössbauer γ-quantum is registered within 100 ns after implantation of the ^{57}Fe ion, leaving no time for diffusion.
Recent investigations by Schlögl et al.[10] showed that in fullerenes oxygen can be absorbed in air. Therefore we cannot completely exclude the influence of oxygen in the data obtained by Mössbauer spectroscopy. Further investigations with other type of irradiation, in particular with various high-energy heavy ions, varying energy and fluence systematically, will be performed in order to shed more light on the mechanism of the irradiation-induced phase transition. Very careful sample preparation will be done to obtain very pure material in order to get more information about the second site of the iron atoms.
We thank W. Krätschmer and S. Klaumünzer for valuable discussions. We are particularly indebted to P. Szimkowiak for his most skillful preparation of the C_{70} foils.

This work was supported by the Austrian Fonds zur Förderung der wissenschaftlichen Forschung, project S5601.

References

[1] J.E. Fischer, P.A. Heiney, A.R. McGhie, W.J. Romanow, A.M. Denenstein, Jr.J.P. McCauley, A.B. Smith III: Science **252**, 1288 (1991).
[2] R. Tycko , G. Dabbagh, R.M. Fleming, R.C. Haddon, A.V. Makhija, S.M. Zahurka: Phys.Rev.Lett. **67**, 1886 (1991).
[3] P.A. Heiney, J.E. Fischer,A.R. McGhie, W.J. Romanow, A.M. Denenstein, Jr.J.P. McCauley, A.B. Smith III, D.E. Cox: Phys.Rev.Lett. **66**, 2911 (1991).
[4] M.A. Verheijen, H. Meekes, G. Meijer, P. Bennema, J.L. de Boer, S. Van Smaalen, G. Van Tendeloo, S. Amelinckx, S. Muto, J. Van Landuyt: Chem.Phys. **166**, 287 (1992).
[5] G. Van Tendeloo, S. Amelincks, J.L. De Boer, S. Van Smaalen, M. A. Verheijen, H. Meekes, G. Meijer: Europhys.Lett. **21**, 329 (1993).
[6] T. Weiske, D.K. Böhme, J.Hrusak, W. Krätschmer, H. Schwarz: Angew. Chemie **103**, 898 (1991).
[7] M. Menningen, R. Sielemann, G. Vogl, Y. Yoshida, H. Bonde-Nielsen, G. Weyer: Europhys.Lett.**3**, 927 (1987).
[8] K. Misof, P. Fratzl, G. Vogl: submitted to Europhys.Lett.
[9] R.L. Collins: Journal of Chem. Physics **42**, 1072 (1964).
[10] R. Schlögl: private communication.

Voltammetric Studies on Fullerenes C_{60} and C_{70} in Low-Polarity Solvents

K. Meerholz, P. Tschuncky, and J. Heinze
Institut für Physikalische Chemie der Universität Freiburg,
Albertstr. 21, and
Freiburger Materialforschungszentrum F.M.F, Stefan-Meier-Str. 31a,
D-79104 Freiburg, Germany

1. Abstract

According to theoretical calculations, the LUMO in C_{60} is three-fold degenerated [1]. Within a short time, the reversible electrochemical formation of increasingly higher charged anions $C_{60}{}^{n-}$ or $C_{70}{}^{n-}$ up to the hexaanion [2] was reported. On the other hand, the fascinating chemical properties of the fullerenes are a point of increasing interest.

The challenge of the work presented here was to study the influence of amines on the electrochemical behaviour of the fullerenes. We carried out measurements in methylene chloride with different contents of dimethylamine (DMA) added up to pure DMA.

Up to now, only chemically irreversible waves in the voltammograms of the electrochemical oxidation of the fullerenes have been reported We obtained proof of the reversible electrochemical generation of (probably tetra-) cations in methylene chloride for the first time.

Our results are already published in J.Electroanal.Chem. 347, 425(1993)

2. Experimental

The fullerenes were produced by a modified Krätschmer-Huffman method [3]. A mass spectroscopy analysis revealed a composition of 83% C_{60} and 17% C_{70}. The experimental techniques used in the voltammetric experiments have been described elsewhere [4]. All potential values are listed in [V] vs. Ag/AgCl.

3. Results and Discussion

Upon reduction of the fullerenes C_{70} and C_{60} in pure CH_2Cl_2 /$TBAPF_6$ as well as at small concentrations of DMA (< 5 vol%) added we observed four reversible one-electron redox steps In front of the tri- or tetraanion step of C_{60}, two additional smaller waves were seen, which can be attributed to the reduction of C_{70} (Fig.1a).

On further addition of DMA, the originally red/violet solution turned green. A UV/VIS spectrum showed two additional absorp-

Springer Series in Solid-State Sciences, Volume 117
Electronic Properties of Fullerenes Eds.: H. Kuzmany, J. Fink, M. Mehring, and S. Roth

tion maxima in the visible region at 620 nm and in the NIR above 850 nm respectively. The voltammetric investigations were carried out shortly after addition of the fullerenes. Within the time required for the measurements, the solution was, as shown by UV spectroscopic control, completely stable.

Under these conditions (25 vol% DMA, Fig. 1b), the original peak at E_{pc_1} = -0.66 V transforms into a kinetically controlled wave, while the reduction peak at E_{pc_2} = -1.03 V increases significantly. These results clearly show that a species B is formed from $C_{60,70}$(A) via a preceding chemical equilibrium. Unconverted A is furthermore reduced at E_{pc1} = -0.66 V while the B/B^- couple appears at E_{pc_2} = -1.03 V. The steps for the higher redox states remain unaltered. This fact indicates the fast chemical transformation of B^- to A^-, which is immediately reduced to A^{2-} (ECE-sequence).

Increasing the amine concentration (50 vol% DMA, Fig. 1c), the wave for the reduction of free $C_{60,70}$ (A) completely disappears, and, in addition to the wave at E_{pc2} = -1.03 V, a new irreversible reduction wave is observed at E_{pc3} = -1.17 V indicating a second product, C, which presumably is formed slowly from B.

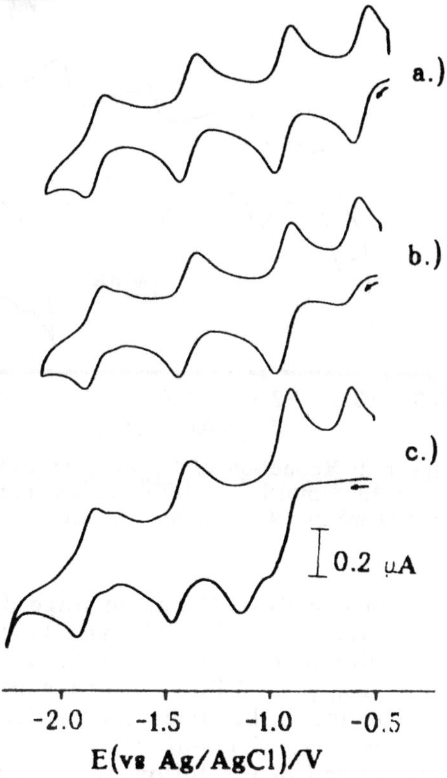

Figure 1: Reduction of C_{60}/C_{70} -20°C, v = 500 mV/s, a) 5 vol%, b) 25 vol%, c) 50 vol% DMA

Figure 2 shows the voltammetric reduction of the fullerenes in pure DMA/TBABr. In the first reduction cycle two cathodic waves are observed at E_{pc2} = -0.93 V and E_{pc3} = -1.12 V which must be assigned to the reduction of the species B or C (in each case an ECE-mechanism).(Fig.2a)

At a scan down to -3.3 V the formation of hexaanions of both fullerenes is observed. The peak current ratio of the respective waves corresponds to the proportion of C_{60}/C_{70} in the fullerene mixture used.

At low scan rates the anodic wave for the reoxidation of free fullerene molecules is almost absent while the peak current of the wave at E_{pa2} = -0.87 V is higher than the other four waves (sums C_{60} and C_{70})(Fig. 2b). This fact shows that the reformation

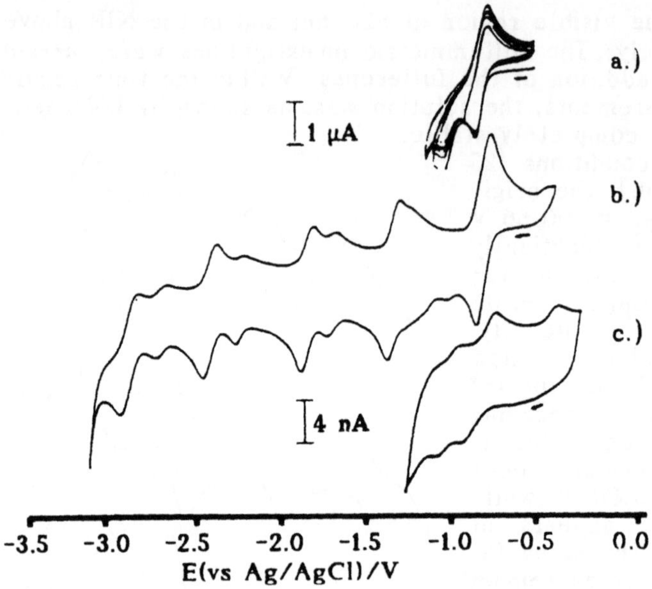

Figure 2: Reduction of C_{60}/C_{70} in DMA/TBABr, T = -40°C, c = $5*10^{-3}$ mol·l^{-1}; a) 200 mV/s, electrode Pt ø = 1mm; b) 200 mV/s, ø = 1 mm; c) 100 V/s, ø = 10 µm.

of B^- proceeds on the anion stage. Fast scan experiments with ultramicroelectrodes showed that the wave at E_{pa1} = -0.45 V increases with increasing scan rate (Fig. 2c) At the same time the wave at E_{pc3} = -1.12 V, which can be assigned to the reduction of the species C, becomes chemically reversible.

The experimental results of our reduction experiments are summarised in the following reaction scheme:

Scheme 1:

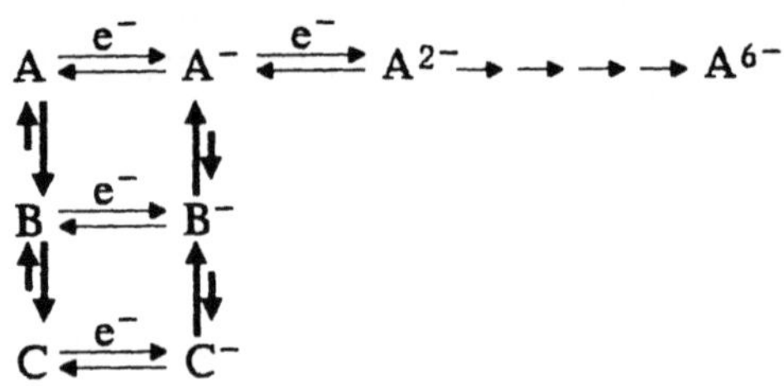

Alongside the free fullerene molecules (A), two species B and C exist in the solution, their quantity depending on DMA content.

Either B and C are coordinative complexes, which differ in the number of DMA molecules involved ($B = C_{60,70}(DMA)_x$, $C = C_{60,70}(DMA)_y$, $y > x$). or only B is a complex, whereas C stands for an adduct with a σ-bond between $C_{60,70}$ and DMA. B and C dissociate on the monoanion stage because of their surplus negative charge, thus freeing DMA and A^-, which at the given potential

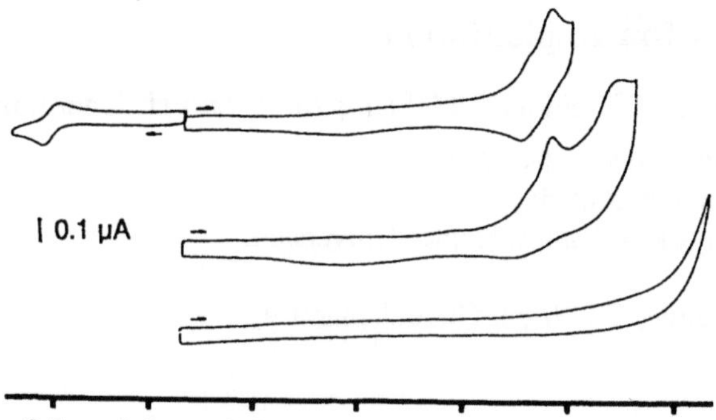

Figure 3: Oxidation and reduction of C_{60}/C_{70} CH_2Cl_2 T = -65°C, v = 100 mV/s, WE(Pt)ø =1mm; a) oxidation and first reduction b) oxidation to E_{λ} = 2.15 V c) background current

immediately reduces to A^{2-}. The total reduction to the dianion thus takes place via a CEE-mechanism for A or an ECE-mechanism for B and C respectively. All further reaction steps are simple one-electron (E) steps of the free fullerene anions A^{n-} (n>2). The oxidation experiments were carried out in CH_2Cl_2/$TBAPF_6$. At 1.65V and 1.75V two oxidation peaks appear with, in each case, approximately doubled peak current in comparison to the first (single-electron) reduction wave.(Fig.3) On the other hand, during the reversal cathodic scan the peak currents are significantly smaller, which can be refered to adsorption effects. We assume that two two-electron transitions occur. If the anodic scan is extended to higher potentials, a further large oxidation wave is observed at about 2.0 V. However, in this case the working electrode is rendered inert by a deposit, which may be attributable to the poor solubility of the cation salt formed.

Acknowledgement

This work was supported by the Fonds der Chemischen Industrie. We would like to thank Mr. W. Schemionek for the preparation of the fullerenes.

References

1 A.D. Haymet, Chem. Phys. Lett., 122 (1985) 421.
2 Q.Xie, E. Perez-Cordero and L. Echegoyen,J.Am.Chem.Soc. 114 (1992) 3978.
3 W.Krätschmer, L.D. Lamb, K. Fostiropoulos, D.R.Huffmann, Nature, 347 (1990) 354.
4 K. Meerholz, J. Heinze, J. Am. Chem. Soc., 111 (1989) 2325.

Buried K-Fullerides by Ion Implantation

J. Kastner[1], *L. Palmetshofer*[2], *P. Bauer*[2], *G. Stingeder*[3], *and H. Kuzmany*[1]

[1]Institut für Festkörperphysik, Universität Wien, Strudlhofgasse 4, A-1090 Wien, Austria

[2]Institut für Experimentalphysik, Johannes Kepler Universität, A-4040 Linz, Austria

[3]Institut für Analytische Chemie, TU Wien, Getreidemarkt 9, A-1060 Wien, Austria

Abstract. We report the formation of buried K-fulleride layers by high dose implantation of 30 keV K^+ at different implant temperatures. Raman scattering showed a transformation of fullerene molecules to amorphous carbon (a-C) within the range of the implanted ions. The microscopic structure of the a-C is discussed. At elevated implant temperatures potassium accumulates underneath the a-C layer. This behaviour is explained in terms of the formation process of a K-fulleride. For comparison, results obtained on Ar^+ implanted fullerene films are presented.

1. Introduction

Ion implantation is commonly used to modify many of the near-surface properties of materials. It is applied in modern microelectronics to the doping of semiconductors and for the formation of buried compounds such as SiO_2 or $CoSi_2$ in silicon [1]. In a recent paper we employed ion implantation for the synthesis of doped fullerene layers [2]. Fullerenes have attracted considerable attention, because they exhibit a series of exciting properties, like phase transitions or superconductivity, when they are doped with alkali metals [3,4]. By using proper implantation conditions we obtained a highly conducting structure consisting of an amorphous carbon (a-C) surface layer and a buried doped fullerene layer.
In this paper we present a detailed analysis of the implantation-induced a-C surface layer by means of Raman scattering and study the formation process of the buried K-fulleride by secondary ion mass spectrometry (SIMS) and Rutherford backscattering spectrometry (RBS).

2. Experimental

Thin films of purified C_{60} with a typical thickness of 300 nm were prepared on silicon by vacuum deposition at 400°C. The $^{39}K^+$ and $^{40}Ar^+$ implantation was carried out at an energy of 30 keV with doses ranging from 1×10^{14} cm^{-2} to 3×10^{16} cm^{-2}. The implant temperature was varied from room temperature up to 300°C. Raman spectra were measured on a Dilor XY spectrometer after excitation with the 514.5 nm line of an Ar^+ laser. SIMS depth profiles were recorded on a Cameca IMS 3f with 5.5 keV O_2^+ as the primary ions. In order to convert the sputtering time into a depth scale, the sputter rate was determined by measuring the depth of the crater by means of optical profilometry. Characterization by RBS was done with 400 keV He^+ as the analysing beam in a standard RBS arrangement.

Springer Series in Solid-State Sciences, Volume 117
Electronic Properties of Fullerenes Eds.: H. Kuzmany, J. Fink, M. Mehring, and S. Roth

3. Results and Discussion

The Raman spectrum of a C_{60} film implanted at 300°C with K^+ (30 keV, $1x10^{16}$ cm^{-2}) is plotted in Fig.1. The Raman spectrum shows the high energy A_g pinch mode of C_{60} at 1468 cm^{-1} and a broad asymmetric line between 1000 cm^{-1} and 1700 cm^{-1} . Very thin films exhibit only this broad asymmetric feature and no signal from the fullerene. The broad asymmetric line is well known and characteristic for a-C [5]. It can be deconvoluted into two Gaussian lines (Fig.1: dotted lines) at around 1340 cm^{-1} and 1540 cm^{-1}, which are designated in literature as the D and G lines [5]. The G line is the broadening of the E_{2g} mode of crystalline graphite when the long range-order is lost. The D line is disorder induced because the k=0 selection rule breaks down at grain boundaries and defects. The intensity of the D line strongly depends on the degree of disorder. The ratio of the integrated intensity of the D line to the G line $R=I_D/I_G$ depends inversely on the in-plane graphite crystallite size L [6]. The dependence of R on L holds over the extended range $0.001 \leq R \leq 1$ and 2.5 nm$\leq$L$\leq$300 nm. In our case R=2.3, which is outside the valid range, but we can conclude that the destruction of C_{60} molecules by the implantation process leads to a-C with a mixture of different sp^2 and sp^3 bondings, where the sp^2 bonded graphitelike carbon clusters are smaller than 2.5 nm.

In order to measure the thickness of the a-C surface layer we prepared 150x150 μm craters with different depths by sputtering with 5.5 keV O_2^+ and measured the Raman spectra at the bottoms of the craters. The penetration depth of O_2^+ into the a-C layer is less than 10 nm, so that the thickness of the a-C layer generated by the sputtering O_2^+ ions is almost negligible, as compared to the thickness of the a-C layer prepared by 30 keV K^+ implantation. The ratio of the peak height of the asymmetric a-C line and the pinch mode of C_{60} (I_{a-C}/I_{C60}) versus the depth x of the sputtered crater is plotted in Fig.2. The obtained data were fitted by the equation [6]

$$\frac{I_{a-C}}{I_{C60}} = k\frac{d-x}{exp[-2\alpha(d-x)]}$$

where d is the thickness of the a-C layer, α is the absorption coefficient of a-C and k is a constant. As fit parameters we obtained 75 nm for d, and $1.8x10^5$ cm^{-1} for α. The value

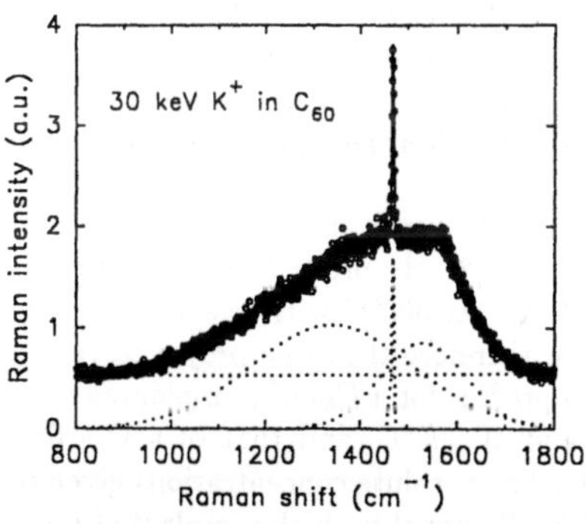

Fig. 1: Raman spectrum of a C_{60} film implanted with K^+ (30 keV,$1x10^{16}$ cm^{-2}) at 300°C. Open circles are experimental data points and dotted lines are the decomposed Raman lines after subtracting a linear background.

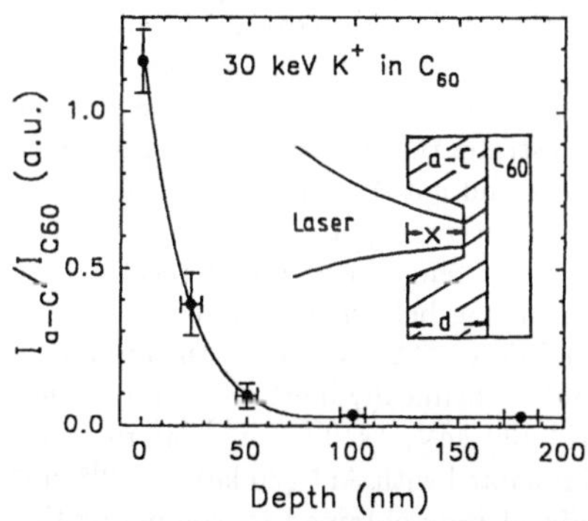

Fig. 2: Ratio of peak height intensities of the a-C line I_{a-C} and the pinch mode of C_{60} I_{C60} as a function of the crater depth. The inset shows that Raman spectra were taken from the bottom of the sputtered craters.

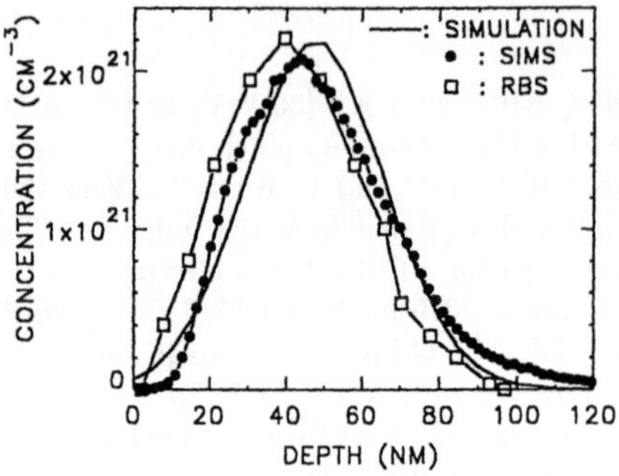

Fig. 3: K depth profiles for a C_{60} film implanted with K^+ (30 keV, $1x10^{16}$ cm^{-2}) calculated with TRIM (full line) and measured with SIMS (full circles) and RBS (open squares).

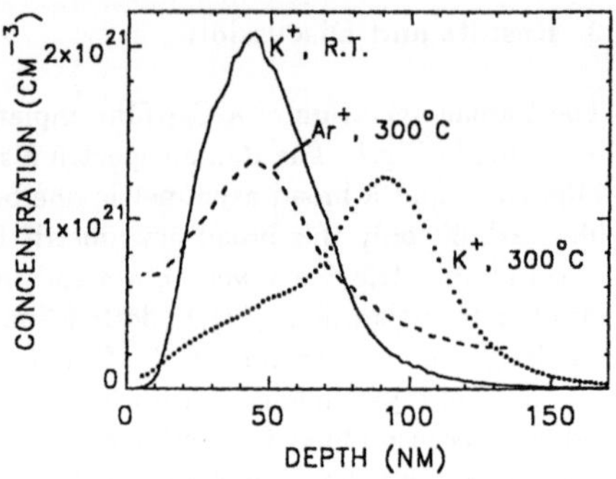

Fig. 4: SIMS depth profiles for C_{60} films implanted with K^+ (30 keV, $1x10^{16}$ cm^{-2}) at room temperature (full line) and at 300°C (dotted line) and implanted with Ar^+ (30 keV, $1x10^{16}$ cm^{-2}) at 300°C (dashed line).

of α is in good agreement with measured absorption coefficients of different a-C films [7,8]. Thus, the a-C layer thickness obtained from the above expression can be regarded as a good estimation.

Fig.3 shows the distribution of K atoms (30 keV, $1x10^{16}$ cm^{-2}) after room temperature implantation calculated with the TRIM code [9] and measured with SIMS and RBS. The profiles are all Gaussian-like, but the measured profiles exhibit a tail at greater depths. This tail may originate from the diffusion of K in the undestroyed C_{60} film, since it is known that K can diffuse in C_{60} even at room temperature [3,4]. Another striking feature is that the maximum of the calculated profile is shifted away from the surface, by 4–8 nm as compared to the experimental results. This behaviour is probably due to the higher density of the a-C layer as compared to C_{60}. C_{60} has a density of 1.65 g/cm^3, which is lower than the density of graphite and most forms of a-C [8]. Thus, it is tempting to assume a higher density for the transformed fullerene molecules, which causes a shallowing of the profile. Moreover, sputtering effects, which shift the maximum of the implantation profile to the surface might play a role as well, although the sputtering yield for carbon is rather low [10]. Comparing the concentration profiles with the thickness of the a-C layer illustrates that the fullerene molecules are only destroyed within the range of the implanted ions, and not deeper inside the film.

In a recent paper we have shown that the implanted K-atoms accumulate at elevated implant temperatures underneath the a-C surface layer [2]. In order to investigate this surprising behaviour in more detail, we implanted $^{40}Ar^+$ at 300°C with the same energy and dose. ^{40}Ar has about the same mass as ^{39}K, and therefore the same projected range, but is chemically inert. Fig.4 shows the SIMS depth profiles for a C_{60} film implanted with K^+ (30 keV, $1x10^{16}$ cm^{-2}) at room temperature and at 300°C, and that of a C_{60} film implanted with Ar^+ (30 keV, $1x10^{16}$ cm^{-2}) at 300°C. The absolute concentrations given in Fig. 4 were obtained by comparing the integrated SIMS signal with the implanted dose. This gives a correct value for K, but is uncertain for Ar, because the Ar-SIMS signal is influenced by contributions of ^{40}Ca, which has the same mass as the implanted $^{40}Ar^+$, but a much higher ionization probability. It is very common that Ca or alkali metals are observed by SIMS measurements due to a slight contamination of the samples. Thus, the plotted Ar profile is the sum of implanted Ar and a certain amount of Ca.

The K-profile after room temperature implantation is the usual Gaussian function with a small tail at greater depths, as was discussed above. The K-profile after implantation at 300°C is unusual. K diffuses out of the implanted region and accumulates at a depth of about 100 nm. In contrast to this behaviour, Ar implantation at 300°C results in a normal diffusion broadened profile, where the maximum of the Ar-concentration coincides with the maximum of the K-concentration after room temperature implantation. This means that the driving force for the K-accumulation at elevated implant temperatures must be related to a chemical interaction of K with C_{60}. The negative heat of formation in the K-fulleride phase favours the diffusion from a-C to the undestroyed fullerene.
We propose the following model for the formation process. Since it was found that at elevated temperatures ($T \geq 100°C$) and at slight doping levels K_1C_{60} is the thermodynamically stable phase [4], we assume that during the implantation process a K_1C_{60} layer is formed. This layer is embedded in between an a-C surface layer and an undoped fullerene layer. At sufficient high doses the K_1C_{60} layer is continuous. When the sample is cooled down to room temperature after implantation, the K_1C_{60} layer decays to dispersed K_3C_{60} clusters in a C_{60} matrix. Further measurements to prove this model are in progress.

4. Conclusion

We have shown that the implantation of 30 keV K^+ in C_{60} leads to an a-C surface layer with a thickness of about 75 nm. The a-C consists of a mixture of different sp^2 and sp^3 bondings and the sp^2 bonded graphitelike carbon clusters are smaller than 2.5 nm. The density of this kind of a-C is slightly higher than that of C_{60}. Using elevated implantation temperatures leads to the formation of a buried K-fulleride layer. The negative heat of formation of this compound is the driving force, which leads to an accumulation of K at a depth of about 100 nm.

Acknowledgement. This work was supported by the Osteuropa Förderung des BMfWF project GZ 45.212/2-27b91. We are grateful to K.H. Ablinger for performing the implantation and to K. Piplits for assistance in SIMS measurements.

References

1. S. Mantl, Mater. Sci. Rep. **8**, 1 (1992)
2. J. Kastner, H. Kuzmany, L. Palmetshofer, P. Bauer and G. Stingeder, Nucl. Instr. and Meth. B , in print
3. A. F. Hebard, M. J. Rosseinsky, R. C. Haddon, D. W. Murphy, S. H. Glarum, T. T. M. Palstra, A. P. Ramirez, and A. R. Kortan, Nature **350**, 600 (1991)
4. J. Winter and H. Kuzmany, Solid State Commun. **84**, 935 (1992)
5. M.S. Dresselhaus and R. Kalish, Springer Series in Materials Science **22**, "Ion Implantation in Diamond, Graphite and Related Compounds" (Springer, Heidelberg, 1992)
6. W. Richter, Springer Tracts in Modern Physics **78** (Springer, Heidelberg, 1976)
7. J. Osaheni, S. A. Jenekke, A. Burns, G. Du, Z. Wang, A. Epstein and C. Wang, Macromol. **25**, 5828 (1992)
8. J. Robertson, Adv. in Physics **35**, 317 (1986)
9. J. F. Ziegler, J. P. Biersack and U. Littmark, "The Stopping and Range of Ions in Solids" vol 1 (Pergamon, New York, 1985)
10. R. Behrisch (ed), Topics in Applied Physics **47**, "Sputtering by Particle bombardment 1" (Springer, Heidelberg, 1981)

C_{60}-Fullerene in Water Solution, γ-Cyclodextrin Complex and Cluster of C_{60}: An NMR and a Photophysical Study*

T. Andersson, M. Sundahl, O. Wennerström, and G. Westman

Department of Organic Chemistry, Chalmers University of Technology, S-412 96 Göteborg, Sweden

*The NMR investigation was presented as a poster by T. Andersson and the photophysical investigation was presented as a research lecture by M. Sundahl

Abstract. γ-CD has been used to dissolve C_{60} in water. At high γ-CD:C_{60} ratio a complex that is monomeric in C_{60} is stable whereas at lower γ-CD:C_{60} ratio a cluster of several C_{60} molecules surrounded by γ-CD is stable. We propose that clusters of different sizes can be formed. The γ-CD:C_{60} complex have an electronic absorption spectrum similar to that of an organic solution of C_{60} and the cluster have an electronic absorption spectrum similar to that of a thin film of C_{60}. In an NMR experiment on a γ-CD:C_{60} complex solution two sets of signals for the cyclodextrin protons were observed indicating that the exchange of C_{60} between different γ-CD:s is slow on the NMR time scale.

For the γ-CD:C_{60} complex, the photophysical properties are similar to those of C_{60} in an organic solvent with the exceptions for quenching rates of the triplet state by molecular oxygen and annihilation rates of the triplet state.

The properties of the clusters are different from those of an organic solution of monomeric C_{60}. For water solutions of small clusters, decay of the excited state of the cluster is clearly dependent on the intensity of the excitation laser pulse; at high laser intensity there is both a fast and a slow component in the decay process whereas at low laser intensity only the slow process is observed.

1. Introduction

The chemical and physical properties of a compound depend on the state of aggregation and can be modified by a solvent or by complexation. Beautiful examples of host-guest complexes have been provided by Cram and others [1]. The fullerenes do form endohedral metallo derivatives [2]. Our approach, however, is to make C_{60} water soluble by making a complex with γ-cyclodextrin where C_{60} should be regarded as a guest molecule, not the host as in the endohedral compounds. In this paper we report on an NMR and a photophysical investigation of water solutions of γ-CD:C_{60} complexes and a photophysical investigation of a water solution of a cluster of C_{60}.

Springer Series in Solid-State Sciences, Volume 117
Electronic Properties of Fullerenes Eds.: H. Kuzmany, J. Fink, M. Mehring, and S. Roth

2. Results and Discussion

2.1 Preparation of Water Solutions of Complex and Cluster

The γ-CD:C_{60} water solution was prepared by boiling an aqueous solution of γ-CD (0.02-0.1 M) containing solid C_{60} (>99%) [3]. The dissolution of C_{60} is slow and to maximize the concentration of the complex it is necessary to reflux the solution for more than 50 hours. The electronic absorption spectrum of a solution prepared in this way is shown in Fig. 1.

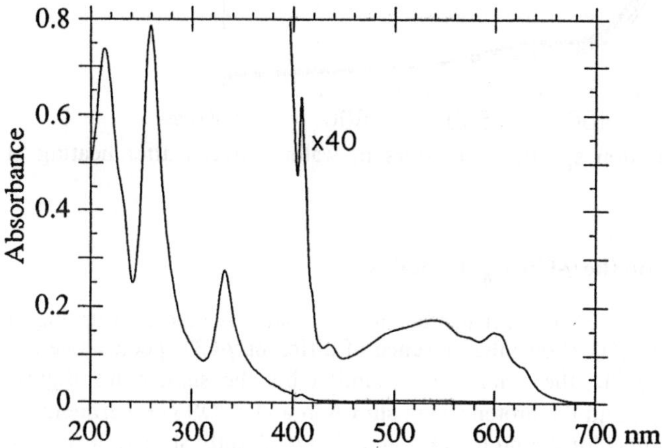

Fig. 1. Electronic absorption spectrum of a water solution of γ-CD:C_{60} complex.

Interestingly, the solution has the characteristic magenta color of C_{60}. There is also a magenta colored precipitate formed from concentrated solutions. On addition of cold water to this precipitate most of the excess of cyclodextrin can be washed away. The remaining precipitate can be dissolved in water by careful warming yielding a magenta solution. On prolonged heating the color changes to yellow (similar to that of a thin film of C_{60}). We propose that a cluster of several C_{60} molecules surrounded by γ-CD is formed [4]. The electronic absorption spectrum of the cluster changes with the time the solution has been heated (Fig. 2). We suggest that larger clusters are formed on prolonged heating. A solution of "large" clusters was extracted with toluene and C_{60} was quantitatively recovered to the toluene solution, assuming that the molar absorptivities for the "332 nm maxima" for a water solution and a toluene solution are the same.

2.2 Selectivity

There are two kinds of selectivity effects of interest when preparing the complex; changing γ-cyclodextrin for α– or β–cyclodextrin and the other, changing C_{60} to C_{70}. Neither α– nor β–cyclodextrin yield any complex with C_{60} under similar experimental conditions as those used with γ-CD. In an experiment using a C_{70} enriched mixture of C_{70} and C_{60} (93:7) an almost pure γ-CD:C_{60} complex was obtained. The selectivity of complexing of C_{60} over C_{70} was estimated to >150:1.

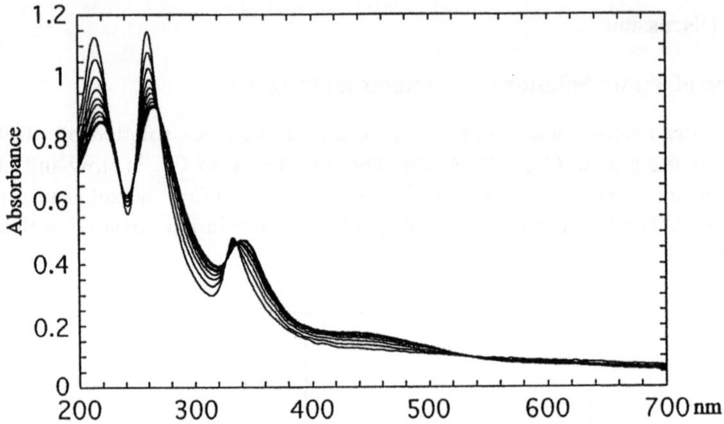

Fig. 2. Electronic absorption spectra of clusters in water solution after heating at different times at 70°C.

2.3 NMR Experiments on the γ-CD:C_{60} Complex

The ^{1}H-NMR spectrum (Fig. 3) of the magenta solution, made by careful warming of the precipitate in water (D_2O), shows the presence of different γ-CD species, one has the same shifts as free γ-CD, the other is very similar but the signals are slightly shifted. The connectivity of all the protons were shown in a DQCOSY-experiment. In the C_{60} complex, the signals for the protons H3 and H6 are slightly shifted downfield whereas small upfield shifts are noted for H1 and H2. The shift for the remaining

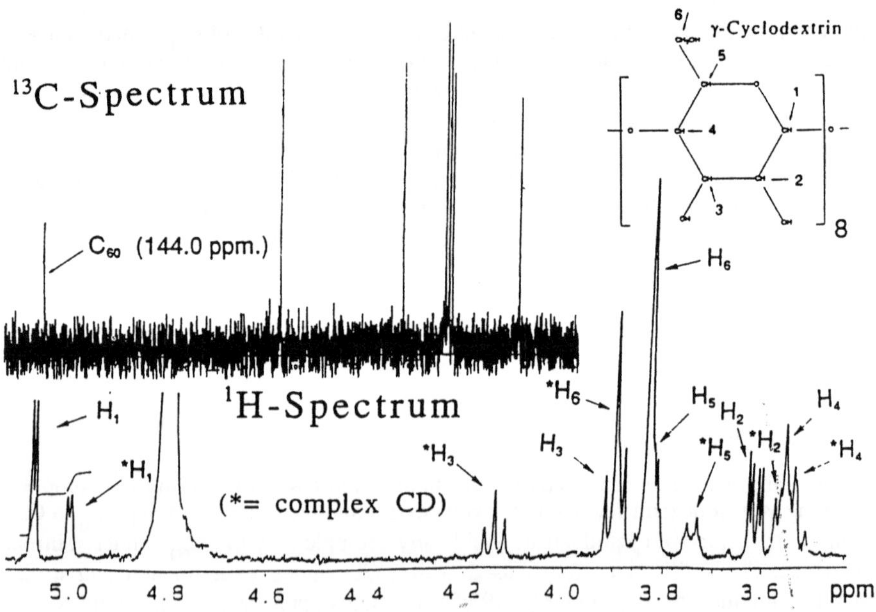

Fig. 3. ^{1}H NMR and ^{13}C NMR of a D_2O solution of the γ-CD:C_{60} complex.

Table 1. Chemical shifts of the protons for different CD-species at 25 °C in ppm.

	H-1	H-2	H-3	H-4	H-5	H-6
free α-CD	5.05	3.63	3.98	3.58	3.86	3.90
free β-CD	5.07	3.65	3.96	3.58	3.88	3.88
free γ-CD	5.10	3.65	3.93	3.59	3.87	3.87
complex	5.03	3.58	4.17	3.55	3.78	3.92

protons, H4 and H5, are unaltered. The largest shift is observed for H3 which is the hydrogen closest to C_{60} in our model structure. The observed ^{1}H-NMR shifts are small and may be due to small conformational changes in the γ-CD upon complexation rather than to an effect of the π-electrons of the C_{60}. For comparison the chemical shifts for the protons in α-, β- and γ-CD are shown in Table 1. Of the three CD:s the α-macrocycle has the more conical shape. On complexation γ-CD should also adopt a more conical shape, which would then explain the changes in the NMR shifts.

The T_1 relaxation times (Tab. 2) for various proton signals were measured and the values shown below were obtained. The T_1 for the protons in the complex is longer than for "free" γ-cyclodextrin.

Table 2. T_1 relaxation measurements in s.

	H-1	H-2	H-3	H-4	H-5	H-6
Free γ-CD	0.96+0.01	1.07+0.07	1.61+0.03	1.07+0.04	0.67+0.00	0.67+0.00
Complex	1.33+0.04	1.34+0.1	1.96+0.08	1.20+0.05	1.22+0.04	1.00+0.06

By ^{13}C-NMR (D_2O) experiment (Fig. 3) C_{60} was identified as a single peak at 144.0 ppm, close to those reported for the ^{13}C-shift of C_{60} in organic solvents. The relaxation time (T_1) for C_{60}-γ-CD in water is much faster than that of C_{60} in toluene. The ^{13}C-NMR spectrum did not show distinct signals from γ-CD bound to C_{60}.

The question whether C_{60} is bound to one or two molecules of γ-CD could not be solved even after quantitative NMR analysis of the amount of free and complexed γ-CD and UV-determination of the C_{60} concentration. Consistent ratios between bound γ-CD and C_{60} was not obtained but rather, ratios varying between 1 and 2. The ratio increases with the amount of free γ-CD and decreases with the concentration of free and complexed γ-CD. In concentrated solutions a 2:1 complex dominates, in dilute solutions a 1:1 complex is more favorable. The main problem is that equilibrium conditions can not be obtained since the host-guest exchange rates are slow and the solubility of the C_{60}-CD complexes are low at room temperature. Heating will cause decomposition of the complexes. The change in solubility as well as the variation in composition of the C_{60}-γ-CD complex with the variation in the concentration of free γ-CD is explained by a slow equilibrium between the 1:1 and the 2:1 complexes.

2.4 Molecular Modelling of γ-CD:C_{60} Complex

To check the size of C_{60} in relation to that of γ-CD we have made molecular modelling using the program BIOGRAPH[5] (BioDesighn Inc. Dreiding II force field). Docking of C_{60} and γ-CD gave a stable 1:1 complex (Fig. 4). Further docking with another γ-CD unit gave a structure in which the C_{60} fits nicely into the cavity of two γ-CD units bond together via the longer rim (Fig. 4). However, the effect of water has not been considered in the modelling experiments.

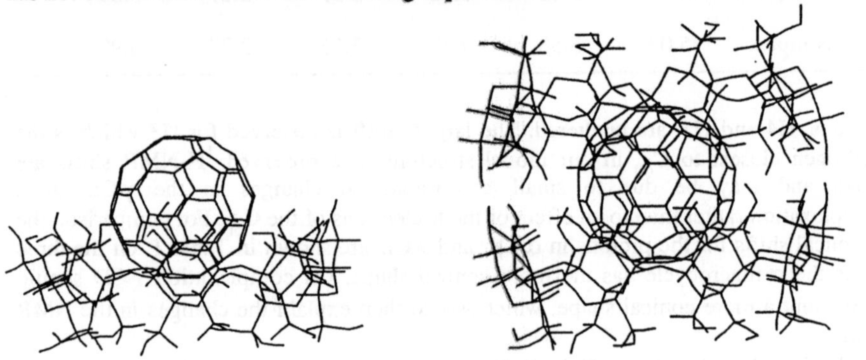

Fig. 4. Models of γ-CD:C_{60} complexes derived from molecular modelling.

2.5 Photophysics of the γ-CD:C_{60} Complex

In Fig. 5 a transient difference absorption spectrum obtained on pulsed laser excitation (7 ns, 355 nm) is shown. At longer wavelengths the spectrum is due to T_1-T_n absorption and at shorter wavelengths there is a combination of ground state bleaching and T_1-T_n absorption. The transient is quenched by molecular oxygen and the experimental quenching rate constant is 6.4×10^8 $M^{-1}s^{-1}$. This rate constant is only 1/10 of the diffusion controlled, whereas in a toluene solution the triplet is quenched

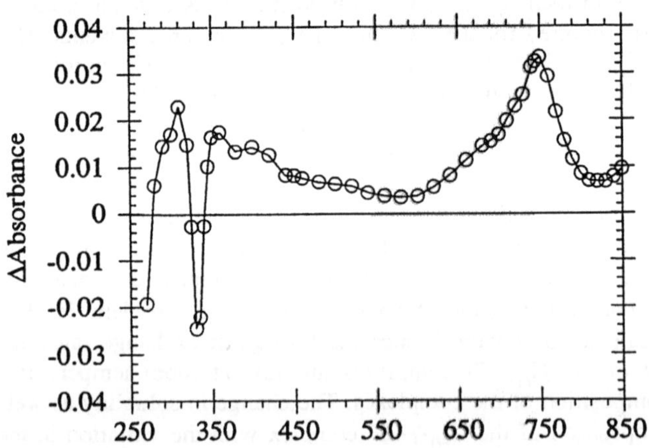

Fig. 5. Difference absorption spectrum obtained on pulsed laser excitation. The spectrum was obtained by slicing traces at different wavelengths.

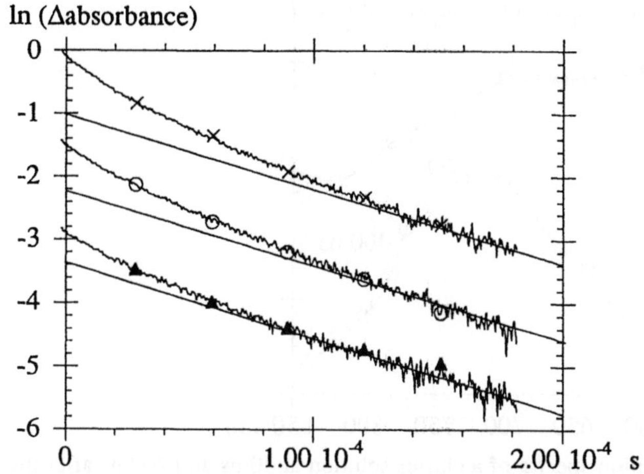

Fig. 6. Decay traces obtained at 750 nm, at different powers of the excitation pulse. The traces are shifted vertically for clarity, Δ absorbance values at 0 s were from top 0.40, 0.22 and 0.15, respectively. The straight lines corresponds to a half life of 83 μs.

with a rate 1/5 of the diffusion controlled. This suggests that half of the surface of $^3C_{60}^*$ in the complex is available for collision with molecular oxygen.

Furthermore, the shape of the decay curve is dependent on the intensity of the excitation laser pulse (Fig. 6). We conclude that this is due to T-T annihilation. Assuming first plus second order decay kinetics and then fitting to the decay curves, we get a triplet state lifetime of 83 μs and a second order decay rate constant 1.2 x 10^9 $M^{-1}s^{-1}$ for a nitrogen flushed solution of the complex. For a toluene solution the second order rate constant was determined to 7.1 x 10^9 $M^{-1}s^{-1}$. The molar absorptivity for the T-T absorption at 750 nm was taken to 2 x 10^4 $M^{-1}cm^{-1}$ [6]. If we correct for the different diffusion rates in water and toluen respectively the annihilation rate in water is only 1/4 of that for a toluene solution, again suggesting that half of the surface of the $^3C_{60}^*$ is available for collision. Thus, oxygen quenching of the excited triplet and T-T annihilation is clearly in favor of a 1:1 γ-CD:C_{60} complex.

2.6 Photophysics of the γ-CD:C_{60} cluster

In Fig. 7 transient difference absorption spectra of a solution of cluster of medium size at different times after the excitation laser pulse (7 ns, 355 nm) are shown. Decay traces were collected at different wavelengths and where then sliced as indicated in Figure 7. The spectrum clearly changes with time. There are different possibilities of assigning the spectra of the transient species. Either, the "short time" transient spectrum is due to an (or a higher) excited state of the cluster and then probably with a delocalized nature, or there are ionic species formed. Another intriguing possibility is that solvated electrons are formed. The spectrum of the "short time" transient species is similar to that of solvated electrons. The spectra after longer times are most probably due to an excited triplet state. Furthermore, the shape of the decay curve is clearly dependent on the intensity of the excitation laser pulse (Fig. 8). At higher

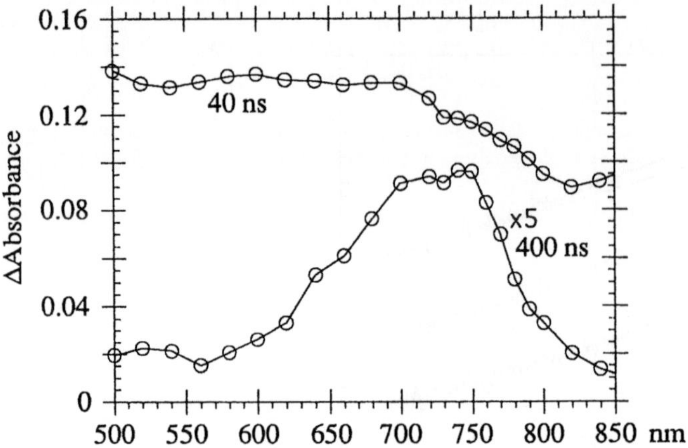

Fig. 7. Difference absorption spectra of a cluster solution at 40 ns and 400 ns after the laser pulse.

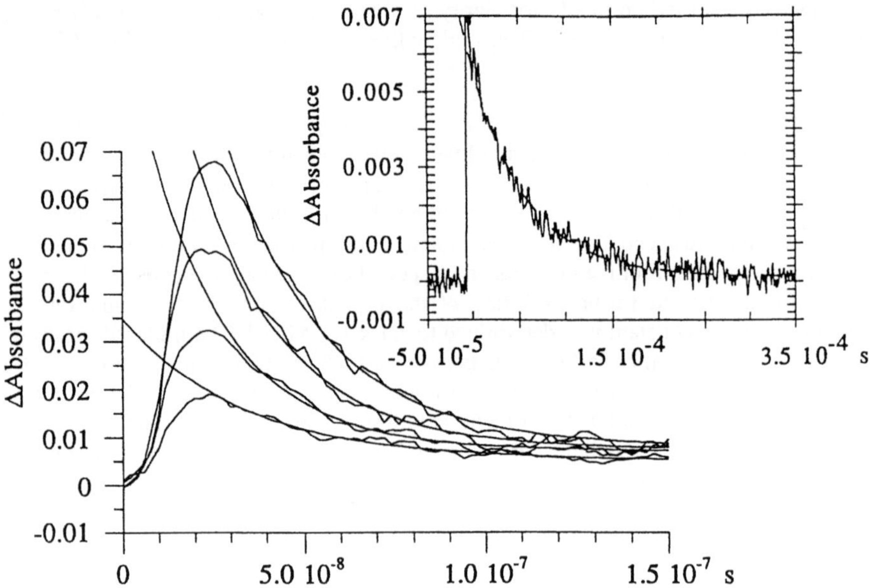

Fig. 8. Decay curves at 720 nm, energy of excitation laser 29, 41, 57, and 74 mJ per pulse. Inset, decay on a long time scale (29 mJ pulse). Excitation wavelength 355 nm, pulse length 7 ns.

intensities we get more of the fast decay process and relatively less of the slower, showing the multiphotonic nature of the formation of the short lived species. Another way of changing the shape of the decay curve is to vary the size of the cluster; for solutions of large clusters we can no longer detect the slow decay process but only the fast one.

3. Conclusions and Outlook

It has been possible to dissolve C_{60} in water with the help of γ-CD. Two different solutions can be formed, one which is monomeric in C_{60} and one where several C_{60} molecules stick together. A ^{1}H NMR investigation show the presence of two sets of cyclodextrin protons for a solution of the complex, The NMR experiments are in favor of an equilibrium between a 2:1 and a 1:1 γ-CD:C_{60} complex. C_{60} was identified by a ^{13}C signal at 144.0 ppm. The photophysical properties of the γ-CD:C_{60} complex are not consistent with a photostable 2:1 γ-CD:C_{60} complex as the dominating species. If the γ-CD:C_{60} ratio is low several C_{60} molecules can stick together forming a solution of a C_{60} cluster. Different sizes of these clusters can be formed. The shape of these clusters might be either spherical or alternatively linear. Dynamic light scattering and other experiments will hopefully reveal the shape and size of these clusters. Further experiments on the photophysical properties are also under way.

References

[1] Cram, D. J.; Tanner, M. E.; Thomas, R. *Angew. Chem.* Intern. Ed. 30 (1990) 1024.

[2] Johnson, R. D.; de Vries, M. S.; Salem, J.; Bethune, D. S.; Yannoni, C. S.; *Nature* 355 (1992) 239.

[3] T. Andersson, K. Nilsson, M. Sundahl, G. Westman, O. Wennerström, *J. Chem. Soc. Chem. Com.* (1992) 604; also presented as a poster at the 14th IUPAC symposium on photochemistry, K.U. Leuven, Belgium, 19-25 july, 1992.

[4] M. Sundahl, T. Andersson ,K. Nilsson , O. Wennerström, G.Westman *Synthetic Metals* in press.

[5] S.L. Mayo, B.D. Olafson, W. A. Goodard, III, *J. Phys. Chem. 94* (1990) 8897.

[6] J.W. Arbogast, A.P. Darmanyan, C.S. Foote, Y. Rubin, F.N. Diedrich, M.M. Alvarez, S.J. Anz, R.L. Whetten, *J. Phys. Chem., 95* (1991) 11; Y. Kajii, T. Nakagawa, S. Suzuki, Y. Achiba, K. Obl, K. Shibuya, *Chem. Phys. Lett., 181* (1991) 100; T.W. Ebbesen, K. Tanigaki, S. Kuroshima, *Chem. Phys. Lett., 181* (1991) 501; R.R. Hung, J.J. Gabrowski, *J. Phys. Chem., 95* (1991) 6073.

Pulse Radiolytic Redox and Alkylation Studies on C_{60}

D.M. Guldi, H. Hungerbühler, and K.-D. Asmus
Hahn-Meitner-Institut Berlin, Bereich S, Abt. Strahlenchemie,
Glienicker Str. 100, D-14109 Berlin, Germany

Abstract. The radiation chemical method of pulse radiolysis is introduced as a suitable tool for the analysis of time-resolved optical changes in radiolytically induced redox and alkylation reactions. A pseudo-first order reaction observable in the radiolysis of C_{60} in 2-propanol leads to the formation of the radical anion in the reaction of C_{60} with solvated electrons and dimethylhydroxymethyl radicals. One-electron oxidation in 1,2-dichloroethane (1,2-DCE) yielded $C_{60}^{\bullet+}$, achieved via interaction of C_{60} with the radiolytically generated oxidizing $(1,2\text{-DCE})^{\bullet+}$radical cations. The reaction of methyl radicals with C_{60} resulted in absorption changes which are very different from redox reactions and are ascribed to the methylation of the fullerene. Corresponding experiments in aqueous solutions have been carried out by incorporating C_{60} into artificial bilayer membranes and larger organic molecules. Several absolute rate constants for the radical induced reactions have been determined.

Since the first successful synthesis of the fullerenes by Krätschmer et al. [1] a steadily increasing number of studies have been conducted with these molecules. Redox chemistry studies, particularly with electrochemical methods [2-6], have demonstrated the ability of fullerenes to undergo up to six subsequent reversible reduction steps [4]. Oxidation on the other hand appears to be limited to a single one-electron step, leading to the formation of the radical cation $C_{60}^{\bullet+}$ with characteristic absorption features in the near IR [7,8]. A further significant property of fullerenes is their capacity to add radicals to their conjugated double bond system, as has been shown recently in ESR [9-11] and pulse radiolysis investigations [12,13]. An interesting aspect is also the possibility to embed C_{60} into host-guest-complexes, e.g. with γ-cyclodextrine [14]. This resulted in a considerable slow-down of the reduction rate of C_{60} by $(CH_3)_2C^{\bullet}(OH)$ radicals compared to a homogeneous 2-propanol solution, and stimulated us to further investigation in this direction, including incorporation of C_{60} into even more organized assemblies such as vesicular and micellar membranes in aqueous environment [15].

Pure C_{60} was purchased from Kaesdorf (Geräte für Forschung und Industrie, München, Germany). Solutions were generally freshly prepared. The pulse experiments were performed by utilizing either 0.5-2 μs pulses of 1.55 MeV, or 500 ns pulses of 3.8 MeV electrons from two van de Graaff accelerator facilities, respectively, described elsewhere [16,17]. The radical concentration generated per pulse was typically (1-3) x 10^{-6} M. All experiments were carried out at room temperature (22±2°C).

Pulse radiolysis of a C_{60} solution in N_2O-saturated 2-propanol solution leads to changes in the optical absorption, which are stable over the experimental time scale (up to 10 ms). The differential absorption spectrum, displayed in Figure 1a and 1b,

Springer Series in Solid-State Sciences, Volume 117
Electronic Properties of Fullerenes Eds.: H. Kuzmany, J. Fink, M. Mehring, and S. Roth

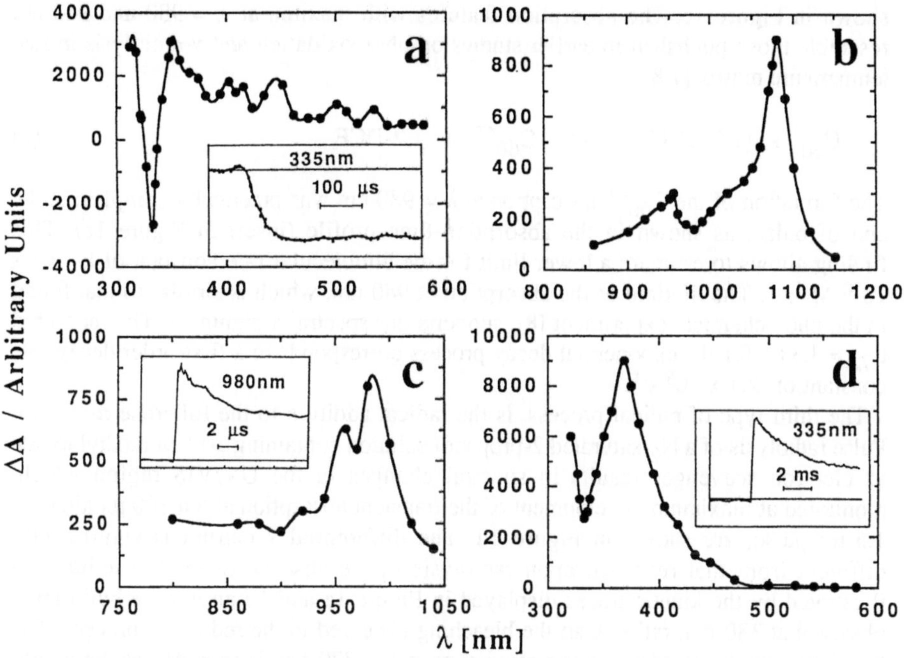

Fig.1: Differential absorption spectrum for a) the formation of the $C_{60}^{\bullet-}$ in the UV/VIS, b) the formation of the $C_{60}^{\bullet-}$ in the near IR, c) the formation of the $C_{60}^{\bullet+}$ in the near IR, d) the formation of the $(C_{60}\text{-}CH_3)^{\bullet}$ in the UV/VIS.

have been recorded 150 µs after the pulse. The spectrum in the UV/VIS (Figure 1a) resembles that obtained for a thin layer spectro-electrochemical reduction [3], while the spectral features in the near IR are identical to those found earlier upon γ-irradiation at low temperature [7]. Since the only reactive species formed under the above conditions is the reducing radical resulting from the radiolysis of 2-propanol, namely $(CH_3)_2C^{\bullet}(OH)$ [16], our finding has to be attributed to the formation of the $C_{60}^{\bullet-}$ radical anion (reaction 1). The absolute rate constant for this reduction of $k_1 = 5\ (\pm 2) \times 10^8\ M^{-1}s^{-1}$ was derived from the time-resolved bleaching of the starting material at 335 nm (insert in Figure 1a) and respective traces at other wavelengths.

$$C_{60} + (CH_3)_2C^{\bullet}(OH) \longrightarrow C_{60}^{\bullet-} + (CH_3)_2CO + H^+ \qquad (1)$$

The same overall changes have been observed upon radiolysis of a N_2-purged 2-propanol solution. In this case the solvated electrons reduce C_{60} directly. The reaction with solvated electrons appears to be diffusion controlled and from the respective absorption time profile a rate constant of $k \geq 10^{10}\ M^{-1}s^{-1}$ can be estimated.

The formation of the $C_{60}^{\bullet+}$ radical cation was observed upon pulse-radiolysis of the fullerene in a N_2-saturated 1,2-dichloroethane solution. Irradiation results primarily in the formation of the highly oxidizing solvent radical cation $(1,2\text{-DCE})^{\bullet+}$ [18-20], which then reacts with C_{60} according to reaction 2. The differential absorption spectrum, recorded in the near IR region almost immediately after a 500 ns pulse, is shown in Figure 1c. The absorption features with maxima at λ = 980 and 960 nm resemble those published in earlier studies on photooxidation and γ-radiolysis in low temperature matrix [7,8].

$$C_{60} + (1,2\text{-DCE})^{\bullet +} \longrightarrow C_{60}^{\bullet +} + 1,2\text{-DCE} \quad (2)$$

The formation of the $C_{60}^{\bullet +}$ absorption at λ = 980 nm was practically complete at the end of pulse, as shown in the absorption-time profile (insert in Figure 1c). This finding allows to estimate a lower limit for the bimolecular rate constant of $k_2 \geq 2 \times 10^{10}\ M^{-1}s^{-1}$. The lifetime of the absorption at 980 nm, which is similar to that found in the photochemical experiment [8], supports our spectral assignment. The measured $t_{1/2}$ = 3.3 µs for the exponential decay process corresponds to a first-order decay rate constant of $2.1 \times 10^5\ s^{-1}$.

The third type of radical process, is the radical addition to the fullerene π-system. Pulse radiolysis of a N_2-saturated 2-propanol solution containing 10^{-1} M methyl iodide as electron scavenger results in spectral changes in the UV/VIS region which, monitored at maximum development of the transient absorption about 300 µs after the 1.5 µs pulse, are shown in Figure 1d. The differential spectrum is significantly different from that recorded upon radiolysis in the absence of methyl iodide. As illustrated by the kinetic traces displayed in Figure 1a and 1d an intense build-up is observed at 330 nm, rather than the bleaching observed in the reduction process. The transient, which exhibits a strong maximum at λ = 370 nm, is regarded as the mono-methyl-adduct of C_{60}, generated via the following processes.

$$CH_3I + e^-_{sol} \longrightarrow {}^{\bullet}CH_3 + I^- \quad (3)$$

$$^{\bullet}CH_3 + C_{60} \longrightarrow (C_{60}\text{-}CH_3)^{\bullet} \quad (4)$$

A bimolecular rate constant of $k_4 = 4.1\ (\pm 2) \times 10^9\ M^{-1}s^{-1}$ has been derived from the build-up of the transient absorption at 370 nm. The radical adduct decays by second-order kinetics with a halflife of 2.0 ms. This decay process is suggested to be dimerization of the radical adduct, in accord with a recent ESR study [11].

The radiolytical reduction of a water soluble host-guest complex, formed upon embedding of C_{60} into γ-cyclodextrine (γ-CD) [14], resulted in a slow-down of the reduction rate. The rate constant for the reduction of C_{60}- γ-CD by $(CH_3)_2C^{\bullet}(OH)$ has been evaluated to $k = 2.7\ (\pm 0.5) \times 10^8\ M^{-1}s^{-1}$. The decrease in rate constant as compared to that for the reduction of the uncomplexed C_{60} might be ascribed, at least in part, to the steric shield provided by γ-CD. Nevertheless, other parameters such as solvation and polarity of the solvent may also have to be considered. A further decrease of the rate was observed when 2-propanol was substituted by ethanol or methanol, which also serves as radical source for reducing species. Rate constants of $1.4\ (\pm 0.5) \times 10^8\ M^{-1}s^{-1}$ and $5\ (\pm 2) \times 10^7\ M^{-1}s^{-1}$ have been determined for the reduction of C_{60}- γ-CD by $CH_3C^{\bullet}H(OH)$ and $C^{\bullet}H_2OH$ radicals, respectively.

Recently we have reported on the incorporation of C_{60} into artificial lipid membranes [15]. Among micellar systems the reduced form of Triton X-100 and BRIJ 35 (which are derivatives of a polyethyleneglycol unit with a hydroxyl head group and a non-polar extended hydrocarbon chain) provided the best yields for the incorporation of C_{60}. The concentration of the incorporated fullerene was determined spectrophotometrically.

The pulse radiolytically induced reduction of the fullerene has been carried out in a N_2O-saturated aqueous solution containing 1% of the micelle and 10% 2-propanol. In such systems, similar to the homogenous solution, the only reactive species is the $(CH_3)_2C^{\bullet}(OH)$ radical. The differential absorption spectrum recorded 100 μs after the pulse resembles that obtained in neat 2-propanol (Figure 1a). It appears to be stable over the experimental time scale of the pulse radiolysis. Hence the reaction can be ascribed to the formation of the $C_{60}^{\bullet-}$ radical anion generated by the reaction of the incorporated C_{60} with $(CH_3)_2C^{\bullet}(OH)$. A bimolecular rate constant of $k = 5 (\pm 2) \times 10^8$ $M^{-1}s^{-1}$ has been derived from the respective absorption time-profile at 332 nm (insert in Figure 2a). The rate resembles that in homogeneous solution; nevertheless an uncertainty for the determination of the C_{60} concentration has to be taken into account.

In summary, redox and free radical reactions with C_{60} can successfully be studied in either homogenous or heterogeneous solution, by radiation chemical methods, and in particular by the time-resolved technique of pulse radiolysis.

References.

[1] W. Krätschmer, L. D. Lamb, K. Fostiropoulos and D. R. Huffmann, Nature, 347, 354 (1990).

[2] C. Jehoulet, A. J. Bard and F. Wudl, J. Am. Chem. Soc., 113, 5456 (1991).

[3] D. Dubois, K. M. Kadish, S. Flanagan, R. E. Haufler, L. P. F. Chibante and L. J. Wilson, J. Am. Chem. Soc., 113, 4364 (1991).

[4] Q. Xie, E. Perez-Cordero and L. Echegoyen, J. Am. Chem. Soc., 114, 3978 (1992).

[5] R. E. Haufler, J. Conceicao, L. P. F. Chibante, Y. Chai, N. E. Byrne, S. Flanagan, M. M. Haley, S. C. O'Brien, C. Pan, Z. Xiao, W. E. Billups, M. A. Ciufloini, R. H. Hauge, J. L. Margrave, L. J. Wilson; R. F. Curl and R. E. Smalley, J. Phys. Chem., 94, 8634 (1990).

[6] D. Dubois, K. M. Kadish, S. Flanagan and L. J. Wilson, J. Am. Chem. Soc., 113, 7773 (1991).

[7] T. Kato, T. Kodama, T. Shida, T. Nakagawa, Y. Matsui, S. Suzuki, H. Shiromaru, K. Yamauchi and Y. Achiba, Chem. Phys. Letters, 180, 446 (1991).

[8] S. Nonell, J. W. Arbogast, and C. S. Foote, J. Phys. Chem., 96, 4169 (1992).

[9] J. R. Morton, K. F. Preston, P. J. Krusic, S. A. Hill and E. Wasserman, J. Phys. Chem., 96, 3576 (1992).

[10] P. J. Krusic, E. Wasserman, B. A. Parkinson, B. Malone, E. R. Holler, J. R. Keizer and K. F. Preston, J. Am. Chem. Soc., 113, 6274 (1991).

[11] J. R. Morton, K. F. Preston, P. J. Krusic, S. A. Hill and E. Wasserman, J. Am. Chem. Soc., 114, 5454 (1992).

[12] N. M. Dimitrijevic, Chem. Phys. Letters, 194, 457 (1992).

[13] D.M. Guldi, H. Hungerbühler, E. Janata, and K.-D. Asmus, J. Chem. Soc., Chem. Commun. 84 (1993).
[14] T. Andersson, K. Nilsson, M. Sundahl, G. Westman and O. Wennerström, J. Chem. Soc. Chem. Commun., 604 (1992).
[15] H. Hungerbühler, D.M. Guldi and K.-D. Asmus, J. Am. Chem. Soc. (in press).
[16] K.-D. Asmus, Methods Enzymol., 105, 167 (1984).
[17] E. Janata, Rad. Phys. Chem., 40, 217 (1992).
[18] N. E. Shank and L. M. Dorfman, J. Chem. Phys., 52, 4441 (1970).
[19] J. Grodkowski and P. Neta, J. Phys. Chem., 86, 2078 (1984).
[20] J. Mönig, K.-D. Asmus, L. W. Robertson and F. Oesch, J. Chem. Soc. Perkin Trans. 2, 861 (1986).

Photoionization Dynamics of C_{60}

T. Drewello[1], *W. Krätschmer*[2], *M. Fieber-Erdmann*[3], *and A. Ding*[3]

[1]Hahn-Meitner-Institut Berlin, Bereich S., Abt. Strahlenchemie, Glienicker Str. 100, D-14109 Berlin, Germany
[2]Max-Planck-Institut für Kernphysik, D-69117 Heidelberg, Germany
[3]Optisches Institut, Technische Universität Berlin, Straße des 17. Juni 135, D-10623 Berlin 12, Germany

Abstract. Photoionization of C_{60} by means of synchrotron radiation is reported. The present investigation covers an energy range of 15 - 120 eV. The findings are compared with other recent reports focusing on the ionization and fragmentation behaviour of C_{60} as a result of certain energy transfer processes to the cluster. The actual ionization mechanism, the temperature dependence on the ion yield curve and the fragmentation behaviour of ionic C_{60} are discussed.

1. Introduction

Since the discovery of fullerenes[1] and their successful synthesis[2], numerous properties of these fascinating molecules have been reported[3]. Gas-phase experiments have been of very special interest in revealing the intrinsic properties of free fullerenes. [4-6]

The present study is focused on the investigation of the photoionization dynamics of free fullerenes. C_{60} was ionized by a single photon in the energy range between 15 and 120 eV. The photoionization process leading to the doubly charged ions and the temperature dependence on the ionic yield curve are discussed.

2. Experimental

The experimental set-up has been described elsewhere[7,8]. A short description is given below.

A simple electrically heated quartz oven filled with a C_{60}/C_{70}-mixture (approx. 10:1) was used to produce an effusive fullerene beam. The Berlin synchrotron facility BESSY served as the light source. The photon energy was varied from 15 - 120 eV, yielding photon fluxes of about 10^{10} photons/s, which made multiple photon ionization events rather unlikely. The resolution of the monochromator was $E/\Delta E=500$. The detection of ions was achieved by time-of-flight mass spectrometry (TOF). Moreover, on account of the time dependent light flux provided by the electron storage ring, the signal intensities were calibrated against the signals of residual background water[9] and He which was introduced at a constant pressure.

3. Results and Discussion

In addition to $C_{60}^{+\cdot}$ and $C_{70}^{+\cdot}$ the corresponding di-cations were observed, for which the ionization energies were recently determined.[10-13] In the energy interval of 65 eV -

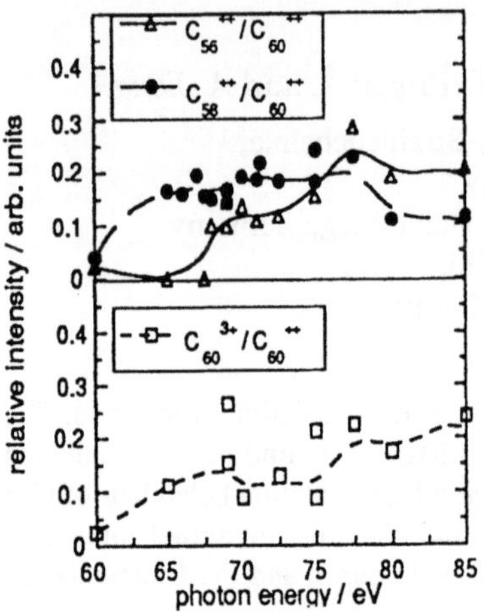

Figure 1: Top: The ratio of $C_{56}{}^{2+}/C_{60}{}^{2+}$- and $C_{58}{}^{2+}/C_{60}{}^{2+}$-intensities as a function of the photon energy. Bottom: The intensity ratios of $C_{60}{}^{3+}/C_{60}{}^{2+}$ versus the photon energy.

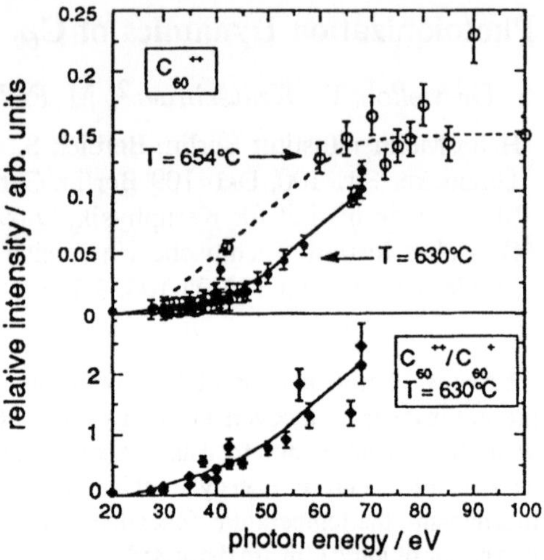

Figure 2: Top: The relative ionization yield for $C_{60}{}^{++}$ versus the photon energy for two different oven temperatures. Bottom: The ratio of the intensities for $C_{60}{}^{2+}/C_{60}{}^{+}$ as a function of the photon energy for an oven temperature of 630°C.

85 eV, the doubly charged species $C_{56}{}^{2+}$ and $C_{58}{}^{2+}$ were efficiently detected (Fig 1, top), as well as the triply charged $C_{60}{}^{3+}$ (Fig. 1, bottom). The observed fragmentation into $C_{58}{}^{2+}$ and $C_{56}{}^{2+}$ was either caused by the loss of C_2 and $2C_2/C_4$ from $C_{60}{}^{2+}$, respectively, or was the result of coulomb fragmentation from $C_{60}{}^{3+}$ into $C_{58}{}^{2+} + C_2{}^{+}$ or $C_{56}{}^{2+} + C_4{}^{+}$ (or $C_2{}^{+} + C_2$). The corresponding singly charged ion of the above scheme could not be detected. The latter process, however, could not be ruled out due to the expected large kinetic energy of the $C_2{}^{+}/C_4{}^{+}$ which reduced detection efficiencies for these ions significantly.

In contrast to the doubly charged ions, $C_{56}{}^{2+}$ and $C_{58}{}^{2+}$, no singly charged fragments could be observed. Berkowitz et al.[14] also reported on the difficulty to detect singly charged fragments for the quasi-single photon ionization of C_{60} using light from a helium discharge lamp. The lack of singly charged fragments for the entired energy range strongly suggests that a large part of the excess energy of the photoionization process contributes to the kinetic energy of the electron released from C_{60}.

The ion yield curve of $C_{60}{}^{2+}$, as well as the intensity ratio of the di- to the mono-cation of C_{60}, is given in Fig. 2. According to refs. [15,16] the idealized direct double photoionization should exhibit a linear increase in photoion yield above the threshold. The measured relative cross section for the production of $C_{60}{}^{2+}$ shows, however, a significant deviation from this behaviour.

The cross section for double ionization increases insignificantly between the threshold at 19.0 eV[12] and approximately 30 eV. Above this energy it rises steeply to a value similar to and even exceeding that of single ionization, depending on the oven temperature. Finally, the slope of the curve decreases for higher energies

($E_{h\nu}$ > 80 eV). This cross section is very sensitive to the internal energy of the C_{60} (as determined by the oven temperature); to such a degree which cannot be explained simply by the large amount of internal energy stored in the molecule at elevated temperatures. Considering the π–σ-plasmon observed above 15 eV, it is tempting to speculate that the steep increase around 35 eV could be due to the double excitation of this π–σ-plasmon (~ 36 eV).

Moreover, the double ionization yield is affected by the temperature of the fullerenes, as already discussed in ref. [18]. An increased oven temperature is accompanied by a strong increase in cross section of the doubly charged C_{60}. This behaviour could be due by the thermal excitation of C_{60}, which seems to indicate that the vibrational excitation might lead to a change in the Franck-Condon factors promoting the double ionization.

In conclusion, we have found a strong increase of the cross section for double ionization of C_{60} at approximately 30 eV, which could be interpreted as a electronic excitation (π–σ-plasmon) and although a temperature dependence of the ratio of $C^{2+}{}_{60}/C^{+}{}_{60}$. Furthermore any fragmentation of singly charged C_{60} were detected, but in contrast to that, doubly charged fragments were clearly visible in the mass spectra.

Acknowledgement. This financial assistance of and the German Ministry of Research and Technology (BMFT) is gratefully acknowledged. T. D. acknowledges the mental support provided by Prof. K.-D. Asmus (HMI-Berlin).

References

1. H. W. Kroto, J. R. Heath, S. C. O'Brien, R. E. Curl, R. E. Smalley: Nature **318**, 162 (1985)
2. W. Krätschmer, L. D. Lamb, K. Fostiropoulos, D. R. Huffman: Nature **347**, 354 (1990)
3. The March issue of Acc. Chem. Res. (1992, 25) is entirely devoted to various aspects of fullerene research.
4. J. P. Hare, H. W. Kroto, R. Taylor: Chem. Phys. Letters **177**, 394 (1991)
5. S. H. Yang, C. L. Pelliette, J. Conceicao, O. Chesnovsky, R. E. Smalley: Chem. Phys. Letters **139**, 233 (1987)
6. E. E. B. Campbell, G. Ulmer, I. V. Hertel: Phys. Rev. Letters **67**, 1986 (1991)
7. I. V. Hertel, H. Steger, J. d. Vries, B. Weisser, C. Menzel, B. Kamke, W. Kamke: Phys. Rev. Letters **68**, 784 (1992)
8. E. Holub-Krappe, G. Ganteför, G. Bröker, A. Ding: Z. Phys. D **10**, 314 (1988)
9. G. Bröker, M. Fieber, T. Drewello, G. Dujardin, E. Holub-Krappe, J. Lehmann, A. Ding: BESSY Jahresbericht 167 (1990)
10. M. S. Banna, B. H. McQuaide, R. Malutzki, V. Schmidt: J. Chem. Phys. **84**, 4739 (1986)
11. J. de Vries, H. Steger, B. Kamke, C. Menzel, B. Weisser, W. Kamke, I. V. Hertel: Chem. Phys. Letters **188**, 159 (1991)
12. S. Petrie, G. Javaher, J. Wang, D. K. Böhme: J. Phys. Chem. **96**, 6121 (1992)
13. H. Steger, J. d. Vries, B. Kamke, W. Kamke, T. Drewello: Chem. Phys. Letters **194**, 462 (1992)

14. D. L. Lichtenberger, K. W. Nebesny, C. D. Ray, D. R.Huffman, L. D. Lamb: Chem. Phys. Letters 176, 203 (1991)
15. R. K. Yoo, B. Ruscic, J. Berkowitz: J. Chem. Phys. 96, 911 (1992)
16. G. Dujardin, L. Hellner, B. J. Olsson, M. J. Besnard-Ramage, A. Dadouch: Phys. Rev. Letters 62, 745 (1989)
17. M. Stankiewicz, P. A. Hatherly, L. J. Frasinski, K. Coldling, D. M. P. Holland: J. Phys. B 22, 21 (1989)
18. T. Drewello, W. Krätschmer, M. Fieber-Erdmann, A. Ding: Int. J. Mass and Mass Spec. 124, R1 (1993)

Application of Fullerenes as Sensor Materials

A.W. Synowczyk[1] *and J. Heinze*[2]

[1]Institut für Physikalische Chemie, Universität Freiburg, Albertstr. 21, D-79104 Freiburg, Germany

[2]Materials Research Center at Freiburg (F.M.F.), Stefan-Meier-Str. 31a, D-79104 Freiburg, Germany

Fullerenes can be used as sensor materials to detect volatile polar substances, such as gaseous NH_3 or amines including organic vapours. The concentration of such analytes are measured by changes in the conductivities in fullerene films. In all cases, a linear correlation between concentration and conductivity of the film has been observed.

1. Introduction

Since 1990, the unusual properties of the fullerenes caused a worldwide boom in the investigation of these systems [1]. However, although theoretical considerations as well as basic experimental results predict a large number of applications, most of these projects are still far from realization. We started our electrochemical studies on fullerenes one year ago. During the voltammetric reduction of these systems we found that C_{60} and C_{70} reversibly react with primary and secondary amines. Similar observations were also made by Wudl et al. [2]. On the other hand, it is well known that fullerenes become conductive on reductive n-doping. As the reaction of fullerenes with amines involves some kind of electron-transfer processes [3, 4], this finding stimulated us to test the application of fullerenes as sensors.

2. Results and Discussion

In order to investigate the influence of NH_3, DMA and other electron-donating compounds on the reduction behavior of fullerenes, we carried out voltammetric measurements in methylene chloride and liquid DMA (NH_3,

Springer Series in Solid-State Sciences, Volume 117
Electronic Properties of Fullerenes Eds.: H. Kuzmany, J. Fink, M. Mehring, and S. Roth

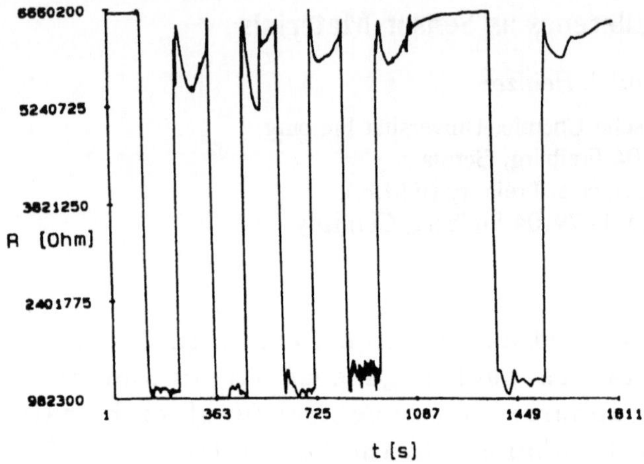

Fig. 1 NH_3 detection with C_{60} films formed via dipcoating, plot of resistance vs. time.

PH_3) and mixtures with different ratios of these two solvents. All electrochemical data proved a complex formation between the electron-donor and the fullerene. At low concentrations of the donor, the complex dissociated after reduction into its free components. Raising the donor content to 50 vol% and more produced a further follow-up product, that was more stable than the precursor-complex. In both species, the electron density of the fullerene unit is apparently higher than that of the free fullerene molecule, since the formal reduction potentials for the complexes are more negative than for pure C_{60} and C_{70}.

This "n-doping" effect of such electron-rich substances led us to the assumption that also the electrical properties of fullerenes should vary during their interaction with electron-donors. Indeed, we observed a dramatic increase of conductivity during exposure of a fullerene film to volatile polar substances.

In order to obtain a more detailed insight into this reaction we investigated the change of the electronic properties of C_{60} films during exposure to electron donors under in-situ conditions. We studied the change of conductivity of fullerene films in contact with gaseous ammonia. The fullerene films were produced by dipcoating. The

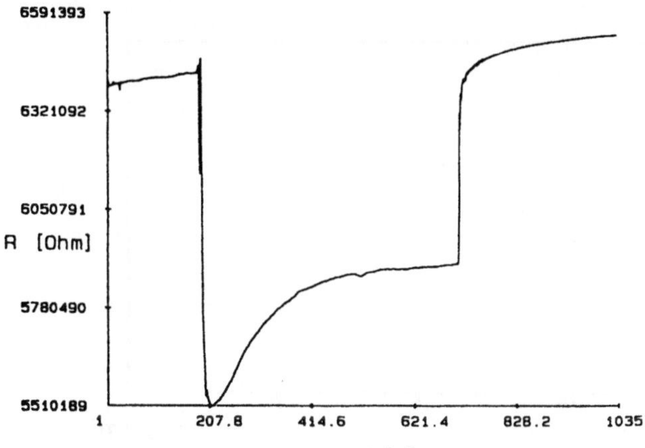

Fig. 2 Resistance vs. time plots for the detection of ammonia with C_{60} (electrochemicaly deposited)

change in the conductivity of sensor-active substrates can be determined using a Wheatstone bridge. Fig. 1 shows the behaviour of a C_{60} sensor. The C_{60} film changes its conductivity reversibly and rapidly, when exposed to ammonia at high concentrations. The change is about two orders in magnitude. When the ammonia stream is cut the sensor returns to its original conductivity within a few seconds. For the detection of ammonia in the ppm-scale high humidity is necessary.

A further method to obtain sensor-active C_{60}-films on arrays was the electrochemical deposition. The deposition was performed in methylene chloride/$TBAPF_6$ at a potential of -1.4 V vs. Ag/AgCl. We cannot explain the mechanism of this electrodeposition yet, but further research is in progress. In term of their sensor-activity, films obtained under these conditions differ strongly from films made by dipcoating. Electrochemical deposition of fullerene produces a red layer. This substrate changes its conductivity reversibly and reproducibly under the influence of ammonia with concentrations in air down to 5 ppm. The response of this layer is very rapid and reaches the maximum value within a few seconds (Fig. 2).

Within two minutes the conductivity decreases again and then reaches a stable plateau value.

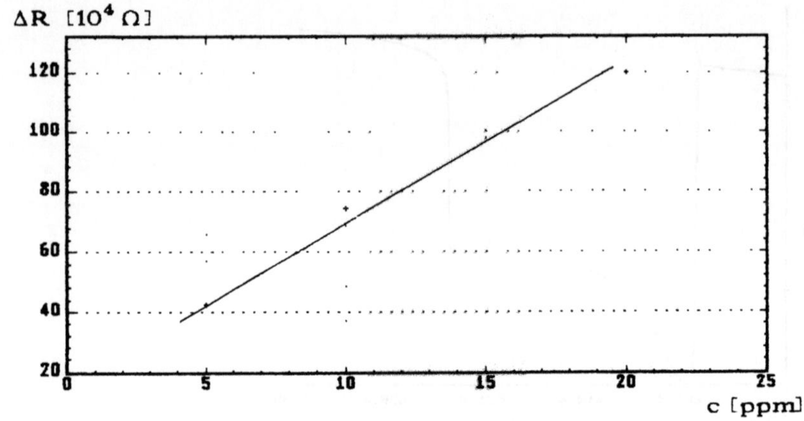

Fig. 3 Plot of minimum resistance (when exposed to gas) for a electrochemicaly formed C_{60} film

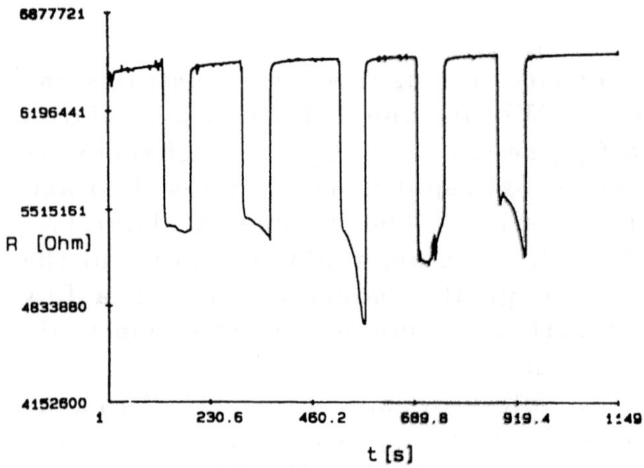

Fig. 4 Tetrachlorethylene detection: resistance vs. time plot during repeated exposure to gas

A linear correlation between maximum conductivity and ammonia concentrations is observed (Fig. 3).
Furthermore, this substrate is sensor-active when treated with organic solvents, e.g. benzene or tetrachlorethylene (Fig. 4).
With tetrachlorethylene there is also a linear correlation between gas concentration in air and conductivity (Fig. 5).
We suppose that swelling of the layers is responsible for the measured effect. Problems with this sensor material are:

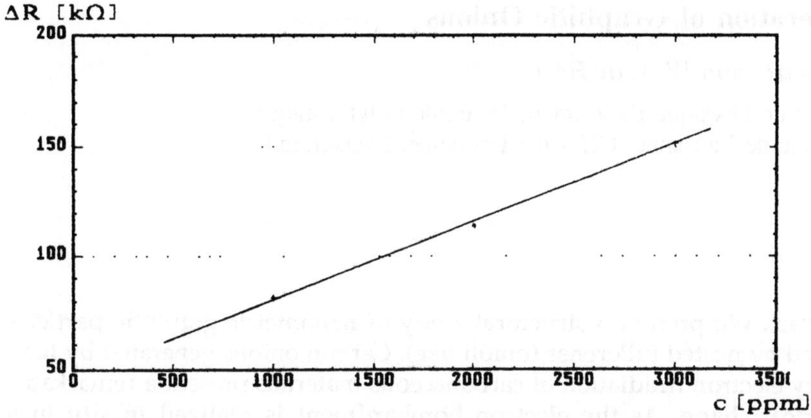

Fig. 5 Plot of resistance vs. gas concentration (ppm) for the detection of tetrachlorethylene with C_{60} film

- its dependency on humidity and
- its limited stability for storage (a few days).

There is still a lot of research to be done in this area and, we cannot fully explain the effects observed. Nevertheless, these preliminary results prove the unique properties of C_{60} as a sensor component.

3. References

[1] H Schwarz, Angew. Chem. 104 (1992), 301.
[2] A.Hirsch, Q. Li, F. Wudl, Angew. Chem. 103 (1991), 1339.
[3] K. Meerholz, P. Tschuncky, J. Heinze, J. Electroanal. Chem. 347 (1993), 425.
[4] K. Meerholz, P. Tschuncky, J. Heinze, Synth. Metalls 56 (1993), 3098.

Generation of Graphitic Onions

D. Ugarte and W.A. de Heer

Institut de Physique Expérimentale, Ecole Polytechnique
Fédérale de Lausanne, CH-1015 Lausanne, Switzerland

Abstract. We present a structural study of nanometric graphitic particles formed by nested fullerenes (onion-like). Carbon onions generated by high energy electron irradiation of carbonaceous materials present a remarkable spherical shape. As the electron bombardment is realized in situ in a electron microscope, the dynamic of the formation may be followed in real time. Macroscopic quantity of onion-like graphitic particles (3-10 nm in diameter) may be obtained by a simple thermal treatment of carbon soot.

1. Introduction

The synthesis and characterization of the fullerenes (buckyballs) [1,2] have generated a great deal of interest in the scientific community. The standard arc-discharge experiment has proved capable of generating fullerenes in macroscopic quantities [2], as well as other fullerene related structures (nanotubes [3], or buckyonions [4]).

Here, we report an experimental study in which we produce onion-like graphitic particles, with a remarkable quasi-spherical shape, by subjecting carbon soot to an intense electron irradiation [5,6]. Furthermore, we present an attempt to produce macroscopic quantities of nanometric onion-like particles through the usual thermal annealing method.

2. Materials and methods

The carbon soot used in this study was generated in the discharge apparatus composed of two 6 mm diameter graphite electrodes sharpened to 4 mm at the ends, which are heated by an alternative current (15 V, 150-200 A). After the arc discharge, we have scraped the surface of contact between electrodes and deposited the powder obtained on holey carbon grids. We have irradiated carbon soot in a 300 kV HREM microscope (Philips EM430 ST), using an electron dose up to 10-20 times higher (100-200 A/cm^2) than under normal operating conditions.

The heat treatment of carbon soot produced with a carbon arc method was performed in a tube rolled from a tantalum foil, which was resistively heated in vacuum (10^{-5} Torr). Details have been reported elsewhere [7].

Springer Series in Solid-State Sciences, Volume 117
Electronic Properties of Fullerenes Eds.: H. Kuzmany, J. Fink, M. Mehring, and S. Roth

3. Electron irradiation experiments

Under the conditions of observation in an electron microscope, amorphous carbon films usually develop a slight graphitization under the electron beam. In particular, irradiation usually produces heating of the sample, bond rupture through electron excitations, and furthermore momentum transfer to the nuclei may result in atom displacements to interstitial lattice sites (knock on). If this experiment is realized in a high resolution electron microscope (HREM), the evolution of a sample can be observed, even up to the atomic details under favorable conditions.

Figure 1 shows a sequence of images of a group of irradiated graphitic particles after 15 minutes of intense electron irradiation. The original soot, collected in an arc-discharge apparatus, contains mostly nanometric needles formed by coaxial graphitic tubes, and also polyhedral graphitic particles [8] (fig. 1a). Finally, in fig. 1b, the electron annealing led to a sample composed nearly entirely of spherical particles. Detailed examination of the particles shows that they consist of an assembly of concentric spherical graphitic cages (see fig. 2b), the distance between layers agreeing well with that of bulk graphite (d_{002}= 3.4 Å).

The final spherical structures correspond to the formation of a shelled spheres with a very small central cage (fig. 2b). The reduced dimension of the final inner shell (0.6-1 nm in diameter) allows an increase in the number of shells [5], and therefore the spherical graphitic onions must have a density very close to bulk graphite.

The change from an ordered structure like a bucky-tube or a polyhedral particle to another well ordered structure, the quasi-spherical onion-like particle, is made through an intermediate state where the ordering of the graphitic structures is seriously reduced [6]. The initial step is the collapse

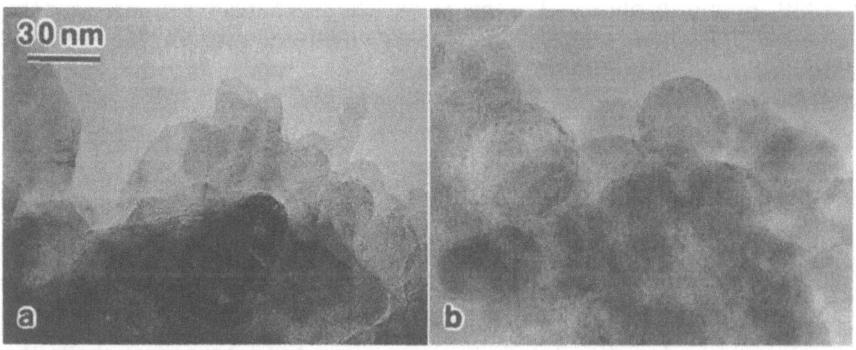

Figure 1. Sequence of TEM images of carbon soot subjected to intense electron irradiation. a) original soot obtained from the surface of a graphitic electrode, containing tubular and polyhedral graphitic particles; b) after 15 minutes of intense irradiation the sample is composed nearly exclusively of spherical particles.

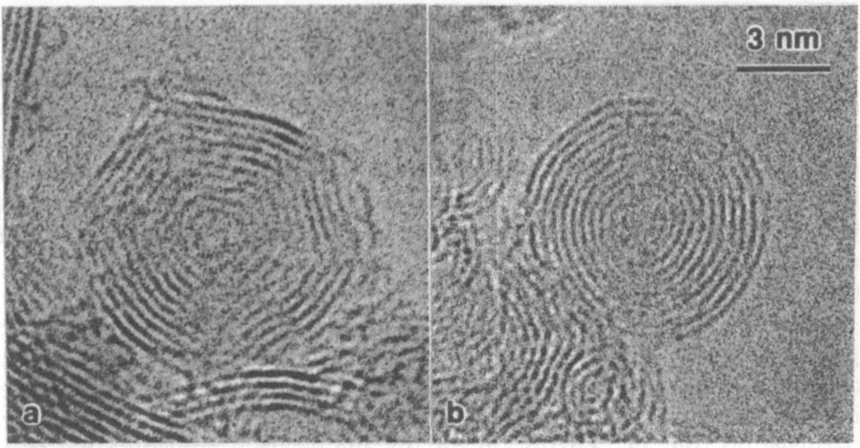

Figure 2. Comparison between an arc-discharge produced polyhedral graphitic particle (a) and an electron irradiation generated onion-like particle (b). Note the remarkable sphericity of the second one.

of the inner shell, which eliminates the hollow space at the center of the particle. The second step consists in the regraphitization of several surface layers. Subsequently the particle's shape evolves into a sphere, and simultaneously there is an internal epitaxial growth of the inner shells up to the center [6].

The growth mechanism described, is analog to the expected solidification of liquid carbon droplets [9]. A free liquid volume has a spherical shape due to its isotropic surface tension, in consequence a graphitization from the surface towards the center, would generate a graphitic onions as observed in this work. The mechanism proposed for the formation of shelled particles [10,11], based on spiral growth, has not been observed, due to the situation in our experiment, which is rather different from the accretion one originally considered.

The sequence shown in fig. 1, clearly reveals the robustness of the spheroidal shelled configuration, which has been proposed as the most stable structure of large carbon clusters [12,13].

One of the most fascinating characteristics of the irradiation generated shelled particles, is their remarkable sphericity, even for very large sizes (≈ 50 nm and 70 concentric shells, see fig. 2 of Ref. [5]). Particles composed of nested truncated icosahedra would show a marked faceting for increasing sizes (see fig. 2a). In the case of nested icosahedral shells the bond lengths can be conserved, but in spherical shells (composed only of 5- and 6-member rings), the bonds must be strongly distorted. Then, it can be expected that the faceted configuration should be preferred. The observed sphericity may be caused by a random orientation of the icosahedral shells [12], leading to an attenuation of the protuberances where the pentagons are located, or to the presence of other defects (associated 7- and 5-member rings).

4 . Thermally produced carbon onions

The fact that the onion configuration represents a very stable form of finely divided carbon, leads us to speculate that they could be produced by a current thermal annealing of carbon soot (fig. 3).

We have annealed Krätschmer soot, which is mostly composed of disordered particles with a minute graphitic component. Annealing at 1700 C produces glassy carbon soot. Further increase of the temperature generates closed shells particles within a typical size range of 3-10 nm [7]. The number of shells composing the onions may be controlled up to some extent, and samples containing mostly 2-3 or 5-8 layers may be produced.

At temperatures of about 2400 C, an important evaporation of the soot is observed, and also the generation of larger graphitic structures: polyhedral onions or bucky tubes (≥ 20-30 nm). Typically, several tenth of milligrams of material are produced per run.

Figure 3. High resolution electron micrograph of nanometric onion-like particles produced by thermal annealing of carbon soot (Treatment: 1700 C 60 min., 2250 C 45 min.).

5. Summary

From our observations, we speculate that spheroidal concentric-shelled particles should be the most stable form of large carbon clusters containing several hundreds of atoms. Further work is necessary to answer whether the shelled particle are faceted or spherical.

Macroscopic quantity production of monodispersed carbon onions, would allow the preparation of new crystalline solid forms of carbon formed by the packing of these particles, interacting through van der Waals forces.

The composition of interstellar matter is not fully understood and a prominent absorption peak at 217 nm was attributed to graphite particles [14]. Preliminary results indicate that the optical properties of graphitic onions are very close to the astronomical data [7,15].

Acknowledgments

We are most grateful to H. Kroto and W. Krätschmer for helpful discussions. We thank R. Monot and A. Châtelain for their encouragment. Supported by the Swiss National Science Foundation.

References

[1] H.W. Kroto, J.R. Heath, S.C. O'Brien, R.F. Curl and R.E. Smalley, Nature **318**, 162 (1985).
[2] W. Krätschmer, L. Lamb, K. Foristopoulos and D. Huffman, Nature **347**, 354 (1990).
[3] S. Iijima, Nature **354**, 56 (1991).
[4] S. Iijima, J. Cryst. Growth **50**, 675 (1980).
[5] D. Ugarte, Nature **359**, 707 (1992).
[6] D. Ugarte, Chem. Phys. Lett., in press.
[7] W.A. de Heer and D. Ugarte, Chem. Phys. Lett., in press.
[8] D. Ugarte, Chem. Phys. Lett. **198**, 596 (1992).
[9] K. Yamada, H. Kunishige and A.B. Sawaoka, Naturwissenschaften **78**, 450 (1992).
[10] Q.L. Zhang, S.C. O'Brien, J.R. Heath, Y. Liu, R.F. Curl, H.W. Kroto and R.E. Smalley, J. Phys. Chem. **90**, 525 (1986).
[11] H.W. Kroto and K. McKay, Nature **331**, 328 (1988).
[12] D. Ugarte, Europhys. Lett., in press.
[13] D. Tomanek, W. Zhong and E. Krastev, submitted for publication.
[14] T.P. Stechter and B. Donn, Astrophys. J. **142**, 1681 (1965).
[15] L. Henrard, A.A. Lucas and Ph. Lambin, Astrophys. J. (accepted).

Part II

Electronic Structure and Electron Spectroscopy

Electronic Properties of Fullerenes in the Molecular and Solid Phases: A Brief Introductory Course to Theoretical Approaches

W. Andreoni

IBM Research Division, Zurich Research Laboratory,
CH-8803 Rüschlikon, Switzerland

Abstract. A very brief introduction is given to the study of the electronic and structural properties of C_{60} and its derivatives, with the aim of providing the reader with the necessary ingredients to follow the more specific literature to which reference is given.

1. Introduction

In this Lecture, I shall present a brief general introduction to the main theoretical approaches used to calculate the structural and electronic properties of C_{60} and its derivatives. The high symmetry of the molecule and the effects of curvature on the chemistry of three-fold coordinated carbon atoms are key features for the understanding of its electronic behavior. Structural optimization has been obtained with *ab initio* methods, i.e. the Hartree-Fock (HF) scheme, also followed by the Møller-Plesset MP2 correction for the electron-electron correlation, and the density functional theory (DFT) in the local density approximation (LDA) for the exchange-correlation functional. LDA is the method commonly used to calculate the electronic structure of the solid phases of C_{60} and of the mixed compounds. Molecular dynamics (MD) simulations are the theoretical tool used to investigate the structural and dynamical properties of these materials. I shall present some of the results obtained with Car-Parrinello (CP) *ab initio* simulations, in particular on doped systems. Two types of doping are considered: intercalation of the solid phase with alkali metals and molecular encagement of metal atoms.

2. The C_{60} Molecule

The carbon atoms are at the vertices of a truncated icosahedron (a semi-regular polyhedron in the classification of Leonardo da Vinci). The closure takes place according to Euler's condition for a closed surface of three-fold coordinated vertices, thus giving rise to 12 five-membered rings (5-MR). In the remaining (20) 6-MRs, the atoms are disposed as in benzene and in graphite. This type of topology (12 5-MRs and 6-MRs only) characterizes an entire class of molecules called fullerenes, of which C_{60} is the prototype. The presence of the 12 5-MRs and the curvature at the carbon atoms are an effect of the three-dimensionality of the molecule and make all the difference between the chemistry of C_{60} (and in general of the known fullerenes) and that of the benzenoid structures [1]. In fact, the curvature introduces a rehybridization of the s and p-states of carbon, so that

Springer Series in Solid-State Sciences, Volume 117
Electronic Properties of Fullerenes Eds.: H. Kuzmany, J. Fink, M. Mehring, and S. Roth

neither the σ nor the π-orbitals are any longer "pure" as in the planar structures. In particular, the π-orbitals acquire a certain percentage (estimated to be ~12%) of s-character from the mixing of the 2p atomic states with the low-lying 2s ones, which tends to increase the electron affinity (EA). This is also increased by the pentagonal rings, bearing low-lying π-states which are electron eager. Measured to be 2.65 eV for C_{60} [2], the EA decreases on passing to the higher fullerenes, in agreement with the relative decrease of the 5-MRs and the weakening of the curvature.

Virtually every theoretical approach has been applied to C_{60}. Since the molecular symmetry is as high as that of the icosahedron, obtaining an electron energy diagram of the molecule within the Hückel approach is easy and at the same time provides a very instructive model [3]. The angular momentum l is a very good quantum number for the classification of the molecular orbitals, although l-mixing starts at the middle of the occupied band. In particular, the states responsible for the chemistry of the molecule and active in the electron doping, i.e. the HOMO (h_u), the LUMO (t_{1u}) and the LUMO+1 (t_{1g}), are five-, three- and three-fold degenerate, respectively, and derive from the splitting of the 11 doubly-occupied $l = 5$ states of the full rotational symmetry group. The LUMO and LUMO+1 are similar in nature, and correspond to the six low-lying π-states associated with the presence of the 12 non-adjacent 5-MRs. More sophisticated approaches reproduce essentially the same sequence of levels as Hückel calculations.

The parameter-free approaches generally used to calculate the electronic and structural properties of C_{60} are the HF or self-consistent-field (SCF) method [4,5] and the DFT in the LDA for the exchange and correlation functional [6-8]. Corrections to the HF approximation for the electron-electron correlation have been introduced within the Møller-Plesset approximation, at the lowest level of the perturbation treatment (MP2) [9]. The various LDA calculations differ among themselves in two respects: one is the all-electron (AE) description versus the use of an atomic pseudopotential accounting for the interaction between the valence states and the 1s core; the other is the choice of the basis set in the description of the one-electron wavefunctions, i.e. either localized Gaussian orbitals [6] or plane waves [7,8]. For the study of the chemical properties, the influence of the core states is indirect, and non-local pseudopotentials are very accurate. However, the pseudopotential approach renders a number of properties inaccessible, such as the determination of the chemical shifts and hyperfine coupling, which are crucial to the understanding of nuclear magnetic resonance (NMR) experiments. Different choices of basis functions also distinguish different SCF calculations. The use of the most sophisticated schemes such as MP at higher levels of perturbation (MP3 or MP4) and configuration interaction (CI) has so far been precluded by the large number of atoms.

Given its high symmetry, the structural optimization of C_{60} reduces to a two-variable problem, e.g. the two different bond lengths: one (d_1) on the pentagonal ring and the other (d_2) at the frontier of two hexagons. d_1 corresponds to a "single" bond (sb) d_2 to a "double" bond (db). Bond alternation is accompanied by an energy gain of the π-electron system. In LDA this turns out to be ~1.5 eV [10]. The results of a few (but not all!) calculations are reported in Table I. The

Table I. C_{60} bond lengths in Å

bond length	HF [5]	MP2 [9]	LDA [6a, 6b, 7, 8]	MNDO [13]	NMR [11a]	XRD [11b]
d_1	1.453	1.446	1.45, 1.445, 1.45, 1.444	1.474	1.45±0.015	1.455
d_2	1.369	1.406	1.39, 1.387, 1.39, 1.382	1.400	1.40±0.015	1.391

experimental value [11] of the difference $d_1 - d_2$ is ~0.05 Å (see Table I) and all calculations are able to give reasonable values for it. However, differences of the order of few percent in the C-C bond lengths are known to be significative. It is well established that HF calculations, even when using a basis as accurate as the 3-21G, systematically underestimate the db length in organic molecules and reproduce, instead, the sb length within 0.01 Å. The lengthening of d_2 found in MP2 is also not surprising, since the first-order correction to the HF description of the electronic configuration contains mainly the excitation to the antibonding states. The values determined in LDA are also consistent with calculations on other carbon systems [12], and show an overall agreement. Many calculations using semi-empirical methods have been reported. Table I gives the bond lengths obtained with MNDO [13], which is commonly used in calculations of the vibrational spectrum.

Similar comments can be made when comparing the different calculated geometries of C_{70}. In particular, among the parameter-free approaches, only the HF [14a,b] and the pseudopotential LDA [14c] methods have so far been applied.

3. C_{60} Solid: Fullerite

Since C_{60} is a very stable molecule, it forms a weakly bound molecular solid at standard pressure [15] (other lectures in this volume are devoted to the discussions of its structure and dynamics). Calculations are generally made within the LDA scheme [8,16,17] and for the fcc structure, which is the room-temperature modification. Experimentally, the molecules are observed to be orientationally disordered [18]. In the calculations, one specific orientation has to be chosen for the sake of feasibility, and generally that (T_h^3) with the highest possible symmetry (space group $F3m$) is adopted. At the equilibrium intermolecular distance, an optimization of the molecular structure in the solid shows that the effect of the confinement on the bond lengths is less than 0.01 Å [10].

The molecular levels provide an entirely sound basis for the understanding of the electronic energy structure of the solid, as seen both in Fig. 1 (the Γ DOS) and Fig. 2. C_{60} is found to be an insulator with a direct gap of ~1.2 eV at the X point of the Brillouin zone. We recall that the Kohn-Sham eigenvalues are not the quasi-particle energies, which in many-body theory describe the electronic excitations. The similarity is largely an assumption and a result of experience on many systems. Comparison with LDA-band structures and experiment must therefore be made with caution. Figure 1 shows to what extent the density of states (DOS) of the occupied part of the spectrum reproduces the experimental (direct) photoemis-

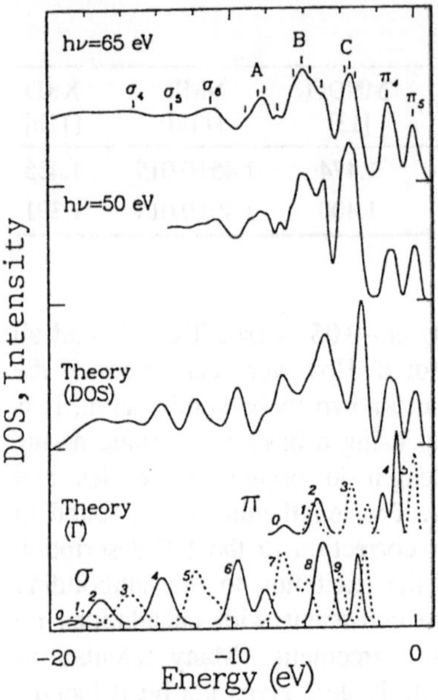

Fig. 1. Solid C60: DOS compared with photoemission. From [8].

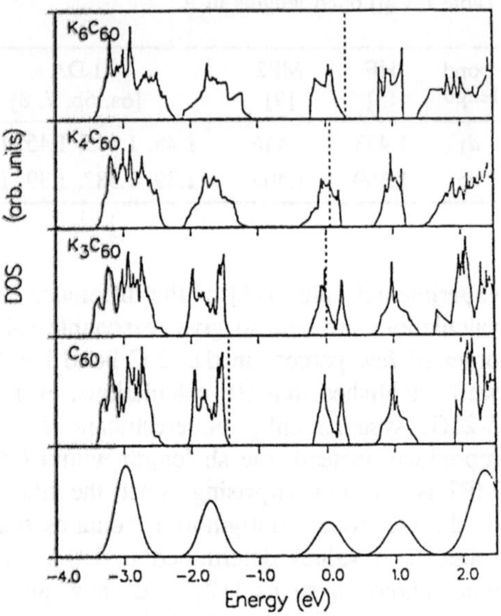

Fig. 2. DOS of solid C60 and K_xC_{60} and broadened molecular levels. From [16].

sion data [19]. It is apparent that the DOS allows a classification of most of the observed structure, and also that — as is usually the case — while reproducing well both the positions and relative heights of the first peaks, it becomes a less reliable approximation for increasing ionization energies. Explicit account of matrix elements effects [6] does not improve this agreement, which requires e.g. the inclusion of electron relaxation and final states effects. Electron-electron correlation beyond mean-field theory must be included to better account for photoemission as well as optical properties and electron-energy loss spectra, as discussed in several of the lectures in this volume [20]. Consistent with the known general trend [21], the LDA energy gap significantly underestimates the experimental one of ~2.3 eV [22].

The structure of C60 at low temperature is simple cubic with *Pa*3 space group and four molecules per unit cell. Accurate calculations of the band structure are not feasible. Within the tight-binding model, however, such a calculation has been performed [17], thus elucidating the effect of the orientational ordering. Figure 3 shows the DOS of the t_{1u}-derived conduction bands (CBs) for the uni- and quadri-directional structure. The main characteristics are similar, e.g. the bandwidth (*W*) and the valley between the two lowest subbands and the upper one.

4. Alkali-Metal Fullerides $(AM)_xC_{60}$

By intercalating the fullerite with alkali metals, a variety of new compounds have been obtained whose structural and electronic properties are strongly dependent on

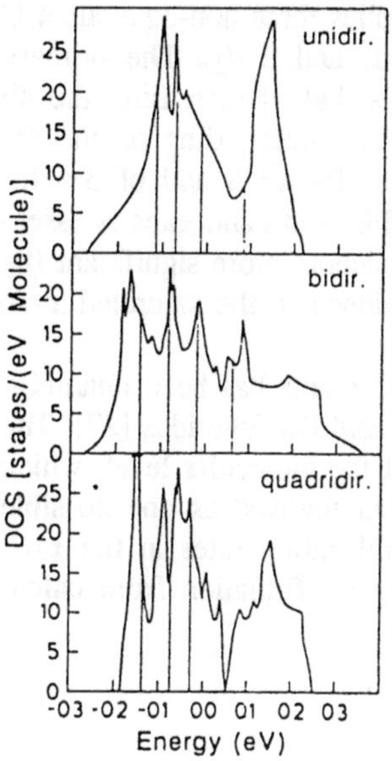

Fig. 3. C_{60}: CB-DOS (see text). Vertical lines correspond to the Fermi energy for $x = 1$ - 4. From [17].

the composition x and on the specific AM [1]. Several calculations are available for the band structure of the K and Rb compounds. The molecule is kept in the same structure as in the gas phase, and the AM atoms are in the crystallographic positions known from experiment. Figure 2 shows the evolution of the DOS [16] from $x = 0$ to $x = 3, 4$ and 6 for the K compounds, whose structures change from fcc ($x = 3$) to bct ($x = 4$) to bcc ($x = 6$). An overall similarity exists and the electronic nature (metal or insulator) depends on the degree of filling of the t_{1u}-derived CBs. However, a discrepancy with the experimental reality exists for $x = 4$, which, in contrast to the LDA results, has been observed to be an insulator [23]. This is not an anomalous finding in the comparison of LDA calculations (which tend to underestimate energy gaps significantly [21]) with experiment (a well-known case being that of semiconducting Ge, for which accurate LDA calculations would predict a metallic ground state). Increasing the electron doping drives a broadening of the occupied bands, a decrease of the HOMO-LUMO band gap of C_{60} as well as of the bandwidth of the t_{1g}-derived states, but keeps that of the t_{1u} states invariably at ~0.5 eV. Experimentally, however, the situation appears to be more complicated [24]. In particular, the width of the occupied t_{1u} band appears to be substantially larger, i.e. 1-2 eV for $x = 3$ and 1.2 eV for $x = 6$. For $x = 3$ this has been explained by Gunnarsson [20] as originating from the combined effects of the electron-phonon and electron-plasmon couplings.

Within the LDA scheme, two more effects have been studied: that of the structural relaxation of the molecule and that of the bidirectionality of the C_{60} units, which has been observed experimentally for the $x = 3$ phase [25]. The latter has been considered in [17] and is illustrated in Fig. 3 for the lowest CBs. For $x = 3$ the change in W clearly is very small. The former effect has been calculated by using the (pseudopotential-LDA based) CP method as first step in the determination of the molecular structure followed by traditional total energy and band structure calculations [26]. The symmetry lowering implies three non-equivalent C positions, thus splitting the C-C bonds into five (3 d_1 and 2 d_2). The primary effect of electron doping on the molecular structure is that of increasing the db length, consistent with the fact that t_{1u} states, which are antibonding on the db, become filled. We find an average lengthening of 1% for $x=3$ and of 3% for $x=6$ [26b]. The main effect on the (relevant) bandwidths and band gaps is essentially a rigid decrease of ~0.1 eV for $x=3$. Changes become more significant (by up to ~0.3 eV) and acquire a non-negligible k-dependence in the saturated $x=6$ phase.

Experimentally, the non-equivalence of the three C atoms has been unambiguously detected with NMR spectroscopy in several K- and Rb-fullerides [27]. The non-homogeneity of the conduction electron density at the molecular level, which results from the deformation of the molecule, has been invoked as one possible contribution to the "anomalous" (non-Korringa-like) relaxation rates in the $x=3$ phases [28]. This still awaits, however, a quantitative confirmation from calculations.

5. Metals Encaged in Fullerenes: Me@C_n

Encapsulating metal atoms in the cage has become a widely pursued way of doping C_{60} as well as higher fullerenes [29]. The hope is that eventually new materials with special and well controlled electronic properties can be created. In addition, the possibility of encaging small metal clusters that, if bare, are very reactive makes it possible to study them e.g. with spectroscopic means. A few calculations exist for single metal atoms in icosahedral C_{60}, and most of them assume the atom at the center of the cage [30]. An elegant and simple model in [16] explains that whenever the dopant atom is ionized (this is the case e.g. for AMs), the center of the cage is an energy maximum and its equilibrium position is displaced off-center. In C_{60}, for a given charge state (Q), the energy difference between the off-center and the central position is considered to be a function of only one variable, the distance d from the center, and is written as the sum of three contributions: the change in the EA of the cage due to the polarization of the electron cloud on the molecule, the change in the direct ion-cage interaction energy, and the increase in the orbital overlap between the electron states localized on the ion and on the cage. The first two terms favor the off-center displacement, while the latter acts as a repulsive factor. The result is shown in Fig. 4.

Ab initio MD is especially suited for determining the low-energy configurations since it fully accounts for the symmetry lowering in the molecular structure. The result of a series of CP simulations [31] largely confirm Erwin's model for one-electron donors such as Na and Al, although the correct energetics and displacements obviously depend on the specific characteristics of the dopant atom. As it is not highly sensitive to the chemistry of the specific system, the model seems to fail in predicting the charge state of La@C_{60}. In contrast to the predicted value of $Q=2$, *ab initio* MD shows [32] that the equilibrium position of La corresponds to ~ $d=1.2$ Å and to a charge state close to 3. In the case of La, in fact, the charge state itself is position-dependent, being 2 at the central position

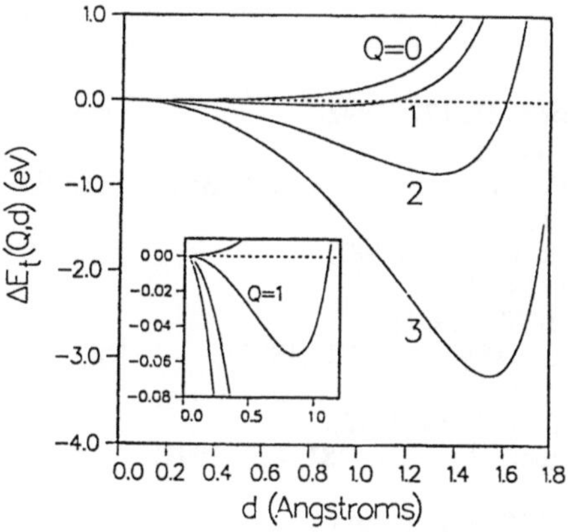

Fig. 4. M@C_{60}: Erwin's model for the energy change due to the off-center displacement. From [16].

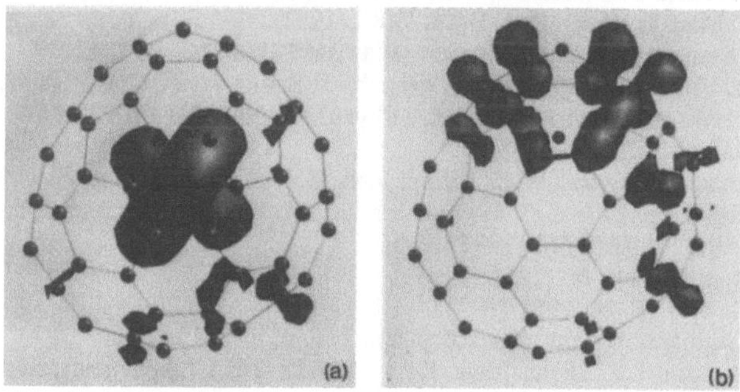

Fig. 5. La@C_{82}: Probability density (one high level) of the HOMO (see text). From Ref. 33, © 1992 by AAAS.

(in agreement with previous HF calculations). However, this configuration is very improbable, as it is ~2.9 eV higher than the off-center ones. The same picture applies to La@C_{82}. There the *ab initio* MD result [33] of a charge state close to $Q=3$ and a HOMO that is not localized on the La atom agrees with the experimental evidence [34]. The probability density distribution of the HOMO is illustrated in Fig. 5 for configurations with La at the center and off-center. In both of the fullerenes considered, the large energy gain associated with the off-center position is accompanied by an increase of the coordination number. For C_{82}, configurations with La outside the cage have also been considered in *ab initio* MD [35], and again turn out to be a few eV higher in energy than the low-lying ones with the encapsulated metal atom. Experimentally, it is clear that La@C_{82} can be obtained (extracted from either organic solvent [29] or directly from the soot using temperature-gradient sublimation [36]) in larger quantities than La@C_{60} or La@C_{70}. However, the reason for this is far from being clear yet.

References

1. R.C. Haddon, Accounts of Chemical Research **25**, 127 (1992).
2. L.S. Wang *et al.*, Chem. Phys. Lett. **182, 5** (1991).
3. The first calculation of a molecular level diagram goes back to D.A. Bochvar and E.G. Gal'pern, Proc. Acad. Sci. USSR **209**, 239 (1973).
4. H.P. Lüthi and J. Almlöf, Chem. Phys. Lett. **135**, 357 (1987).
5. G.E. Scuseria, Chem. Phys. Lett. **176**, 423 (1991).
6. a) J.W. Mintmire *et al.*, Phys. Rev. B **43**, 14281 (1991)
   b) L. Ye, A.J. Freeman and B. Delley, Chem. Phys. **160**, 415 (1992).
7. B.P. Feuston *et al.*, Phys. Rev. B **44**, 4056 (1991).
8. N. Trouiller and J.L. Martins, Phys. Rev. B **46**, 1754 (1992).
9. M. Häser, J. Almlöf and G.E. Scuseria, Chem. Phys. Lett. **181**, 497 (1991).
10. W. Andreoni (unpublished).
11. a) C.S. Yannoni *et al.*, J. Am. Chem. Soc. **113**, 3190 (1991)
    b) W.I.F. David *et al.*, Nature **353**, 147 (1991).
12. a) W. Andreoni, D. Scharf and P. Giannozzi, Chem. Phys. Lett. **173**, 449 (1990);
    b) M.C. Shabel and J.L. Martins, Phys. Rev. B **46**, 7185 (1992).
13. M.D. Newton and R.E. Stanton, J. Am. Chem. Soc. **108**, 2469 (1986).
14. a) G.E. Scuseria, Chem. Phys. Lett. **180**, 451 (1991); b) J. Baker *et al.*, Chem. Phys. Lett. **184**, 182 (1991); c) W. Andreoni *et al.*, Chem. Phys. Lett. **189**, 241 (1992).
15. W. Krätschmer *et al.*, Nature **347**, 354 (1990).
16. S.C. Erwin, in: *Buckminsterfullerenes*, W.E. Billups and M.A. Ciufolini (eds.) (VCH Publishers, New York, 1993), p. 217.
17. S. Satpathy *et al.*, Phys. Bev. B **46**, 1773 (1992).
18. P.A. Heiney *et al.*, Phys. Rev. Lett. **66**, 2911 (1991).
19. P.J. Benning *et al.*, Phys. Rev. B **44**, 1962 (1991).
20. See the lectures by O. Gunnarsson, G.A. Sawatzky, and J. Fink in this volume.
21. R.W. Godby, M. Schlüter and L.J. Sham, Phys. Rev. Lett. **56**, 2415 (1986) and references therein. For diamond see p. 453 in [12a].
22. R.W. Lof *et al.*, Phys. Rev. Lett. **68**, 3924 (1992).
23. R.F. Kiefl *et al.*, Phys. Rev. Lett. **69**, 2005 (1992); D.W. Murphy *et al.*, J. Phys. Chem. Solids **53**, 1321 (1992).
24. J. Fink *et al.*, in: *Proc. First Italian Workshop on Fullerenes: Status and Perspectives*, C. Taliani *et al.*, Eds. (World Scientific, Singapore, 1992), p. 161.
25. P.W. Stephens *et al.*, Nature **351**, 632 (91) and Phys. Rev. B **45**, 543 (1992).
26. a) W. Andreoni, F. Gygi and M. Parrinello, Phys. Rev. Lett. **68**, 823 (1992)
    b) W. Andreoni, P. Giannozzi and M. Parrinello, to be published.
27. C.S. Yannoni *et al.*, Synthetic Metals (1992); P.P. Bernier *et al.*, this volume.
28. K. Holczer *et al.*, Europhys. Lett. (in press) and the lecture by K. Holczer in this volume; see also K. Holczer and R.L. Whetten, Carbon **30**, 1261 (1992).
29. See e.g. C. Jin *et al.*, in: *Proc. First Italian Workshop on Fullerenes: Status and Perspectives*, C. Taliani *et al.*, Eds. (World Scientific, Singapore, 1992), p. 21; M.E. Johnson *et al.*, Acc. Chem. Res. **25**, 169 (1992); M.M. Alvarez *et al.*, J. Phys. Chem. **95**, 10561 (1991); C.S. Yannoni *et al.*, Science **256**, 1191 (1992).
30. A. Rose'n and B. Wästberg, J. Am. Chem. Soc. **110**, 8701 (1988); A.H.H. Chang *et al.*, J. Chem. Phys. **94**, 5004 (1991); J. Ciolowski and E.D. Fleischmann, *ibid.* **94**, 3730 (1991)
31. W. Andreoni, Nanostructured Materials (in press) and to be published.
32. W. Andreoni, K. Laasonen and M. Parrinello, to be published.
33. K. Laasonen, W. Andreoni and M. Parrinello, Science **258**, 1916 (1992).
34. M.E. Johnson *et al.*, Nature **355**, 239 (92); J.H. Weaver *et al.*, Chem. Phys. Lett. **190** 460 (1992).
35. K. Laasonen, W. Andreoni and M. Parrinello, to be published.
36. C. Yeretzian *et al.*, J. Phys. Chem. (in press).

Experimental Investigation of the Electronic Structure of Gas-Phase and Solid C_{60}

S. Krummacher, M. Biermann, M. Neeb, A. Liebsch, and W. Eberhardt

Institut für Festkörperforschung der Forschungszentrum Jülich GmbH,
P.O. Box 1913, D-52425 Jülich, Germany

Abstract. We have investigated the electronic structure and electron correlation effects of gas phase and solid C_{60} by C 1s photoabsorption, photoelectron, and Auger spectroscopies. Direct *in situ* comparison between C 1s absorption spectra taken in the gas phase and in the solid reveals a close similarity between the unoccupied levels of the two phases. The same is true for the shake up satellites in the C 1s XPS spectrum. The satellite due to plasmon excitations, on the other hand, appears at about 6 eV higher binding energy in the gas phase than in the solid, due to the contribution of higher multipole moments. Solid state effects play only a minor role in this new material, and electron correlation effects are predominantly intra molecular.

1. Introduction

One of the missing links in the studies of the fascinating new material fullerene has been a detailed comparison of the electronic structure of a single C_{60} "molecule" or cluster with that of the solid. The same question arises with respect to correlation effects, which have recently been suggested to be involved in the explanation of the (super)conducting properties of doped fullerene material [1]. This prompted our investigation of the electronic properties of gas phase and solid C_{60} using photoabsorption by electron yield and photoemission spectroscopy with synchrotron radiation.

2. Experimental details

Fullerite powder usually containing about 90% C_{60} was evaporated from a Knudsen furnace at roughly 960° K. Only the powder used for the photoabsorption measurements had been purified by liquid chromatography to reduce the contribution of higher fullerenes below the detection limit; for the other measurements unpurified material was used. At a distance of about 2 mm from the 0.5 mm diameter nozzle the vapor beam was crossed at right angles by the synchrotron radiation beam from the X1B soft X-ray undulator beam line at the National Synchrotron Light Source (NSLS), Brookhaven, which is described in detail elsewhere [2,3]. The photoelectrons were energy analyzed in a commercial CMA-type electron spectrometer whose axis was oriented perpendicular to both the light and vapour

Springer Series in Solid-State Sciences, Volume 117
Electronic Properties of Fullerenes Eds.: H. Kuzmany, J. Fink, M. Mehring, and S. Roth

beam. The C_{60} vapor was collected on a stainless steel plate which could be moved into the interaction region after termination of the gas phase measurements. This guaranteed an *in situ* comparison of gas phase and solid spectra using the same energy calibration. Absolute energy calibration for the photon energy was obtained using the $1s \rightarrow \pi^*$ transitions of gas phase CO and CO_2 [4] taken immediately before and after the C_{60} studies. Similarly, the kinetic energy scale was calibrated against the K-LL Auger transition of O_2 [5].

3. Results and discussion

The upper panel of figure 1 shows a comparison of the onset of the C 1s excitations into unoccupied electronic states in gas phase and solid C_{60}. These spectra were obtained by measuring the secondary electron yield at kinetic energies between 2 eV and 10 eV for the solid species, and the Auger electron yield at 250 eV for the gas phase species. They were normalized using the signal from a semiconductor photodiode located behind the interaction region in the direct photon beam in order to eliminate the variations of the incident photon flux in the region of the C K-

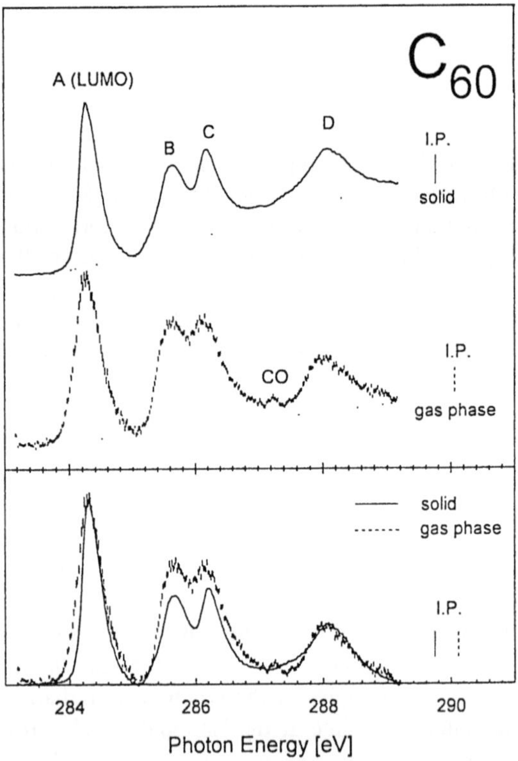

Figure 1: C 1s near edge absorption spectra of gas phase (error bars) and solid C_{60} (solid line).

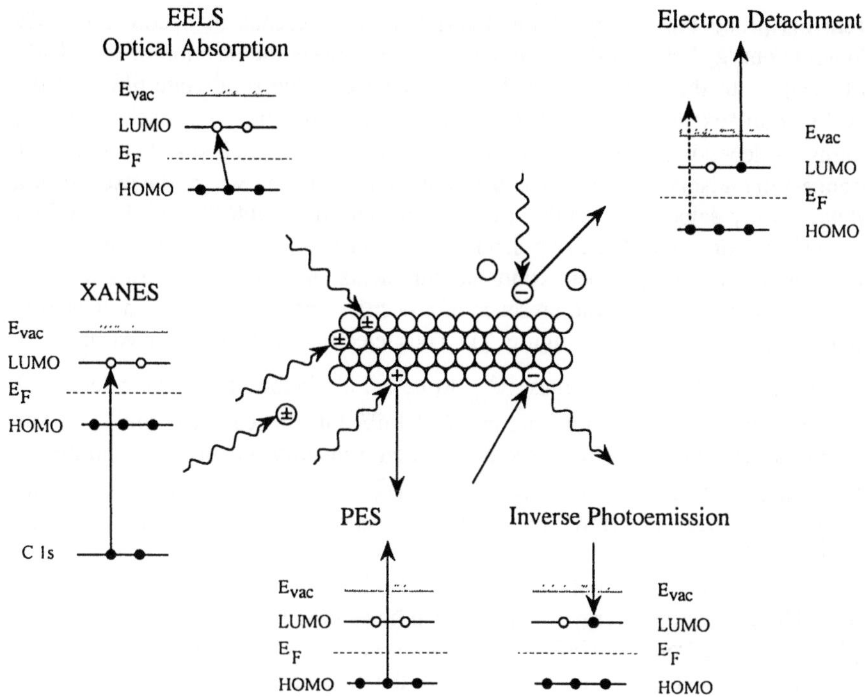

Figure 2: Comparison of different methods to probe the energy of the LUMO.

edge. The band pass of the monochromator was set to 0.06 eV for the gas phase measurements, and 0.03 eV for the measurements on the deposited film.

While for the gas phase no core absorption spectrum has been reported previously, several spectra already exist in the literature for the solid case, obtained either by photoabsorption [6,7] or by electron energy loss spectroscopy [8,9]. Our spectrum of solid C_{60} agrees well with those obtained with purified C_{60} [7-9]. The peak observed at 287.3 eV is due to the 1s $\rightarrow \pi^*$ transition in CO traces left in the chamber from the calibration measurement.

In order to focus on the very subtle differences between the two spectra, the continuous part of the cross section has been subtracted as a two segment linear "background" in the lower panel. The energy positions of all four features are identical within the accuracy of the experiment, but the amplitudes of peaks B and C in the solid spectrum relative to peak A are reduced by 26% and 18%, respectively. The slight broadening on the low energy side of peaks A and C in the gas phase spectrum by about 0.1 eV compared to the solid state spectrum is due to the population of higher vibrational levels in the ground state at 960° K during the gas phase measurements. No additional broadening and/or peak shift due to solid state band dispersion can be recognized.

Nevertheless, the accurate energy positions of peaks A in figure 1 do contain some information on the degree of localisation in the two systems, as will be explained now. As figure 2 demonstrates, the peak positions in the XANES measurements probe the energy of the LUMO in the presence of a core hole. The Coulomb energy between these two charges then lowers the position of the LUMO with respect to the vacuum level. In the language of the solid state physicist one speaks of an "excitonic state". The amount of this lowering — or exciton binding energy — depends on the degree of (de)localisation of the excited state orbital. It is strongest in an isolated molecule and zero in an ideal metal, where the excited state orbital is fully delocalized, so that the electron can diffuse into the solid away from the positive core hole. It can be determined from a comparison with electron detachment or inverse photoemission measurements, in which the position of the LUMO is probed without the presence of a positive core hole, so that no Coulomb interaction is invoked (see figure 2). Thus electron detachment measurements showing the LUMO in gas phase C_{60}^- to be 2.7 eV below the vacuum level [10] yield a Coulomb interacttion energy of 3.1 eV for the core exciton. The corresponding value in the solid, based on reported measurements of direct and inverse photoemission [1,11,12], is only (2.2 ± 0.1) eV. This reduction in interaction energy is a direct manifestation of the increased delocalization of the excited electron in the solid.

A similar but much smaller effect can also be observed for the valence band states. The correlation energy U between two holes in the valence band can in principle be inferred from the shift of the Auger spectrum against the self convolution of the valence band density of states, or from the difference between the onset of the Auger spectrum and the maximum Auger energy calculated from the relation

$$KE^{\max}(Auger) = BE(1s) - 2\times BE(HOMO) - U$$

for $U = 0$. In the case of C_{60}, however, the Auger spectrum is so broad and structureless, that these determinations are difficult to make without ambiguity. The relative shift between the Auger spectra of gas phase and solid C_{60}, however, can be determined with much higher accuracy. Thus we measure the kinetic energy at the steepest slope of the solid Auger spectrum to be about 1.4 eV higher than that of the gas phase spectrum. From the relationship

$$\Delta U^{\text{gas-solid}} = \Delta BE(1s)^{\text{gas-solid}} - 2\times\Delta BE(HOMO)^{\text{gas-solid}} - \Delta KE^{\text{gas-solid}}$$

and the differences in the binding energies we thus determine a difference between the correlation energies in the gas phase and in the solid of only 0.23 eV. This is much smaller than the difference between previous estimates (3.4 eV for the gas phase and 1.6 eV for the solid [1]). However, at present it can not be excluded that the differences between the gas phase and solid spectra observed by us are reduced by some small C_{70} contributions, since only the photoabsorption measurements were performed with purified material.

We have also investigated the satellite structure in the C 1s photoelectron spectrum of gas phase and solid C_{60}. While the discrete shake up structure in the region within about 16 eV of the main line shows again no differences between the two phases, the broad peak due to collective plasmon excitation of all 240 valence electrons is observed at 34 eV in the gas phase compared to 28 eV reported for the solid [11]. Corresponding excitations have been observed in the optical absorption spectra of gas phase C_{60} at a photon energy of 19 eV [13], while in the electron energy loss spectrum of solid C_{60} this structure appears again at about 28 eV [14,15]. Based on simple model calculations, these differences in the energy position can be explained by the contribution of higher multipole moments and different symmetry selection rules. More details are given in a separate publication [16].

In conclusion, electron correlation effects are mainly a property of the individual C_{60} cluster, and it seems unlikely that they reflect upon the mechanism of macroscopic conductivity or even superconductivity of the solid.

Acknowledgements

It is a pleasure to acknowledge the skillful help of F.P. Johnen, F. Loeb, and the staff of the National Synchrotron Light Source. Fruitful discussions with B. Kessler, T. Schober, and F. Hudson are also gratefully acknowledged. The experiments were carried out on beam line X1B at the NSLS Brookhaven, which is supported by the DOE, Office of Basic Energy Sciences.

References

[1] R.W. Lof, M.A. van Veenendaal, B. Koopmans, H.T. Jonkman, and G.A. Sawatzky, Phys. Rev. Lett. **68** (1992) 3924.

[2] W. Eberhardt, K.J. Randall, J. Feldhaus, A.M. Bradshaw, R.F. Garrett, and M.L. Knotek, Physica Scripta **41** (1990) 745.

[3] K. J. Randall, J. Feldhaus, W. Erlebach, A. Bradshaw, W. Eberhardt, Z. Xu, Y. Ma, and P.D. Johnson, Rev. Sci. Instrum. **63** (1992) 1367.

[4] G.R. Wight, C.E. Brion, J. Electr. Spectr. **3** (1974) 191; R.P. Hitchcock and C.E. Brion, J. Electr. Spectr. **18** (1980) 1.

[5] W.E. Moddeman, T.A. Carlson, M.O. Krause, B.P. Pullen, W.E. Bull, and G.K. Schweitzer, J. Chem. Phys. **55** (1971) 2317.

[6] S.L. Molodtsov, A. Gutierrez, M. Domke, and G. Kaindl, Europhys. Lett. **19** (1992) 369.

[7] L.J. Terminello, D.K. Shuh, F.J. Himpsel, D.A. Lapiano-Smith, J. Stoehr, D.S. Bethune, and G. Meijer, Chem. Phys. Lett. **182** (1991) 491.

[8] G. Gensterblum, J.J. Pireaux, P.A. Thiry, R. Caudano, J.P. Vigneron, Ph. Lambin, A.A. Lucas, and W. Krätschmer, Phys. Rev. Lett. **67** (1991) 2171.

[9] E. Sohmen, J. Fink, and W. Kraetschmer, Z. Phys. B **86** (1992) 87.

[10] R.E. Haufler, L.S. Wang, L.P.F. Chibante, C. Jin, J. Conceicao, Y. Chai, and R.E. Smalley, Chem. Phys. Lett. **179** (1991) 449.

[11] J.H. Weaver, J.L. Martins, T. Komeda, Y. Chen, T.R. Ohno, G.H. Kroll, N. Troullier, R.E. Haufler, and R.E. Smalley, Phys. Rev. Lett. **66** (1991) 1741.

[12] J.H. Weaver, J. Phys. Chem. Solids **53** (1992) 1433.

[13] I.V. Hertel, H. Steger, J. de Vries, B. Weisser, C. Menzel, B. Kamke and W. Kamke, Phys. Rev. Lett. **68** (1992) 784.

[14] G. Gensterblum, J.J. Pireaux, P.A. Thiry, R. Caudano, J.P. Vigneron, Ph. Lambin, A.A. Lucas, and W. Krätschmer, Phys. Rev. Lett. **67** (1991) 2171.

[15] E. Sohmen, J. Fink, and W. Kraetschmer, Z. Phys. B **86** (1992) 87.

[16] S. Krummacher, M. Biermann, M. Neeb, A. Liebsch, and W. Eberhardt, submitted to Phys. Rev. B.

Low-Energy Electron Diffraction and High-Resolution Electron Energy-Loss Spectroscopy of Heteroepitaxial C_{60} Thin Films on GeS(001)

G. Gensterblum, L.-M. Yu, J.-J. Pireaux, P.A. Thiry, and R. Caudano

Laboratoire Interdisciplinaire de Spectroscopie Electronique,
Facultés Universitaires Notre-Dame de la Paix,
61, rue de Bruxelles, B-5000 Namur, Belgium

Abstract. *The heteroepitaxial growth of well-ordered C_{60} fullerite films on GeS(001) has been studied by low-energy electron diffraction (LEED) for different film thicknesses and substrate temperatures. The peculiar corrugations of the basal plane of this layered compound together with favorable lattice matching conditions allow the growth of uniaxially-oriented $C_{60}(111)$ layers up to a thickness of 1000 Å, as evidenced by the sharp LEED patterns. The vibrational structure of the C_{60}/GeS interface is analyzed in detail by infrared high-resolution electron energy loss spectroscopy (HREELS) in specular reflection geometry. The structural order is confirmed by the clear detection in the specular beam of all four T_{1u} dipole-active vibrations of the C_{60} molecule and by the small angular width of the elastic peak.*

1. Introduction

The small typical size (0.5 x 0.5 mm²) of C_{60} single crystals [1] compared to the spot size of a low energy electron or photon beam, together with the critical charging effect on thick insulating crystals makes them unsuitable for studies by low-energy electron spectroscopies such as High-Resolution Electron Energy Loss Spectroscopy (HREELS), photoemission (UPS) and inverse photoemission (IPES). For these spectroscopies epitaxial thin films on conducting substrates are more appropiate. The epitaxial growth of C_{60} has already been observed by Sakurai *et al.* [2] on a MoS_2 substrate using RHEED, and by Schmicker *et al.* [3] and Krakow *et al.* [4] on mica using helium atom scattering and electron diffraction, respectively. Here, we report on the epitaxial growth of $C_{60}(111)$ fullerene films on GeS(001) [5], as evidenced by sharp LEED diffraction patterns, and on measurements of the HREELS spectrum in the infrared enegy domain [6].

2. Experimental set-up

The GeS single crystal (7x7 mm²) grown by the vacuum sublimation method was freshly cleaved in UHV before C_{60} deposition. The crystallographic quality of the substrate was checked by LEED prior to deposition. Fig. 1a shows a sharp LEED pattern typical of a freshly cleaved GeS(001) surface at an incident energy of 80 eV.

The chromatographically purified C_{60} powder was provided by J.E. Fischer and G.A. Sawatzky. The fullerene powder was loaded into a graphite crucible of a Knudsen cell and carefully outgassed for 24 h. It was then sublimed at temperatures of 400 to 420°C at constant pressures below $2x10^{-9}$ mbar on the GeS substrate. The deposition rate ranged from 0.25 to 5Å/min, as estimated with a quartz crystal oscillator. Thin films of different thicknesses were grown for different substrate temperatures from room temperature up to 200°C.

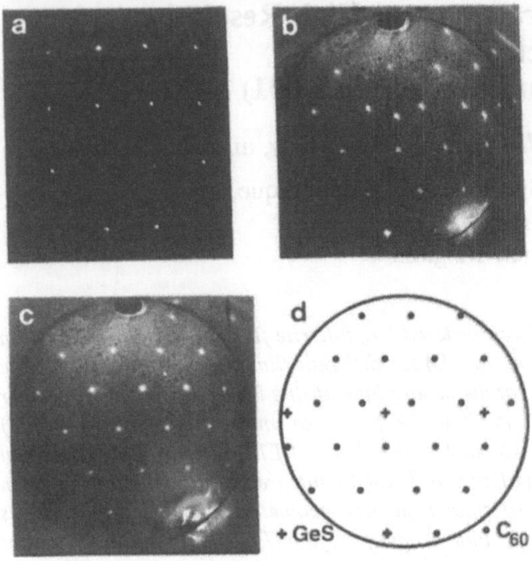

Fig. 1a-d. LEED patterns of (a) a cleaved GeS(001) substrate with E_p=80eV; (b) 1ML of C_{60} on GeS with E_p=24 eV, both C_{60} and substrate spots are visible: they are identified by dots and crosses respectively in (d); (c) 4ML of C_{60} on GeS with E_p=26 eV; only spots of the fcc C_{60}(111) hexagonal arrangement are visible.

The HREELS spectra were recorded with a commercial electron energy loss spectrometer (ISA-Riber), and for the LEED study we used a rear-view LEED diffractometer (Riber) using a standard electron gun placed at an incident angle of 45°.

3. LEED results

Fig. 1b,c show the LEED patterns of 1 and 4 monolayers (ML) of C_{60} sublimed on a GeS substrate held at about 180°C during deposition, at an incident energy of 24 eV and 26 eV, respectively. The LEED pattern revealing the C_{60} hexagonal arrangement corresponding to the (111) plane of the room temperature fcc phase is of very good quality, with sharp spots devoid of fine structure and a rather low background. In the case of the 1ML deposition, spots characteristic of both C_{60} and GeS are visible, the larger rectangular mesh (identifid by crosses in Fig. 1d) originating from the substrate. LEED patterns of the same quality, as the one shown in Fig. 1c, have been obtained for films up to a thickness of 1000 Å.

From the inspection of Fig. 1b, it is evident that in the reciprocal space the $\bar{\Gamma}-\bar{K}$ ($\bar{\Gamma}-\bar{M}$) direction of the first C_{60} surface Brillouin zone (SBZ) is aligned with the $\bar{\Gamma}-\bar{D}-\bar{S}$ ($\bar{\Gamma}-\bar{G}-\bar{U}$) direction of the GeS surface BZ. In the direct space this means that the hexagonal layer of the (111) face of the C_{60} crystal is aligned with the substrate so that one of the six equivalent C_{60} $<\bar{1}01>$ direction is parallel to the GeS **b**-axis and the corresponding C_{60} $<1\bar{2}1>$ directions is parallel to the GeS **a**-axis, as shown in Fig.2. The fact that the GeS substrate acts as such a perfect template for the growth of the ordered fullerite films stems from the close match between twice the lattice parameter of GeS along its **a**-axis (2 x 4.30 Å = 8.6 Å) and the separation between rows of C_{60} clusters along $<\bar{1}01>$ or the equivalent directions (8.68 Å). As sketched in Fig. 3, which presents another schematic view of our model structure, the GeS(001) surface exhibits a peculiar corrugation in its basal plane. As a matter of fact, in a single layer of the orthorhombic structure, each atom is covalently bonded to three nearest neighbors, forming zigzag chains along the **b**-axis with alternating positions perpendicular to the basal plane. The cleaved surface has the aspect of a corrugated roof due

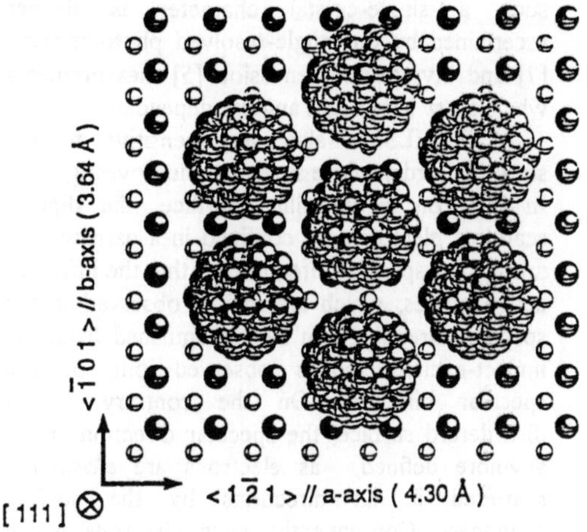

Fig.2. Top view of the model structure of a C_{60} layer grown on a GeS(001) substrate. Small grey (white) spheres represent the surface germanium (sulphur) atoms of GeS.

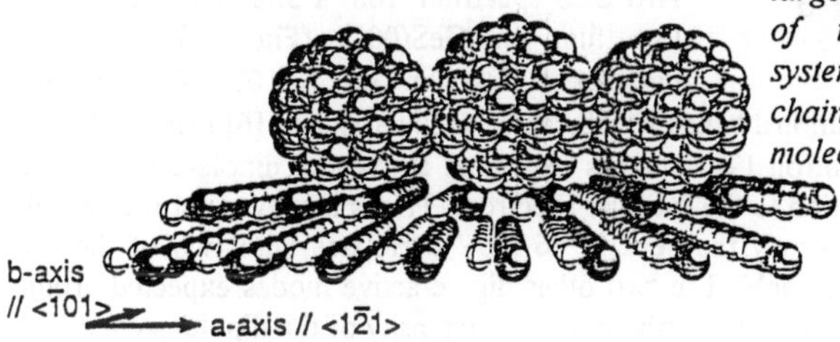

Fig.3. Side view of the model structure of a C_{60}(111) monolayer on a GeS(001) substrate showing the large corrugation in the a-direction of the basal plane and the systematic skipping of one inter-chain groove by the big C_{60} molecules.

to the alignment of well-separated zigzag chains at the selvedge, which provide a preferential adsorption direction in the inter-chain grooves along the **b**-axis. This one-dimensional corrugation with a periodicity of 4.3 Å allows the easy alignment of C_{60} $< \bar{1}01 >$ rows along the substrate **b**-axis. Completion of the first monolayer is accomplished if the C_{60} cluster skip one inter-chain streak systematically, thereby allowing near-perfect arrangement at the interface.

4. HREELS results

The observation of a sharp LEED pattern ascertains a domain size of the order of the coherence length of the unmonochromatized electron beam (typically 100 Å). In HREELS one uses a highly monochromatized electron beam ($\Delta E < 10$ meV) which consequently has a coherence length larger than 1000 Å. If the real domain size were smaller than this coherence length, the angular width of the elastic peak would be broadend with respect to the intrinsic angular width of the electron gun. In Fig. 4 we show the angular profile of the elastic peak for the GeS substrate and for a 500 Å thick C_{60} film. The FWHM of 1.2° for both profiles corresponds to the angular divergence of the electron beam. We can therefore state that the C_{60} films present a majority of single domains well oriented over large areas. The existence of

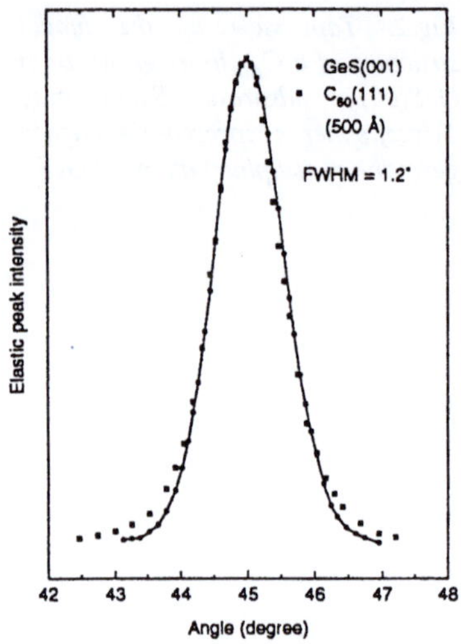

Fig.4. Angular distribution curve around the specular direction of the elastic peak intensity recorded on the GeS substrate and on a 500 Å thick C_{60} film.

such a single-crystal character is further ascertained by our angle-resolved photoemission [7] and inverse-photoemission [5] measurements which show significant angular dependance.

HREELS is also very sensitive to the structural order of the surface under investigation. In fact, on a crystalline surface, the dipole-scattered electrons are confined in a narrow lobe around the specular direction, so that the infrared-active modes, which are mainly observed in the specular direction, can be distinguished from the impact-excited modes observed out of the specular direction. On the contrary, on a disordererd surface, the specular direction is not anymore defined, as electrons are *elastically* scattered in all directions by the surface roughness. Consequently, even electrons which have excited a dipole-active mode are now scattered into a broad lobe, and the spectrum becomes isotropic.

Fig. 5 shows a comparison between a HREELS spectrum from a 350 Å thick epitaxial C_{60} film on GeS(001) (Fig. 5a) and of a disordered C_{60} film on Si(100) [6] (Fig. 5b). Both spectra show a lot of vibrational bands which can be easily assigned [6] using published infrared [8], Raman [8] and neutron [9] data. As explained above the dipole-active modes should make the largest contribution to the specular spectrum. This is however not the case for the C_{60}/Si(100) spectrum, where only the peak at 66 meV correlates with two of the four dipole-active modes at 66 and 72 meV. The two other dipole-active modes expected around 146 and 178 meV are not observed at all. Furthermore, the intensity of the dipole-excited band at 66 meV is comparable to that of the strong impact excited bands at 94, 156 and 194 meV. Turning now to the C_{60}/GeS(001) spectrum, it differs from that of C_{60}/Si(100) in different aspects which all point to the highly ordered structure of the present film. The strong peak at 66 meV due to the two lowest T_{1u} dipole modes of C_{60} is very prominent, suggesting that the dipolelobe becomes very sharp. All the impact-excited bands are drastically reduced in intensity. The other two T_{1u} dipole modes are now clearly detected at 147 and 178 meV. Although they become dominant in the upper part of the spectrum, their intensity is of the same order of magnitude as observed for impact-excited modes. This is related to the high symmetry of the C_{60} molecule which makes it very difficult to induce a dipole into the molecule, so that the oscillator strength of the dipole-active modes is very small. Indeed, the strongest dipole mode at 66 meV has an oscillator strength of only 0.07 [10], and the strength of the two weakest T_{1u} modes at 147 and 178 meV is even 20 times smaller. Under such circumstances the cross sections for impact and dipole scattering become comparable. Keeping this in mind, it becomes than understandable why the two highest dipole-active modes are not observed in the C_{60}/Si(100) case, where the dipole lobe is very broad.

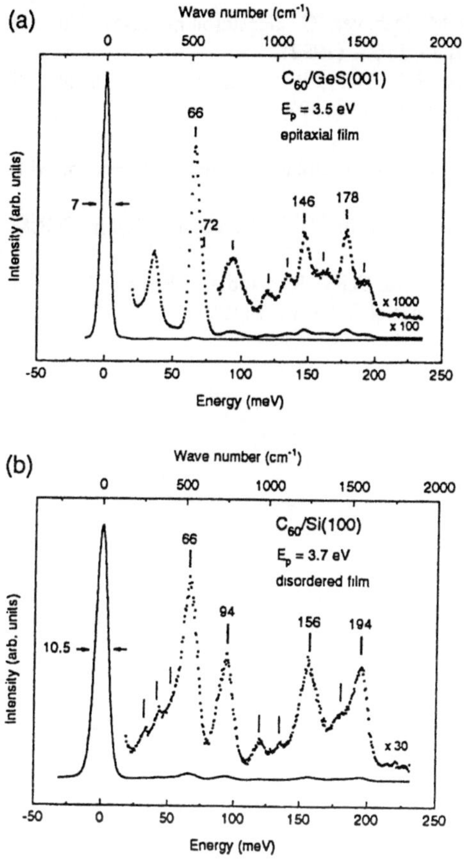

Fig. 5. (a) HREELS spectrum of a 350 Å thick epitaxial C_{60} film on GeS(001); (b) HREELS spectrum of a disordered C_{60} film on Si(100).

Acknowledgements

This work was funded by the Belgian national program of Interuniversity Research Projects by the State Prime Minister Office (Science Policy Programming).

References

[1] R.L. Meng, D. Ramirez, X. Jiang, P.C. Chow, C. Diaz, K. Matsuishi, S.C. Moss, P.H. Hor, C.W. Chu: Appl. Phys. Lett. 59, 3402 (1991).
[2] M. Sakurai, H. Tada, K. Saiki, A. Koma: Jpn. J. Appl. Phys. **30**, L565R (1991).
[3] D. Schmicker, S. Schmidt, J.G. Skofronick, J.P. Toennies, R. Vollmer: Phys. Rev. B **44**, 10995 (1991).
[4] W. Krakow, N.M. Rivera, R.A. Roy, R.S. Ruoff, J.J. Cuomo: IBM Research Report, RC 17345 (#76621) 10/30 (1991).

[5] J.-M. Themlin, S. Bouzidi, F. Coletti, J.-M. Debever, G. Gensterblum, L.-M. Yu, J.-J. Pireaux, P.A. Thiry, R. Caudano: Phys. Rev. B **46**, 15602 (1992).
[6] G. Gensterblum, L.-M. Yu, J.-J. Pireaux, P.A. Thiry, R. Caudano, J.-M. Themlin, S. Bouzidi, F. Coletti, J.-M. Debever: Appl. Phys. A 56, 175 (1993).
[7] G. Gensterblum *et al.*: to be published.
[8] D.S Bethune, G. Meijer, W.C. Tang, H.J. Rosen, W.G. Golden, H.Seki, C.A. Brown, M.S. Devries: Chem. Phys. Lett. **179**, 181 (1991).
[9] K. Prassides, T.J.S. Dennis, J.P. Hare, J. Tomkinson, H.W. Kroto, R. Taylor, D.R.M. Walton: Chem. Phys. Lett. 187, 455 (1991).
[10] G. Gensterblum, L.-M. Yu, J.-J. Pireaux, P.A. Thiry, R. Caudano, Ph. Lambin, A.A. Lucas, W. Krätschmer, J.E. Fischer: J. Phys. Chem. Solids **53**, 1427 (1992).

Electron Energy-Loss Studies on Rb-and Ca-Doped Fullerenes

M. Roth, H. Romberg, E. Sohmen, P. Adelmann, and J. Fink

Kernforschungszentrum Karlsruhe, Institut für Nukleare Festkörperphysik, Postfach 3640, D-76021 Karlsruhe, Germany

Abstract: The electronic structure of undoped, alkali (Rb) and alkaline earth (Ca) metal doped C_{60} has been studied by high-energy electron energy-loss spectroscopy (EELS). Valence band excitations yield information on the charge-carrier plasmon as well as on π-π^* transitions. From measurements of the C1s absorption egdes information on the unoccupied π^* and σ^* states could be derived. In the superconducting compound Rb_3C_{60} the LUMO-derived bands are half filled and completely filled for the insulating compound Rb_6C_{60}. In the superconducting compound Ca_5C_{60} the LUMO+1 derived bands are partially filled.

1. Introduction

The synthesis of macroscopic amounts of fullerenes [1] has allowed the preparation of a new class of solids with new and interesting properties. Doping of C_{60} with alkali metals produces conductors [2] and for A=K and Rb the A_3C_{60} phase is superconducting [2-5]. Kortan et al. showed [6] that Ca intercalates in the C_{60} lattice and that Ca_5C_{60} is a superconductor with a transition temperature of T_c=8.4 K. The mechanism leading to those superconducting properties is not clear and it is a prerequisite to understand the electronic structure in the normal state in order to understand the superconducting behaviour. Here we present electron energy-loss spectroscopy (EELS) studies on the electronic structure of Rb and Ca doped fullerenes.

2. Experiment

The fullerenes were synthesized by the standard method [1]. Thin, free-standing films with a thickness between 500 Å and 2000 Å are needed for EELS measurements in transmission. Those samples were prepared by evaporation of C_{60} on NaCl and KBr crystals. The crystals were dissolved in distilled water and the fullerite film floated onto standard electron microscopy grids. Subsequently the films were annealed under UHV conditions at 300°C and doping was achieved by evaporating various amounts of alkali resp. alkaline earth metals on the films. The electron energy-loss data were recorded with a 170 KeV Spectrometer [7] in transmission. The energy resolution was set to 0.18 eV and the momentum

Springer Series in Solid-State Sciences, Volume 117
Electronic Properties of Fullerenes Eds.: H. Kuzmany, J. Fink, M. Mehring, and S. Roth

resolution was chosen to be 0.04 Å for the low energy spectra and 0.2 Å for the C1s absorption edges, respectively.

3. Alkali metal doping

Fig. 1a shows the change in the low energy spectra of a C_{60} sample upon doping. In the undoped sample one clearly sees the bandgap of 1.9 eV. At 2.2 eV a structure appears which can be assigned to a dipole forbidden $h_{1u} \rightarrow t_{1u}$ transition. The following features at 2.7 eV and 3.6 eV correspond to allowed transitions from the h_u to the t_{1g} level. The plasmon of all valence electrons is observed at 6.2 eV and the plasmons of the π and σ electrons has its maximum at 25.5 eV (not shown).

Upon doping two features appear in the bandgap. For x=3 the peak at 0.5 eV reaches its maximum and can be assigned to a plasmon caused by an intraband transition within the partially filled lowest-unoccupied-molecular-orbital-derived (LUMO) conduction band. The second peak at 1.2 eV is caused by a transition from the LUMO to the next group of π^* bands derived from the molecular t_{1g} level. Consequently, while the first peak decreases upon further doping and disappeares for the fully doped case, the second peak approximately doubles its intensity, both indicating a complete filling of the t_{1u} derived band. The data shown in the figure are in qualitative agreement with calculations of the loss function [8].

The peak appearing at 3.6 eV in the undoped system has shifted to 2.9 eV in the fully doped case. This gradual shift indicates a shrinking of the energy difference between the π and π^* bands. Calculations of the electronic structure predict a reduction in the energy difference between the π and π^* bands due to a stronger delocalisation of the π bands in the doped case compared to the undoped one, where the electrons are localized in every second bond of the hexagon [9].

In Fig. 1b we show the C1s absorption spectra for the undoped, half, and fully doped case. Neglecting core hole effects, transitions from the C1s core level are a direct measure of the density of states of the π^* and σ^* bands. Therefore the spectra clearly show the filling of the t_{1u} derived band upon doping by the gradual decrease of the first peak, having about half its original intensity for x=3 and disappearing in the Rb_6C_{60} samples. When looking at the spectra one should note the shift of all structures to lower energies with increasing dopant concentration. This is caused partly by a shift of the C1s level to lower binding energy and by a lowering of the π^* bands.

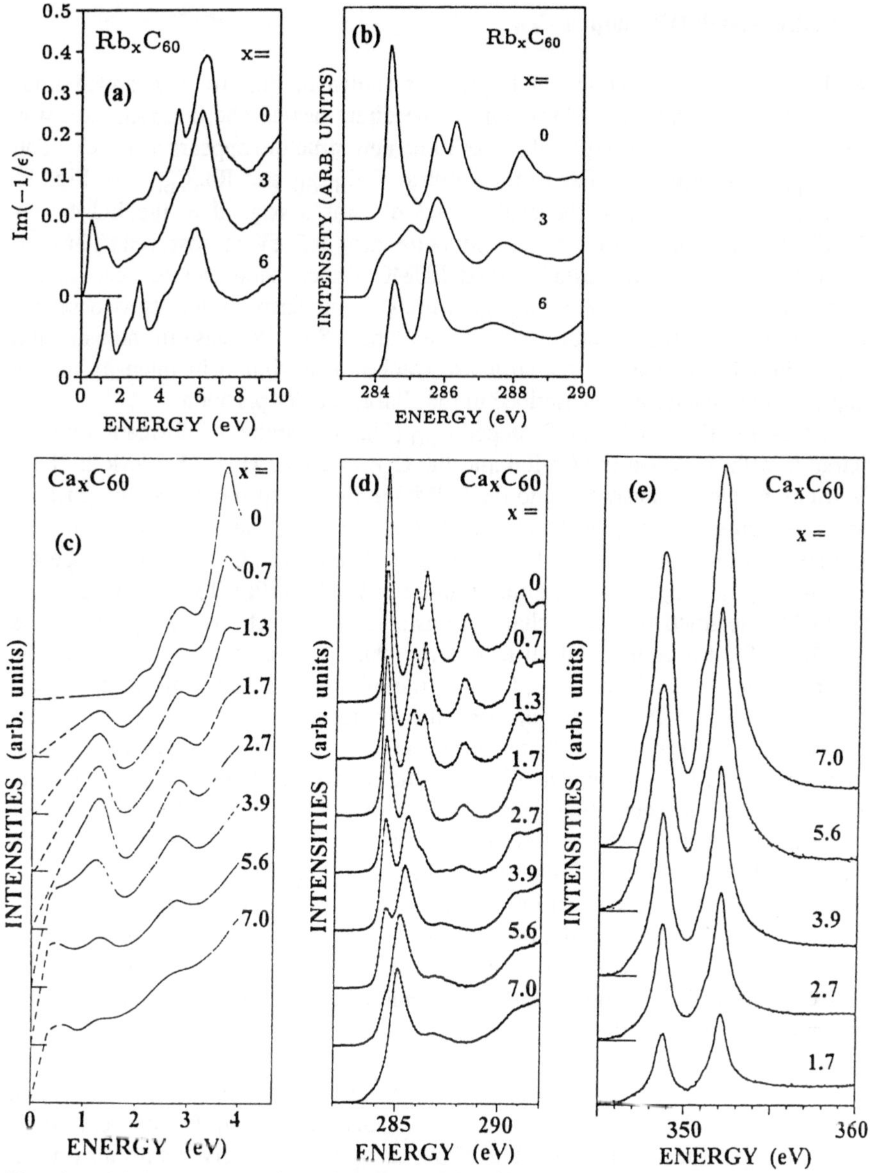

Fig. 1: **(a)** Low-energy loss function of Rb_xC_{60}. **(b)** C1s absorption edges of Rb_xC_{60}. **(c)** Low-energy loss function of Ca_xC_{60}. **(d)** C1s absorption edges of Ca_xC_{60}. **(e)** Ca2p absorption edges of Ca_xC_{60}.

4. Alkaline-earth (Ca) doped C_{60}

The low energy loss function of Ca_xC_{60} is shown in Fig. 1c. The dotted lines below 0.5 eV indicate the subtraction of the direct beam. The undoped case was discussed above. For doping levels $0<x\leq3$ one new structure appears at 1.3 eV. The $Ca_{2.7}C_{60}$ spectrum is similar to the spectra of K_6C_{60} and Rb_6C_{60} (see Fig. 1a and Ref. 10-12) in which the peak at 1.3 eV was asssigned to the LUMO → LUMO+1 transitions. However, no structure near 0.5 eV is observed. Thus no intraband-plasmon of a partially filled LUMO drived band can be seen in the spectra of Ca_xC_{60} ($x<3$). For Ca_xC_{60} with $x>3$, increasing intensity is observed near 0.5 eV while the structures near 1.3 eV and 2.8 eV decrease in intensity and become broadened. The 0.5 eV structure reaches a maximum in intensity near a Ca_5C_{60} stochiometry and is ascribed to the charge-carrier plasmon.

The C1s absorbtion edge of Ca doped C_{60} (Fig. 1d) contains information on the charge transfer between the Ca-ion and the C_{60} molecule [13]. The peak at 284.4 eV which is due to transitions into the LUMO in the undoped case has disappeared for a Ca-doping with x=3. At this concentration the two remaining strong peaks are assigned to transitions into the higher lying π^* bands. The C1s absorption spectra of Ca_3C_{60} (Fig. 1 d) are very similar to those of Rb_6C_{60} (Fig. 1b), thus in Ca_3C_{60} the LUMO is filled due to a charge transfer of two electrons per Ca atom. As mentioned for Rb doping, all structures are shifted towards lower energies. The spectra for $0<x<3$ can be simulated by appropriatly weighted amounts of the spectra for x=0 and x=2.7 therefore indicating the existence of only two phases for Ca concentrations lower than x=3, i.e. x=0 and x=3. This phase separation is also supported by the low-energy data (Fig. 1c) and the recorded diffraction pattern (not shown). The phase separation would explain the lack of metallic conductivity or even superconductivity [6] as well as the lack of a Fermi edge in photoemission [14], though disagreeing with the results of Kortan et al. [6].

Upon doping with Ca concentrations higher than x=3 the situation changes. Here the structure at 284.5 eV, assigned to LUMO+1, decreases upon doping and moves continuously to lower energies. This corresponds to a continuous filling of the LUMO+1 derived band and explains the metallic behavior observed for $x>3$, being also in agreement with the appearance of a charge carrier plasmon at 0.5 eV in the low energy spectra.

Fig. 1e shows the Ca2p core level excitations which can be used to obtain information the electronic structure of the Ca ions in Ca_xC_{60}. For concentrations lower than x=5 the spin-orbit doublet of the Ca2p level is observed with maxima at 348.7 eV and 352.0 eV, with widths of 0.9 and 1.0 eV (full width at half maximum), repespectively. The spectra for Ca_xC_{60} ($x>4$) show widths increasing to about 1.5 eV for Ca_7C_{60}, while additional shoulders appear at 347.7 eV and 350.9 eV. The spectra of Ca_xC_{60} ($x<4$) are similar to those of CaF_2 thus providing evidence for a fully ionised Ca^{2+} at those doping levels. The new features for higher doping levels might be explained by Ca atoms occupying interstitial sites with lower symmetry or by the onset of hybridisation of the Ca (4s,4p) orbitals with the LUMO+1 and the LUMO+2 levels.

Electrons from a monochromatic source (ΔE = 30 meV) pass through an interaction region and are analyzed in a tandem electron spectrometer. In order to maintain high resolution, the electron energies inside the electron spectrometers were kept at a low level (2 eV to 5 eV).

To perform the experiments in the gas phase fullerene samples were vaporized in an oven in the interaction region of the spectrometer. Oven temperatures ranged between 400 and 600°.

3. Results and Discussion

Fig. 1 displays EELS-spectra of pure C_{60} (left column) and pure C_{70} (right column) taken at low and high primary electron energies. In the case of C_{60} four distinct features (denoted A to D) arise in the energy interval between 2 and 10 eV. The higher transiions(B,C) are optically allowed; they coincide approximately with the absorption spectra measured by Krätschmer et al.[2]. According to Rosén et al. [11] they are assigned

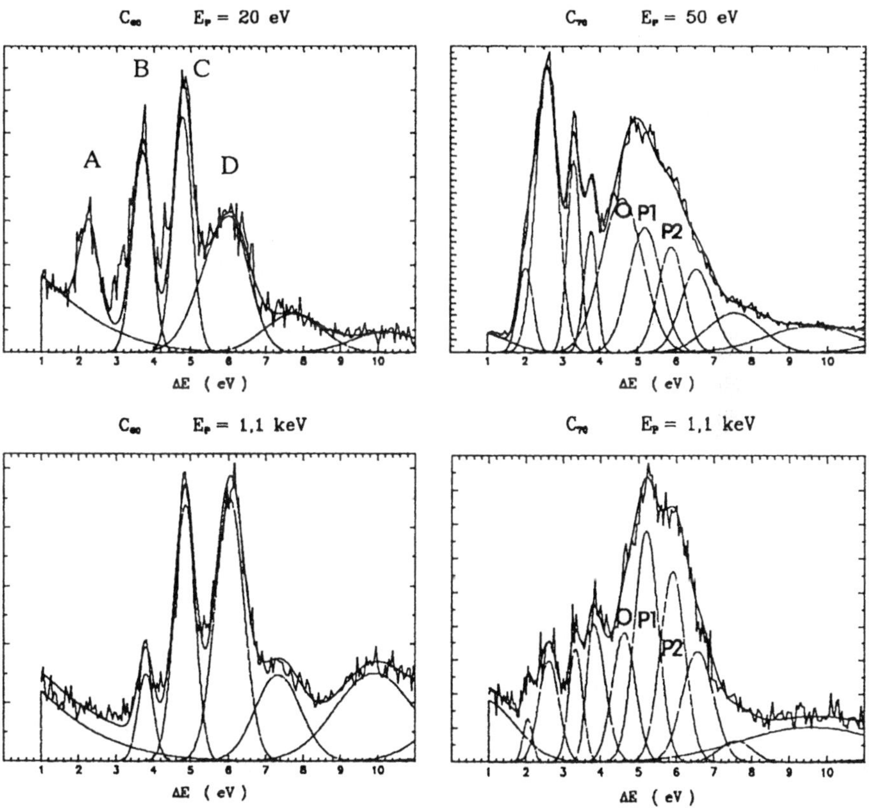

Fig. 1: EEL-spectra of C60 (left column) and C70(right column) taken at high and low primary electron energies. Solid lines represent the simulation using Gauss functions.

to the h_u-t_{1g} (HOMO→(LUMO+1)), h_u-t_{2g}/h_g (HOMO→LUMO+2), and h_g-t_{1u} ((HOMO-1)→LUMO) transitions while the lowest one h_u-t_{1u} (HOMO→LUMO) corresponds to an optically forbidden transition. The latter transition (noted A in Fig.1) increases in intensity when lowering the energy of the primary electrons, leading to a breakdown of the optical selection rules. It emerges as the strongest transition at E_p=10eV.

The peak at D increases strongly with electron energy and is therefore assigned to a collective resonance of the π-electrons[5].

In order to obtain insight into the excitation dynamics 2-step spectral simulations were performed. The experimental distributions were approximated using a set of Gauss functions of variable width placed at different energies. In the first step, line position, width, and area were optimized independently for each of the measured spectra . In the second step the same procedure was applied again, but using average positions, identical for all spectra, to calculate new linewidth and peak areas. Thus we have identified transitions at 2.26, 3.68, 4.77, 6.01, 7.72, 10.2, 12.3, 16.4, and 21.8 eV, the latter 3 being asssigned to the π-σ-collective excitation.

EEL-spectra of C_{70} are more complicated because of its lower symmetry. The right column of Fig.1 displays such spectra for E_p= 50 eV and 1200 eV. It shows a similar behaviour to that of C_{60} with transitions at ΔE=2.01, 2.58, 3.30, 3.77, 4.58, 5.17, 5.87, 6.54, 7.54, 9.59, 13.0, 16.3 and 20.7eV. The spectra were simulated in a similar way to that used for C_{60}. Collective excitations in C_{70} should be split because of the spheroidal (rather than spherical) distribution of the charge around the molecule. Varying the primary electron energy E_p, results in a change of the relative intensities except for the lines at 5.2 eV and 5.9 eV which keep their intensity ratio regardless of E_p. This and the fact that the energy of these peaks is near that of the π-plasmon of both C_{60} and graphite leads us to the conclusion that these two lines represent the two split levels of the π-plasmon of C_{70}. The splitting is a direct result of the deviation from the "quasi"-spherical geometry and is caused by the varying curvature of the 2-dimensional electron cloud in the C_{70} particle.

We have compared our experimental results with the calculations of Bulgac et. al.[14] , who obtained the oscillator strength for different multipole excitations. Therefore we converted our data to oscillator strength. For this we use a relationship between the general oscillator strength $f_n(K,E_P)$ and the differential inelastic cross section for inelastic excitation $d\sigma/d\omega$ first developed by Bethe[9, 10]:

$$\left.\frac{\partial\sigma}{\partial\omega}\right|_{\Theta=0} = 16\, a_0^2\, R^2 \frac{E_P}{\Delta E^3} f_n \qquad (1)$$

E_p is the kinetic energy of the interacting electrons, ΔE is the energy loss, a_o and R are the Bohr radius and the Rydberg constant, respectively.For our experimental conditions ($\hat{\Theta}<0.5^o$, $\Delta E < 0.1 \cdot E_P$) this results in a scale factor proportional to $(\Delta E)^3$. Fig 2 shows the data treated in such a way. Using the calculated oscillator strength of Bulgac et al. it seems that the best results are obtained using a combination of 30% L=0(monopole) and 70% L=1(dipole) excitation. Fig. 2 (left) shows the C_{60} spectra treated this way

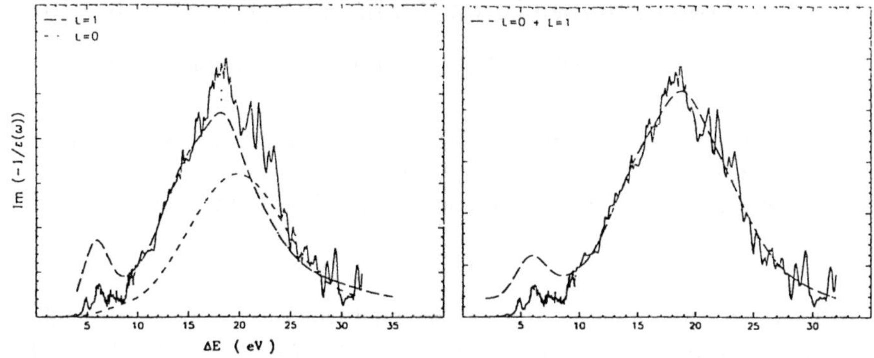

Fig.2: Comparison of the oscillator strength for C_{60} with calculations for L=0 and L=1 excitations [6]. Best fit is achieved using 30% L=0 and 70% L=1 excitation (right).

together with the calculations for L=0 and L=1. The right part of the figure shows a comparison of the experimental with the simulated oscillator strength optimized for the best fit. This implies that a significant amount of the electrons are passing through the sphere and influence the electron cloud in such a way that a breathing mode is excited.

We greatfully acknowledge the financial assistance of the German Research Council (DFG Sonderforschungsbereich 337) and of the German Ministry for Research and Development (BMFT).

References:

1.H. W. Kroto, J. R. Heath, S. C. O'Brian, R. F. Curl, R. E. Smalley: Nature, **318**, 162 (1985)

2.W. Krätschmer, L. D. Lamb, K. Fostiroupolos, D. R. Huffman, : Nature, **347**, 354 (1990)

3.C. Becker, A. Burose, A. Ding: Z.Phys.D **20**, 35, (1991)

4.A. Burose, T. Dresch, A. M. Ding, Z.Phys.D, in print

5.E. Sohmen, J. Fink, W. Krätschmer: Z.Phys.B **87**, 87 (1992)
G. Gensterblum, J. J. Pireau, P. A. Thiey, R. Caudano, J. P. Vigneron, P. Lambin,A. A. Lucas: Phys.Rev.Lett. **67** , 2171 (1991)

6.A. Bulgac, N. Ju: Phys.Rev. B 46, 4297(1992)

7.M. Inokuti: Rev.Mod.Phys. **43**, 297 (1971) and references therein

High-Resolution Photoemission Studies of A_3C_{60} (A = K, Rb)

M.S. Golden[1], *M. Knupfer*[1], *M. Merkel*[1], *J. Fink*[1], *O. Gunnarsson*[2], *and V.P. Antropov*[2]

[1] Kernforschungszentrum Karlsruhe, Institut für Nukleare Festkörperphysik, Postfach 3640, D-76021 Karlsruhe 1, Germany
[2]Max-Planck-Institut für Festkörperforschung, D-70569 Stuttgart, Germany

Abstract. We report high resolution photoemission spectra of A_3C_{60} (A=K, Rb) recorded at low temperature. The spectra clearly show near Fermi-edge fine structure, as well as an anomalous width of the conduction band spectral weight and density of states at E_F. Calculations of the photoemission spectra show that these features can be understood in terms of the presence of satellites due to both phonons and coupling to the charge carrier plasmon.

1. Introduction

Despite the intense effort expended on their study, there still remain a number of questions regarding the detailed electronic structure of the A_xC_{60} materials. It has been shown [1] that the correlation energy in solid C_{60} is as high as 1.6eV. Taken together with the small bandwidths in these materials this means that the effects of electron correlation are important. Consequently, the question arises as to whether the half-doped A_3C_{60} materials are indeed good metals (and superconductors) as bulk studies suggest, or whether stoichiometric, ordered A_3C_{60} is a Mott-Hubbard insulator [1]. The effect of phonons on the electronic structure of the fullerides also cannot be ignored, as some of the intramolecular phonon modes are of the same order as the Fermi energy.

Photoemission represents a powerful tool for the study of the occupied electronic structure of the fullerides, but despite publication of photoemission studies of A_xC_{60} with high resolution and accurate, absolute determination of the overall doping level [2,3], there persist a number of anomalies in the data. In particular, the observed width of the occupied conduction band in A_xC_{60} ($x \approx 3$) and the low density of states (DOS) at E_F have still not been fully understood.

In this contribution we show that the fine structure observed in the low temperature spectra of A_3C_{60} (A=K, Rb) is consistent with the observation of satellites in the photoemission spectral weight due to phonons and the charge carrier plasmon. We briefly discuss the significance of this result with reference to the effects of phonons and

Springer Series in Solid-State Sciences, Volume 117
Electronic Properties of Fullerenes Eds.: H. Kuzmany, J. Fink, M. Mehring, and S. Roth

electronic correlation on the electronic structure of the half-doped K and Rb fullerides.

2. Experimental

Polycrystalline A_3C_{60} films of thickness ~100Å were prepared and characterized as described elsewhere [2]. Photoemission was carried out using He I radiation (hν=21.22eV) with a total energy resolution of 25meV.

The calculations of the photoemission spectra were carried out using a simple, exactly soluble Hamiltonian in which photoemission from a discrete energy level includes coupling to the A_g and H_g phonon modes of C_{60} and to the charge carrier plasmon observed at 0.5eV in EELS spectra [4]. Unfortunately, space considerations preclude a detailed description of the calculation method here, however more details can be found in Ref. [8]. The electron-phonon coupling constants were obtained from *ab initio* calculations [5] and correspond to the coupling strength $\lambda=0.057N(0)$ in superconductivity, where N(0) is the electron density of states per eV, spin and C_{60}. The plasmon coupling was obtained from a tight-binding model [6] of the doped C_{60} solid by calculating the strength of the pole in the inverse dielectric function. The frequency and width of the resulting plasmon were modified to agree with experiment. This resulted in the coupling $(g_{pl}/\omega_{pl})^2=1.0$ for the plasmon, where g_{pl} and ω_{pl} represent the coupling to and the frequency of the plasmon, respectively.

The calculated spectrum is convoluted with a DOS representing the occupied part of the conduction band of A_3C_{60}. For simplicity, a band of the shape shown as an inset to Fig. 1 was used. The width of this band is reduced in comparison with the value obtained from band structure, calculations [7], in order to take a polaronic type of narrowing into account. The actual bandwidth used (0.19eV) represents an optimal value for agreement of the calculated spectra with experiment, particularly in the region of the two low energy peaks (i.e. up to ~0.4eV BE). The calculations include a broadening of 25meV, representing the experimental energy resolution of the photoemission experiments.

3. Results and discussion

Figure 1 shows a comparison of the low temperature (10K) photoemission spectra of Rb_3C_{60} and K_3C_{60} with a series of simulated photoemission spectra of A_3C_{60} (A=Rb or K). In the simulated spectra, the plasmon coupling, $(g_{pl}/\omega_{pl})^2$, and the coupling to the low lying phonon

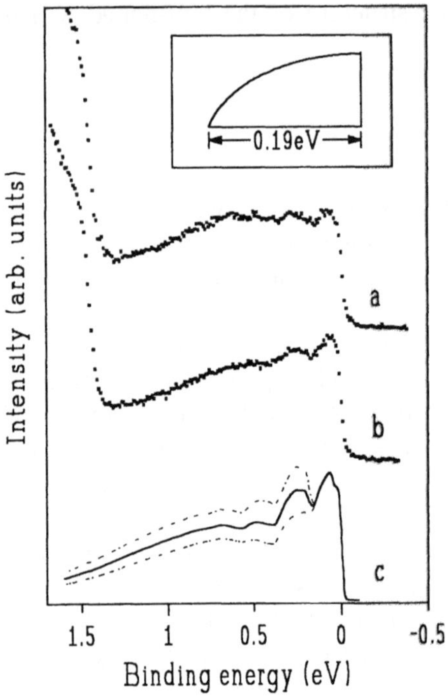

Fig. 1. Photoemission spectra of a) K_3C_{60} and Rb_3C_{60} recorded at 10K. Also shown in c) are calculated PES spectra of A_3C_{60} (A=Rb or K) with electron phonon coupling constants, $\lambda/N(0)$, of 0.069 (dotted-dashed line), 0.091 (solid line) and 0.114 (dashed line). The inset shows the form of the half-filled conduction band used in the calculation.

modes are equal to the calculated results while the coupling to the high lying modes (i.e. the two highest energy H_g modes and the highest energy A_g mode) are multiplied by the empirical constants 1.25, 1.75 and 2.25. The corresponding values of $\lambda/N(0)$ are 0.069, 0.091 and 0.114. The experimental Rb_3C_{60} spectrum has a clear peak centred at ~0.06eV below E_F and is followed by a dip centred at 0.15eV binding energy (BE), with a further peak at 0.25eV and another dip at 0.45eV BE. There are also signs of an additional broad peak at ~0.6eV BE. The spectrum of K_3C_{60} is very similar. For both materials, the spectra show a sharp Fermi edge and onset due to emission from the HOMO-derived band of C_{60} at ~1.5eV BE. The width of the occupied conduction band-derived spectral weight is ~1.3eV. It is clearly seen from Fig. 1 that the best agreement between the calculation and the experimental spectra is achieved with a value of $\lambda=0.091N(0)$, which gives values of $\lambda=0.77$ and 0.91, taking [7] $N(0)=8.5$ and 10 for K_3C_{60} and Rb_3C_{60}, respectively. Thus λ is a factor of 1.6 larger than the value arrived at from the *ab initio* calculation. We note, however, that the low BE part of the calculated spectrum (up to ~0.4eV) is sensitive to the assumed shape of the conduction band DOS, and thus care is needed in the determination of λ from the photoemission data. The presence of phonon satellites has also been suggested in a recent low temperature photoemission study [3] of K doped $C_{60}(111)$, together with transfer of spectral weight from the quasiparticle peak at E_F attributed to the effects of electron correlation.

Analysis of the occupied conduction band spectra shown in Fig. 1, assuming the total spectral weight represents three electronic states, leads to a DOS at E_F of 3.1 and 4.1 states/eV-C_{60} for K_3C_{60} and Rb_3C_{60}, respectively. These figures are lower than those from band structure calculations [7] and suggest that the spectra shown in Fig. 1 may not be a true reflection of the conduction band density of states, but rather the spectral function including phonon and plasmon satellites. An estimation of the relative weight of the so-called 'zero phonon line', or coherent part of the spectral function which gives rise to the sharp Fermi edge seen in Fig. 1, yields a value of around 20%. Thus, although it is evident from the calculated spectra that the effects of the molecular phonons and the charge carrier plasmon are strongly felt by the system during photoemission, the influence of these excitations, as well as the effects of electron correlation, are not great enough to result in a spectrum consisting of incoherent spectral weight alone. This rules out, for example, a scenario in which the electron-phonon coupling is great enough to result in a complete condensation into polaronic states.

4. Conclusions

The low temperature photoemission spectra of A_3C_{60} show clear signs that the system couples strongly to the molecular phonons. Calculations show that the detailed shape of the spectral weight can be explained in terms of electron-phonon and electron-plasmon coupling. About 20% of the spectral weight remains in coherent states at E_F. This suggests that although phonons and electron correlation may be important in A_3C_{60}, they do not completely dominate the determination of the electronic structure.

M.S.G. is grateful for the award of a SERC/NATO Advanced Postdoctoral Fellowship.

References

[1] R. W. Lof, M. A. von Veenendaal, B. Koopmans, H. T. Jonkman and G. A. Sawatzky, Phys. Rev. Lett, **68**, 3924 (1992)

[2] M. Merkel, M. Knupfer, M. S. Golden, J. Fink, R. Seemann and R. L. Johnson, Phys. Rev. B, in print

[3] P. J. Benning, F. Stepniak, D. M. Poirier, J. L. Martins, J. H. Weaver, L. P. F. Chibante and R. E. Smalley, Phys. Rev. B, to be published
[4] E. Sohmen, J. Fink and W. Krätschmer, Europhys. Lett., **17**, 51 (1992)
[5] V. P. Antropov, O. Gunnarsson and O. K. Andersen, (unpublished)
[6] O. Gunnarsson and G. Zwicknagel, Phys. Rev. Lett., **69**, 957 (1992)
[7] S. Satpathy, V. P. Antropov, O. K. Andersen, O. Jepsen, O. Gunnarsson and A. I. Lichtenstein, Phys. Rev. B, **46**, 1773 (1992)
[8] M. Knupfer, M. Merkel, M. S. Golden, J. Fink, O. Gunnarsson and V. P. Antropov, (unpublished)

Electronic Structure of Alkaline-Earth Fullerides

I. Turek and J. Hafner*

Institut für Theoretische Physik, TU Wien,
Wiedner Hauptstr. 8–10, A-1040 Wien, Austria
*On leave of absence from Institute of Physical Metallurgy,
Czech Academy of Sciences, Brno, Czechia

Abstract. We present selfconsistent calculations of the electronic structure of alkaline–earth doped fullerides.

1. Introduction

The discovery of conductivity and superconductivity in C_{60} films exposed to alkali- [1,2] and alkaline earth metals [3,4] has stimulated a great deal of interest in the properties of the metal fullerides. The atomic and electronic structures of the alkali–metal fullerides are now quite well understood. The metal atoms diffuse into the tetrahedral and octahedral holes of the face–centred cubic C_{60} lattice. The fcc A_3C_{60} phases (A = K, Rb,Cs) exhibit superconductivity [2]. At higher metal concentrations, a body–centred tetragonal A_4C_{60} phase [5] and an insulating body–centred cubic A_6C_{60} phase [6] are found. Electronic structure calculations [7–11] show that the electronic density of states of the alkali–metal fullerides around the valence band maximum is very similar to that of the pure C_{60} phase. The transition to a metallic state and the return to a semiconducting behaviour arise from the gradual filling of the conduction band derived from the LUMO–band of pure C_{60}.

Less is known about the alkaline–earth fullerides. Chen et al [4] showed that the fcc solid–solution phases Ca_xC_{60} with $x \leq 3$ have non–metallic character. Larger dopant concentrations lead to metallic behaviour, and Ca_5C_{60} has been shown to be superconducting [3]. In the Ca_5C_{60} phase calcium clusters are formed in the octahedral sites, and different variants of the structure have been proposed [3,12]. Photoemission experiments[4,13] suggest an appreciable degree of hybridization of the alkaline earth s–states and the LUMO and LUMO+1 bands of pure C_{60}. At $x = 3$ the Fermi level falls into a narrow gap between the LUMO and LUMO+1 derived bands. For Ca_3C_{60} rather preliminary electronic structure calculations [13,14] have been presented, producing qualitative agreement with the observed photoelectron spectra.
In this work we present selfconsistent calculations of the electronic structure and of the photoemission and inverse photoemission spectra of the alkaline–earth fullerides AE_xC_{60} with AE = Ca,Sr, and Ba.

2. Method of calculations

We used the linear muffin–tin orbital (LMTO) method in the atomic sphere approximation (ASA) [15]. Exchange and correlation are described in the local–density approximation, using the Barth–Hedin parametrization of the exchange–correlation functional [16]. Since the fcc phase of C_{60} has a very open structure, empty spheres must be placed into the voids of the lattice. We choose a minimum number of 76 atomic spheres: 60 carbon spheres (radius r = 0.90 Å), one large sphere in the center of the C_{60} molecule (r = 2.96 Å), one sphere in the octahedral interstice (r = 3.10 Å), two spheres in the tetrahedral interstices (r = 2.62 Å), (these spheres are occupied by the metal ions in the doped fullerides) in addition twelve spheres placed halfway between the pentagon and hexagon faces of nearest–neighbours C_{60} molecules (r = 1.42 Å), all radii refer to pure C_{60}. We used the full set of s,p,d–LMTO's in all atomic spheres except the sphere inside the C_{60} molecule. In the innermost sphere all electrons were down–folded [17]. The selfconsistent calculations of the potential and charge density were performed with a mesh of 20 $\vec{k}$–points, the calculation of the electronic density of states was performed using the tetrahedron method.

In addition to the electronic structure we have calculated the intensities of photoemission (PES) and inverse (IPS) photoemission spectra in terms of the local, angular–momentum decomposed DOS $n_{i,l}(E)$ weighted with the partial cross section $\sigma_{i,l}(E, \hbar\omega)$ for the transition of the electron between its initial and final states. $\sigma_{i,l}(E, \hbar\omega)$ may be calculated in terms of the selfconsistent potential using a dipole approximation for the photon field, neglecting wave vector conservation and multiple scattering, and averaging over photon directions and polarizations and the orientation of the wave vector of the outgoing photoelectron in PES, respectively of the incoming electron in IPS [18,19]. Although the plane–wave approximation and the neglect of wave vector conservation will be better justified at high photon energies, we generally find good agreement between theory and experiment even at low photon energies in previous systems studied.

3. Electronic structure of alkaline–earth fullerides

Fig.1 shows the electronic density of states of the alkaline–earth fullerides Ca_3C_{60}, Sr_3C_{60} and Ba_3C_{60}, for comparision the DOS of the pure fullerite C_{60} and of the potassium fulleride K_3C_{60} are also shown. Our results for the fcc C_{60} phase are in good agreement with previous band structure calculations [20–22]. The HOMO–LUMO gap is 1.1 eV, i.e. smaller than the gap width of E_g = 1.6 – 1.9 eV estimated from the electrical, microwave and optical conductivities and distinctly smaller than the apparent gap of E_g = 2.3 – 2.6 eV in combined PES and IPS experiments [23]. This discrepancy is well understood in terms of electron correlation effects neglected in the LDA and must be considered in comparison of the theoretical and experimental spectra. The band structure of K_3C_{60} may be derived from that of pure C_{60} by a rigid shift of

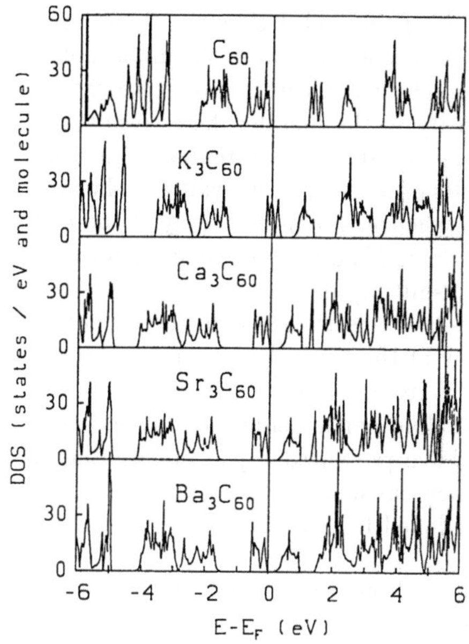

Fig.1: Electronic density of states of C_{60}, K_3C_{60}, Ca_3C_{60}, Sr_3C_{60} and Ba_3C_{60} in the vicinity of the Fermi level.

the Fermi level, only at energies < 3 eV above E_F distinct differences are observed. The DOS at the Fermi level is $n(E_F) = 17.4$ states per eV per molecule; again our results are in good agreement with previous calculations [8–11]. In the alkaline–earth fullerides, the Fermi energy is shifted to the upper edge of the LUMO–derived band. Ca_3C_{60}, Sr_3C_{60} and Ba_3C_{60} are predicted to be narrow gap semiconductors with a gap varying between $E_g = 0.2$ eV in Ca_3C_{60} and $E_g = 0.1$ eV in Ba_3C_{60}. The structure of the valence bands conforms more or less with a rigid band picture, apart from a slight broadening of all bands arising from the more extended character of the C–p states of the now charged C_{60}^{6-} molecular ion. The structure of the conduction bands, however, is substantially modified: in Ca_3C_{60} the lowest conduction band is only slightly modified, but a new band containing two electrons appears 1.3 eV above the Fermi level, and the bands at E > 1.6 eV are formed by C–p states strongly hybridized with Ca-s,p and d-states. The two–electron band is formed by an atomic–like Ca-4s states centred at the octahedral site. The Ca-4s states centred in the smaller tetrahedral holes are more strongly hybridized with the C-p states. The electronic structure of Sr_3C_{60} is very similar to that of the Ca–fulleride, but that of Ba_3C_{60} shows distinct differences: the two–electron band has merged with the second conduction band showing strong Ba-5d character. The difference is due to the larger size of the Ba–atom and to relativistic effects which tend to lower the Ba-5d level relative to the Ba-6s level (remember that pure bcc Ba has to be considered as a transition metal with ~ 0.8 d-electrons per atom [24]). Although the lowest alkaline–earth states lie above the Fermi level, an analysis of the local DOS shows that the atomic spheres of the metal-sites are by no means completely depleted of electrons. However, the local DOS in the metal-spheres has a strongly mixed angular momentum character with a strong weight for $\ell = 2$. This suggests that the charge density in the metal-spheres results from overlapping C-states and is dominated by the strong C-potential. Our results for Ca_3C_{60} are at variance with the results of Saito and Oshiyama [14] showing a band derived from Ca-4s states at the octahedral site immediately at the Fermi level. This is in disagreement with the PES and IPS results [4] demonstrating the insulating character of Ca_3C_{60}. Only by shifting the Ca-atom to an off-center position of the octahedral hole, the Ca-4s states is moved to higher energies and a gap opens at E_F. The discrepancy of the two calculations is eventually due to the restricted Gaussian basis in Saito and Oshiyamas approach which contains only s- and p-states. For an alkaline-earth element this is necessarily of questionable validity. Chen et al [4] have presented a pseudopotential calculation for Ca_3C_{60}. However, since they calculated the eigenvalues only at two special $\vec{k}$-points, the Gaussian-broadened DOS is too crude to allow for a detailed comparison.

Preliminary results for the Ca_5C_{60} phase show that the Fermi-level moves into the LUMO-band of Ca_3C_{60}. The partially filled conduction band is separated by a gap of 0.3 eV from a C-p – Ca-s,d band complex.

4. Photoemission and inverse photoemission spectra

Fig.2 shows the calculated PES intensities for C_{60} for photon energies varying from the XPS to the UPS regime, compared with available data (refs. 25-28).

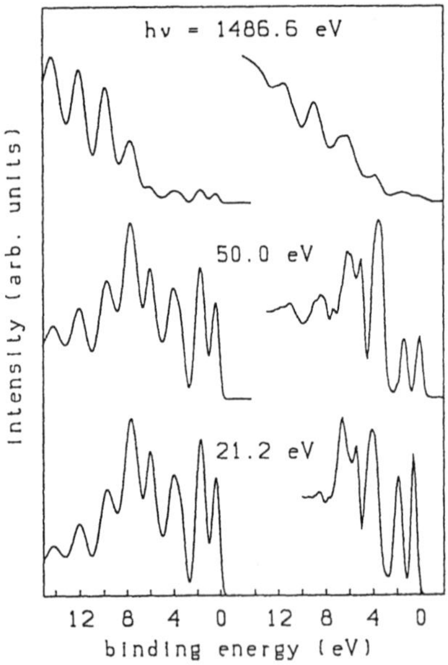

Fig.2: Photoemission intensities for C_{60} at various photon energies. Left column – calculated, right column – experiment (refs. 25–28).

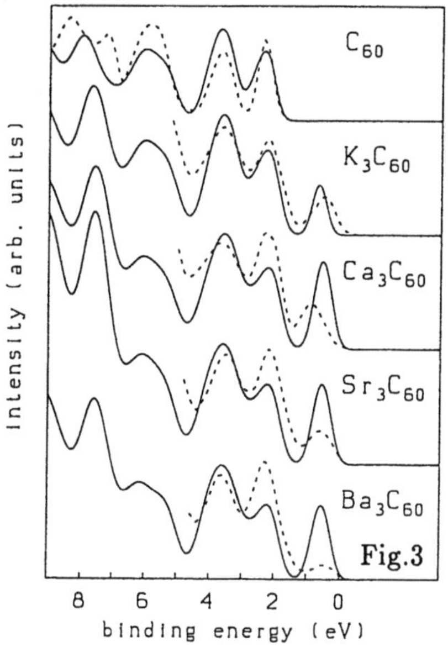

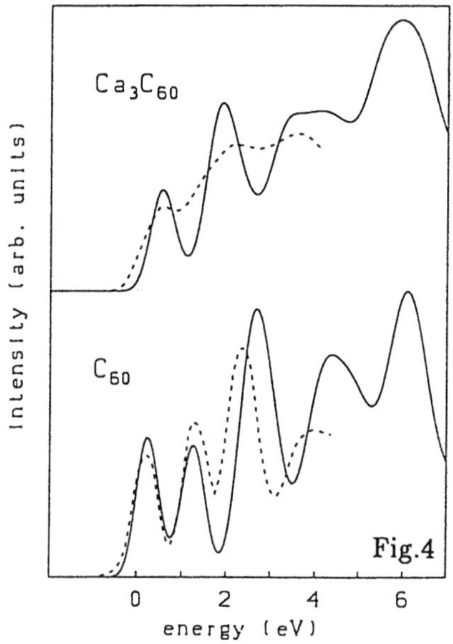

Fig.3: Photoemission intensities for the alkaline–earth fullerides, theory (solid lines) vs. experiment (dashed lines) (refs. 4, 13).

Fig.4: Inverse photoemission spectra for C_{60} and Ca_3C_{60}, theory (solid lines) vs. experiment (dashed lines).

We find that although the positions of the most prominent features in the PES agree quit well with the predicted position of the bands, quit important changes occur in the relative intensities of the PES features. These variations are well described by the calculated variation in the photoionization cross sections with the energy of the incident photon. However, for bin ding energies > 6 eV dthe peaks observed in the PES are shifted to lower energy compared to the theoretical prediction. This is probably due to the neglect of the self–energy correction associated with the hole in the valence band.

Fig.3 shows the calculated and measured PES for the alkaline–earth fullerides. The comparison with experiment is complicated by uncertainties in the exact stoichiometry, but altogether both theory and experiment agree on a complete filling of the LUMO–derived band. The changes in the PES at higher binding energy compared to pure C_{60} reflect the changes in the bonding induced by the interaction of the C–states with the potential of the metal ions. The inverse photoemission spectra are shown in Fig.4. For the alkaline–earth fullerides the IPS demonstrates the existence of a narrow gap at the Fermi level and of many new states at higher energy. However, the much lower resolution does not allow for a critical evaluation of the calculated DOS.

5. Conclusions

We have presented selfconsistent calculations of the electronic structure, PES and IPS spectra of C_{60} fullerite and of the alkaline–earth fullerides Ca_3C_{60}, Sr_3C_{60}, and Ba_3C_{60}. Our results show that the alkaline–earth fullerides are narrow–gap semiconductors ($E_g \sim 0.1 - 0.2$ eV), the valence electrons of the alkaline–earth elements fill the LUMO–band of the fullerite. The interaction of the C-p states with the potential of the metal–ions leads to a broadening of the valence–band, the conduction bands are strongly modified by hybridization of C-p and alkaline–earth s- and d–states.

Acknowledgement:

This work was supported by the Austrian Science Foundation, project no. S59/08.

References:

|1| R.C. Haddon et al., Nature (Lonon) 350, 320 (1991).
|2| A.F. Hebard et al., Nature (London) 350, 600 (1991).
|3| A.R. Kortran et al., Nature (London) 355, 529 (1992).
|4| Y. Chen et al., Phys.Rev. B46, 7961 (1992).
|5| R.M. Fleming et al, Nature (London), 352, 701 (1991).
|6| O. Zhou et al, Nature (London), 351, 462 (1991).
|7| S.C. Erwin and M.R. Pederson, Phys.Rev.Lett. 67, 1610 (1991).
|8| D.L. Novikov, V.A. Gubanov, and A.J. Freeman, Physica C191, 399 (1992).
|9| J.L. Martins and N. Trouiller, Phys.Rev. B46, 1766 (1992).
|10| S. Saito and A. Oshiyama, Phys.Rev. B44, 11536 (1991).
|11| S.C. Erwin and W.E. Pickett, Science 254, 842 (1992).
|12| Y.Z. Li et al, Phys.Rev. B46, 12914 (1992).
|13| Y. Chen et al, Phys.Rev. B46, 8845 (1992).
|14| S. Saito and A. Oshiyama, Sol.State Comm. 83, 107 (1992).
|15| O.K. Andersen, O. Jepsen, and D. Glötzel, in 'Highlights of Condensed Matter Theory', ed. by F. Bassani et al. (North Holland, Amsterdam 1985).
|16| U. von Barth and L. Hedin, J.Phys. C5, 1629 (1972).
|17| W.R.L. Lambrecht and O.K. Andersen, Phys.Rev. B34, 2439 (1986).
|18| J. Redinger, P. Marksteiner, and P. Weinberger, Z.Phys. B63, 321 (1986).
|19| W. Jank and J. Hafner, J.Phys. CM2, 5065 (1990).
|20| B. Szpunar, Europhys.Lett. 18, 445 (1992).
|21| S. Satpathy et al., Phys.Rev. B46, 1773 (1992).

|22| N. Trouiller and J.L. Martins, Phys.Rev. B46, 1754 (1992).
|23| J.H. Weaver, J.Phys.Chem.Solids 53, 1433 (1992).
|24| Y. Chen, K.M. Ho, and B.M. Harmon, Phys.Rev. B37, 283 (1988); W. Jank and J. Hafner, Phys.Rev. B42, 6926 (1990).
|25| R.W. Lof et al., Phys.Rev.Lett. 68, 3924 (1992).
|26| G.K. Wertheim et al., Science 252, 1419 (1991).
|27| P.J. Benning et al., Phys.Rev. B44, 1962 (1991).
|28| J.H. Weaver et al., Phys.Rev.Lett. 66, 1741 (1991).
|29| M.B. Jost et al., Phys.Rev.Lett. 44, 1966 (1991).

Jahn-Teller Distorted Excited States of C_{60}

P.R. Surján[1], K. Németh[1], and L. Udvardi[2]

[1]Dept. of Theoretical Chemistry, Eötvös University, P.O.B. 32, H-1518 Budapest 112, Hungary

[2]Quantum Theory Group, Institute of Physics, TU Budapest, Budafoki út 8, H-1111 Budapest, Hungary

The geometrical and electronic structure of various excited states of the C_{60} molecule is determined by a semi-empirical extended Hubbard model. The Jahn-Teller distortions are found to be small in terms of bond length changes (max. 0.02 Å). Triplet excited states of symmetries D_{5d} and D_{2h} have been found to be 0.089 eV and 0.040 eV below the icosahedral state. Other symmetries as D_{3d} can also be located but they are of higher energies. The triplet ground state in solution is predicted to have D_{5d} symmetry, while the D_{2h} state is likely the lowest triplet in the solid phase.

1 Introduction

The geometry and the electronic structure of the isolated C_{60} molecule in its ground electronic state has been a subject of many quantum chemical investigations (see e.g. [1-4]). Similarly, the ionized structures C_{60}^- and C_{60}^+ have recently been studied in detail [5-8]. While the ground state of this molecule is known to have icosahedral nuclear configuration, the geometry will be distorted in the degenerate ionized and excited states. The present paper aims to discuss the energetics and geometry of excited states of C_{60}. The singlet and triplet low-lying states will be evaluated in various distorted symmetries.

We have adopted a π-electron model at the extended Hubbard level [9], that is, nearest-neighbor hopping, on-site and first-neighbor electron repulsion are taken into account. The β(r) hopping and γ(r) repulsion integrals for each bond depend on the bond distance

$$\beta(r) = Ae^{-\frac{r}{\rho}} \qquad \gamma(r) = 1/(\epsilon r) \tag{1}$$

where A, ρ and ϵ are adjustable parameters having the values 179.7003 eV, 0.32629 Å and 2.24 $\text{Å}^{-1}eV^{-1}$ respectively. The on-site repulsion integral U was taken as 3.536 eV. The bond lengths r have been optimized through Coulson's bond order - bond length relationship [10]

$$r = r_0 - \kappa P \tag{2}$$

where the bond orders for the bonds between sites μ and ν are defined as the offdiagonal elements of the first order density matrix

$$P_{\mu\nu} = \sum_{\sigma} \langle \Psi \mid a^+_{\nu\sigma} a_{\mu\sigma} \mid \Psi \rangle \tag{3}$$

Here Ψ is either a ground-state or an excited state many electron wave function, and σ is the spin label. In (2), we used r_0 = 1.54 Åand κ = 0.21 Å for carbon - carbon bonds. The parameters were obtained by fitting the bond lengths and the gap of polyacethylene, and the lowest singlet and triplet excitation energies of ethylene to experiments. As we checked, this set of parameters gives reliable results for butadiene, benzene and naphtalene. The ground state wave function $^{1,3}\Psi_0$ is approximated by a single determinant of 30 doubly occupied molecular orbitals, that is, the extended-Hubbard model is solved at the Hartree-Fock level. The singlet and triplet excited state wave functions $^{1,3}\Psi_J$ are written as linear combinations of all possible singly excited spin adapted configurations:

$$|^{1,3}\Psi_J\rangle = \frac{1}{\sqrt{2}} \sum_i^{occ} \sum_{k^*}^{virt} C^J_{ik^*} (a^+_{k^*\alpha} a_{k^*\alpha} \pm a^+_{k^*\alpha} a_{k^*\alpha}) \mid \Psi_0 \rangle \tag{4}$$

with $C^J_{ik^*}$ optimized variationally. The positive sign holds for singlets while the negative is for triplets. The asterisk indicates virtual orbitals. The solution of the above model involves a double iteration

Springer Series in Solid-State Sciences, Volume 117
Electronic Properties of Fullerenes Eds.: H. Kuzmany, J. Fink, M. Mehring, and S. Roth

procedure: (i) one starts with some initial geometry at which the extended-Hubbard orbitals optimized self-consistently, and evaluates ground and excited state bond orders; then (ii) one modifies the geometry according to (2) and reoptimizes the orbitals. Similar methods in the one-electron (Hückel) approximation were introduced a long time ago [11] and have been applied recently to a variety of conjugated polymers [12-16] and C_{60} [17,18] with success. Generalization from Hückel to extended-Hubbard level was motivated by the high degeneracies of C_{60} where one electron excitations are inappropriate.

Monitoring the Jahn-Teller distortion is possible by reducing the degenerate subspace of excited-state wave functions according to the irreducibile representations of the relevant subgroups by appropriate projections. We considered the subgroups D_{5d}, D_{3d} and D_{2h} which have one-dimensional (i.e. Jahn-Teller inactive) irreducibile representations contained in the degenerate icosahedral excited states. Depending on the selected particular subgroup (and irreducibile representation), one will arrive at different excited state wave functions, excited densities and thus different excited state geometries.

2 Results

Geometries. The optimized bond length of various C_{60} structures are colleted in Table 1. For the states of icosahedral symmetry there are only 2 different bond lengths. The experimental results for the ground state geometry is 1.40 Åand 1.45 Å[19] in perfect agreement with the present calculations.

Table 1. Lengths of inequivalent bonds in C_{60} in various states [Å]. The letters a,b,c,... refer to Fig. 1.

bond	I_h		D_{2h}		D_{5d}		D_{3d}	
	ground	excited	^3B_g	^1B_g	$^3A_{2g}$	$^1A_{2g}$	$^3A_{2g}$	$^1A_{2g}$
a	1.4463	1.4436	1.4108	1.4084	1.4447	1.4447	1.4183	1.4003
b	1.4023	1.4099	1.4310	1.4504	1.4052	1.4052	1.4425	1.4438
c			1.4154	1.4053	1.4451	1.4451	1.4382	1.4521
d			1.4466	1.4399	1.3997	1.3996	1.4481	1.4392
e			1.4346	1.4508	1.4523	1.4524	1.4097	1.4109
f			1.4466	1.4399	1.4229	1.4226	1.4507	1.4390
g			1.4019	1.4154	1.4247	1.4248	1.4006	1.4160
h			1.4154	1.4053			1.4482	1.4399
i			1.4505	1.4402			1.4109	1.4116
j			1.4505	1.4402			1.4373	1.4454
k			1.3974	1.4152				
l			1.4505	1.4402				
m			1.4505	1.4402				
n			1.4154	1.4053				
o			1.4310	1.4504				
Figure	1b	1b	1c	1c	1b	1b	1a	1a

On exciting the system to the (instable) I_h state the two bond lengths are only slightly affected, the pentagons become somewhat smaller while the size of hexagons increase.

Allowing the geometry to relax freely, we arrived at one of the geometries given in the last 6 columns of Table 1. The distortions are still small, the largest change is cca. 0.02 Å.

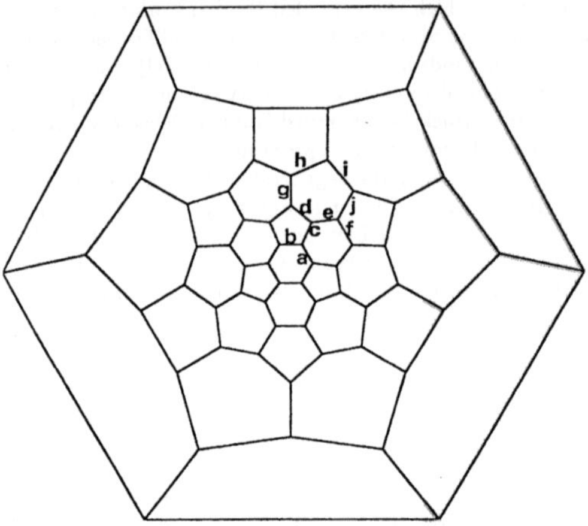

Fig. 1a

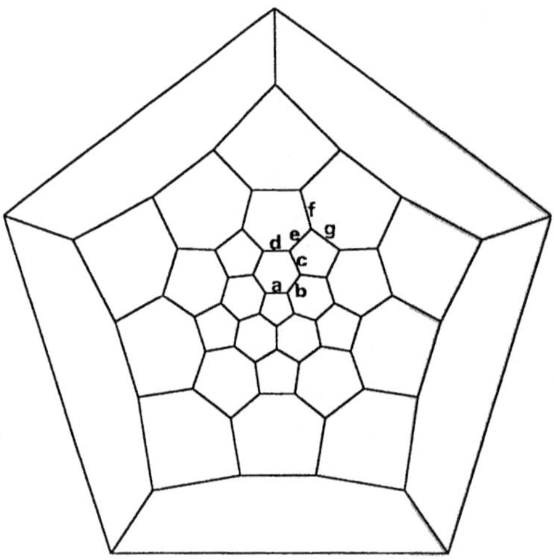

Fig. 1b

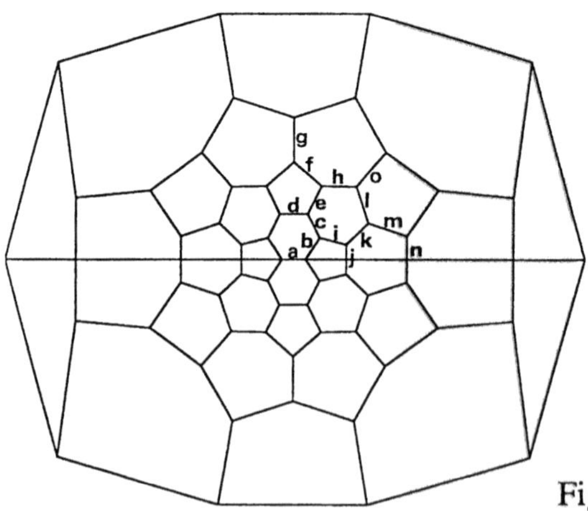

Fig. 1c

Energetics. At the ground state geometry, the lowest energy (forbidden) vertical transitions are found to be at 2.1 eV (^1G_g state) and 1.9 eV ($^3T_{2g}$ state). They show a singlet-triplet splitting of ~ 0.2 eV which is consistent to recent experimental results [20]. The singlet energy is close to UV–VIS results [28], CNDO/S [29] and INDO/S [30] calculations. The lowest symmetry allowed transition occurs at 2.8 eV ($^1T_{1u}$) and 2.5 eV ($^3T_{1u}$). These results are somewhat too low [28,29]. In Table 2 we calculated Jahn-Teller distortion energies which are obtained from a calculation of transition energies by the present model at the distorted geometries, and estimating the energy of the geometry distortion from CNDO/2 semi-empirical all-valance-electron calculations using a scaling factor 0.3 for the force constants. The energy difference between the ground and excited I_h states was found to be 0.1 eV. It is clearly seen from the table that there are only 2 stable distorted states: the one of D_{2h} and another one with D_{5d} symmetry. (Note that a negative Jahn-Teller energy indicates that the corresponding state is predicted to be above the icosahedral state.)

Table 2. Distortion and relative energies of various states of C_{60} [meV]

Subgroup	irrep.	J-T energy		relative energy	
		S	T	S	T
D_{3d}	A_{2g}	-64	-86	191	175
D_{2h}	B_{1g}	-37	40	164	49
D_{5d}	A_{2g}	127	89	0	0

3 Discusssion

The Jahn-Teller distortion energies for the symmetries D_{5d} and D_{2h} are in the order of magnitude of 50-100 meV, i.e., 1-3 kcal/mol. This energy is roughly the same as those obtained for the distortion energies for C_{60} ions [8].

The greatest stabilization energy is obtained in the D_{5d} symmetry. This holds also for singlets and for triplets, the latter being more interesting from the experimental side [21-25]. The present calculations predict that the symmetry of the C_{60} molecule in the lowest triplet state is D_{5d}. This is consistent to experimental results in solution [26]. However, experiments in the solid phase make D_{2h} probable as the lowest triplet [27]. This can easily be understood by symmetry arguments. The C_{60} molecules embedded in the crystal do not have icosahedral symmetry, rather T_h. This group does not contain D_{5d} as as subgroup, just D_{2h}. Thus, when reducing the degenerate excited states according to the irreducible representations of various subgroups, the D_{5d} state plays no role.

In this work we have been looking for the states which have highest possible symmetries but which are not subjected to further Jahn-Teller distortions. Besides from the solutions which have been found so far, there can be several other states which meet the above conditions. A more complete search will be reported for in a forthcoming paper.

Acknowledgement. This work was supported by the grant OTKA 517/1990.

References

[1] A.D.J.Haymet, J.Am.Chem.Soc. 108, 319 (1986), M.D.Newton, R.E.Stanton, ibid., 108, 2469 (1986)
[2] I.László and L.Udvardi, Chem.Phys.Lett.,136 (1987) 418
[3] I.László and L.Udvardi, Journal of Molecular Structure (Theochem),183 (1989) 271
[4] H.P.Lüthi, J.Almlöf, Chem.Phys.Letters 135 357 (1986)
[5] K.Tanaka, M.Okada, K.Okahara and T.Yamabe, Chem.Phys.Letters 193 101 (1992);
[6] R.D.Bendale, J.F.Stanton, M.C.Zerner, Chem.Phys.Letters 194 467 (1992)
[7] F.Negri, G.Orlandi, F.Zerbetto, Chem.Phys.Letters, 144 31 (1988);
[8] N.Koga, K.Morokuma Chem.Phys.Letters 196 191 (1992)
[9] P.R.Surján: Second Quantized Approach to Quantum Chemistry, Springer, Berlin, 1989
[10] C.A.Coulson, Proc.Roy.Soc A207 91 (1951)
[11] H.C.Longuet-Higgins and L.Salem, Proc.Roy Soc. A251 172 (1959)
[12] M.Kertesz and P.R.Surjan, Solid State Commun. 39 611 (1981)

[13] P.R.Surjan and H.Huzmany, Phys.Rev. B33 2615 (1986)
[14] J.Kürti, P.R.Surján, J.Chem.Phys. 92 3247 (1990); J.Math.Chem. 10 313 (1992)
[15] C.Kuhn, Phys.Rev. B40 7776 (1989)
[16] P.R.Surjan and K. Nemeth, Synth.Metals (in press);
[17] M.Ozaki, A.Takahashi, Chem.Phys.Letters 127 242 (1986)
[18] C.Kuhn (to be published)
[19] D.L. Lichtenburger, K.W. Nebesny, C.D. Ray, D.R. Huffman,D.L. Lamb Chem. Phys. Letters 176 (1991) 203
[20] T.Thomas, private communication
[21] M.Bennati, A.Grupp, M.Mehring, K.P.Dinse, J.Fink, Chem.Phys.Letters 200 440 (1992)
[22] R.V. Benasson, T. Hill, C. Lambert, E.J. Land, S. Leach, T.G. Truscott, to be published
[23] P.H.M. van Loosdrecht, M.A. Verheien, P.J.M. van Bentum, G.Meijer, this conference
[24] A.Grupp, M.Bennati and M.Mehring, this conference
[25] T.N.Thomas, D.Mihailovic, J.F.Ryan, R.A.Taylor, R. Zamboni, this conference
[26] A. Grupp, private communication
[27] G.Meijer, private communication
[28] H. Aije, M. M. Alvarez, S. J. Anz, R. D. Beck, F. Diederich, K. Fostiropoulos, D. R. Huffman, W. Kratschmer, Y. Rubin, K. E. Schriver, K. Sensharma, R. L. Whetten, J. Phys. Chem. 94, 8630-8633 (1990).
[29] F. Negri, G. Orlandi, F. Zerbetto, J. Chem. Phys. 97 6496 (1992)
[30] R.D. Bendale, J.D. Baker, M.C. Zerner, Int. J. Quant. Chem, Quantum Chemistry Symposium 25 557–568 (1991)

The C_{60} Molecule with the Step-Potential Model

C. Kuhn

Laboratoire Léon Brillouin, C.E. de Saclay, Bât. 563,
F-91191 Gif-Sur-Yvette Cédex, France

Abstract. The geometry of the C_{60}-molecule, determined by electron-phonon coupling, is obtained by finding the selfconsistency between bond length and π-electron density in the middle of the bond. The π-electrons are considered to move along a one-dimensional step-potential forming a network of bond potentials. The one-particle Schrödinger equation is solved numerically fulfilling branching conditions at each site and the π-electron density distribution is evaluated. The bond length pattern and the electronic spectrum of the ground state of the C_{60}-molecule is investigated as well as the charge localization in $C_{60}{}^{n-}$.

1. Introduction

The nearly spherical conjugated system of C_{60} is composed of 12 regular pentagon rings fused with 20 nearly regular hexagons and with all 60 (formally sp^2 hybridized) carbon atoms at equivalent sites. The 60 valence electrons occupy π-orbitals that have mostly p-like character and project radially from the surface [1-5].

As zeroth-order approximation we can consider these π-electrons in terms of free electrons [6] on the sphere with radius r=3.6Å [7]. The orbitals are spherical harmonics with energies $E=\gamma L(L+1)$, angular momentum quantum number $L=0,1,2...$ and $\gamma = \hbar^2/2mr^2 = 0.3$ eV. Their occupation for the neutral C_{60} is $(L=0)^2(L=1)^6(L=2)^{10}(L=3)^{14}(L=4)^{18}(L=5)^{10}$ with an open L=5 shell. In the following we treat the C_{60}-molecule on the basis of the step-potential model, which

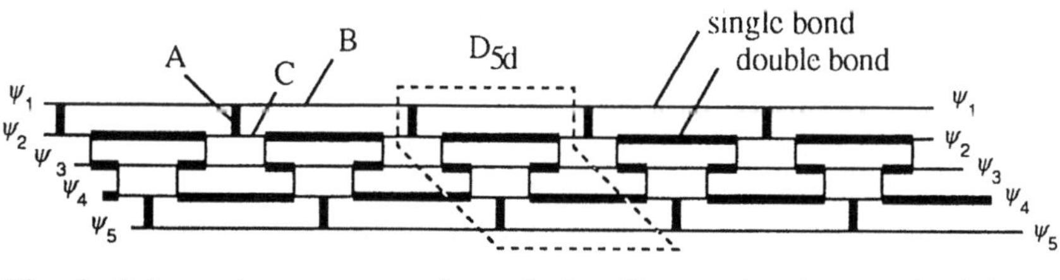

Fig. 1. Schematic representation of the C_{60}-molecule emphasizing the reduced symmetry of D_{5d} (pentagons at top and bottom). The network is cut open showing five horizontal paths ("main streams") of potential steps B and C and potential steps A ("branches") connecting two main streams (see Fig.2c).

Springer Series in Solid-State Sciences, Volume 117
Electronic Properties of Fullerenes Eds.: H. Kuzmany, J. Fink, M. Mehring, and S. Roth

considered a nearly free π-electron (NFE) in a one dimensional potential composed of successive bond potentials in case of the planar linear hydrocarbons [8-12]. The Schrödinger equation of a nearly free π-electron in a one dimensional step-potential forming the network of bond potentials corresponding to the C_{60}-molecule (Fig.1) is solved numerically with the transfer matrix method by an inverse eigen value problem of a 10*10 matrix (without any symmetry restrictions):

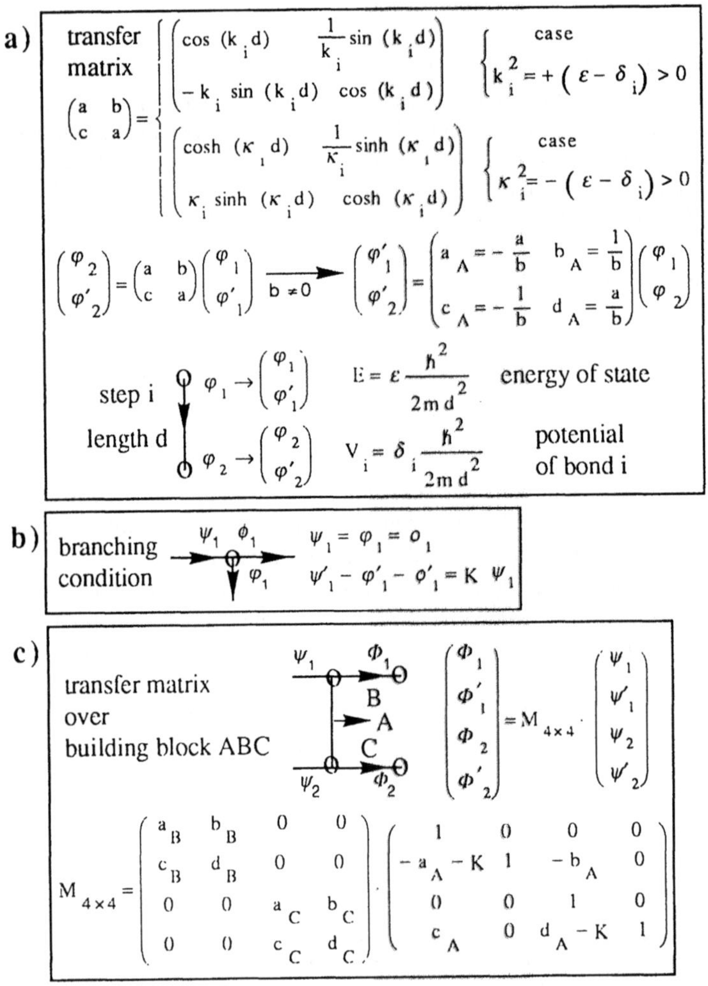

Fig. 2. a) The matrix transfers the amplitude of the wave function and its derivative at one end of the potential step onto the other end. Alternatively, the derivatives at both ends as a function of the amplitudes at both ends are given by a transformed matrix. **b)** Branching condition [13-15] with Kd=0 (value Kd=1.56 does not change the results in this case). **c)** The amplitude of two wave functions and their derivatives are transfered over a building block ABC by the matrix $M_{4\times4}$.

$$M^{network}_{10\times10} = \ldots \cdot \begin{pmatrix} E_{6\times6} & 0 \\ 0 & M_{4\times4} \end{pmatrix} \cdot \begin{pmatrix} E_{4\times4} & & 0 \\ & M_{4\times4} & \\ 0 & & E_{2\times2} \end{pmatrix} \cdot \begin{pmatrix} E_{2\times2} & & 0 \\ & M_{4\times4} & \\ 0 & & E_{4\times4} \end{pmatrix} \cdot \begin{pmatrix} M_{4\times4} & 0 \\ 0 & E_{6\times6} \end{pmatrix}$$ is a product of 30 of $M_{4\times4}$ matrices (Fig. 2) and transfers the wave functions of the fife main streams at the dissection $\vec{\psi}_{10} = (\psi_1, \psi'_1, \psi_2, \psi'_2, \psi_3, \psi'_3, \psi_4, \psi'_4, \psi_5, \psi'_5)$ over the network onto themselves: $M^{network}_{10\times10}(\varepsilon_n) \cdot \vec{\psi}_{10} = \vec{\psi}_{10}$ (matching condition). The coupling of the π-electron system to the σ-bonded elastic lattice is considered by relating the potential V_i in a given bond i with the bond length d_i and the density ρ_i in the middle of the bond. By iteration to selfconsistency between ρ_i and $V_i = \alpha(2/3 - \rho_i d)\hbar^2/2md^2$ for each bond ($\hbar^2/2md^2$=2.0eV, d=1.38Å), the molecular lattice converges to the equilibrium with bond lengths $d_i = d(1 + 0.1(1 - \rho_i d))$. The electron-phonon coupling is supposed to be larger in this case of a non-planar conjugated system [5], the value α=3.4 is thus used rather α=1.95, which was scaled to the experimental bond lengths of butadiene and benzene for planar hydrocarbons.

2. Results and Discussion

Figure 3a shows the energy levels of the free π-electron in the network of the C_{60}-molecule (equal bonds of length 1.42Å of all 90 bonds. In contrast to the free electron on the sphere, where the L=5 shell is 11 times degenerate, we find three seperate levels due to the icosahedral symmetry: the 5-times degenerate $1h_u$ HOMO (highest occupied molecular orbital) of closed shell, the triply degenerate $2t_{1u}$ LUMO (lowest unoccupied molecular orbital) and the triply degenerate t_{2u} orbital at higher energy. Simlarly, in the case of the L=6 shell, the 13 times degeneracy is removed and we find the triply degenerate $1t_{1g}$ LUMO+1 and states of higher energy. Relaxing

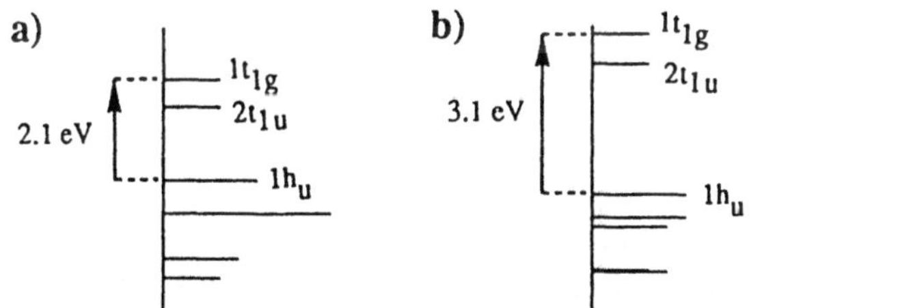

Fig. 3. Energy levels vs. degeneracy of the C_{60}-molecule calculated by the **a)** free π-electron in the icosahedral network (all bond lengths 1.42Å), **b)** nearly free π-electron in the network of the relaxed geometry (bond lengths 1.38Å and 1.45Å).

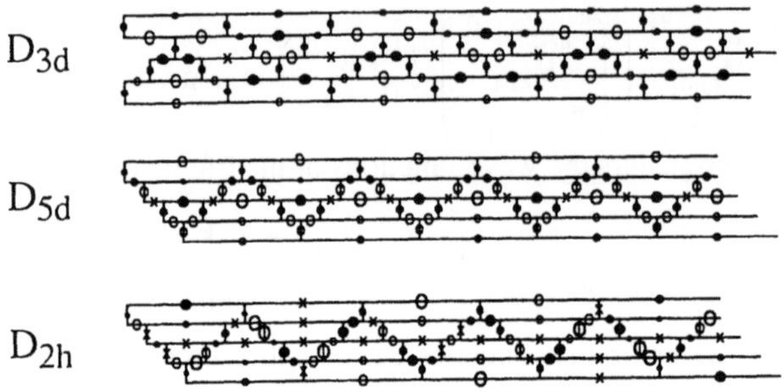

Fig. 4. $2t_{1u}$ LUMO (Fig. 3b) emphasizing D_{3d}, D_{5d} and D_{2h} reduced symmetry. Values in the middle of each bond are proportional to the diameter of black dots = positive value, circles = negative value, X = zero.

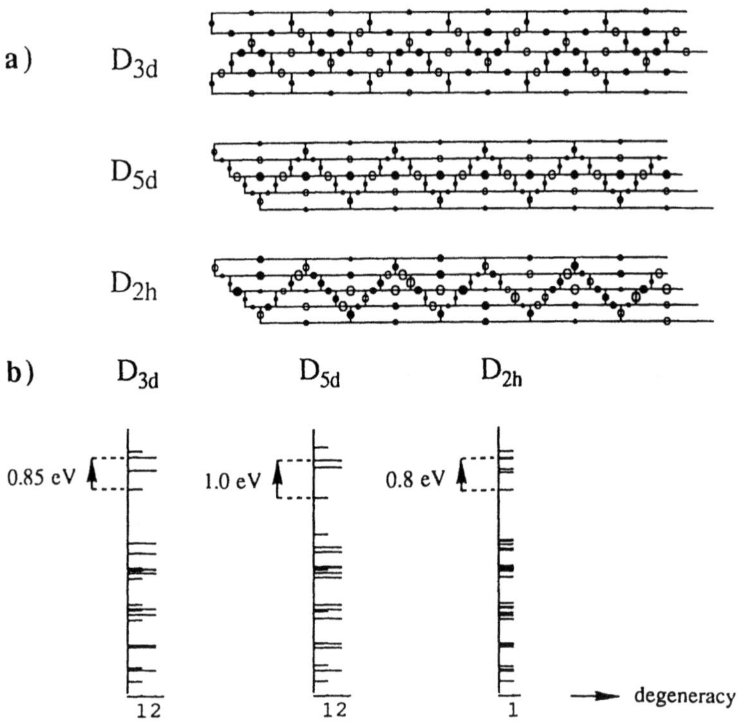

Fig. 5. a) Additional charge density of (singlet) C_{60}^{2-} with respect to the neutral molecule in cases of D_{3d}, D_{5d}, D_{2h} symmetry. **b)** Corresponding level schemes.

the lattice by electron-phonon coupling, the 60 bonds building up the pentagons become "single bonds" of length 1.45Å, whereas the hexagons become alternating with the "double bonds" of length 1.38Å in excellent agreement with experiment (NMR [16] values of 1.45Å and 1.40Å respectively and x-ray-diffraction [17] values of 1.43Å and 1.39Å respectively). The (HOMO-LUMO) gap widens to 2.5eV and the optically allowed (HOMO-LUMO+1) gap widens to 3.1eV (Fig. 3b) in good agreement with photoelectron spectra [18,19] and results from Hückel calculations [20-22].

The partial occupation of the $2t_{1u}$ LUMOs (Fig. 4) by one or two electrons leads by electron-phonon coupling to lower symmetry of either D_{3d} or D_{5d} or D_{2h} (Fig. 5, the case of singlet C_{60}^{2-}) and to the localization of the charge along the equatorial line with lowest potential energy in the case of D_{5d} symmetry (all bonds equalized to length of 1.42Å along this equatorial line, Fig. 5a case D_{5d}, doubly occupied HOMO wave function within the gap, Fig. 5b case D_{5d}, more localized along this equatorial line than the original $2t_{1u}$ LUMO , see Fig. 4 case D_{5d}). Similar results of this Jahn-Teller deformation [23-25] were obtained within the density functional formalism [26], MNDO calculations [27,28] and the SSH-model [29].

3. Acknoledgements

The author is grateful to Prof. S. Aubry and to Drs. R. Papoular and J. Szeftel for many valuable discussions.

References

1. A.F. Hebard, Physics Today Nov., 26 (1992)
2. M. Schlüter, et. al., J. Phys. Chem. Solids **53**, 1473 (1992)
3. F.C. Zhang, M. Ogata, T.M. Rice, Phys. Rev. Lett. **67**, 3452 (1991)
4. S. Charkavarty, M.P. Gelfand, S. Kivelson, Science **254**, 970 (1991)
5. C.M. Varma, J. Zaanen, K. Raghavachari, Science **254**, 989 (1991)
6. H. Kuhn, J. Chem. Phys. **16**, 840 (1948) and **17**, 1198 (1949)
7. G.A. Gallup, Chem. Phys. Lett. **187**, 187 (1991)
8. C. Kuhn, Phys. Rev. B **40**, 7776 (1989)
9. C. Kuhn, Synth. Met. **41-43**, 3681 (1991)
10. C. Kuhn, Synth. Met. **57**, 4350 (1993), **57**, 4357 (1993)
11. C. Kuhn, W.F. van Gunsteren, Solid State Comm., submitted
12. H. Kuhn, C. Kuhn, Chem. Phys. Lett. **204**, 206 (1993)
13. H. Kuhn, Helv. Chim. Acta **32**, 2247 (1949)
14. H. Kuhn, Z. Naturforsch. **9a**, 989 (1954)
15. H. Kuhn, J. Chem. Phys. **22**, 2098 (1954), **25**, 293 (1956)
16. C.S. Yannoni, et. al., J. Am. Chem. Soc. **113**, 3190 (1991)
17. J.M. Hawkins, et. al., Science **252**, 312 (1991)
18. S.H. Yang, et. al., Chem. Phys. Lett. **139**, 233 (1987)
19. L.-S. Wang et al., Chem. Phys. Lett. **179**, 449 (1991)

20. M. Ozaki, A. Takahashi, Chem. Phys. Lett. **127**, 242 (1986)
21. R.C. Haddon, L.E. Brus, K. Raghavachari, Chem. Phys. Lett. **125**, 459 (1986)
22. Y. Deng, C.N. Yang, Phys. Lett. A **170**, 116 (1992)
23. M. Lannoo, G.A. Baraff, M. Schlüter, Phys. Rev. B **44**, 12106 (1991)
24. G. Stollhoff, Phys. Rev. B **44**, 10998 (1991)
25. R.A. Jishi, M.S. Dresselhaus, Phys. Rev. B **45**, 2597 (1992)
26. V. de Coulon, J.L. Martins, F. Reuse, Phys. Rev. B **45**, 13671 (1992)
27. K. Tanaka, et. al., Chem. Phys. Lett. **193**, 101 (1992)
28. A. Tachibana, S. Ishikawa, T. Yamabe, Chem. Phys. Lett. **201**, 315 (1993)
29. K. Harigaya, Phys. Rev. B **45**, 13676 (1992)

Part III

Structure and Structural Phase Transitions

Structure and Dynamics of C_{60} and C_{70}

K. Prassides

School of Chemistry and Molecular Sciences, University of Sussex,
Falmer, Brighton BN1 9QJ, UK

Abstract. The structure and dynamics of C_{60} and C_{70} fullerenes in the solid state are briefly reviewed. Solid C_{60} shows a first-order phase transition at 260 K from a face-centred to a simple cubic structure. The isotropic molecular reorientations change abruptly to quasi-random jump motion between nearly-degenerate orientations differing in energy by 11.4(3) meV. A transition to an orientation glass state occurs at 85 K. Static disorder dominates the crystal chemistry of C_{70}. The orientational ordering transition of the face-centred-cubic phase at 280 K is accompanied by severe hysteresis effects. The rotational dynamic behaviour changes progressively on cooling from isotropic to anisotropic to uniaxial reorientations about the unique molecular axis. High pressure diffraction experiments reveal that C_{70} has a smaller compressibility than C_{60}.

1. Introduction

Fullerene research has advanced at an unprecedented pace since the discovery of a simple, efficient method to produce bulk crystalline samples [1]. The original [2] proposed structures for C_{60} and C_{70} were immediately confirmed: truncated icosahedral (I_h) and ellipsoidal (D_{5h}) respectively [3]. Notwithstanding the stability which led to their original detection, fullerenes are found to participate readily in a plethora of chemical reactions. Reactions can lead among others to the formation of molecular derivatives $C_{60}X_n$ (for X= Br, bromides with stoichiometries $C_{60}Br_6$, $C_{60}Br_8$ and $C_{60}Br_{24}$ have been characterised [4]), charge-transfer salts (reaction with ferrocene affords [5] the solid $[Fe(C_5H_5)_2]_2C_{60}$) and essentially ionic salts (reaction with alkali metals [6,7] leads to the formation of superconducting solids of stoichiometry A_3C_{60}). Moreover, it soon became apparent that a variety of small molecules can be trapped in the interfullerene spacing in solid fullerenes, making the isolation of highly pure samples not a routine procedure. For instance, $C_{60}.(C_6H_6)_4$ crystals can be isolated from benzene solutions [8]. Solvent molecules are retained in the interfullerene spacing even after sublimation and long annealing of the solid samples. In our experience, we find that the hydrogen content of both sublimed and annealed C_{60} and C_{70} solids is no less than 1 hydrogen atom per ~2.5-3 fullerene molecules.

In this paper, we briefly review the evolution with temperature of the structural and dynamical properties of C_{60} and C_{70} fullerenes in the solid state, drawing from the results of a number of experimental techniques. The structural phase transitions accompanying orientational ordering of the molecules are sensitively related to deviations from isotropy arising through the presence of more than one type of

Springer Series in Solid-State Sciences, Volume 117
Electronic Properties of Fullerenes Eds.: H. Kuzmany, J. Fink, M. Mehring, and S. Roth

carbon-carbon bonds which give rise to an anisotropic π electronic distribution on the fullerene surface.

2. Structure and Dynamics of C_{60}

The icosahedral symmetry of the C_{60} molecules and its incompatibility with periodic translational symmetry make the determination of the structure of solid C_{60} of fundamental importance. Fleming et al. [9] have shown that the C_{60} I_h molecular symmetry is reduced to at most *m3* by a cubic crystalline field. The two standard orientations of the C_{60} molecule placed at the origin of a cubic lattice are shown in Fig. 1. They are related by a 90° rotation about [100] axes which are now coincident with rotational 2-fold symmetry axes. The crystal symmetry can only be raised through disorder; superposition and equal fractional occupancy of the two standard orientations of Fig. 1 results in raising the symmetry to *m3m* with the [100] axes now having 4-fold rotational symmetry.

Early X-ray diffraction work [1] revealed that at room temperature the C_{60} crystalline powder consisted of spheroidal molecules of diameter 7.1 Å, forming a random mixture of hexagonal close-packed (hcp) and face-centred cubic (fcc) arrays. However, elimination of solvent molecules trapped in interstitial cavities leads only to a fcc crystal structure in which each C_{60} molecule is orientationally disordered. Powder neutron [10] and X-ray [11] diffraction data are consistent with the space group *Fm3m*, with all four quasi-spherical C_{60} units being symmetry equivalent; the cubic lattice constant at 290 K is 14.1569(5) Å. A variety of experimental techniques revealed that in this plastic 'rotator' phase, the C_{60} molecules rotate effectively freely and randomly. Deviations from a uniform charge spherical distribution are however still present with excess electron density present along the <111> and a deficiency along the <110> directions [12]. Rotational correlation times τ_{ROT} of 9-12 ps at room temperature are only 3-4 times longer than expected for unhindered gas phase rotation and faster than any other known solid state rotor [13]. The high temperature description of pristine C_{60} that has emerged is of a prototypical plastic crystal (cf. adamantane, norbornane, etc) with a well-defined translational order and a smooth rotational potential with many shallow minima (E_a=35±15 meV [14]), leading to continuous small-angle diffusive motions.

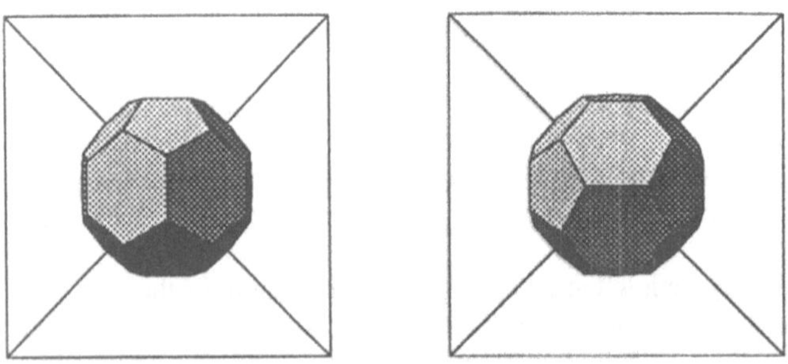

Figure 1. The two standard orientations of a C_{60} molecule placed at the origin of a cubic lattice.

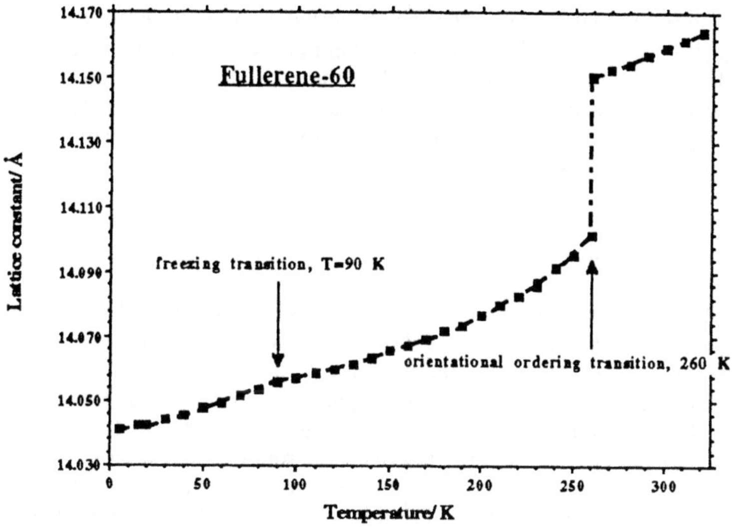

Figure 2. Temperature evolution of the cubic lattice constant of solid C_{60}.

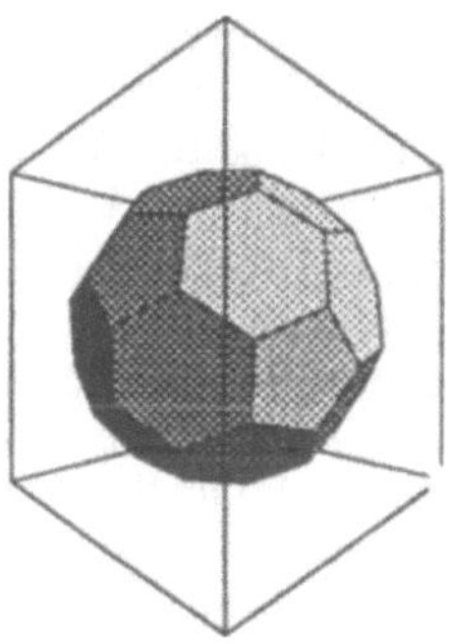

Figure 3. View of the C_{60} molecule at the origin along one of the <1$\bar{1}$0> axes straddling short C-C bonds.

As the temperature is lowered, the rotational motion abruptly changes. The C_{60} units are no longer symmetry-equivalent because of orientational ordering, leading to a simple-cubic structure [10,11,15]. The low-temperature structure of C_{60} can be traced to its anisotropic electronic distribution which arises from the presence of two types of carbon-carbon bonds: long (fusions of five- and six-membered rings, 1.448(10) Å) and short (fusions of two six-membered rings, 1.404(10) Å). The electrostatic contribution to the predominantly van der Waals intermolecular bonding is then optimised by aligning the "electron-rich" regions (short carbon-carbon bonds) of one molecule over the "electron-deficient" regions (pentagonal faces) of its neighbours. The first-order orientational ordering transition at 260 K (Fig. 2) is accompanied by a change in the nature of the rotational dynamics to jump reorientational motion, as the molecules shuffle between two *nearly*-degenerate orientations differing in energy by 11.4(3) meV [16] (short C-C bonds facing pentagonal or hexagonal faces). The two orientations are accessible via either 60° hops about <111> or 42° hops about <1$\bar{1}$0> axes straddling short C-C bonds. An estimate of the jump angle of ~36-42° can be derived [17] by assuming a sinusoidal hindrance potential and using the magnitude of the rotational barrier (~250-300 meV) and of the librational energy (~2.5 meV). Furthermore, both recent avoided-level-crossing μSR [18] and 2D ^{13}C NMR [19] studies are consistent with the quasi-isotropic random nature of the motion implied by the hops about the <1$\bar{1}$0> axes of Fig. 3 and discard the possibility of uniaxial 60° jumps about <111>. Finally, in the vicinity of 85 K, the molecular jump motion essentially freezes and a transition to an orientational glass phase occurs.

3. Structure and Dynamics of C_{70}

Early X-ray diffraction and electron microscopy studies [20] have revealed that both fcc and hcp phases coexist while a large concentration of defects is also present. At high temperatures, solid C_{70} is found to adopt a fcc crystal structure. A rhombohedral distortion (α~85.6(1)°) accompanies an orientational ordering transition at ~280 K as the molecules orient themselves with their long axis towards the unit cell diagonal. This is reflected in both an abrupt expansion along <111> (~1.9 Å, Fig. 4) as the long axis of the molecules is now pointing towards this direction and an abrupt contraction along the close packing <110> direction (~0.4 Å) reflecting their smaller size on the equatorial plane [21]. A small volume change is also observed at the phase transition while the rhombohedral angle α does not change with temperature. The fcc and rhombohedral phases coexist over a large temperature range. A lower symmetry crystallographic phase (possibly monoclinic) is present below 180 K. The structural properties of the fcc structural modification of C_{70} are found to be dominated by a high concentration of defects, associated with static and dynamic disorder, leading to extensive "undercooling" for the high-temperature and "overheating" behaviour for the low-temperature phase. However, the presence of substantial disorder in the sample does not make it possible to present definite conclusions on the lower symmetry phases. A puzzling feature also relates to the reports from many experimental measurements of the thermal properties of solid C_{70} that a phase transition also occurs in the vicinity of 350 K. It is tempting to note that it is precisely at this temperature that on heating the

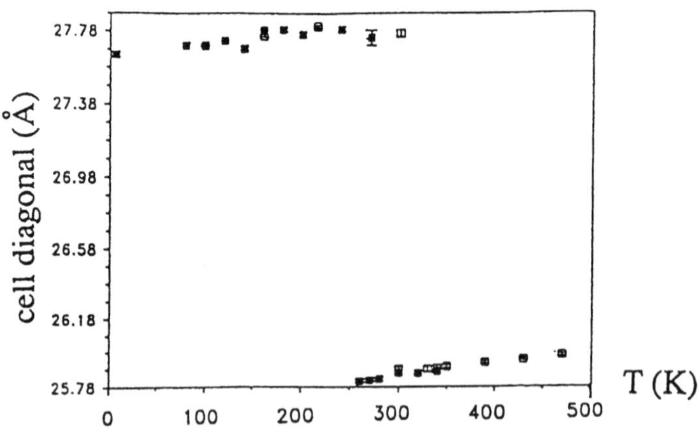

Figure 4. Temperature evolution of the unit cell diagonal in solid C_{70} (<111> direction).

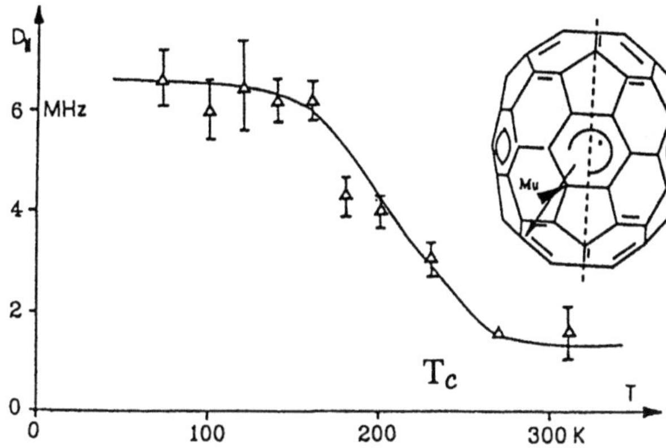

Figure 5. Temperature evolution of the axial hyperfine anisotropy of the MuC_{70} muonated radical shown in the inset.

"overheated" rhombohedral phase is finally converted to the fcc phase as the hindrance potential barrier associated with microstrains and defects appears to be of the order of 350 K. Modest pressure at ambient temperature is also enough to cause an ordering transition at ~0.35 GPa [21], the resulting rhombohedral structure (a~85.4(1)°) being slightly less compressible (κ=4.0±1.4 $\times$ 10^{-2} GPa^{-1}) than C_{60} (κ=5.5±0.6 $\times$ 10^{-2} GPa^{-1}). Large hydrostatic pressures (18 GPa) lead to the formation of an amorphous carbon phase.

The rotational dynamics of C_{70} were probed by μSR and neutron scattering measurements. Zero-field μSR [22] of the endohedral muonium adduct Mu@C_{70} reveals a single frequency oscillation (ν=0.70(2) MHz) at the orientational ordering transition (~270 K), signaling the presence of an axially symmetric hyperfine interaction in the orientationally ordered phase. An avoided level crossing μSR study [23] of one of the exohedral MuC_{70} radiacals reveals that C_{70} is not in an entirely frozen state at 73 K but it probably still undergoes uniaxial rotation about its long molecular axis. As the temperature increases from 160 K to 270 K, there is an increasing degree of rotational disorder of the rotational axis itself. A small anisotropy is found to persist across the phase transition to the plastic fcc phase above 270 K. Neutron scattering studies [24] reveal the presence of well-defined librations in the ordered phase of solid C_{70}. These broaden and soften with increasing temperature until they collapse into a single quasielastic line above the phase transition. The motion is somewhat anisotropic at 300 K but becomes quasi-isotropic on heating to T=525 K with a rotational diffusion coefficient D_r=1.5 $\times$ 10^{10} s^{-1}. The temperature dependence of the width of the quasielastic line between 300 and 640 K is consistent with a thermally activated process having an activation energy of 32.0±6.5 meV. Diffuse scattering arising from both static and dynamic disorder is present in the diffraction profiles down to 100 K. Its intensity changes little with temperature in sharp contrast with C_{60} in which the diffuse scattering is reduced substantially upon orientational ordering [14].

Acknowledgement

I am indebted to my numerous coworkers and collaborators whose contributions made the work presented here possible.

References

[1] W. Krätschmer, L.D. Lamb, K. Fostiropoulos, and D.R. Huffman, *Nature* **347**, 354 (1990).
[2] H.W. Kroto, J.R. Heath, S.C. O'Brien, R.F. Curl, and R.E. Smalley, *Nature* **318**, 162 (1985).
[3] R. Taylor, J.P. Hare, A.K. Abdul-Sada, and H.W. Kroto, *J. Chem. Soc. Chem. Commun.* 1423 (1990).
[4] P.R. Birkett, P.B. Hitchcock, H.W. Kroto, R. Taylor, and D.R.M. Walton, *Nature* **357**, 479 (1992).
[5] J.D. Crane, P.B. Hitchcock, H.W. Kroto, and D.R.M. Walton, *J. Chem. Soc. Chem. Commun.* (1992).
[6] A.F. Hebard, M.J. Rosseinsky, R.C. Haddon, D.W. Murphy, S.H. Glarum, T.T.M. Palstra, A.P. Ramirez, and A.R. Kortan, *Nature* **318**, 600 (1991).
[7] K. Holczer, O. Klein, S.-M. Huang, R.B. Kaner, K.-J. Fu, R.L. Whetten, and F. Diederich, *Science* **252**, 1154 (1991).
[8] M. Meidine, P.B. Hitchcock, H.W. Kroto, R. taylor, and D.R.M. Walton, *J. Chem. Soc. Chem. Commun.* 1534 (1992).
[9] R.M. Fleming, et al., in 'Fullerenes: Synthesis, Properties, and Chemistry of Large Carbon Clusters', eds. G.S. Hammond and V.J. Kuck, ACS Sympos. Series **481**, 25 (1991).
[10] W.I.F. David, R.M. Ibberson, T.J.S. Dennis, J.P. Hare, and K. Prassides, *Europhys. Lett.* **18**, 219 (1992); 735 (1992).
[11] P.A. Heiney, J.E. Fischer, A.R. McGhie, W.J. Romanow, A.M. denenstein, J.P. McCauley, A.B. Smith, and D.E. Cox, *Phys. Rev. Lett.* **66**, 2911 (1991).
[12] P.C. Chow, X. Jiang, G. Reiter, P. Wochner, S.C. Moss, J.D. Axe, J.C. Hanson, R.K. McMullan, R.L Meng, and C.W. Chu, *Phys. Rev. Lett.* **69**, 2943 (1992).
[13] R.D. Johnson, C.S. Yannoni, H.C. Dorn, J.R. Salem, and D.S. Bethune, *Science* **255**, 1235 (1992).
[14] J.R.D. Copley, D.A. Neumann, R.L. Cappelletti, and W.A. Kamitakahara, *J. Phys. Chem. Solids* **53**, 1353 (1992).
[15] W.I.F. David, R.M. Ibberson, K. Prassides, T.J.S. Dennis, J.P. Hare, H.W. Kroto, R. Taylor, and D.R.M. Walton, *Nature* **353**, 147 (1991).
[16] K. Prassides, H.W. Kroto, R. Taylor, D.R.M. Walton, W.I.F. David, J. Tomkinson, M.J. Rosseinsky, D.W. Murphy, and R.C. Haddon, *Carbon* **30**, 1277 (1992).
[17] K. Prassides, *Int. J. Mod. Phys. B* **6**, 4007 (1992).
[18] K. Prassides, unpublished.
[19] R. Blinc, J. Seliger, J. Dolinsek, and D. Arcon, to be published.
[20] G.B.M. Vaughan et al., *Science* **254**, 1350 (1992).
[21] C. Christides, I.M. Thomas, T.J.S. Dennis, and K. Prassides, *Europhys. Lett.* in press.

[22] K. Prassides, T.J.S. Dennis, C. Christides, E. Roduner, H.W. Kroto, R. Taylor, and D.R.M. Walton, *J. Phys. Chem.* **96**, 10600 (1992).
[23] T.J.S. Dennis, K. Prassides, E. Roduner, L. Cristofolini, and R. deRenzi, to be published.
[24] C. Christides, T.J.S. Dennis, K. Prassides, R.L. Cappelletti, D.A. Neumann, and J.R.D. Copley, to be published.

K-C_{60}: Growth Structures, Phase Formation, and Electronic Properties

J.H. Weaver, D.M. Poirier, and Y.B. Zhao

University of Minnesota, Department of Materials Science and Chemical Engineering, Minneapolis, MN 55455, USA

This paper focuses on K-C_{60} thin films. Scanning tunneling microscopy shows that high quality films can be grown epitaxially on GaAs(110), ultraviolet photoemission shows the complex nature of the electronic states of the K_3C_{60} phase, and x-ray photoemission reveals the phase diagram for this system.

1. Introduction

Many investigations have examined the properties at room temperature or below for fullerides grown by vapor phase deposition of alkali metals onto fullerene thin films [1]. In this paper, we first use scanning tunneling microscopy (STM) to demonstrate the very high quality of thin films of C_{60} grown on GaAs(110) and the persistence of highly ordered surfaces after doping to produce the K_3C_{60} phase. With ultraviolet photoemission, we characterize the valence band emission for the K_3C_{60} phase, finding that the result is poorly described by independent-electron calculations because of coupling to vibrational modes and probable electron-correlation effects. We then use x-ray photoemission to investigate the K-C_{60} phase diagram. This last study was motivated by the report of a K_1C_{60} phase at high temperature [2], analogous to that observed at room temperature for Rb and Cs fullerides [3]. Given the existence of a K_1C_{60} phase, it is important to understand how it transforms into the lower temperature solid solution or α-phase of K in C_{60} and the K_3C_{60} phase. We find that there is a eutectoid transformation at ~425 K that dictates the lower temperature microstructure of samples having stoichiometries, x, between the solubility limit and about 3 [4].

2. K-C_{60} Film Structures

Figure 1 shows an STM image of a 10 monolayer film produced by the deposition of C_{60} onto cleaved GaAs(110) in ultrahigh vacuum. The image measures 660×680 Å^2. Growth was done at 180°C. Under these conditions, the C_{60} film forms an incommensurate structure bound to the substrate by van der Waals forces. Figure 1 reveals a perfectly formed overlayer with a single close-packed layer exposed. In other images, steps of 8 Å height were observed,

Springer Series in Solid-State Sciences, Volume 117
Electronic Properties of Fullerenes Eds.: H. Kuzmany, J. Fink, M. Mehring, and S. Roth

Fig. 1. STM image of 10 ML of C_{60} grown on a step-free region of GaAs(110) showing a high degree of perfection. The image is ~660×680 Å^2.

representing incompletely formed layers or islands. The quality of such fullerene films reflects the quality of the starting surface itself. In regions where there are no GaAs(110) substrate steps, the overlayer is like that of Fig. 1. In all cases, the high degree of perfection reflects the high surface mobility of C_{60} on the GaAs substrate and on the layers of C_{60} that form.

STM images at submonolayer coverage show that GaAs(110) steps produced during cleaving act as nucleation sites for stable groups of C_{60} molecules. Once nucleation is initiated at a step, molecules diffusing on the terrace find low energy bonding sites and grow out from that step. Such step flow is characteristic of fullerene film formation for any of the fullerenes, from C_{60} to C_{84}, as shown recently by Li *et al.* [5]. Because of these steps, there will be an offset for layers forming on the two sides of the step, and this offset propagates to the surface of the multilayer film where it is readily detected with STM. For growth on GaAs(110) where the step density is very low, there is also nucleation on the terraces themselves. Such growth is expected if the diffusion length is small compared to the terrace width. Under these conditions, two-dimensional islands reach a critical radius and then grow laterally from the "gas" of diffusing fullerenes. The image of Fig. 1 was obtained from a large, step-free region.

For growth on GaAs(110), it is important to note that the lattices of GaAs(110) and C_{60}(111) have different symmetry so that pseudomorphic growth (growth in registry with the substrate) should produce a strained overlayer and, ultimately, strain-relieving dislocations would form when the thickness exceeds a critical value. Fortunately, this can be avoided for C_{60}-GaAs(110) since interface bonding is not so strong that the overlayer is locked into registry. Instead, the overlayer can adopt its own lattice constant, starting from the very first layer. This incommensurate growth is readily achieved at elevated temperatures. Li *et al.* [6] showed an STM image for 1 ML C_{60}/GaAs(110)

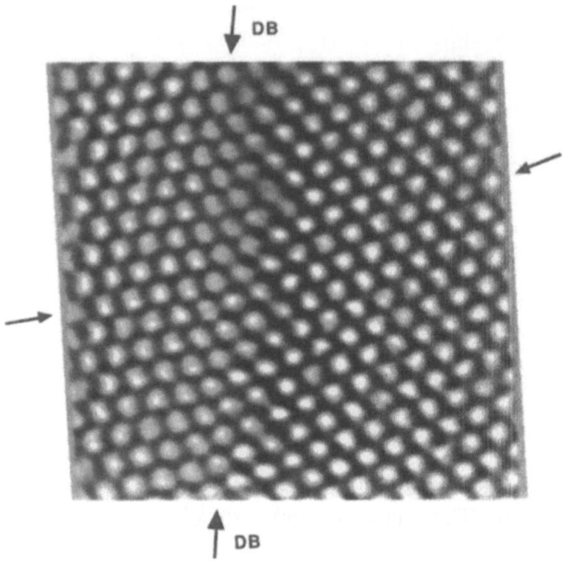

Fig. 2. STM image of a K_3C_{60} surface grown by exposing the 10 ML film characterized by Fig. 1 to a K flux. Tunneling conditions demonstrate metallic character. Arrows point to a domain boundary (DB) between two large grains of K_3C_{60} that are rotated about 7° relative to one another. The boundary is viewed clearly by holding the picture horizontally and looking down the row of molecules indicated by arrows on the sides of the figure. The image is ~150×160 Å^2.

formed at 150°C where there was a long wavelength modulation of the apparent height of the C_{60} molecules. In such a situation, the $C_{60}(111)$ layer is oriented with its $[\bar{1}10]$ direction approximately parallel to the GaAs[001] direction so that the fullerenes will fall over equivalent surface sites every 45 Å. The modulation in fullerene height is ~0.8 Å. This incommensurate monolayer acts as an ideal template that facilitates strain-free homoepitaxial growth of subsequent C_{60} layers.

There are two orientations for C_{60} on GaAs(110) that are energetically equivalent. These correspond to having the $[\bar{1}10]$ of $C_{60}(111)$ oriented ±3.5° with respect to [001] of GaAs(110). Hence, when islands form on the terraces, they can do so with slightly misaligned domains. When the islands grow together, there can be residual domain boundaries that are kinetically trapped in the structure. As an example of such a small angle domain boundary, we show in Fig. 2 an STM image for C_{60} that was obtained after the fullerene film was doped with K to give $K_3C_{60}(111)$. The location of the boundary is emphasized by the arrows labeled DB. The offset in atomic rows is evident by viewing along the rows, emphasized by the arrows on the left and right sides of the figure. This image was obtained with a sample bias relative to the tip of 11 mV, confirming the metallic character of the $K_3C_{60}(111)$ grains. For comparison, the bias of Fig. 1 was 2 V, a voltage typical for imaging pure fullerene films [6]. The persistence of the domain boundary after the growth of 10 layers of C_{60} demonstrates that the crystalline quality of the film is very high and is dictated by lattice matching conditions at the boundary of the first layer with the substrate.

3. Electronic Properties of K_3C_{60}

Photoemission studies for K-doped C_{60} yield the spectral signature shown in Fig. 3 when the stoichiometry is near 3 and before the K_4C_{60} phase starts to nucleate [9]. Measurements at low temperature and high resolution reveal that the band of states near the Fermi level is much broader than that predicted by 1-electron band calculations [7]. This band is derived from the lowest unoccupied molecular orbital (LUMO) of the fullerenes, a band that becomes occupied by electrons donated from the alkali metals, as first shown by Benning *et al.* [8]. For this band, experiments show a sharp Fermi level onset, a shoulder shifted 0.2 eV from the peak, and a second feature centered 0.7 eV below E_F [9, 10]. In contrast, band calculations have postulated a partially occupied LUMO-derived

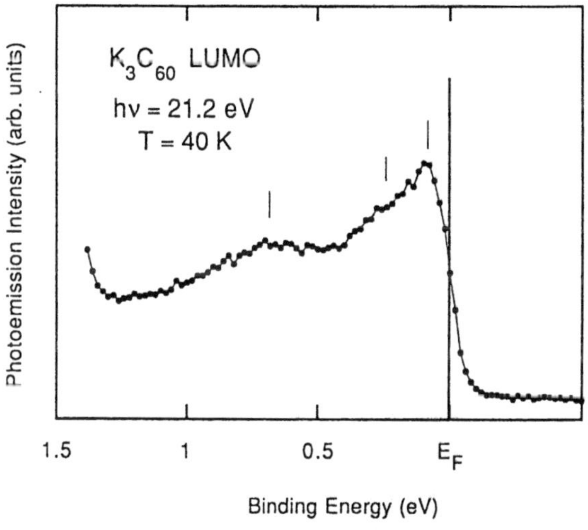

Fig. 3. High resolution photoemission spectrum acquired at 40 K for the band derived from the lowest unoccupied molecular orbitals of K_3C_{60} showing a sharp Fermi level cut-off and spectral features at 0.3 and 0.7 eV. Band calculations predict a narrow band at E_F but the experimental results indicate a redistribution of spectral weight. The feature at 0.3 eV is probably due to coupling with vibration modes of the molecule. That at 0.7 eV reflects a redistribution of spectral weight due to electron correlation and/or coupling to a plasmon mode.

band with a width of ~0.3 eV, much smaller than the ~1.2 eV width observed experimentally. Part of this spectral broadening can be explained by considering that the photoelectrons couple to the on-ball vibrational modes of the fullerene during excitation, producing a broad side band or satellite structure, as proposed by Benning *et al.* [9]. Recent calculations by Knupfer *et al.* have confirmed the importance of phonon coupling [10]. The origin of the second feature at 0.7 eV is less certain. According to Benning *et al.*, it reflects the incoherent part of the spectral function, shifted in energy by electron correlation for K_3C_{60}. Alternatively, Knupfer *et al.* considered the possibility of coupling to intrinsic charge carrier plasmons. While it remains uncertain as to whether the 0.7 eV feature is correlation-derived or plasmon-derived (or both), it is clear that the spectral signature does not represent simply the density of states of the fullerides.

4. K-C_{60} Phase Diagram

To explore the ordering of K ions in tetrahedral and octahedral sites of the C_{60} fcc lattice, we used temperature-dependent x-ray photoemission, a technique particularly useful because of the very different bonding energies for K ions in the two interstices, as demonstrated by Poirier *et al.* [3,4,11]. Figure 4 shows the K 2p core level emission measured for x=0.7 at 25°C and 180°C. These results show octahedral and tetrahedral site occupancy with features separated by 1.2

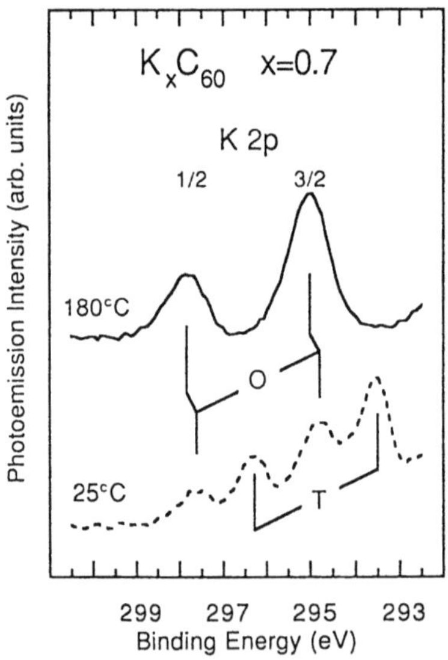

Fig. 4. XPS core level spectra showing inequivalent K 2p binding energies associated with octahedral (O) and tetrahedral (T) site occupancy. At 180°C, there is only octahedral site occupancy as the K_1C_{60} phase forms. At 25°C, the ratio of O to T occupancy is the same as that for K_3C_{60}.

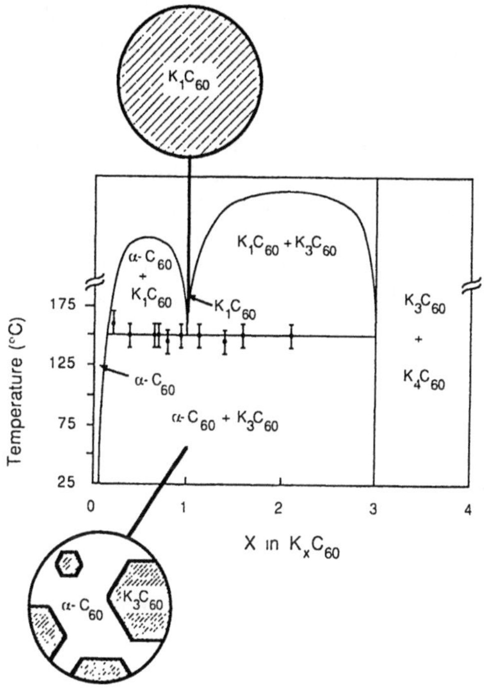

Fig. 5. Phase diagram for K-C_{60} showing an invariant eutectoid reaction at about 150°C and the fields of coexistence of the solid solution phase, the K_1C_{60} phase, and the K_3C_{60} phase. For simplicity, the K_3C_{60} and K_4C_{60} phases are drawn as line compounds because the extent of vacancy formation is not known. The microstructure sketch indicates the coexistence of two phases, as discussed in the text.

eV because of the different Madelung energies for ions in the two sites. The octahedral-to-tetrahedral ratio is 1:2. The underlying background is due to the satellite structure associated with the main C 1s line at ~284.5 eV. With increased doping, the room temperature spectra show minimal change in appearance (except for reduction in the relative importance of the background) as the octahedral to tetrahedral site occupancy remains unchanged at 1:2. As discussed by Poirier and Weaver [4], this behavior, and the corresponding behavior of the C 1s line, can be understood in terms of phase separation for the K-C_{60} sample.

Figure 5 reproduces the phase diagram that we have proposed for K-C_{60}, with an inset that depicts the microstructure expected for a sample derived from a mixture of the α-phase solid solution and K_3C_{60}. For any global stoichiometry, C_0, between the phase boundaries K_3C_{60} (denoted C_3) and α-C_{60} (denoted C_α) and any temperature below the transformation temperature, the fractions of the two phases would be given by the lever rule for a two phase system, namely $f_\alpha = (C_3 - C_0)/(C_3 - C_\alpha)$ and $f_3 = (C_0 - C_\alpha)/(C_3 - C_\alpha)$.

Changing x, i.e. C_0, changes the relative amounts of the two phases but not their stoichiometries. This phase diagram is in agreement with the NMR results of Tycko *et al.* [12] and the recent Raman results of Winter and Kuzmany [2].

Figure 4 shows that the lineshape of the K 2p emission changes dramatically between 25 and 180°C. These results reflect the fact that K ions are redistributed so that they occupy only octahedral sites at 180°C. For any stoichiometry below 1, the spectroscopic results show complete depletion of the tetrahedral sites upon heating from 25°C to 180°C. For a global stoichiometry of x>1, however, there is partial occupation of tetrahedral sites since all of the octahedral sites are filled, but the ratio of these sites is not the same as that at 25°C. In particular, this ratio is 1:2 for 0<x<3 at 25°C because the sample is two-phase with the K occupancy dictated overwhelmingly by the K_3C_{60} phase.

In Fig. 5 the data points correspond to stoichiometries where measurements were made, representing temperatures at which the change in O:T population was observed. It is clear that the transformation temperature is invariant for any x between ~0.2 and ~2.5. As noted above, the changes were greatest for x≤1. Measurements near the x=3 phase boundary showed no change, as expected because the sample was dominated by the K_3C_{60} component at any temperature.

The occupation of just octahedral sites for x=1 is indicative of the formation of the K_1C_{60} phase. Additional evidence for such a phase can be found in the Raman scattering results of Winter and Kuzmany [2] and the x-ray diffraction studies of Zhu *et al.* [13]. The invariance in transition temperature implied by Fig. 5 argues that the system is characterized by a conventional eutectoid phase diagram. Briefly, this implies that the K_1C_{60} phase will form for 0<x<3 and T>150°C but that it will transform reversibly and completely into α-C_{60} and K_3C_{60} below 150°C. From thermodynamic considerations, the high temperature sample must be composed of K_1C_{60} plus K_3C_{60}, again with fractions dictated by the lever rule and the phase boundaries of x=1 and x=3. Cooling of a sample with x>1 converts all of the K_1C_{60} into α-C_{60} plus K_3C_{60} and the excess K_3C_{60} simply cools without changing. This transformation also implies a very different microstructure from that produced by slow doping at T<150°C because growth at lower temperature allows K_3C_{60} nucleation and growth for all $x>x_\alpha$. Doping at higher temperature produces K_1C_{60} and K_3C_{60} but cooling disproportionates the K_1C_{60} into, possibly, smaller granular structures of α and K_3C_{60}. Of course, doping should be sufficiently slow from the vapor phase that K can diffuse into the lattice rather than accumulating at the surface layers where the bcc-based K_6C_{60} phase will form. Transformation back from K_6C_{60} to K_3C_{60} by high temperature heating almost certainly introduces large numbers of defects and new grain boundaries. This transformation is not possible *in vacuo* because the film sublimes at the necessary temperature.

5. Acknowledgments

This work was supported by the National Science Foundation and the Office of Naval Research. We thank F. Stepniak and P.J. Benning for helpful discussion and for taking UPS measurements.

6. References

1. For reviews, see J.H. Weaver in Solid State Physics, ed. H. Ehrenreich and F. Spaepen (Academic Press) and the special issue of J. Phys. Chem. Solids dedicated to fullerenes [vol. 53, pp. 1321-1486 (1992)].
2. J. Winter and H. Kuzmany, Solid State Commun. **84**, 935 (1992).
3. D.M. Poirier, T.R. Ohno, G.H. Kroll, P.J. Benning, F. Stepniak, J.H. Weaver, L.P.F. Chibante, and R.E. Smalley, Phys. Rev. B (15 April 1993).
4. D.M. Poirier and J.H. Weaver, Phys. Rev. B (15 May 1993).
5. Y.Z. Li, J.C. Patrin, M. Chander, J.H. Weaver, K. Kikuchi, and Y. Achiba, Phys. Rev. B (15 April 1993).
6. Y.Z. Li, M. Chander, J.C. Patrin, J.H. Weaver, L.P.F. Chibante, and R.E. Smalley, Science **253**, 429 (1991).
7. J.L. Martins and N. Troullier, Phys. Rev. B **46**, 1766 (1992); see also S. Saito and A. Oshiyama, *ibid.* **44**, 11537 (1991); Y.-N. Xu, M.Z. Huang, and W.Y. Ching, *ibid.* **44**, 13171 (1991); M.P. Gelfand and J.P. Lu, Phys. Rev. Lett. **68**, 1050 (1992); S. Satpathy *et al.*, Phys. Rev. B **46**, 1773 (1992); S.C. Erwin and W.E. Pickett, Science **254**, 842 (1992).
8. P.J. Benning, J.L. Martins, J.H. Weaver, L.P.F. Chibante, and R.E. Smalley, Science **252**, 1417 (1991).
9. P.J. Benning, F. Stepniak, D.M. Poirier, J.L. Martins, J.H. Weaver, L.P.F. Chibante, and R.E. Smalley, Phys. Rev. B (15 May 1993); J.H. Weaver, P.J. Benning, F. Stepniak and D.M. Poirier, J. Phys. Chem. Solids **53**, 1707 (1992).
10. M. Knupfer, M. Merkel, M.S. Golden, J. Fink, O. Gunnarsson, and V.P. Antropov, Phys. Rev. B (submitted 2/5/93).
11. D.M. Poirier, T.R. Ohno, G.H. Kroll, Y. Chen, P.J. Benning, J.H.Weaver, L.P.F. Chibante, and R.E. Smalley, Science **253**, 646 (1991).
12. R. Tycko *et al.* Science **253**, 884 (1991)
13. Q. Zhu *et al.*, submitted to Phys. Rev. B.

Structure and Lattice Dynamics of C_{70} Single Crystals

G. Meijer

Department of Molecular and Laser Physics, University of Nijmegen, Toernooiveld, 6525 ED Nijmegen, The Netherlands

Abstract. Single crystals of C_{70} are grown from the vapour phase using chromatographically purified C_{70} powder as starting material. Most of the crystals grown have the fcc habit but a minority fraction shows a perfect hcp morphology. These hcp grown crystals are also microscopically hcp as evidenced from X-ray and electron diffraction studies. On lowering the temperature the ideal hcp crystals undergo two phase transitions due to a sequential freezing out of the rotations of the molecules around their two different axes, resulting in an orientationally ordered low-temperature monoclinic structure. Raman measurements on C_{70} crystals in this low-temperature structure reveal a large number of Raman active inter-molecular librational and vibrational modes, consistent with the selection rules for monoclinic $P2_1/m$-C_{70}. Harmonic lattice dynamics calculations on the low temperature structure of solid C_{70} have been performed and a lattice structure in agreement with the experiments is found. From these calculations additional information on the orientational ordering is obtained and the observed Raman active lattice modes can be quantitatively interpreted.

1 Introduction

Now single crystals of high purity C_{60} and, to a minor extent, C_{70} have become available detailed experimental investigations on the structure, the orientational ordering, and the dynamics of these molecular crystals have been undertaken. It is well established by now that above 260 K C_{60} forms a fcc lattice in which the molecules have almost complete rotational freedom. Below 260 K the crystal structure becomes simple cubic with four molecules per unit cell; the molecules are still at the same relative positions as in the high temperature phase, but the four positions per unit cell have become inequivalent due to orientational ordering [1]. The orientation of the molecules can be described by a single angle ϕ over which the molecules are rotated around the [111] direction with respect to a reference configuration. The two minimum energy orientations that are experimentally observed correspond to $\phi = 98°$ and $\phi = 38°$, yielding double bonds on the one molecule facing pentagons and hexagons, respectively, on its nearest neighbours [2]. More recently, van Tendeloo *et al.* [3] postulated that the two forementioned orientations occur alternatingly in

Springer Series in Solid-State Sciences, Volume 117
Electronic Properties of Fullerenes Eds.: H. Kuzmany, J. Fink, M. Mehring, and S. Roth

the lattice, based on their observation of $2a_o$-superstructure spots in electron diffraction experiments on defect free areas of C_{60} crystals. In triplet EPR experiments Groenen *et al.* [4] confirmed the occurence of C_{60} molecules in two different orientations in the perfect crystal lattice, two orientations that have a difference in ϕ of 60 degrees. As these two types of orientations occur in an exact one-to-one ratio these experiments strongly support the existence of the proposed low temperature $2a_o$-superstructure.

First experimental results on the crystalline form of C_{70} were reported by Vaughan *et al.* [5] from measurements on powder samples. They describe the high temperature structure of crystalline powder grown from the vapour phase as being a mixture of ideal hcp (a=10.62 Å, c=17.31 Å) and fcc (a=15.01 Å), the fcc fraction being larger, especially after repeated annealing. Detailed information on the phase transitions occuring in polycristalline fcc C_{70} powder is reported by Prassides *et al.* elsewhere in this issue.

Here we report on the structure and dynamics of high purity C_{70} *single crystals* with the hcp morphology. Single crystals of high purity C_{70} grown from the vapour phase around 900-950 K predominantly have the fcc morphology. A minor fraction of the crystals (several percent), however, has the hcp morphology and, without annealing, these crystals are also microscopically hcp. The hcp C_{70} phase has to be considered to be metastable to the fcc phase. X-ray diffraction and electron diffraction experiments have been performed on these crystals at various temperatures. Upon cooling the hcp grown crystals, two phase transitions occur due to a sequential freezing out of the rotations around the two inequivalent axes. At low temperatures a monoclinic superstructure exists and harmonic lattice dynamics calculations have been performed to understand the orientational ordering in this superstructure in more detail. In addition, the frequencies of the librational and translational modes in the low temperature structure are calculated and the results are compared to the observed Raman active lattice modes.

2 Experimental results and discussion

The C_{70} crystals are grown from the vapour phase in a dynamically pumped quartz tube in which the chromatographically purified C_{70} starting material is first further purified via sublimation transport, an identical procedure as used for the growth of C_{60} crystals [6]. Both C_{70} powder that is commercially obtainable (Symcom, The Netherlands) as well as material produced and purified in the Nijmegen laboratory have been used. The purity of the C_{70} material is $\geq$ 99.5 %, as checked by HPLC, optical spectroscopy and mass spectrometry. In the mass spectrometer fullerenes are brought in the gas phase via laser desorption (N_2 laser), they are entrained in a pulsed supersonic expansion where the internal degrees of freedom are cooled, and they are ionized with an F_2 excimer laser (157 nm) in a differentially pumped detection chamber. The ions are mass selectively detected in a linear Time-Of-Flight detector with a mass resolution $M/\Delta M \geq 1000$. With the F_2 excimer laser single photon ionization of fullerenes is performed and problems with fragmentation upon ionization as well as with delayed ionization are thus circumvented. The presence of residual C_{60} and, for instance, $C_{70}O$ in the C_{70} samples can be detected at the 10^{-4} level.

The crystals are grown by cooling the saturated vapour from 950 K to room temperature in a couple of hours. No repeated heating-cooling cycles are used as this will cause the metastable hcp grown crystals to anneal to the fcc phase. C_{70} crystals as large as 0.2 mm^3 are obtained in this way. Most of the crystals have a fcc morphology, many of them being severly twinned. About 5 to 10 % of the crystals grown in this way have the hcp morphology, and are used for further investigation.

Both X-ray diffraction and electron diffraction experiments have been performed on crystals with the hcp morphology [7]. Somewhat above room-temperature the C_{70} molecules are free to rotate in a nearly ideal hcp structure with lattice constants a=b=10.56 Å and c=17.18 Å (Z=2). Upon cooling the hcp grown crystals, two phase transitions due to a sequential freezing out of the rotations around the two inequivalent axes occur. First, the C_{70} molecules arrange themselves in a deformed hcp lattice, in which the molecules have their long axes parallel and aligned along the hexagonal c-axis but are still free to rotate around this long axis, which has the smallest moment of inertia. The lattice parameters of this hexagonal phase are a=b=10.11 Å and c=18.58 Å (Z=2). This first phase transition occurs somewhat above room-temperature but is rather sluggish, evidenced by the coexistence of the ideal and the deformed hcp phases at room temperature. Then, around 270 K, the C_{70} molecules can no longer rotate freely around the long axis either and a low-temperature ordered superstructure with four molecules per unit cell results. The molecules do not rotate in this ordered structure but only perform small amplitude librations around a well-defined equilibrium orientation [7]. The experimentally observed low-temperature structure is a monoclinic structure ($P2_1/m$) with the unique b_m-axis along the former hexagonal c-axis and an angle β close to, but not necessarily identical to, 120°. The a_m-axis is identical to the former hexagonal a-axis and the monoclinic unit cell is doubled along the c_m direction (corresponding to one of the former a-axes), due to the orientational ordering in the former hexagonal plane [7]. The experimental results for the lattice parameters in this structure are given in Table I, together with the results from a harmonic lattice dynamics calculation, to be discussed later.

Unpolarised Raman spectra of C_{70} single crystals in this low-temperature ordered structure have been recorded with a DILOR XY multichannel Raman spectrometer using a Ti:sapphire laser at 771 nm as excitation source [8]. This excitation wavelength is choosen because of its relatively large penetration depth and to avoid optically induced changes in the structural and electronic properties of C_{70}. In the low frequency part of the Raman spectra, shown in figure 1(a), a large number of librational and translational lattice modes are observed. The two experimental spectra shown originate from two different crystal faces and although the relative intensities for the various modes are dif-

Table 1: Experimentally determined and calculated lattice parameters a, b, c and β (α and γ were found to be 90°) for monoclinic $P2_1/m$ C_{70}. Calculations are performed using a Lennard-Jones potential (LJ) given by $V = Ar^{-12} - Br^{-6}$, with A in kcal/mol Å^{12} and B in kcal/mol Å^6. In addition the calculated results for the orientational ordering angle ϕ_0, as discussed in the text, and the lattice energy E_{lat} (in kcal/mol.) is given

	Exper.	LJ
A	-	531070
B	-	343.9
a	9.98	9.90
b	19.96	20.04
c	18.51	18.34
β	~120	119.65
ϕ_0	-	6.57
E_{lat}	-	-41

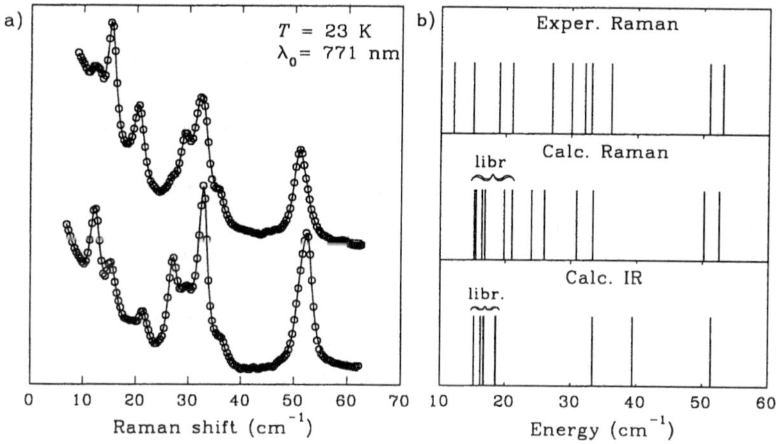

Figure 1: Experimental Raman spectrum [8] and calculated Raman and IR spectra. The upper experimental spectrum in (a) is recorded in a back scattering geometry from a (101) plane and the lower spectrum is taken from a (100) plane. In (b) the experimental Raman frequencies deduced from the spectra in (a) are shown together with the calculated Raman and IR frequencies.

ferent, the modes are found at the same frequencies. It should be noted that the spectra do not depend on the spot position on a certain crystal face, and that only a weak polarisation dependence is found. The solid curves through the data points in both spectra are fits of Lorentzian lines centered at the frequencies indicated in the figure with a typical FWHM of $\approx$2 cm^{-1}, superimposed on a Gaussian background. The number of modes observed is consistent with the selection rules for the proposed monoclinic superstructure. The librational

(l) and translational (t) inter-molecular modes in $P2_1/m$ C_{70} are classified as $\Gamma^l = 2A_g^l + 4B_g^l + 2B_u^l + 4A_u^l$ and $\Gamma^t = 4A_g^t + 2B_g^t + 2B_u^t + A_u^t$. Of these 21 modes only the 12 gerade modes are Raman active, whereas the IR spectrum will originate from the 9 ungerade modes.

3 Lattice dynamics calculations

Harmonic lattice dynamics calculations have been performed on the low temperature monoclinic phase of solid C_{70}, both to understand the orientational ordering in more detail and to make an assignment of the observed lattice modes possible [9]. These calculations have been performed using a standard harmonic lattice dynamics program for molecular crystals, details of which can be found elsewhere [10]. Rigid C_{70} molecules are used, with the atomic positions within each molecule deduced from the five, experimentally determined, non-equivalent atomic positions [11] by applying the symmetry operations of the D_{5h} point group. The calculations are performed for zero temperature where an orientational ordering exists. The interaction between the molecules is modeled as the sum over atom-atom interactions, using a Lennard-Jones interaction potential. It should be explicitly noted that no Coulomb interaction terms have been included. The lattice sum is performed over all molecules within a 25 Å range. The lattice energy is minimized with respect to the lattice parameters and with respect to the positional coordinates and the Euler angles of the four molecules in the unit cell. The experimental values are used as the initial values of the 30 minimization parameters. No symmetry restrictions have been imposed on any of these parameters.

The energetically most favorable lattice structure thus obtained is in perfect agreement with the experimental observations, as can be seen in Table I. In the low temperature structure all C_{70} molecules are aligned with their long axes parallel to the b_m-axis. Therefore, the orientation of the molecules can be described by the angle ϕ by which the molecules are rotated around their long axis. At $\phi = 0°$ the corner points of the top (and bottom) pentagons of the C_{70} molecules are directed along the minus a_m-axis. It is important to note that the calculated lattice structure obeys the experimentally observed monoclinic symmetry; the angles ϕ_i of molecules i (i=1,.,4) are related via $\phi_3(\phi_4) = 180°[\text{mod}.72°] + \phi_2(\phi_1)$. The $2_1/m$ screw axis is parallel to the b_m-axis and intersects the a_m-c_m plane at position (x, z) with x=0.1663 and y=0.3390, still reflecting the apparent hexagonal symmetry of the high temperature phase. The spacegroup is therefore $P2_1/m$. The molecules are located at positions with site symmetry m. The mirror planes perpendicular to the screw axis cut through the middle of the C_{70} molecules at (0,0,0) and (0.322,0.178,0.500).

For this lattice structure the frequencies and eigenvectors of the librational and translational inter-molecular modes are computed. The results calculated for wave vector $\mathbf{k}$=**0** can be directly compared with the experimental Raman

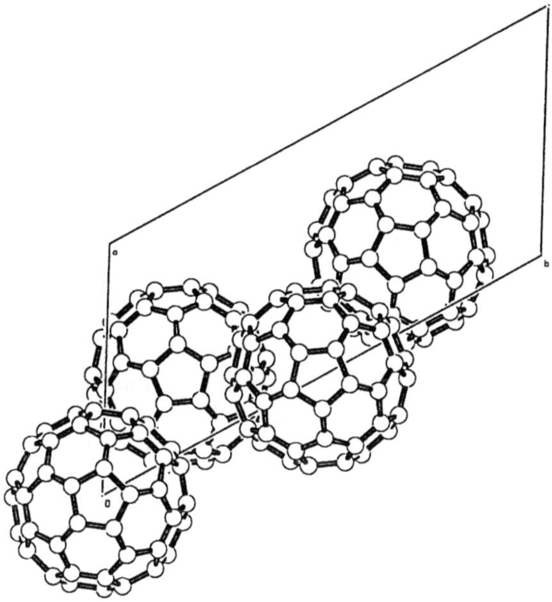

Figure 2: Representation of the orientational ordering in solid C_{70} as calculated for a Lennard-Jones potential. A view along the b_m-axis is shown. The orientations of the molecules around their long axes can be described by a single angle ϕ_0: $\phi_1 = \phi_0$, $\phi_2 = 36° - \phi_0$, $\phi_3 = -\phi_0$, $\phi_4 = 36° + \phi_0$, with $\phi_0 = 6.6°$.

data. The transformation properties of the eigenvectors under the symmetry operations of the $P2_1/m$ space group are used to make an unambiguous symmetry assignment of the modes. In figure 1(b) the calculated Raman spectrum is compared to the stick spectrum deduced from the observed Raman spectra and a good agreement is found. The frequencies of the 9 complementary IR active lattice modes are predicted and the spectrum is given in the lower part of the figure.

From the calculations some additional conclusions on the orientational ordering can be drawn. The ordering found by the calculation is schematically shown in figure 2.

Considering only the $P2_1/m$ symmetry two independent angles, ϕ_1 and ϕ_2, are needed to unambiguously describe the structure. In the model of Verheijen *et al.* [7] it was assumed that these angles are $\phi_1 = 0°[\text{mod.}72°]$ and $\phi_2 = 180°[\text{mod.}72°]$. From the calculations it follows, however, that the difference $\phi_1 - \phi_2$ is not equal to $180°[\text{mod.}72°]$. Instead, it is found that the (in principle) independent angles ϕ_1 and ϕ_2 are related via $\phi_2 = 180°[\text{mod.}72°] - \phi_1$. Therefore the structure can be described by the single angle ϕ_0 given in Table I.

The present calculations indicate that including the van der Waals interactions alone is sufficient to correctly describe the orientational ordering in solid C_{70}. The theoretically obtained crystal structure agrees with the experimentally determined low temperature structure; the symmetry is found to be $P2_1/m$. The calculations yield more insight in the orientational ordering, which involves a fixed rotation of the four molecules in the unit cell around their long axes, and which can be described by a single angle ϕ_0. The calculated frequencies of the 12 Raman active lattice modes agree quantitatively with the recently observed Raman spectrum. The frequencies of the 9 complementary IR active lattice modes are predicted.

4 Acknowledgement

This work has been supported by the Dutch Foundation for Fundamental Research of Matter (NWO/FOM-SON), and would not have been possible without the collaboration of M.A. Verheijen, H. Meekes, P.H.M. van Loosdrecht, P.J.M. van Bentum, B.J. Nelissen, A. van der Avoird, G. van Tendeloo, S. Amelinckx, S. van Smaalen and L.J. de Boer.

References

[1] P.A. Heiney *et al.*, Phys. Rev. Lett. 66, (1991), 2911
R. Sachidanandam and A.B. Harris, *ibid.* 67, (1991), 1467

[2] W.I.F. David, R.M. Ibberson, T.J.S. Dennis, J.P Hare, and K. Prassides, Europhys. Lett. 18, (1992), 219

[3] G. van Tendeloo, A. Amelinckx, M.A. Verheijen, P.H.M. van Loosdrecht, and G. Meijer, Phys. Rev. Lett. 69, (1992), 1065

[4] E.J.J. Groenen, O.G. Poluektov, M. Matsushita, J. Schmidt, J.H. van der Waals, and G. Meijer, Chem. Phys. Lett. 197, (1992), 314

[5] G.B.M. Vaughan *et al.*, Science 254 (1991) 1350

[6] M.A. Verheijen, H. Meekes, G. Meijer, E. Raas, and P. Bennema, Chem. Phys. Lett. 191, (1992), 339

[7] M.A. Verheijen *et al.*, Chem. Phys. 166 (1992) 287
G. van Tendeloo *et al.*, Europhys. Lett. 21, (1993), 329

[8] P.H.M. van Loosdrecht, M.A. Verheijen, H. Meekes, P.J.M. van Bentum, and G. Meijer, Phys. Rev. B (in press)

[9] B.J. Nelissen, P.H.M. van Loosdrecht, M.A. Verheijen, A. van der Avoird, and G. Meijer, Chem. Phys. Lett. (in press)

[10] A.P.J.M. Jongenelis, T.H.M. van den Berg, A.P.J. Jansen, J. Schmidt, and A. van der Avoird, J. Chem. Phys. 89, (1988) 4023
T.H.M. van den Berg and A. van der Avoird, J. Phys.: Condens. Matter 1, (1989) 4047

[11] P. Procacci, G. Cardini, P.R. Salvi and V. Schettino, Chem. Phys. Lett. 195, (1992) 347

Inelastic Neutron Scattering Study of the External Vibrations in C_{60}

L. Pintschovius[1,2], *S.L. Chaplot*[1], *R. Heid*[1], *M. Haluška*[3], *and H. Kuzmany*[3]

[1]Kernforschungszentrum Karlsruhe, Institut für Nukleare Festkörperphysik, Postfach 3640, D-76021 Karlsruhe, Germany

[2]Laboratoire Léon Brillouin, CEN Saclay, F-91911 Gif-sur-Yvette Cédex, France

[3]Institut für Festkörperphysik der Universität Wien, Strudlhofgasse 4, A-1090 Wien, Austria

*On leave of absence from Solid State Physics Division, Bhaba Atomic Research Centre, Bombay, India

Abstract. The external vibrations in solid C_{60} were studied by inelastic neutron scattering on single crystals. In the room temperature phase the phonons can be very well described by simple models. In the low temperature phase the phonon spectrum is much more complex, not only because of the appearance of librational phonon branches, but also because of a considerable mixing between some of the translational and librational vibrations. A new bond charge model describes the observed phonon branches very well.

1. Introduction

Considerable effort has been spent to understand the ground state and the structural phase transitions in solid C_{60}. In contrast to what might be expected from the large minimum distance between two C atoms on different molecules (~ 3 Å), a simple Lennard-Jones potential does not give the correct ground state [1]. Therefore, different models have been proposed which include secondary interactions to make the observed low temperature structure stable [2,3]. The most stringent test of these models is a comparison of the predicted translational and librational phonon dispersion to experiment. By inelastic neutron scattering on single crystal C_{60} we were able to obtain very detailed information on the external vibrations, and that is for both the translational and the librational vibrations. A first account has been given in [4]. Since then, progress has been achieved on the experimental as well as the theoretical side. In the following, we will focus on this progess.

Springer Series in Solid-State Sciences, Volume 117
Electronic Properties of Fullerenes Eds.: H. Kuzmany, J. Fink, M. Mehring, and S. Roth

2. Experimental

The experimental details were given in [4] Compared to our first data we achieved progress by using longer counting times and a slightly bigger sample (V_{max} = 6 mm). Longer counting times not only improved the statistics, but allowed to use a set-up with better resolution for a part of the scans.

3. Phonons in the room temperature phase

The room temperature dispersion of the translational modes is shown in Fig. 1. Lattice dynamical models with 1 to 3 adjustable parameters were fitted to the data. In the simplest model, a single force constant represents a central interaction between nearest neighbor molecules. Such a model is expected to hold for fully spherical, i.e. freely rotating, molecules. If, however, the molecules are not freely rotating, but jump between different orientations., a more complex force field is expected. Similar to what we reported in [4], tensor forces (3 parameters) gave a better fit than the simplest model, but the improvement was not very substantial. This means that the phonons at room temperature reflect only marginally the non-spherical nature of the molecules. This finding is in line with the results of diffraction experiments which showed that 90 % of the diffraction intensity can be explained by a uniform spherical shell [5].

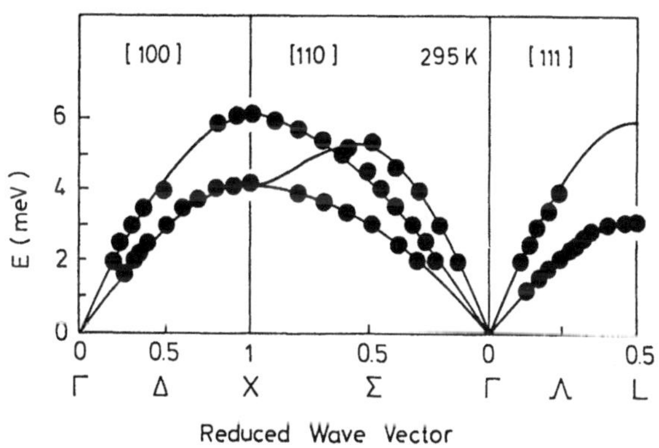

Fig. 1: Phonon dispersion of C_{60} at room temperature.

4. Phonons in the low temperature phase

Whereas the room temperature data can be explained in terms of simple models based on an fcc-structure, the interpretation of the low temperature data is less straight-forward. In particular, there is no simple criterion to

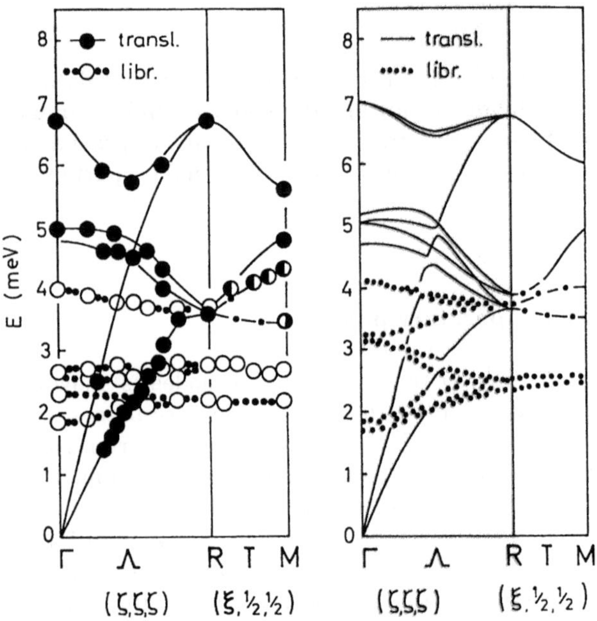

Fig. 2: Phonon dispersion of C_{60} in the low temperature phase. Left: Experimental data taken at T=200 K. Lines are a guide to the eye only. Right: Dispersion Curves calculated from the bond charge model described in the text.

distringuish librational modes from translational ones. Lattice dynamical calculations showed that even in the Q-range, which is optimum for librations, there may be equally large contributions from the translations to the scattering intensity. By measurements in many different zones we finally could assign most of the observed phonon modes with confidence. These assignments differ in some details from those given in [4].

The picture which emerged is illustrated in Fig. 2. It shows the phonon dispersion in the (ξ, ξ, ξ) - and in the (ξ, 0.5, 0.5) - direction, which are the directions with the largest number of degeneracies. The lines were drawn as a guide to the eye in accordance to what is expected from lattice dynamical models. Different symbols denote modes of translational and librational character. The branches around 4 meV in the T-direction can only be understood on the assumption of an appreciable mixing between the translational and librational degrees of freedom. We found clear evidence for such a mixing also in other directions. The most prominent example is shown in Fig. 3.

Unlike in [4], we now assign phonon peaks around 5 meV to translational modes. Some neutron groups reported in [4] at about 3.2 meV were later separated with better resolution into peaks at ~ 2.7 meV and ~ 4 meV, respectively. A previously unobserved librational mode was found at

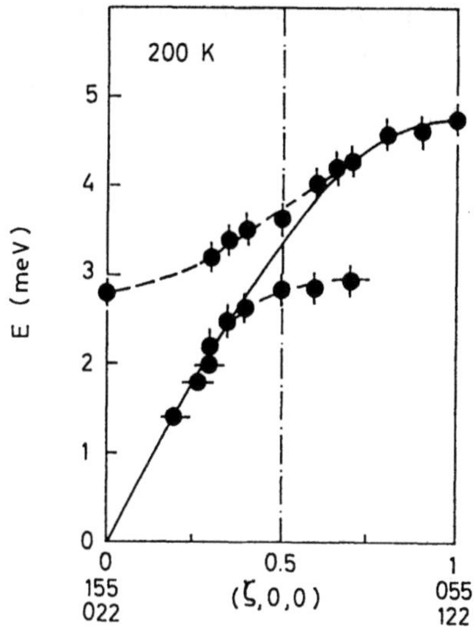

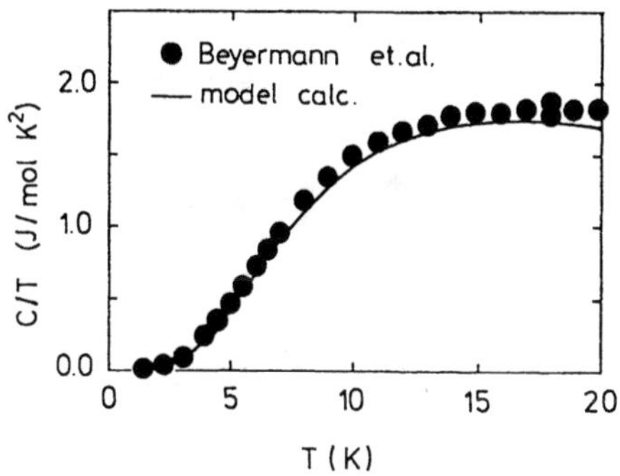

Fig. 3 (left): Dispersion of phonon peaks observed in the (ξ,0,0)-direction at T=200 K. The data are plotted in an extended zone scheme to emphasize the analogy to the TA (ξ,0,0)-branch of the room temperature phase. The full line is expected in the absence of any coupling between translational and librational motions.

Fig. 4 (right): Comparison of the low temperature specific heat calculated from the neutron results to experimental data of Beyermann et al. [6].

a comparatively low energy, i.e. E = 1.85 meV at q = 0 at T=200 K. In total, we have identified 4 out of the expected 5 librational zone center modes. From model calculations we expect the missing zone center mode in the range E=2.5 ... 3 meV. An overall check of our analysis can be performed by calculating the low temperature specific heat $c_p(T)$ from our results and comparing it with published data [6]. Fig. 4 shows that the agreement is very good.

Selected phonons were investigated as a function of temperature (Fig. 5). At present, the measurements had to be restricted to T≥90 K for intensity reasons. On approaching T_s from below, all the phonons soften somewhat, but no real soft mode behavior was found for any of the phonons investigated - in accordance with the fact that the phase transition is of first order.

It was stated in [4], that the intermolecular potentials proposed in the literature do not give a satisfactory description of the external vibrations in solid C_{60}, as the librational frquencies come out too low. A reassignment of peaks observed at E = 5 meV to translational modes removes part of the discrepancy, but the disagreement remains substantial, since we

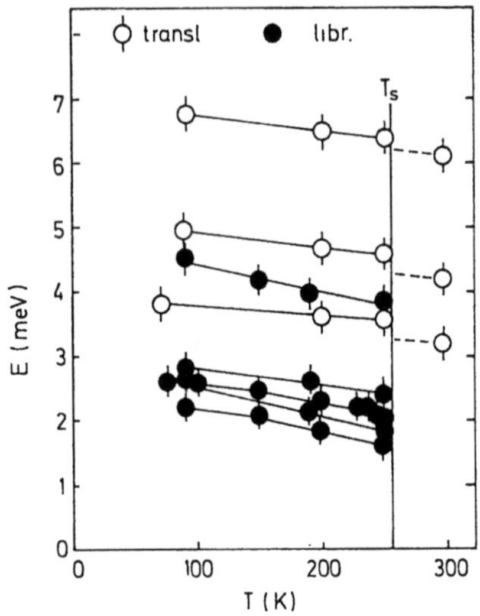

Fig. 5: Temperature dependence of selected phonon frequencies.

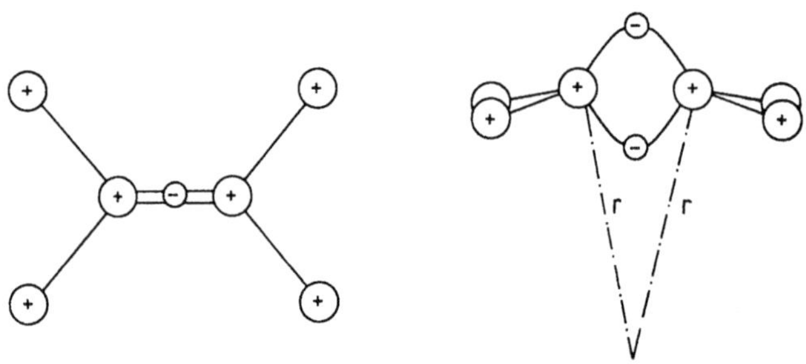

Fig. 6: Illustration of the charge distribution assumed in our lattice dynamical model as seen from top (left) or from aside (right).

confirmed librational frequencies up to E = 4.5 meV. When looking for a better model, we started from the idea that the four bonds of the carbon atoms would like to point to the corners of a tetrahedron. We therefore propose a bond charge model where the charge on the double bond is split in the radial direction (see Fig. 6). The total potential is the sum of van der Waals interactions between C atoms and electrostatic interactions between C atoms (+ 0.25 e) and double bonds (2x-0.25e). This model gives the correct ground state of solid C_{60} and a very good account of the observed phonon frequencies (see Fig. 2). It will now be used to elucidate problems linked to the orientational disorder. In particular, we would like to learn why the librational phonon peaks remain fairly sharp even close to T_s, in spite of having about 1/3 of the molecules not sitting in their ground state.

5. Acknowledgements

S.L.C. thanks the Alexander von Humboldt Foundation for the award of a research fellowship. M.H. and H.K. thank the BfWF, A. for financial support under Project No. 45212.

6. References

1. Y. Guo, N. Karasawa, and W.A. Goddard III, Nature (London) **351**, 464 (1991)
2. J.P. Lu, X.P. Li, and R.M. Martin, Phys. Rev. Lett. **68**, 1551 (1992)
3. M. Sprik, A. Cheng, and M.L. Klein, J. Phys. Chem. **96**, 2027 (1991)
4. L. Pintschovius, B. Renker, F. Gompf, R. Heid, S.L. Chaplot, M. Haluska, and H. Kuzmany, Phys. Rev. Lett. **69**, 2662 (1992)
5. R.J. Papoular, G. Roth, G. Heger, and H. Kuzmany, this volume
6. W.P. Beyermann, M.F. Hundley, S.D. Thompson, F.N. Diederich, and G. Grüner, Phys. Rev. Lett. **68**, 2046 (1992)

Is K_3C_{60} a Line Phase?

Q. Zhu[1], *J.E. Fischer*[1], *and D.E. Cox*[2]

[1]Department of Materials Science and Engineering, University of Pennsylvania, Philadelphia, PA 19104, USA

[2]Physics Department, Brookhaven National Laboratory, Upton, NY 11973, USA

Abstract. We present x-ray diffraction evidence that K_xC_{60} with x as low as 1.6 exhibits phase-pure fcc structures over some temperature range. These results have implications for the equilibrium binary phase diagram, and for the possibility of testing the hypothesis that fullerene superconductors are doped Mott-Hubbard insulators.

1. Introduction

Solid C_{60} intercalated with alkali metals is of interest for two main reasons: the evolution of lattice symmetries with concentration, temperature and hydrostatic pressure; and the existence of metallic and superconducting behavior which is believed to occur at only one special composition, x = 3. The general features of structural evolution with increasing K and Rb concentration have been discussed by Fleming *et al.* [1] and Zhu *et al* [2]. Thus far, stoichiometric phases with x = 3, 4 and 6 have been identified as face-centered cubic (fcc), body-centered tetragonal (bct) and body-centered cubic (bcc) respectively. A number of experiments give results consistent with phase separation at 300K for global compositions lying between these values [3-5]. As for the electronic properties, rigid band arguments and early conductivity vs. concentration data [6] indicated that metallic behavior should be the rule rather than the exception, the exception being x = 6 with a completely filled LUMO-derived band. On the other hand, the single-particle gap of solid C_{60} is about 1.5 eV while the difference between the HOMO of C_{60}^{+} (photoemission) and the LUMO of C_{60}^{-} (inverse photoemission) is more like 3.5 eV, implying a correlation energy U of at least 2 eV [5]. The measured and calculated bandwidths are of order 0.5 eV. These results taken together suggest that one-electron band structure might not be an appropriate framework for understanding the electronic properties. In particular, it has been argued that all phases with integral charge transfer (x = 1, 2, 3, etc.) are Mott-Hubbard insulators, and that the observed metallic and superconducting behavior near x = 3 is due to non-stoichiometry [7].

Recent results show that the binary (T,x) phase diagram is more complicated than originally thought [5,8,9]. Furthermore, a definite resolution of the correlation energy question requires much better knowledge of the relationships among composition, structure and superconductivity than has been obtained to date. The experiments described in this paper represent our initial attempts to address both these issues. High-resolution x-ray powder diffraction profiles and detailed Rietveld refinements demonstrate that for K doping, single-phase fcc materials can be obtained with x = 1.6, 2.2 and 2.9 which are stable over various ranges of T. These are lattice gas phases with random vacancies in

Springer Series in Solid-State Sciences, Volume 117
Electronic Properties of Fullerenes Eds.: H. Kuzmany, J. Fink, M. Mehring, and S. Roth

tetrahedral and/or octahedral sites, and are therefore isostructural with the superconducting phase of nominal composition x = 3 in which both sets of sites are fully occupied.

2. $K_{2.89}C_{60}$

A sample of nominal K_3C_{60} was prepared by direct reaction of preweighed amounts. X-ray powder diffraction patterns were recorded at 12K and 298K at beamline X7A at the NSLS with a Ge(111) channel-cut monochromator, flat Ge(220) analyzer and a NaI:Tl scintillation detector. Figure 1 shows the 12K profile (dots) and the Rietveld refinement (solid curve) in space group fm3m (merohedrally disordered C_{60} orientations [10]), from which we obtain **a** = 14.1559(3) Å, R(I) = 0.069, R(wp) = 0.065, and S(wp) = 1.32. There is no evidence for second-phase reflections to better than 0.1% of the strongest fcc peak intensity. Isotropic temperature factors were assigned to C, K(tet) and K(oct), and the C atoms were placed on ideal positions with a single scaling variable corresponding to the radius of the C_{60} pseudo-sphere. Refined occupancy factors showed a significant deviation from stoichiometry accommodated by tetrahedral K vacancies, N(tet) = 0.946(16), corresponding to an

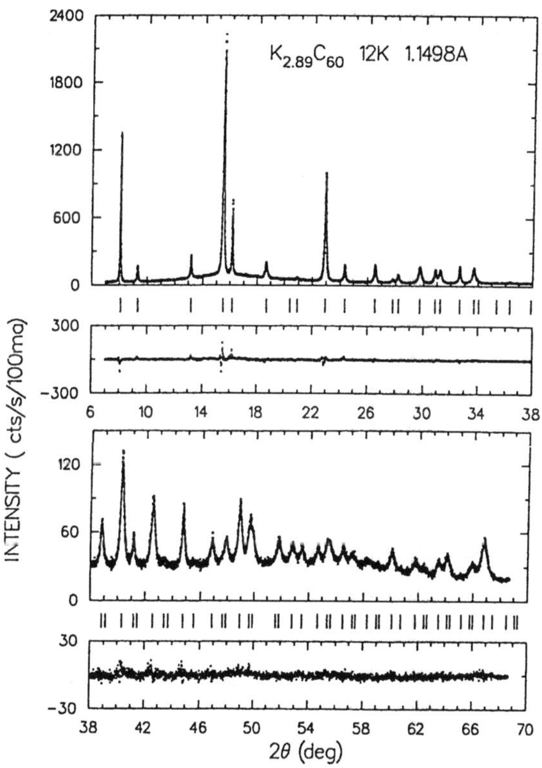

Figure 1. $K_{2.89}C_{60}$: X-ray profile (dots) and refinement (solid curve).

overall composition $K_{2.89(3)}C_{60}$. The quoted standard deviation is based on the integrated intensity goodness of fit and is therefore conservative; the values normally quoted are based on S(wp) and are a a factor of four smaller. Refined temperature factors for C, K(oct) and K(tet) are 0.9(2), 4.1(4) and 0.9(2) Å^2 respectively. The larger thermal factor for K(oct) indicates the presence of residual static disorder at 12K, while the 298K data are consistent with large thermal fluctuations. The high overall quality of the refinement insures that correlations between thermal factors and occupancies do not significantly affect the refined stoichiometry. DC magnetization of a 2 mg. packed powder sample showed a 100 ± 20% shielding fraction with $T_c = 18.7 \pm 0.5$ K.

3. $K_{2.2}C_{60}$

A sample from the same batch was dedoped in an uncontrolled manner by heating in vacuum. The 298K profile again revealed a single fcc phase, plus a few very weak peaks tentatively identified as a reaction product of K and O_2 or H_2O. It was immediately obvious from the relative intensities that x was substantially less than 3. Refinement based on the same model (Figure 2) yielded **a** = 14.245 Å, R(I) = 0.108, R(wp) = 0.069, and S(wp) = 2.06, and a substoichiometric composition $K_{2.20(5)}C_{60}$ accommodated by both tetrahedral and octahedral vacancies, N(tet) = 0.730(18) and N(oct) = 0.744(29). Thermal factors are 3.5(4) and 13(2) Å^2 for K(tet) and K(oct) respectively, much larger than found in the previous sample at 12K. The observed single-phase lattice gas behavior at x = 2.2 may not be representative of equilibrium at 298K since the sample was cooled rapidly from the (unknown) dedoping temperature. Magnetization measurements are in progress.

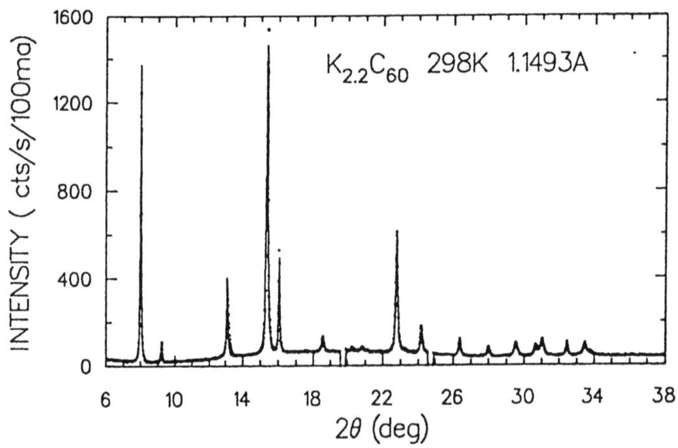

Figure 2. $K_{2.2}C_{60}$: X-ray profile (dots) and refinement (solid curve).

4. $K_{1.6}C_{60}$

A sample of nominal composition $K_{1.4}C_{60}$ prepared by the dilution method [11] exhibits single-phase fcc patterns throughout the range 353 < T < 423K. Below 353K, peaks from a lower-symmetry lattice appear, while above 423K we observe a stoichiometric K_1C_{60} rocksalt phase (N(tet) = 0) plus a small amount of K_3C_{60} [9]. From Rietveld refinement of a 393K profile, we find R(I) = 0.143, R(wp) = 0.145 and **a** = 14.27 Å, comparable to that of K_3C_{60}) at 300K but with distinctly different relative intensities. For this sample N(tet) = 0.43(3) and N(oct) = 0.78(5), yielding a composition $K_{1.6(2)}C_{60}$; the corresponding thermal factors are 5 and 14 Å^2. The absence of K_3C_{60} reflections implies that the single-phase lattice gas (or solid solution) regime identified from the two previous samples extends to at least x = 1.6 in the range 353K < T < 423K. DC magnetization reveals a maximum shielding fraction, ~ 15%, when the sample is quenched to 4.2K from within this range of T, with a T_c very close to that of nominal K_3C_{60} [9].

5. Conclusions

Our x-ray results clearly show that the ideal x = 3.00 fcc structure can support at least ~ 50% metal vacancies, over some x-dependent range of T, without phase separating. Tetrahedral vacancies are favored if x is slightly less than 3. We emphasize that pure phases with x = 1.6, 2.2 and 2.9 exist, and are isostructural with superconducting fcc K_3C_{60}. Analogous phases have not been observed with Rb or Cs doping [2,9]. Bulk superconductivity in single-phase samples with x in the range 2.8-2.9 has been clearly established (see also [12]). Superconductivity at smaller x values is problematic since quenching is required to preserve the fcc lattice gas structure. However, we believe that the diamagnetic shielding observed in the quenched x = 1.6 sample cannot be attributed to K_3C_{60}; one would have to argue that the small, highly crystalline x = 3 fraction which is observed at higher T dissolves at the same time that nanocrystalline x = 3, undetectable by x-rays, precipitates. By carefully adding small amounts of K_6C_{60} to the x = 2.9 sample, it should be possible to find out if superconductivity vanishes for x sufficiently close to 3.00, and thus to test the doped Mott-Hubbard insulator hypothesis.

We thank N. Bykovetz and C. L. Lin for help with the magnetization measurements. This work was supported by the Department of Energy, Grant Nos. DE-FC02-86ER45254 and DE-FG05-90ER75596, and by the NSF MRL Program under Grant No. DMR91-20668. The work at Brookhaven was also supported by DOE, Division of Materials Sciences, DEAC02 76CH00016.

References

1. R. M. Fleming, M. J. Rosseinsky, A. P. Ramirez, D. W. Murphy, J. C. Tully, R. C. Haddon, T. Siegrist, R. Tycko, S. H. Glarum, P. Marsh, G. Dabbagh, S. M. Zahurak, A. V. Makhija and C. Hampton, Nature **352**, 701 (1991).

2. Q. Zhu, O. Zhou, N. Coustel, G. Vaughan, J. P. McCauley Jr., W. J. Romanow, J. E. Fischer and A. B. Smith III, Science **254**, 545 (1991).

3. D. M. Poirier, T. R. Ohno, G. H. Kroll, P. J. Benning, F. Stepniak, J. H. Weaver, L. P. F. Chibante and R. E. Smalley, Phys. Rev. (in press).

4. R. Tycko, G. Dabbagh, M. J. Rosseinsky, D. W. Murphy, R. M. Fleming, A. P. Ramirez and J. C. Tully, Science **253**, 884 (1991).

5. J. Winter and H. Kuzmany, Solid State Comm. **53**, 1321 (1992).

6. R. C. Haddon, A. F. Hebard, M. J. Rosseinsky, D. W. Murphy, S. J. Duclos, K. B. Lyons, B. Miller, J. M. Rosamilia, R. M. Fleming, A. R. Kortan, S. H. Glarum, A. V. Makhija, A. J. Muller, R. H. Eick, S. M. Zahurak, R. Tycko, G. Dabbagh and F. A. Theil, Nature **350**, 320 (1991).

7. R. W. Lof, M. A. van Veenendaal, B. Koopmans, H. T. Jonkman and G. A. Sawatzky, Phys. Rev. Lett. **68**, 3924 (1992).

8. D. M. Poirier and J. H. Weaver, Phys. Rev. (submitted).

9. Q. Zhu, O. Zhou, N. Bykovetz, J. E. Fischer, A. R. McGhie, W. J. Romanow, C. L. Lin, R. M. Strongin, M. A. Cichy and A. B. Smith III, Phys. Rev. (submitted).

10. P. W. Stephens, L. Mihaly, P. L. Lee, R. L. Whetten, S-M. Huang, R. B. Kaner, F. Diederich and K. Holczer, Nature **351**, 632 (1991).

11. J. P. McCauley Jr., Q. Zhu, N. Coustel, O. Zhou, G. Vaughan, S. H. J. Idziak, J. E. Fischer, S. W. Tozer, D. M. Groski, N. Bykovetz, C. L. Lin, A. R. McGhie, B. H. Allen, W. J. Romanow, A. M. Denenstein and A. B. Smith III, J. Am. Chem. Soc. **113**, 8537 (1991).

12. R. Schlögl *et al*, this volume.

Structural and Morphological Characterization of Fullerite Crystals Prepared from the Vapor Phase

M. Haluška[1], *P. Fejdi*[2], *M. Vybornov*[3], *and H. Kuzmany*[1]

[1]Institut für Festkörperphysik, Universität Wien, Strudlhofgasse 4, A-1090 Wien, Austria

[2]Department of Mineralogy and Petrology, Comenius University, Bratislava, Slovakia

[3]Institut für Physikalische Chemie, Universität Wien, A-1090 Wien, Austria

Abstract. Crystal structure, habits and surface structures of fullerite crystals prepared from vapor phase were characterized by X-ray analysis, interfacial angle measurements and optical and scanning electron microscopy (SEM). The study of selected C_{60} crystals confirmed the fcc structure at room temperature. The crystal habit is determined by two types of morphological faces, namely {100} and {111}. SEM was used for the observation of thermal etched surfaces.

1. Preparation of crystals

Since C_{60} and C_{70} sublime at relatively low temperatures [1], crystals can be prepared with moderate effort by a sublimation-condensation method in a closed evacuated glass tube placed in a double-temperature-gradient furnace. From chromatographically purified fullerene powder aromatic solvents were removed by heating at 250 - 300°C. Then molecules with lower vapor pressures were removed by 2 successive sublimation processes. During all temperature treatments (heating of material and sealing off the glass tube) an active vacuum of $\sim 10^{-7}$mbar was maintained. Closed evacuated glass tubes with pure source material on one end were placed in the horizontal furnace with two controlled and oppositely oriented temperature gradients, which created a V- or U-shaped temperature profile. For the preparation of C_{60} crystals the source material was kept at $T_2 \sim 590°C$, and the region where the crystals were expected to grow was kept at a temperature $T_1 \sim 550°C$. For C_{70} we used $T_2 \sim 615°C$ and $T_1 \sim 570°C$. These conditions were maintained for several days. In this way we obtained crystals of various sizes up to 12 mg mass. Most of them had well-expressed smooth and shiny faces; the crystal habit was a truncated octahedron.

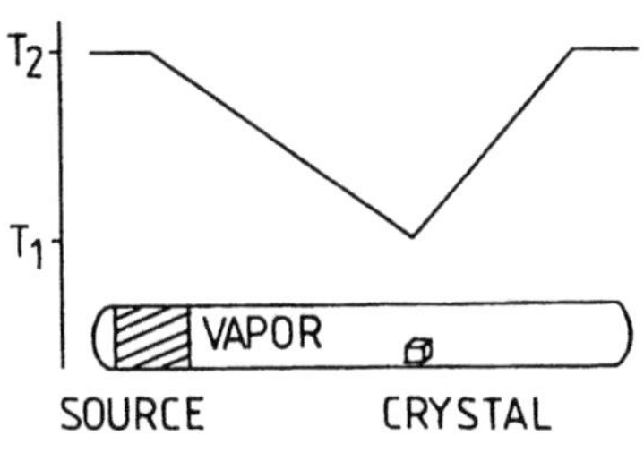

Fig. 1: The temperature profile for the crystal growth process by the double-gradient technique.

Springer Series in Solid-State Sciences, Volume 117
Electronic Properties of Fullerenes Eds.: H. Kuzmany, J. Fink, M. Mehring, and S. Roth

2. Morphology

The crystal form is characterized by the shapes and areas of crystal faces. It may vary from specimen to specimen, depending on the velocity of growth. We observed two basic forms of crystals, "bulk" crystals and "planar" crystals. In the latter two parallel {111} faces are dominating. The high quality of the crystal faces allowed goniometric studies of C_{60} crystals by a two-circle optical goniometer. Crystals exhibit two different types of faces: octahedral {111} (with hexagonal or triangular shape) and hexahedral {100} (with rectangular shape). The angles between octahedral and hexahedral faces, two octahedral and two hexahedral faces were 125.38(21)°, 109.48(12)°, and 90°, respectively. A strong tendency to twinning was observed. Most of the twinned crystals had a (111) twinning plane. Forms of observed twinning [2] are in accordance with computed forms for fcc materials [3].

3. Crystal structure

X-ray diffraction patterns were obtained at room temperature for a number of C_{60} and C_{70} samples from different growth batches using a Weissenberg camera (R=27.53 mm, Cu-K_α radiation, λ = 0.154056 nm). Lattice parameters were calculated on the basis of the X-ray data from powdered samples obtained in a Debye-Scherrer camera (R_k= 114.59 mm, Cr-K_α radiation λ = 0.228970 nm) and in a Guinier-Huber camera (R_k = 114.59 mm, Cu-$K_{\alpha 1}$ radiation (λ = 0.154056 nm). For the latter data an internal standard of 99.9999% pure Ge a_{Ge}=0.5657906 nm) was employed. The single crystals of C_{60} and C_{70} were characterized by reconstruction of the reciprocal lattice on the basis of X-ray rotation photographs obtained in a Weissenberg camera. All X-ray photographs of C_{60} were successfully indexed on the basis of an fcc structure. C_{70} crystals from different growth batches showed continuing differences in crystal structure. More detailed investigations are going on. The lattice parameters and cell volumes calculated by a least-squares refinement for selected samples of C_{60} and C_{70} are presented in Table 1.

Table 1. Lattice parameters obtained for selected single crystals of C_{60} and C_{70}.

Sample	Camera/radiation	a,[nm]	V,[nm^3]
1	Debye-Scherrer/Cr-K_α	1.4194(4)	2.859(1)
2	Debye-Scherrer/Cr-K_α	1.4195(6)	2.860(3)
C_{60} 3	Debye-Scherrer/Cr-K_α	1.4175(4)	2.848(2)
4	Guinier-Huber/Cu-$K_{\alpha 1}$	1.4193(5)	2.859(2)
5	Guinier-Huber/Cu-$K_{\alpha 1}$	1.4179(6)	2.850(2)
C_{70} 1	Guinier-Huber/Cu-$K_{\alpha 1}$	1.4851(6)	3.2754(3)

4. Surface structure

Different kinds of surface structures were observed. Most of the surfaces were smooth and shiny. However, in some cases expressive features developed at the surfaces [2]. They could be characterized as dendrites, screw dislocations, barriers at the edges of faces,

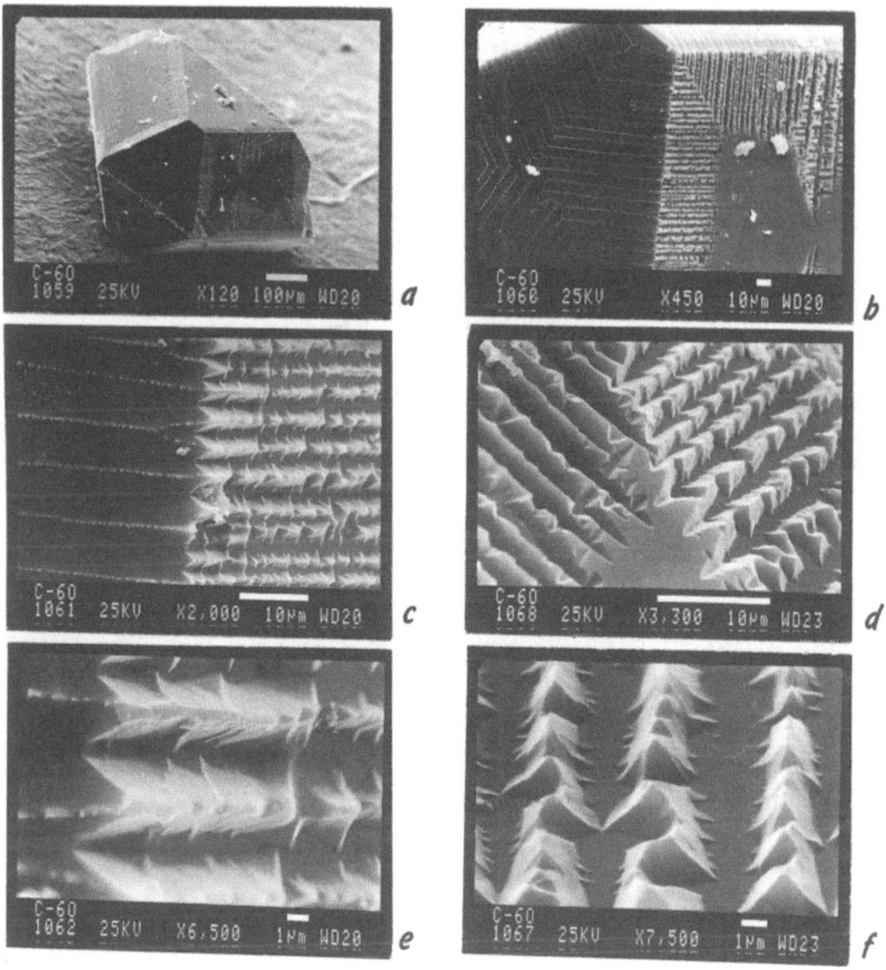

Fig. 2: SEM photographs of surface stuctures for a C_{60} crystal after thermal etching in vacuum. Macroscopic view of the crystal (a); more detailed views (b-f); view of the edge between {111} (left side) and {100} (right side) faces (b). (c), (e); detailed views of {100} face (d), (f).

lamellar growth and cavity formation. Other kinds of surface structures can be developed by thermal etching. This method is very useful and informative for the study of line defects, and it can be used to reveal the growth history of crystals [4]. When a crystal is heated, evaporation of the surface takes place in a manner reverse to growth. The crystal shown in Fig 2 was etched in vacuum, and shows a hill and valley surface structure formed on the {100} faces formed by facets of higher-index planes. The process of evaporation starts at the edge between faces and continues perpendicular to the edges in the direction towards the midle of the face. Surface structures obtained in this case are crystallographically oriented. The thermally etched reliefs have twofold symmetry (Fig. 2b, right side) and 3 fold symmetry (Fig. 2b, left side) on {100} faces and on {111} faces, respectively.

In Fig. 2b,d part of the original smooth surface can be seen. Since the reticular density of the {111} faces is higher than that of the {100} faces, more material was evaporated from {100} faces.

5. Conclusion

We prepared fullerite crystals (C_{60} and C_{70}) with fcc structure and with a habit approaching the ideal shape (Wulf's polyhedron for fcc material). The quality of the surface is strongly dependent on the growth conditions. For careful preparation atomically smooth and shiny surfaces can by obtained. During thermal etching, striations at the surface occur, and low-index faces break up into very small facets (micro facets) of higher index.

Acknowledgment. This work was supported by the Osteuropa Förderung des BMfWF project GZ 45.212/2-27b91. We are grateful for valuable discussions with P.Rogl, B.Burger and R.Winkler.

References

1. J. Abrefah., D. R. Olander, M. Balooch, W. J. Siekhaus: Appl. Phys. Lett. **60**, 1313 (1992)
2. M. Haluška, H. Kuzmany, M. Vybornov, P.Rogl, P. Fejdi: Appl. Phys.A **56**, 161 (1993)
3. L. D. Marks: J. Crystal Growth **61**, 556 (1983)
4. K. Sangwal: Etching of Crystals, In Defects in Solids, ed. by S. Amelinckx, J. Nihoul (North-Holland, Amsterdam 1987) vol.15

Low-Frequency Elastic Response in Single-Crystal C_{60}

W. Schranz[1], *A. Fuith*[1], *P. Dolinar*[1,*], *H. Warhanek*[1], *M. Haluška*[2], *and H. Kuzmany*[2]

[1]Institut für Experimentalphysik der Universität Wien, A-1090 Wien, Austria
[2]Institut für Festkörperphysik, Universität Wien,
Strudlhofgasse 4, A-1090 Wien, Austria
*Permanent address: J. Stefan Institute, University of Ljubljana,
61111 Ljubljana, Slovenia

Abstract. The complex elastic constant of sublimation grown C_{60} single crystals is measured between 100K and 300K in the frequency range 0.6 Hz - 50 Hz. The order disorder phase transition at $T_I = 260$K shows up in a negative dip - anomaly and a peak maximum in the real and imaginary parts of the complex elastic constant, respectively. A very slow relaxation ($\tau = 0.2$s) was measured near T_I. In addition the elastic behaviour around the frequency dependent freezing transition at T_{II} is measured and described by a single activated relaxation time τ. From our data and the data of previous measurements we find, that the frequency shift of T_{II} from 180 K to 80 K can be described by a simple Arrhenius behaviour over a frequency range of 11 decades.

1.Introduction

The synthesis of the icosahedral carbon molecule C_{60} [1], the preparation of large quantities of material [2] and the growth of single crystals [3,4] have stimulated and enabled a large number of investigations leading to exciting results. At $T_I = 260$K the crystal undergoes an order disorder phase transition from a face centered cubic structure to a simple cubic structure, which is orientationally ordered [5]. The phase transition is accompanied by a discontinuous decrease of the cubic lattice constant [5,6] of about 0.3% at T_I. This demonstrates that the C_{60} lattice is not rigid, but there is a coupling between the orientational and the elastic degrees of freedom in crystalline C_{60}. In order to study the effect of this coupling we have measured the temperature dependence of the complex elastic constant [7]. Only one measurement of the elastic constant has been reported up to now [8] showing a positive jump of 8% at T_I. In the second part of our paper we report on the elastic behaviour around the freezing transition at T_{II}.

2. Experimental

The C_{60} single crystals were grown by a sublimation technique [3]. The typical size of the samples was $0.6 \times 0.6 \times 1mm^3$. The temperature dependence of

Springer Series in Solid-State Sciences, Volume 117
Electronic Properties of Fullerenes Eds.: H. Kuzmany, J. Fink, M. Mehring, and S. Roth

the complex elastic response to a time dependent stress was measured using a commercial dynamical mechanical analyzer (DMA 7, Perkin Elmer) in the frequency range between 0.6 Hz and 50 Hz. We used as grown crystals with parallel natural [111] - faces and applied a longitudinal stress perpendicular to these faces resulting in an effective elastic constant:

$$\frac{1}{C_{eff}} = \frac{1}{3}\left(\frac{1}{C_{11} + 2C_{12}} + \frac{1}{C_{44}}\right)$$

3. Results

3.1. Order - Disorder Phase transition at T_I

Figure 1 shows the temperature dependence of the real and imaginary part of the complex elastic constant $C_{eff} = C'_{eff} + iC''_{eff}$ in the vicinity of the first order phase transition at $T_I = 260$K.

The most important features of Fig. 1 are the negative jump of about 25% in C_{eff} at T_I and the corresponding maximum in C''_{eff} appearing somewhat higher in temperature. These anomalies are preceded by a rounding in C'_{eff} and a corresponding finite damping C''_{eff} which start already 5K above T_I. In addition a thermal hysteresis of about 3K was observed demonstrating the first order nature of the phase transition in agreement with recent theoretical [9] and experimental results [5,6]. The behaviour of C_{eff} of Fig.1 is in agreement with recent calculations performed on the basis of the Pippard-Buckingham-Fairbank relations [10]. These relations connect the compressibility with the thermal expansion coefficient and the pressure dependence of the phase transition temperature and lead to a negative jump in the elastic constant at T_I which should be larger than 60%. The variation of the elastic constant near the phase transition is determined by the coupling terms between the order parameter and the strain variables appearing in the free energy of the system. The general form of the free energy expansion F(η) in terms of symmetry adapted

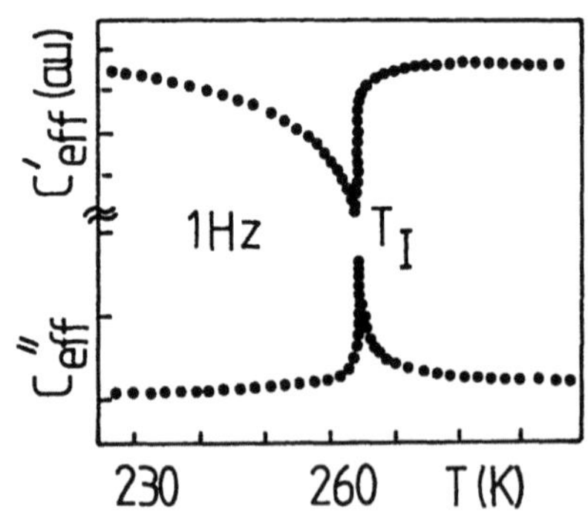

Figure 1: Temperature dependence of the real part C'_{eff} and the imaginary part C''_{eff} of the complex elastic constant measured at 1 Hz.

rotator functions was calculated on the basis of a microscopic model [9] and was recently extended to include the strain coupling $F_c(\eta, \epsilon)$ [11] and to derive the temperature dependence of the elastic constants [12]. Applying symmetry arguments the lowest order coupling terms between the order parameter η and the strains ϵ are of the type $F_c = a\eta^2\epsilon + b\eta^2\epsilon^2$. The first term in F_c leads to a negative jump in the elastic constant at T_I similar as it is shown in Fig.1 [13]. The second term in F_c induces an anomaly which is proportional to the square of the order parameter. This term may explain the previous vibrating reed measurements of Shi, et al which were performed at 10 kHz [8]. In these measurements a positive jump of about 8% was obtained in contrast to our findings. The difference may be due to the different measurement frequencies ω. It is well known, that the $\eta^2\epsilon$-anomaly can depend on the applied frequency [13]. Assuming a relaxational process with a relaxation time τ we can discuss two limits: If $\omega\tau \ll 1$ the full jump is obtained at T_I in the elastic constant, whereas for $\omega\tau \gg 1$ the stress varies too fast for the order paramter motion to follow and the jump at T_I is diminished by a factor of $(1+\omega^2\tau^2)^{-1}$.

In the relaxational model of above the real and imaginary parts of the complex elastic constant are related according to the Debye-equations. Then the relaxation time τ can be calculated from the data of Fig.1 leading to $\tau \approx 0.2s$ just above T_I. Further work is necessary to clarify the microscopic origin of this large relaxation time.

3.2 Freezing transition at T_{II}

It is a widely adopted picture, that below T_I the molecules shuffle between at least two nearly degenerate states. With decreasing temperature the population of the energetically less favorable orientation decreases, but depending on the time scale of the experiment a different amount of misorientations can be quenched at different temperatures. This freezing of the misoriented molecules shows up in different quantities like thermal expansion [6], elastic constants [8], and dielectric permittivity [14]. We have measured the temperature dependence of the complex elastic constant for frequencies between 0.6Hz and 50Hz [7]. Fig.2 shows an example for 0.6Hz. With increasing frequency the knee in C'_{eff} and the maximum in C''_{eff} shift to higher temperatures: T_{II}=121K for 0.6Hz and 140K for 50Hz. The data can be fitted using a Debye relaxation with a single relaxation time $\tau = \tau_0 exp(E/k_BT)$, yielding $\tau_0 = 4 \pm 2\ 10^{-14}$ s and $E = 300 \pm 10$ meV. These values are quite consistent with previous measurements [6,8,14] implying that the anomalies observed around 160K in sound velocity [8] and around 80K in thermal expansivity [6] are of the same microscopic origin (Fig.3).

Quite recently a broad distribution of relaxation times was reported from dielectric measurements [14]. A simulation of our data with a Gaussian distribution of activation energies showed that the width must be smaller than 10meV (Fig.2). Work to clarifiy this difference is in progress.

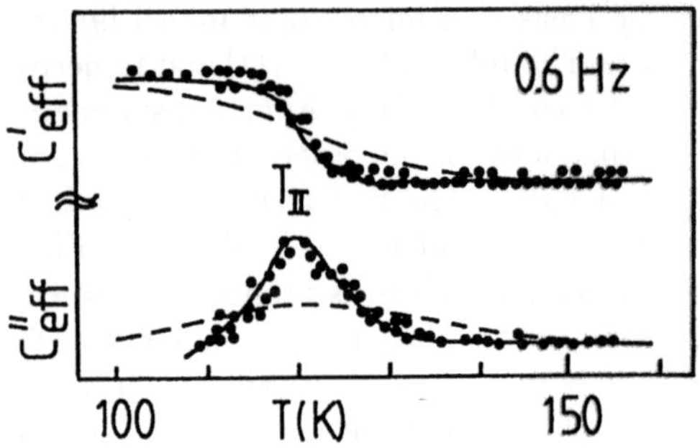

Figure 2: Temperature dependence of the elastic constant C'_{eff} and C''_{eff} measured at 0.6Hz (o) and Debye-fit (—): Gaussian distribution of activation energies (– – –) with W=40meV, according to ref. [14]

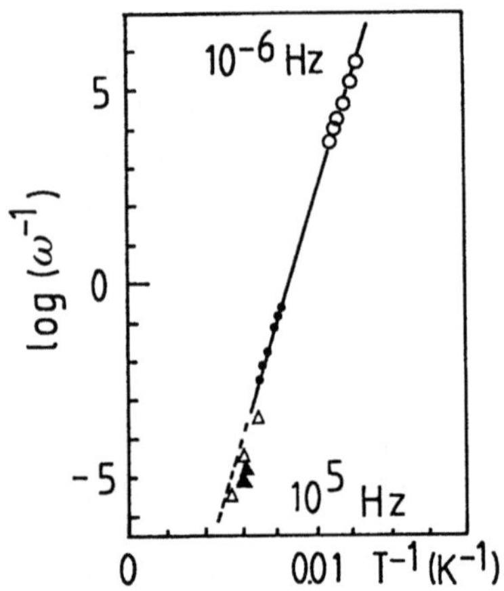

Figure 3: $log(\omega^{-1})$ vs T_{II}^{-1} for different measurement frequencies: o...thermal expansivity [6], •...our data ▲...sound velocity [8], △...dielectric permittivity [16]

4.Summary

In summary very slow relaxations have been found from low frequency elastic measurements on C_{60} single crystals. Around the order-disorder phase transition at T_I=260K they may originate from correlated motions of C_{60} molecules, i.e. order parameter fluctuations, phase front motions, or twin boundary motions. The elastic anomaly around T_{II} occurs due to the freezing in of molecular disorder. From our data and the mesurements of thermal expansivity [6], sound velocity [8] and dielectric permittivity [14] we find that the frequency shift of the freezing temperature T_{II} from 180K to 85K follows an Arrhenius law over a frequency range of 11 decades.

Acknowledgement. The present work was supported by the Österreich. Fonds zur Förderung der wissenschaftl. Forschung under Project No P 8285 and by the BMWF Project GZ 45.212/2-27b91.

References

1. H.W. Kroto, J.R. Heath, S.C. O'Brian, R.F. Curl and R.E. Smalley, Nature **318**, 162 (1985)
2. W. Krätschmer, L.D. Lamb, K. Fostiropoulos, D.R. Huffman, Nature **347**, 354 (1990)
3. M. Haluska, H. Kuzmany, M. Vybornov, P. Rogl and P. Fejdi, Appl. Phys. **A56**, 161 (1993)
4. R.L. Meng, D. Ramirez, X. Jiang, P.C. Chow, C. Diaz, K. Matsuishi, S.C. Moss, P.H. Hor, C.W. Chu, Appl. Phys. Lett. **59**, 3402 (1991)
5. W.I.F. David, R.M. Ibberson, T.J.S. Dennis, J.P. Hare and K. Prassides, Europhys. Lett. **18**, 219 (1992)
6. F. Gugenberger, R. Heid, C. Meingast, P. Adelmann, M. Braun, H. Wühl, M. Haluska and H. Kuzmany, Phys. Rev. Lett. **69**, 3774 (1992)
7. W. Schranz, A. Fuith, P. Dolinar, H. Warhanek, M. Haluska and H. Kuzmany, submitted to Phys. Rev. Lett.
8. X.D. Shi, A.R. Kortan, J.M. Williams, A.M. Kini, B.M. Savall and P.M. Chaikin, Phys. Rev. Lett. **68**, 827 (1992)
9. K.H. Michel, J.R.D. Copley and D.A. Neumann, Phys. Rev. Lett. **68**, 2929 (1992)
10. C. Meingast, P. Adelmann, M. Braun, G. Czijek, B. Renker, G. Roth and E. Sohmen, to be published
11. D. Lamoen and K.H. Michel, to be published
12. K.H. Michel, private communication
13. W. Rehwald, Adv. Phys. **22**, 721 (1973)
14. G.B. Alers, Brage Golding, A.R. Kortan, R.C. Haddon, F.A. Theil, Science **257**, 511 (1992)

Electron Microscopy of C_{60} and C_{70} Fullerites

G. Van Tendeloo and S. Amelinckx

University of Antwerp (RUCA), Groenenborgerlaan 171,
B-2020 Antwerp, Belgium

Abstract. The crystal structures and their phase transitions in C_{60} and C_{70} molecular crystals are studied by means of electron microscopy and electron diffraction. Lattice defects in FCC crystals of C_{60}, such as twins, stacking faults and dislocations are observed at room temperature using high resolution imaging and compared with models.

1. Introduction

High resolution electron microscopy and electron diffraction have played, and continue to play an essential role in elucidating structures and textures of fullerites [1,2]and of related objects such as whiskers[3], tubes[4,5], onions [6]... Because these techniques require only minute crystal fragments, as compared to other diffraction techniques, they were extensively used in the early structural studies. Moreover the relationship between direct and reciprocal space of the same crystal fragment can easily be established, which is especially helpful for the interpretation of the observed textures and defects.

Cooling and heating the specimens allows to study the structural aspects of phase transitions, usually in a qualitative manner only. This is due to the fact that in an electron microscope the specimen temperature is difficult to measure with a reasonable precision.

In this paper we shall briefly summarize some recent results of such studies on C_{60} and C_{70}

2. C_{60}

a) The room temperature structure

The room temperature crystal structure of the C_{60} fullerite has initially been the subject of some discussion until it was realized that the sample purity is of overall importance. Crystals grown from an organic liquid such as toluene are predominantly faulted hexagonally close packed [7, 1]. However the observation in the microscope vacuum under electron irradiation causes a transformation of such crystals into a faulted cubic close packed structure. It is assumed that the solvent molecules occupy the channels along c, formed by the unoccupied interstitial positions in the HCP structure. Under vacuum these solvent molecules are removed and a shear transformation converts the faulted hexagonal into a faulted cubic structure. Sublimation grown crystals of sufficient purity are always cubic at room temperature.

The isotropically rotating quasi-spherical molecules occupy the node points of a face centered cubic lattice with a_0= 1.41 nm [8, 9]. The molecules can be modelled as spherical shells of diffracting material with an average radius r_0 = 0.353 nm. Due to the fact that fortuitously the lattice parameter is four times the radius of

Springer Series in Solid-State Sciences, Volume 117
Electronic Properties of Fullerenes Eds.: H. Kuzmany, J. Fink, M. Mehring, and S. Roth

the spherical shell, the h00 reflections have vanishingly small intensities for all value of h [10]. Experimentally the kinematical intensities of the h00 reflections are found to be negligibly small, but under conditions of dynamical diffraction this is no longer strictly true [11].

2.2 Lattice defects in the face centered cubic structure [12].

The defects in the room temperature structure can conveniently be studied by means of high resolution images, especially since single molecular columns can easily be resolved along the [110] zone, i.e. along the close packed rows of molecules. The most abundant defects are coherent twin interfaces and intrinsic stacking faults on (111) planes. Occasionally also extrinsic stacking faults may occur. Many of the complex defects, occurring in low stacking fault energy metallic alloys have also been described in C_{60} crystals.

Intrinsic stacking faults, jogging from one (111) plane to the adjacent one, producing a dipole of stair rod dislocations have been revealed. Lomer - Cottrell barriers, consisting of V-shaped arrangements of two stacking faults on intersecting (111) planes, forming a stair-rod dislocation along their intersection line, have been observed as well. Stacking faults were often found to terminate when meeting a twin interface. Different types of dislocations were observed. The image of fig.1 shows for instance a Frank partial dislocation with Burgersvector $1/3[1\bar{1}1]$ and its associated intrinsic stacking fault.

Most of these defects are clearly resolved in C_{60} because of the large size of the molecules as compared to that of atoms. The geometrical features of the defects in such alloys, in C_{60} and in bubble rafts are amazingly similar because the bonding of metallic atoms, molecules and soap bubbles has some common characteristics: it is non-directional and short range.

A line defect, which was so far not observed in a high resolution image of any material is shown in fig.2. It consists of a row of fractional vacancies, formed along the intersection line of two intrinsic faults. In metallic alloys the stacking faults usually do not intersect, but one terminates on meeting the other. In C_{60} the stacking fault energy is very small and moreover the formation energy of vacancies is sufficiently small to allow the formation of such line defects.

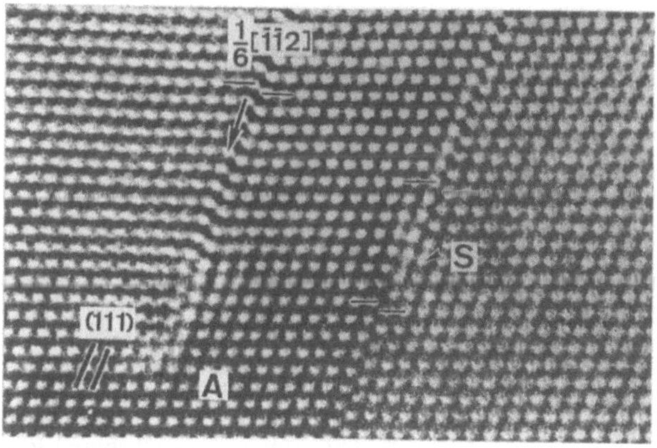

Fig.1: Intrinsic stacking fault limited by a Frank partial dislocation. An intrinsic fault jogging from one lattice plane to the next causing a dipole of stair rod dislocations is also present.

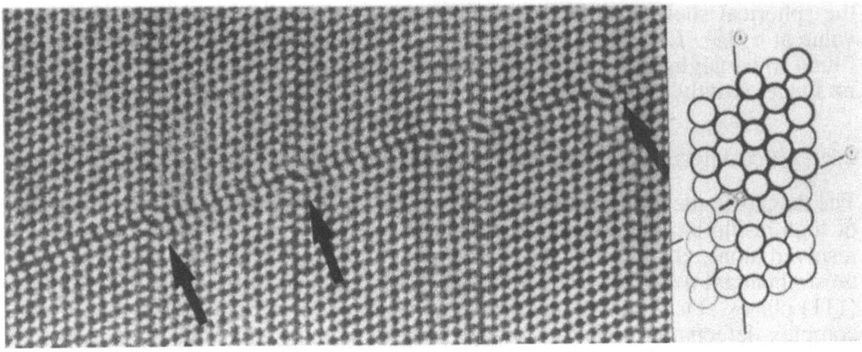

Fig.2: a) Intersection line of stacking faults causing a row of fractional vacancies
b) Schematic representation of a line of fractional vacancies in C_{60}
(according to S. Amelinckx in "Dislocations in Solids Ed. F.R.N. NabarroN. Holland 1979)

The room temperature diffraction patterns sometimes exhibit additional spots at fractional reciprocal lattice positions. They can be shown to be due to the intersection of Ewald's sphere with long <111>* streaks resulting from the presence of a high density of coherent twin interfaces, stacking faults, or both on (111) planes. Phase transitions clearly have to be studied on crystal fragments which do not exhibit such additional spots at room temperature.

2.3 Phase transitions in C_{60}

"In situ" electron diffraction has confirmed the orientational order-disorder transition at around 255 K, from the room temperature face centered cubic structure into a primitive cubic structure, with approximately the same lattice parameter. The diffraction conditions allow to establish unambiguously the space group as $Pa\bar{3}$ [10, 13]. The reduction of the fourfold symmetry axis to a two fold one on transformation gives rise to two orientation variants, related by a 90° rotation about a cube axis. The loss of the symmetry translations of the type $1/2\langle110\rangle_{FCC}$ leads to four translation variants of the primitive structure for each orientation variant. The phase transformation thus causes a severe domain fragmentation of the primitive structure (SC), even when starting with a perfect FCC single crystal. The orientationally ordered structure, proposed in [10] is consistent with all these observations. Fig.3 shows the [001] diffraction pattern of the primitive structure; in (a) and (b) a single orientation variant is present in the selected area, whereas in (c) both variants are present..

The domain fragmentation was studied using dark field diffraction constrast imaging in a superlattice reflection.[2]. It confirmed the presence of the two types of orientation domains; domain walls also often coincide with twin interfaces.

At a somewhat lower temperature certain crystal fragments which were free of additional spots at room temperature, were found to exhibit not only the spots due to the primitive lattice, but moreover also spots characteristic of a face centered cubic structure with a lattice parameter $2a_0$[13]. This pattern was attributed to a structure in which the successive molecules in rows along the cube directions of the simple cubic structure adopt alternatingly two different orientations. Whereas the observation of the additional spots is unambiguously established, the interpretation

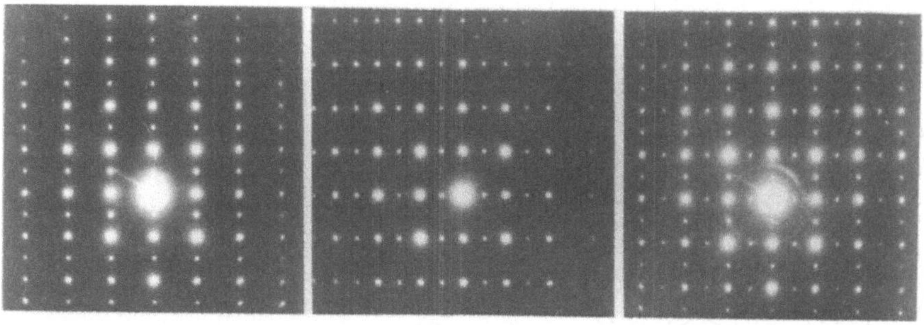

Fig.3: [001] zone diffraction pattern of the single cubic phase of C_{60} exhibiting the diffraction conditions of the Pa$\bar{3}$ space group.

a) one variant in the selected area

b) second variant related by a 90° rotation to (a)

c) two variants are both present. when the selector aperture is put across the 90° boundary.

is still controversial, because the additional spots are not always observed in X-ray or neutron diffraction experiments. On the other hand low temperature e.p.r. experiments confirmed the doubling of the lattice parameters [14]. Different reasons for the absence of additional spots in some diffraction experiments can be imagined; the most probable one being the small size of coherent domains in this phase.

Some recent theoretical work [15] indicates the existence at room temperature of minima of the interaction energy for two different orientations of the C_{60} molecules. Starting from the standard orientation of the molecule with respect to the crystal axis these two orientations are obtained by rotations of 38° and 98° about the [111] axis i.c. the two orientations differ by 60°, which is a quasi-symmetry operation. At room temperture one of these energy minima is an absolute minimum (98°) whilst the other one (38°) is only a shallower relative minimum. It is possible that at low temperature this second minimum becomes deeper, thus leading to two roughly equally occupied sublattices, associated with each of the <111> type rotation axis. It is clear that the occurrence of two sublattices would be consistent with the proposed $2a_o$ structure. Low temperature neutron diffraction evidence also suggested the occurrence of two different molecular unequally occupied orientations on the same sublattice of the primitive structure, but no ordering was found [15].

3. C_{70} [16, 17, 18]

Vapour grown crystals of very pure C_{70} (> 99.9%) exhibit, even within the same batch, two different morphologies: the majority being cubic, only a minority being hexagonal. At room temperature the corresponding predominant crystal structures are face centered cubic and hexagonal close packed. However both types of structures are always faulted; respectively on $(111)_{FCC}$ or $(0001)_{HCP}$ planes. The face centered cubic structure is moreover heavily twinned.

The hcp phase exhibits orientationally ordered superstructures of the hcp stacking (c/a = 1.82) at low temperatures. The reciprocal lattices of the two low temperature superstructures, as deduced from electron diffraction patterns at liquid nitrogen temperature [17] indicate that in both structures one of the a-parameters of

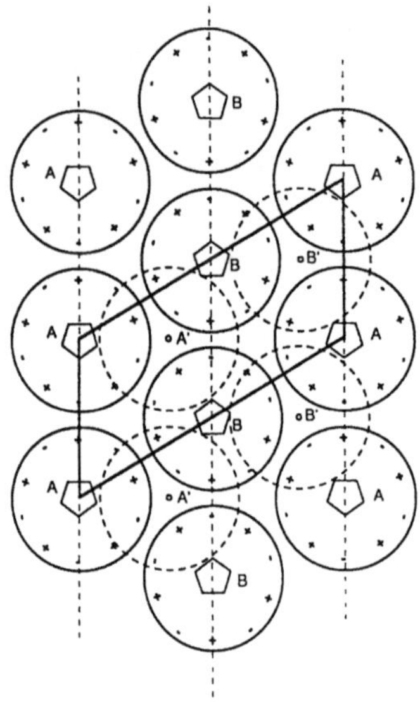

Fig.4: Orientational order of C_{70} molecules in one layer of the monoclinic crystal structures which are based on an ABAB stacking of the molecular layers.

the hcp basic lattice is doubled, breaking the hexagonal symmetry. Moreover in one of them the c-parameter is doubled as well. The two structures can be referred to monoclinic unit cells with angles close to 120°; one unit cell is primitive the second one is C-centered . The corresponding structures are closely related; they contain the same type of layers which are orientationally ordered variants of close packed layers of molecules with their long axis parallel to the c-direction. The relative azimuthal orientation of the molecules within one layer are presumably determined by the same packing principle as in C_{60}. Along the "equator", where the intermolecular separation is smallest, electron rich regions (+ in fig.4), along the 6 - 6 or 6 - 5 edges of the cage, should face electron poor regions (- in fig.4) in the centre of the hexagons of the neighbouring molecules.

Applying the mentionned stacking principle the resulting packing of molecules, with their long axis parallel to c, leads to the layer structure represented in fig.4 . The molecules in neighouring rows thus adopt two different azimuths which differ by 180°, this leads to doubling of one of the edges of the unit mesh in the layer planes. The AB spatial arrangement of such layers can give rise to various orientational stackings with different c-parameters. The structures suggested by the two types of diffraction patterns are represented in fig.5. The stackings only differ by the orientation of the molecules in successive layers; they could be called orientational polytypes. One structure, the most abundant one, has a primitive monoclinic lattice, whereas the other structure is based on a C-face centered monoclinic lattice; both lattices have a monoclinic angle of approximately $\beta = 120°$.

Both structures can be formed in six orientation variants out of one hcp crystal, simulating hexagonal symmetry in polydomain crystals.

On heating the low temperature orientationally ordered phase gradually disorders by the successive excitation of different rotational degrees of freedom [16-18]. In the first transition (276 K) the superstructure spots disappear; the structure becomes hcp, still with a c/a = 1.82-3. This ratio is close to the ideal ratio for a hcp stacking of spheres, corrected for the aspect ratio of the elongated C_{70} molecules. In

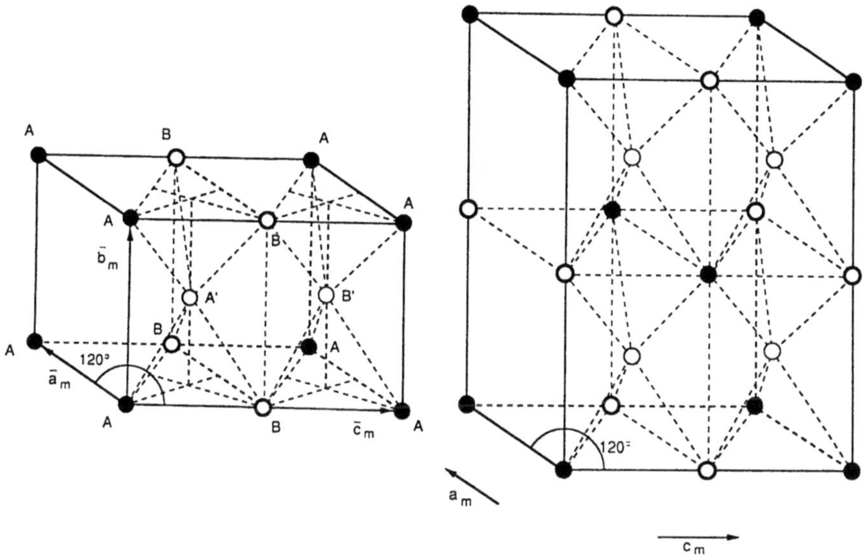

Fig.5: Crystal structures of the low temperature superstructures in C_{70}.
a) primitive monoclinic structure
b) C-centered monoclinic structure

this phase the molecules have cylindrical symmetry since they are nearly freely rotating about their long axis; they are packed with their long axis parallel to c.

At around 337 K the hcp structure acquires a smaller c/a ratio, which is still somewhat larger than, but close to the c/a ratio for spheres (1.63). This indicates that the molecules now also rotate about axes other than the long axis, leading to an effective, time averaged, shape for the molecules which approaches a spherical shell.

It is assumed that slightly above this temperature the hcp stacking changes into an ABC stacking with a rhombohedral lattice. The rhombohedral deformation suggests the persistance of a slightly preferential orientation of the molecules with their long axis along the long diagonal of the elongated rhombohedron.

At temperatures slightly above room temperature, the structure becomes FCC. The transformation rhombohedral - FCC is reversible; it can be driven in one or the other sense by changing the electron beam current. The shear transformation AB $\leftrightarrow$ ABC on the other hand does not seem to be reversible; cooling an ABC stacked crystal does not produce a hcp stacking; this may be related to the fact that the most stable rotation axes of the elongated C_{70} molecule are those perpendicular to the long molecular axis. The formation of the hcp stacking starting from an ABC stacking would require the realignment of the long axis, which seems a difficult cooperative process. The FCC - HCP transformation is furthermore hampered by the fact that the shear can with equal probability take place on four equivalent families of (111) planes, leading to a polycrystalline endproduct.

The authors would like to acknowledge the following scientists for their collaboration: S. Muto, C. Van Heurck, J. Van Landuyt of the University of Antwerp (RUCA), M.A. Verheijen, P.H.M. van Loosdrecht, G. Meijer of the University of Nijmegen.

References

[1] G. Van Tendeloo, M. Op de Beeck, S. Amelinckx, J. Bohr, W. Krätschmer, Europhys. Lett. **15** (1991), 295

[2] G. Van Tendeloo, C. Van Heurck, J. Van Landuyt, S. Amelinckx, M.A. Verheijen, P.H.M. van Loosdrecht, G. Meijer, J. Phys. Chem. **96** (1992) 7424

[3] S. Amelinckx, W. Luyten, T. Krekels, G. Van Tendeloo, J. Van Landuyt, J. Crystal Growth **121** (1992) 543

[4] S. Iijima, Nature 354 (1991) 56

[5] P.M. Ajayan, S. Iijima, Nature, **361** (1993) 333

[6] D. Ugarte, Nature **359** (1992) 707

[7] W. Krätschmer, K. Fostiropoulos, D.R. Huffman, Chem. Phys. Lett. **170** (1990) 167.

[8] R. M. Fleming, T. Siegrist, P. M. Marsh, B. Hessen, A. R. Kortan, D. W. Murphy, R. C. Haddon, R. Tycko, G. Dabbagh, A. M. Mujsce, M. L. Kaplan and S. M. Zahurac, Mat. Res. Soc. Symp. Proc. **206**, (1991), 691

[9] D. Huffman, Physics Today, november (1991), 22.

[10] P. A. Heiney , J.E. Fisher, A.R. Mc Ghie, W.J. Romanow, A.M. Denenstein, J.P. Mc Cauley, Jr., A.B. Smith, III, D.E. Cox, Phys. Rev. Lett. **66** (1991) 2911.

[11] S. Amelinckx, C. Van Heurck, D. Van Dyck, G. Van Tendeloo, Phys. Stat. Sol. (a) **131** (1992) 589.

[12] S. Muto, G. Van Tendeloo, S. Amelinckx, Phil. Mag. (1993) in the press.

[13] G. Van Tendeloo, S. Amelinckx, M.A. Verheijen, P.H.M. van Loosdrecht, G. Meijer, Phys. Rev. Lett. **69** (1992) 1065

[14] E.J.J. Groenen, O.G. Poluektov, M. Matsushita, J. Smidt, J.H. Van der Waals, G. Meijer, Chem Phys. Lett. **197**, 314 (1992)

[15] D. Lamoen, K. H. Michel, Phys. Rev. B (in the press)

[16] G.B.M. Vaughan, P.A. Heiney, J.E. Fischer, D.E. Luzzi, D.A. Ricketts-Foot, A.R. McGhie, Y.W. Hui, A.L. Smith, D.E. Cox, W.J. Romanow, B.H. Allen, N. Coustel, J.P. McCauley, Jr., A.B. Smith,III, Science, **254**, (1991) 1350

[17] G. Van Tendeloo, S. Amelinckx, J.L. De Boer, S. van Smaalen, M.A. Verheijen, G. Mijer, Europhys. Lett. **21** (1993) 329

[18] M. A. Verheijen, H. Meekes, G. Meijer, P. Bennema, J. L. De Boer, S. van Smaalen, G. Van Tendeloo, S. Amelinckx, S. Muto, J. Van Landuyt, Chem. Phys. **166** (1992) 287

Orientational Order and Electron Density of C_{60} at Room Temperature Studied by Combined Neutron and X-Ray Diffraction on Single Crystals

R.J. Papoular[1], *G. Roth*[2], *G. Heger*[1], *M. Haluška*[3], *and H. Kuzmany*[3]

[1]Laboratoire Léon Brillouin, C.E.N. Saclay, F-91191 Gif-sur-Yvette Cédex, France
[2]Kernforschungszentrum Karlsruhe, Box 3640, D-76021 Karlsruhe, Germany
[3]Institut für Festkörperphysik, Universität Wien, Strudlhofgasse 4, A-1090 Wien, Austria

Abstract. Neutron and x-ray diffraction experiments have been performed on sublimation grown high-quality C_{60} single crystals, yielding complete data sets of Bragg intensities. Structure factors were first phased by fitting a spherical model of the C_{60} molecules. High resolution 3-dim densities were then reconstructed using Maximum Entropy, yielding consistent results concerning an orientational order of C_{60} in the high temperature phase. Besides the main effect (about 90%) consisting of an isotropic distribution, we have found clear evidence for an additional modulation of both nuclear and electron densities, and consistent with results first published by Chow *et al.* We propose a simple model to describe this effect.

1 . Introduction

Since the occurence of the seminal work by Kroto *et al* [1], which paves the way for the new field of Fullerenes and Fullerites, subsequent landmarks have been set by allowing the production of these molecules in macroscopic quantities [2] or the elaboration of sizable single crystals [3].

The structural studies on pure C_{60} were first carried out on powders, using either x-rays [4] or neutrons [5-7]. For the high-temperature phase, above $T \approx 260K$, a cubic close-packed structure (space group $Fm\bar{3}m$) was determined, consisting of nearly freely rotating C_{60} molecules. This rotational disorder freezes away when the temperature is lowered, as shown by David *et al* [6,7].

More detailed structural information has recently been obtained by Chow *et al* [8] by making use of both a single crystal sample and a high-flux high-resolution x-ray synchrotron device. This pioneering work demonstrates for the first time the anisotropy of the electron density on the averaged C_{60} molecule obtained by an x-ray diffraction experiment.

The purpose of our work is three-fold: i) to confirm the anisotropy of the electron density by an independent single crystal x-ray study, ii) to look for this anisotropy using yet another crystal and neutrons, which are sensitive to the nuclear density rather than the electron density and iii) to propose a simple model to describe our results.

Springer Series in Solid-State Sciences, Volume 117
Electronic Properties of Fullerenes Eds.: H. Kuzmany, J. Fink, M. Mehring, and S. Roth

2. Experimental

The two crystals used in our room temperature study and mentioned hereafter have been sublimation grown using the procedure described in [3].
A diffraction experiment yields the moduli of the Fourier components of the electron density in the case of x-rays, and of the nuclear density in the case of neutrons. Physically, these two densities are not identical and a joint x-n study is required to distinguish between chemical bonding effects and atomic distributions related to preferred C_{60} orientations. Practically, the electronic cloud extending around each atom results in a higher spatial resolution being achievable for the atomic density when neutron diffraction data are used.

2.1 The X-ray Diffraction Experiment

Our C_{60} single crystal was isometrical, of diameter 0.2 mm, and showed no detectable twins. The cubic lattice parameter a was refined to 14.152(2) Å.

A conventional 4-circle STOE diffractometer, equipped with a graphite (002) monochromator, was operated at a wavelength $\lambda_{MoK\alpha} = 0.71073$ Å.

Using $\omega - 2\theta$ scans, 1894 Bragg intensities were collected, yielding a complete data set up to $\frac{\sin\theta}{\lambda} = 0.65$ Å^{-1}, resulting in 205 unique reflections, with an internal agreement factor $R_{int} = 0.039$. No absorption correction was applied (insignificant $\mu R < 0.005$).

2.2 The Neutron Diffraction Experiment

The rather large single crystal, of about 3 mm^3 in volume, had natural (100) and (111) faces. Absorption is also negligible ($\mu R < 0.0001$) in neutron diffraction.

Two sizable twins could be detected, originating from stacking faults of the cubic close-packed structure along the [111] directions, along with a main domain orientation of our sample, corresponding to volumic contributions of 9.3%, 2.3% and 88.4%, respectively. The experiments were performed in a way to measure only those "pure" Bragg reflections of the main domain orientation which are unaffected by the contributions from the two small twins.

We have used the 5C2 4-circle diffractometer, located at the hot source of the ORPHEE reactor. A neutron wavelength of $\lambda = 0.8308(2)$ Å was selected by a Cu(200) vertically focusing monochromator, and a 0.25 mm thick Erbium filter reduced the $\frac{\lambda}{2}$ contamination to be less than 0.1%.

We have measured 210 non twin-affected Bragg intensities using ω scans up to $\frac{\sin\theta}{\lambda} = 0.51$ Å^{-1}, resulting in a data set of 102 unique reflections. The good agreement between symmetry related reflections ($R_{int} = 0.022$) shows that extinction effects are unimportant.

3. Data Analysis

We have used two distinct procedures, leading to the same qualitative picture: an anisotropic atomic density for the C_{60} molecules.

Firstly, following Chow *et al*, the sought nuclear and electron densities were determined by fitting the measured intensities using symmetry adapted spherical (cubic) harmonics [9]. In this way, a radial averaged density is obtained. Our results, to be detailed elsewhere [10], are in excellent agreement with those in [8]. In particular, our modelled electron density is undistinguishable from Chow's.

Secondly, a preliminary phasing of the structure factors was carried out using simple spherical shell models involving either a gausssian thickness in direct space or an isotropic Debye-Waller factor in reciprocal space. Due to the centrosymmetry of the crystal structure, the phases are merely 0 or π, and were found to be essentially independent of the model used. These signs were subsequently coupled to the measured moduli and full-fledged 3-dim electron as well as nuclear densities reconstructed by the Maximum Entropy Method (MaxEnt) [11,12] using the procedure described in [13]. MaxEnt is well known for doing away with truncation and noise effects. Technically, the latter procedure involved a 64*64*64 discretization of the unit cell and the 3-dim densities were retrieved as sets of 64 sections at constant z, including a section at z=0.

In order to compare with the results by Chow *et al*, the densities were subsequently interpolated on spheres of selected radii. The density on a sphere for a given radius was mapped by a one degree equistepped longitude-latitude gridding in such a way that the pairs (0°, 0°), (0°, 90°), (x°, 90°) and (45°, 35.26°) correspond to the [100], [010], [001] and [111] directions respectively. The obtained density was further smoothed in order to eliminate spurious effects linked to the 64*64*64 pixelization of the unit cell, which we had at first erroneously identified with Carbon atoms. Different shells with increasing radii were finally added up. The resulting averaged densities are in excellent agreement with the previously mentioned results.

4. Our Main Results

Diffraction analyses on sublimation grown single crystals demonstrate that the averaged C_{60} molecules seen by either x-rays or neutrons form very nearly isotropic empty shells, with no evidence for extraneous density either inside or between them. The (001) sections at z=0 from both analyses are displayed in Figure 1. It is clearly seen that the conventional Fourier synthesis does not rule out the occurence of some density at the center of the molecules. MaxEnt, used in this context for the first time, proves this possible endohedral contribution to be not physically significant and due solely to truncation effects (Fig. 1a).

MaxEnt model-free reconstructions confirm (x-rays) and establish (neutrons) the anisotropy of the radial averaged atomic density of the C_{60} molecules in the high temperature phase. Figure 2, where only one octant of molecule is represented, evidences that the anisotropy of the nuclear density exists and is very similar to that of the electron density (Compare with Fig. 1a from [8]).

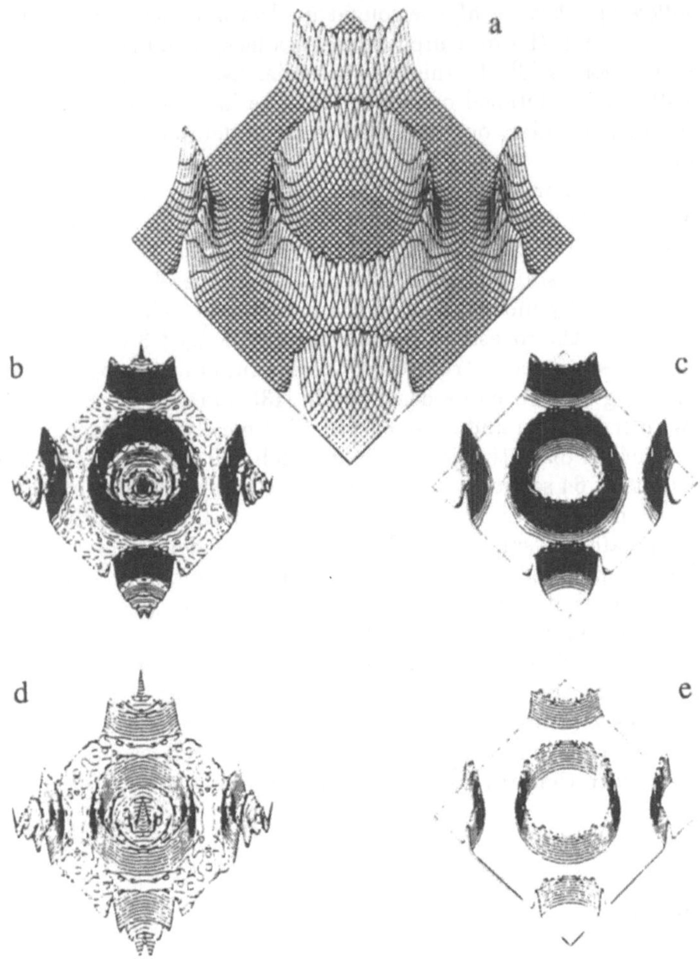

Fig. 1: 3-dim densities of (001) sections (z=0) for C_{60} at room temperature. a) MaxEnt eliminates both noise and truncature effect present in the conventional Fourier synthesis: there are no density contributions inside or between the C_{60} molecules.
In b) and c): Conventional synthesis versus MaxEnt reconstructions from x-ray data. d) and e) are analogous to b) and c) for the neutron case.

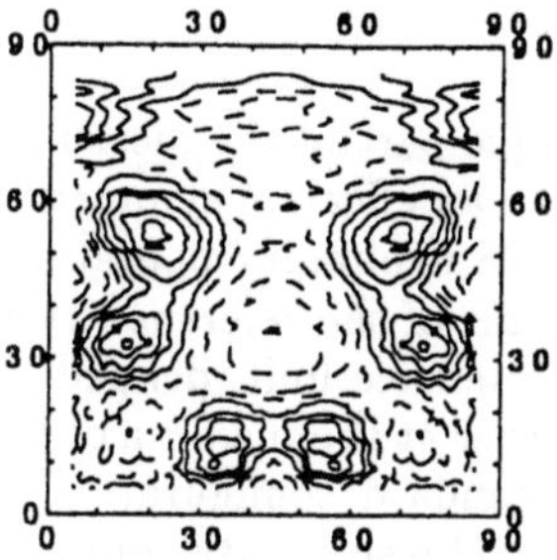

Fig. 2: Anisotropic part of the radial averaged nuclear density for one octant of a C_{60} molecule from MaxEnt and neutron data.
The solid and dotted lines relate to maxima and minima of the density.

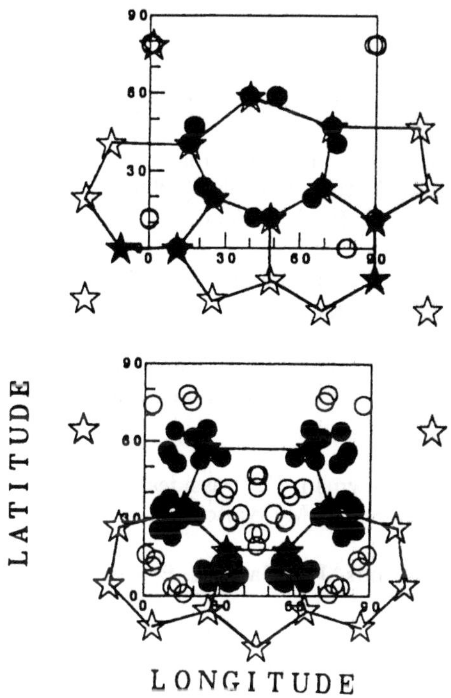

Fig. 3: Modelling of the anisotropic part of the density of C_{60} by symmetry adapted molecular disorder.
Above: Resulting from the standard configuration, where the crystal axes coincide with three 2-fold axes of one C_{60} molecule.
Below: Our model. Hexagons and pentagones are preferentially oriented about 3-fold and 2-fold axes of the crystal, in order to reproduce the observed density distributions.

5. A Simple Model

The simplest possible way to account for the anisotropic part of the measured densities is to introduce a rigid C_{60} unit and generate an effective symmetry-adapted molecule from it, obtained by applying the 48 symmetry operations of the crystal $m\bar{3}m$ point group to the 60 carbon atoms of the reference unit. The end game is to find the best orientation of this reference unit, defined by its three Euler angles, so that the generated atoms gather at the maxima of the observed densities. Figure 3 demonstrate that the standard configuration as defined by Michel [9] fails to reproduce the experimental results. By contrast, our model which keeps hexagons centered about the [111] direction, but also forces pentagons about the [110] direction, yields a much better agreement.

References

[1] H.W. Kroto, J.R. Heath, S.C. O'Brien, R.F. Curl, R.E. Smalley: Nature **318**, 162 (1985)

[2] W. Krätschmer, L.D. Lamb, K. Festiropolous, D.R. Huffmann: Chem. Phys. Lett. **170**, 107 (1990); Nature **347**, 354 (1990)

[3] M. Haluška, H. Kuzmany, M. Vybornov, P. Rogl, P. Fejdi: Appl. Phys. A **56**, 161 (1993)

[4] P.A. Heiney, J.E. Fischer, A.R. McGhie, W.J. Romanow, A.M. Denenstein, J.P. McCauley Jr., A.B. Smith III, D.E. Cox: Phys. Rev. Lett. **66**, 2911 (1991)

[5] J.R.D. Copley, D.A. Neumann, R.L. Cappelletti, W.A. Kamitakahara, E. Prince, N. Coustel, J.P. McCauley Jr., N.C. Maliszewskyj, J.E. Fischer, A.B. Smith III, K.M. Creegan, D.M. Cox: Physica B **180 & 181**, 706 (1992)

[6] W.I.F. David, R.M. Ibberson, J.C. Matthewman, K. Prassides, T.J.S. Dennis, J.P. Hare, H.W. Kroto, R. Taylor, D.R.M. Walton: Nature **353**, 147 (1991)

[7] W.I.F. David, R.M. Ibberson, T.J.S. Dennis, J.P. Hare, K. Prassides: Europhys. Lett. **18**, 219 (1992)

[8] P.C. Chow, X. Jiang, G. Reiter, P. Wochner, S.C. Moss, J.D. Axe, J.C. Hanson, R.K. McMullan, R.L. Meng, C.W. Chu: Phys. Rev. Lett. **69**, 2943 (1992)

[9] K.H. Michel, J.R.D. Copley, D.A. Neumann: Phys. Rev. Lett. **68**, 2929 (1992)

[10] R.J. Papoular, G. Roth, G. Heger, P. Schiebel, W. Prandl, M. Haluška, H. Kuzmany: *In preparation*

[11] J. Skilling, R.K. Bryan: Mon. Not. R. astr. Soc. **211**, 111 (1984)

[12] Program system MEMSYS, by S.F. Gull and J. Skilling: Maximum Entropy Data Consultants Ltd., Meldreth, Royston SG8 6NR, England

[13] R.J. Papoular, W. Prandl, P. Schiebel: in *Maximum Entropy and Bayesian Methods, Seattle, 1991*, ed. by C.R. Smith, G.J. Erickson, P.O. Neudorfer, (Kluwer Academic Publishers, 1992) p. 359

Orientational Glass Transition in Solid C_{60}

C. Meingast[1], F. Gugenberger[1], R. Heid[1], P. Adelmann[1], M. Braun[1], H. Wühl[1], M. Haluška[2], and H. Kuzmany[2]

[1]Institut für Nukleare Festkörperphysik, Institut für Technische Physik, D-76021 Karlsruhe, Germany
[2]Institut für Festkörperphysik, Universität Wien, Strudlhofgasse 4, A-1090 Wien, Austria

Abstract. The glass transition in both single- and polycrystalline C_{60} at about 90 K was studied using high-resolution capacitance dilatometry. This glass transition is most pronounced in samples which also have a sharp orientational-ordering transition at T≈260 K. The kinetics of the relaxation processes near the glass transition were obtained by fitting the experimental results with a simple glass-transition model. The relaxation time obeys an Arrhenius-behavior with $E_a \approx 288$ meV over more than 10 orders of magnitude when including other published data.

1. Introduction

Solid C_{60} is a very interesting example of a molecular plastic crystal. At room temperature the nearly spherical C_{60} molecules rotate almost freely and form a close-packed fcc lattice [1,2]. At lower temperatures, the non-spherical features of the molecules start to play a role and an orientationally ordered phase is formed at a first-order phase transition at 260 K [2,3]. There is a further structural anomaly at T≈90 K, which has been shown by a number of studies to result from a glass transition due to the freezing-in of thermally induced orientational disorder [3-6]. Relaxation studies at higher temperatures (shorter relaxation times) confirm this picture [7-9]. This glass transition is very pronounced in the thermal expansion and we have previously examined it in detail on a high-quality single crystal [6]. In this paper we present some additional high-resolution dilatometry measurements of several different quality solid C_{60} samples between 5 and 300 K. We also compare the relaxation times of the molecular reorientations from the different experiments.

2. Experimental

Pressed pellets (6 mm diameter) were made from C_{60} powder, which was separated from carbon soot by column chromatography. The C_{60} single crystal ($\approx 0.6 \times 1 \times 1$ mm^3) was grown by a sublimation technique [10]. The shiny faces of the crystal indicated a high degree of crystallinity. The linear thermal expansion was measured using a high-resolution capacitance dilatometer [11], in which the temperature is varied continuously at constant rates between ±0.15 and ±20 mK/s.

Springer Series in Solid-State Sciences, Volume 117
Electronic Properties of Fullerenes Eds.: H. Kuzmany, J. Fink, M. Mehring, and S. Roth

The relative length resolution varied between $\Delta L/L \approx 5x10^{-8}$-$5x10^{-9}$, depending on the length of the sample.

3. Results and Discussion

In Fig.1 the temperature dependence of the thermal expansivity $\alpha(T) = \partial \ln L / \partial T$ is shown for two different polycrystalline samples and for the single crystal. The orientational-ordering transition near 250 K is clearly visible for all samples. This transition is broader and is shifted to lower temperatures for the polycrystalline samples than for the single crystal. This is presumably due to residual solvents in the powdered, non-sublimated samples, although it is not apparent why the transition is shifted to lower temperatures. A close up of the expansivity near the glass transition is shown in the inset of Fig. 1 for all three samples upon heating. The size of the glass-transition anomaly correlates well with the sharpness of the '250 K' anomaly; i.e. a sharp upper transition also has the largest anomaly at 90 K. This correlation suggests that the glass transition is an intrinsic phenomena since a sharp transition at 260 K indicates a high-quality and pure sample. The same conclusion was reached by the interpretation of the specific heat data at the glass transition [5,6].

In Fig.2a the single-crystal expansivity data taken at six different cooling (heating) rates where in each case $|q_{heating}| = |q_{cooling}|$ is shown. The anomalies in both heating and cooling curves are shifted to lower temperatures with decreasing rate, as expected for a glass-transition. We have previously modeled this transition with a very simple 'glass-transition model', from which it is then possible to obtain the relaxation times as a function of temperature [6]. The simulation of the model is shown in Fig. 2b and one can see that both the shape of the hysteresis and the shifting of the transition for different q's is reproduced quite nicely by the model.

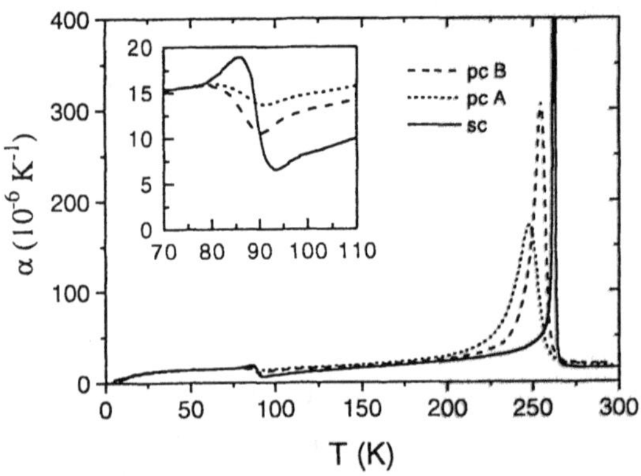

Fig.1 Temperature dependence of the thermal expansivity α(T) for three different quality single- (sc) and polycrystalline (pc) C_{60} samples. The inset shows the details at the glass transition. All curves are heating curves.

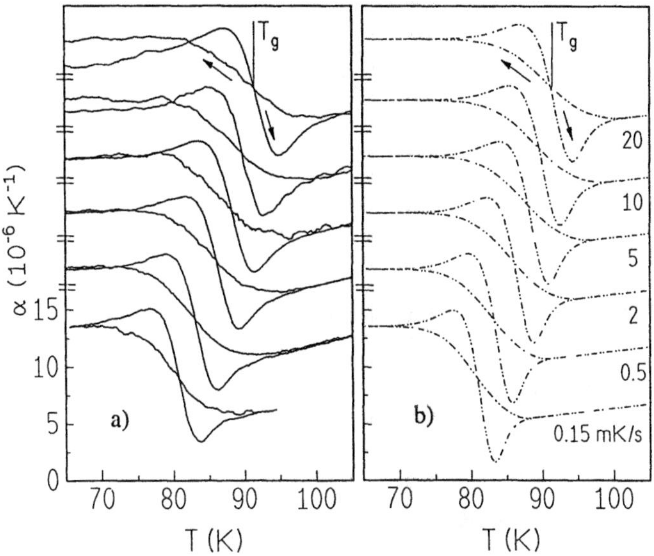

Fig.2 a) Experimental and b) modeled thermal expansivity vs. temperature at the glass transition. The data were taken at six different heating (cooling) rates q. The curves for q≥0.15 mK/s are vertically offset for clarity. The simple glass-transition model reproduces both the shape of the hysteresis and the shifting of the transition for different q's quite well (see Ref. 6 for details of the model).

In this model the relaxation time, t_r, was assumed to follow an Arrhenius temperature dependence, i.e.

$$t_r^{-1}(T) = \nu \cdot e^{-E_a/k_B T} \tag{1}$$

where E_a=288±5 meV and $\nu=10^{13}$-10^{14} s^{-1} are parameters which were determined by fitting the model to the experiment [6]. In Fig. 3 the relaxation times determined from the expansivity measurements are shown in comparison to other published data. The dashed line is an extrapolation of our data to higher temperatures (shorter t_r's) using equation 1. We note that the slope of this extrapolation line is determined quite accurately due to our small error in E_a and that the extrapolation agrees well with the data from sound velocity [7], dielectric [8] and elastic constant [9] measurements at higher temperatures. This demonstrates that the relaxation time can be approximately described by an Arrhenius behavior over a range of over 10 orders of magnitude. There is however considerable scatter in the absolute relaxation times between the different experiments (which may be due to the different physical quantaties being measured, sample differences or experimental uncertainties), and it is therefore not clear whether the behavior is strictly Arrhenius or whether a Vogel-Fulcher description with a slight curvature as seen in 'strong' glasses [12] provides the best description

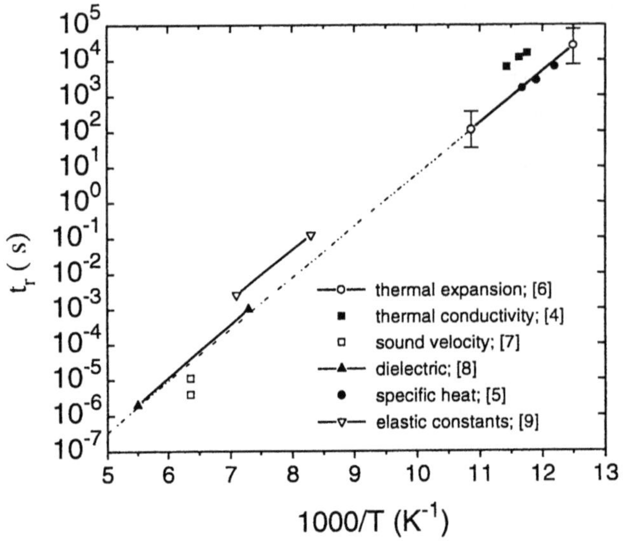

Fig.3 Arrhenius plot of the reorientational relaxation time, t_r, for all published experiments. The dashed line is an extrapolation of t_r using eq. 1 and the values of E_a and ν determined from fitting the glass-transition model to the thermal expansion data. t_r appears to obey an activated behavior over more than 10 decades.

of the relaxation in C_{60}. Further experiments on well defined samples over a wide temperature and frequency range would be useful.

4.Conclusions

The comparison of the thermal expansion behavior of different quality C_{60} samples shows that the glass transition near 90 K is an intrinsic phenomena. Further, the reorientational relaxation process associated with the glassy behavior appears to be quite simple as suggested by the the Arrhenius behavior of the relaxation time and the fact that only a single activation energy is needed to describe the data [6,9]. The microscopic interpretation of David et al. [3], in which two nearly degenerate orientations of the C_{60} molecules occur, provides a simple picture for understanding this glassy behavior. It is however suprising that this 'single-particle' picture describes the data as well as it does, because the temperature dependent degree of misorientation of neighboring molecules [3], which is expected to affect both the difference in energy of orientated and misorientated states and the rotational barrier height, is totally neglected.

References

[1] R.Tycko et al., Phys.Rev.Lett. **67**, 1886 (1991).
[2] P.A.Heiney et al., Phys.Rev.Lett. **66**, 2911 (1991).
[3] W.I.F.David et al., Europhys.Lett. **18**, 219 (1992); Europhys.Lett **18**, 735 (1992).
[4] R.C.Yu et al., Phys.Rev.Lett. **68**, 2050 (1992).
[5] T. Matsuo et al., Solid State Commun. 83, 711 (1992).
[6] F. Gugenberger et al., Phys. Rev. Lett. **69**, 3774 (1992).
[7] X.D.Shi et al., Phys.Rev.Lett. **68**, 827 (1992).
[8] G.B.Alers et al., Science **257**, 511 (1992)
[9] W. Schranz et al. (these proceedings).
[10] M. Haluška et al. Appl. Phys. A**56**, 161 (1993).
[11] C.Meingast et al., Phys.Rev. B **41**, 11299 (1990).
[12] C.A.Angell, J.Non-Cryst.Solids **131-133**, 13 (1991).

Orientational Ordering and Molecular Structure in Solid C_{60}

D. Lamoen and K.H. Michel

Department of Physics, Universiteit Antwerpen (UIA),
B-2610 Wilrijk, Belgium

Abstract. The crystal field in the disordered phase of C_{60}-fullerite is derived from an intermolecular potential model, which takes into account the geometric difference between double bonds and single bonds. The angular dependence of the crystal field, resulting from an anticlockwise rotation of the molecule around the [111] axis, exhibits minima at angles of 98° and 38°. The discontinuity of the order parameter at the first order phase transition is calculated.

1. Introduction

C_{60}-fullerite is a molecular crystal [1]. At room temperature the space group is Fm3m [2], the molecules are orientationally disordered [3,4]. At a transition temperature $T_1 \approx 250$ K, the crystal undergoes a phase change [5,6] to a $Pa\bar{3}$ structure [7,8,9]. The molecules are orientationally ordered on four different sublattices. Neutron powder diffraction studies [8] have revealed the packing configuration of the C_{60} molecules. Electron rich double bonds which fuse the hexagons on the C_{60} molecule [10] face the electron-poor pentagon centers of adjacent C_{60} units [8]. This idea has been implemented in molecular dynamics calculations [11].

A theoretical description of the orientationally disordered phase and the phase transition has been formulated with the use of symmetry adapted multipolar rotator functions [12]. Originally the theory was presented for a model of interacting molecules on a rigid lattice, where the intermolecular potential was a sum of atom-atom Lennard-Jones (L-J) interactions. Subsequently the theory was generalized to a more complex model of the molecular structure, where atoms, centers of double bonds and centers of single bonds contribute to the intermolecular potential [13].

In the disordered phase the molecules are not free rotators, but experience a crystal field [12]. The crystal field, which is of cubic symmetry, depends sensitively on the model of molecular structure. The crystal field determines the orientational probability distribution function and the sign of the orientational order parameter. We will calculate the order parameter at the transition.

Springer Series in Solid-State Sciences, Volume 117
Electronic Properties of Fullerenes Eds.: H. Kuzmany, J. Fink, M. Mehring, and S. Roth

2. Molecular Interactions

We consider N C_{60} molecules on a rigid fcc lattice. Each molecule is described as a rigid structure built up from interaction centers, which are located at the vertices and edges (bonds) of the truncated icosahedron [10]. The intermolecular potential V is a sum of two body interactions between centers belonging to different molecules. It is expanded in terms of rotator functions [14-17]

$$U_l^{\tau}(\omega) = \sum_{n,m} \alpha_{l(I)}^{nl} D_l^{nm}(\omega)\, \alpha_{l(O_h)}^{m\tau}. \tag{1}$$

Here D_l^{nm} are the Wigner matrices. They depend on the Euler angles ω which account for the orientation of the molecule, l is the angular momentum quantum number. The coefficients $\alpha_{l(I)}^{nl}$ account for the icosahedral symmetry of the molecule, thereby l = 0, 6, 12 ... The coefficients $\alpha_{l(O_h)}^{m\tau}$ account for the symmetry of the site. The index τ labels the irreducible representations of the cubic group O_h and their components.

The orientation interaction potential V separates in two contributions

$$V = V^{RR} + V_{CF} \tag{2}$$

Here V^{RR} is the rotation-rotation interaction potential [12] summed over lattice sites $\vec{n}$, $\vec{n}'$:

$$V^{RR} = \frac{1}{2} \sum_{\vec{n}\vec{n}'} \sum_{jj'} J^{j'j}(\vec{n}' - \vec{n})\, U_{10}^{j'}(\vec{n}')\, U_{10}^{j}(\vec{n}). \tag{3}$$

Here $U_{10}^{j}(\vec{n})$, j = 1, 2, 3 are the rotator functions of T_{2g} symmetry and belonging to the manifold l = 10, ϱ = 3. Our classification of representations is the same as in Ref. [18]. The T_{2g} modes with l = 10, ϱ = 3 are dominant in driving the phase transition from Fm3m to $Pa\bar{3}$ Ref. [12]. The interaction matrix J is given by

$$J^{j'j}(\vec{n}'-\vec{n}) = \sum_{P} \sum_{A'A} v_{1010}^{(P)} \begin{bmatrix} A' & A \\ j' & j \\ \vec{n}' & \vec{n} \end{bmatrix} g_{10}^{(A')} g_{10}^{(A)} \mu_A \mu_{A'}. \tag{4}$$

Here the index A = (a, b or s) labels the type of interaction center, with a for atoms, b for double bonds and s for single bond centers. The quantities $g_l^{(A)}$ are the molecular form factors [17]. The numbers μ_A account for the distribution of interaction centers, where μ_a = 1, μ_s = 1 while μ_b is the number of interaction centers along a double bond. While originally [11,13] double bonds were discribed by one single interaction center, we will consider a more general model, where double bonds are described by a distribution of interaction centers (μ_b = 2 or 3) along the bond [19]. Such an interaction model takes into account the spatial extension of the electrons along the double bonds. The matrix elements in (4) are calculated numerically for the case of Lennard-Jones (P = L-J) and Coulomb (P = C) potentials [11,13].

The second term on the right hand side of Eq. (2) stands for the crystal field. This is the potential of a central molecule at any site $\vec{n}$, if the surrounding molecules are taken in spherical approximation. The crystal field is written as

$$V_{CF} = \sum_{\vec{n}} \sum_{l} \sum_{\rho} w_l^{(\rho)} U_l^{A_{1g},\rho}(\vec{n}). \tag{5}$$

The rotator functions have full cubic symmetry (A_{1g}). For the cases l = 6, 10, one has ϱ = 1, while for l = 12, one has ϱ = 1,2. The coefficients are given by

$$w_l^{(\rho)} = 12 \sum_{AA'} v_{l\,0}^{(L-J)} \begin{pmatrix} A'A \\ A_{1g},\rho \end{pmatrix} g_l^{(A')} g_0^{(A)} \mu_A \mu_{A'} \, \mathrm{sgn}(\xi_l^{A'}), \tag{6}$$

where $v_{l\,0}^{(L-J)}$ are calculated numerically [12,13], and where $\mathrm{sgn}(\xi_l^{A'}) = -1$ for l = 6 and 12 and for A' = b, and +1 otherwise (we will restrict us here tot $l \leq 12$). Due to the charge neutrality of the molecule, there are no Coulomb contributions. Restricting ourselves to nearest neighbour interactions, we find that $v_{l\,0}$ is always negative. Due to the sgn factor in (6), there is a subtle interplay between different types of interaction centers in determining the sign of $w_l^{(\varrho)}$ for l = 6 and l = 12 [20].

3. Orientational Distribution

The orientational density distribution in the disordered phase has cubic symmetry and is expanded as [15]

$$f(\Omega) = \sum_{l} \sum_{\rho} \gamma_l^{(\rho)} S^{A_{1g},\rho}(\Omega). \tag{7}$$

Here $\Omega = (\theta,\phi)$ are the polar angles of the direction of observation. The coefficients

$$\gamma_l^{(\rho)} = g_l^{(a)} \left\langle U_l^{A_{1g},\rho} \right\rangle_{CF} \tag{8}$$

are thermal averages over the molecular orientation calculated with the crystal field Hamiltonian V_{CF} [20]. On the other hand the coefficients $\gamma_l^{(\varrho)}$ can be determined experimentally from diffraction experiments by using synchroton X-rays [21] or neutrons [22]. The numerical coefficients $C_{l\,\varrho}$ of [21] correspond to $\gamma_l^{(\varrho)}(10^3\sqrt{4\pi}/16)$. The values of $\gamma_l^{(\varrho)}$ obtained from [21] are quoted in Table 1, second row.

We have calculated from a theoretical model with 210 interaction centers per molecule (60 atoms, 60 single bonds, 3 centers along 30 double bonds) the crystal field coefficients w_l. The results are quoted in Table 2.

Table 1

γ_6	γ_{10}	$\gamma_{12}^{(1)}$	$\gamma_{12}^{(2)}$
- 0,389	0.216	0.159	0.440
-0.188	0.864	0.0035	0.309

Table 2

w_6	w_{10}	$w_{12}^{(1)}$	$w_{12}^{(2)}$
198.4	- 206.9	- 46.5	- 230.4

The corresponding values for $\gamma_l^{(\varrho)}$ are found in Table 1, third row. The signs agree with those of the experimental coefficients. Our calculations show that it is essential to consider a distribution of interaction centers along the double bonds.

Given the theoretical expression of the crystal field, we can study its angular dependence as a function of the orientation of the central molecule. We consider the molecule in its standard position [8] and rotate it counterclockwise by an angle Ψ around the [111] direction. The corresponding variation of V_{CF} (per molecule) is shown in Fig. 1, for the crystal field coefficients of Table 2.

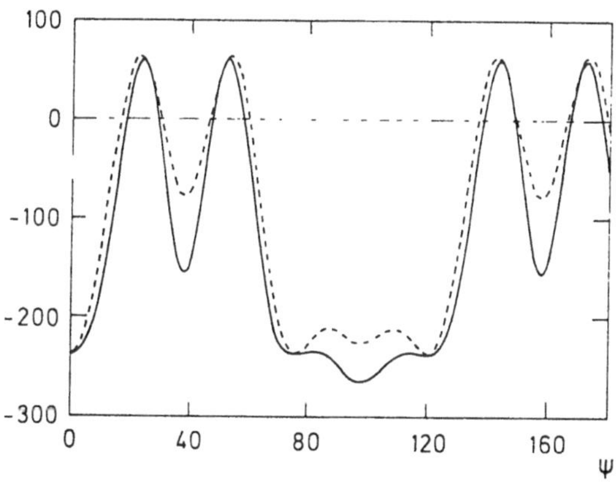

Angular dependence of crystal field V_{CF} for rigid lattice (dotted line). Same for deformable lattice (full line). V_{CF} in units Kelvin.

We observe the appearance of a broad minimum with triple well potential centered at $\Psi = 98°$ and a secondary minimum at 38°. Previously it has been shown by neutron scattering that the most probable molecular orientation at low T in the ordered phase correspond to $\Psi = 98°$. A secondary favorable occupation of the molecule has been found at $\Psi = 38°$ in the low T phase [23,24]. This orientation has been related to the existence of an orientational glass below T = 90K [23]. Our results demonstrate that the remarkable angular dependence is already present in the crystal field potential of the disordered phase.

4. Orientational Order Parameter

The phase transition from the orientationally disordered phase Fm3m to the orientationally ordered phase $\mathbf{Pa\bar{3}}$ is driven by the condensation of the orientational modes U_{10}^{j} at the X point of the Brillouin zone. The condensation scheme reads [12]

$$\begin{aligned} \mathrm{Fm3m} : [U_{10}^{3e}(\vec{k}_x^X) &= U_{10}^{2e}(\vec{k}_z^X) = U_{10}^{1e}(\vec{k}_y^X) = \sqrt{N}\eta \neq 0; \\ U_{10}^{2e}(\vec{k}_x^X) &= U_{10}^{3e}(\vec{k}_y^X) = U_{10}^{1e}(\vec{k}_z^X) = 0] \rightarrow \mathrm{Pa3}. \end{aligned} \tag{9}$$

Here the index e indicates that $U_{10}^{i\,e}$ is a thermal expectation value, η is the amplitude of the order parameter. The wave vectors $\vec{k}_\alpha$, α = x, y, z are defined by $\vec{k}_x = (2\pi/a)(1,0,0)$ etc.

In order to calculate η, we start from the Landau free energy per molecule

$$(F/N) = f_0 + A\eta^2 + B\eta^3 + C\eta^4. \tag{10}$$

Here f_0 is the single molecule free energy, the coefficients A, B and C have been derived from microscopic theory [25]. One has

$$A = \frac{3}{2}\left[\frac{T}{x^{(2)}} - J\right]. \tag{11}$$

$$B = -T(x^{(2)})^{-3}x_{123}^{(3)}. \tag{12}$$

$$C = \frac{T}{8(x^{(2)})^4}\left[9(x^{(2)})^2 - s_{1111}^{(4)} - 6x_{1122}^{(4)}\right]. \tag{13}$$

Here the quantities $x^{(n)}$ are crystal field thermal averages of products of rotator functions

$$\begin{aligned} x^{(2)} &= \left\langle (U_{10}^1)^2 \right\rangle_{CF} \; ; \; x^{(3)} = \left\langle U_{10}^1 U_{10}^2 U_{10}^3 \right\rangle_{CF}, \\ x_{1111}^{(4)} &= \left\langle (U_{10}^1)^4 \right\rangle_{CF} \; ; \; x_{1122}^{(4)} = \left\langle (U_{10}^1)^2 (U_{10}^2)^2 \right\rangle_{CF}. \end{aligned} \tag{14}$$

Notice that $x^{(3)} \neq 0$ follows from the fact that the product of three T_{2g} components is a cubic invariant. Therefore B $\neq$ in Eq. (10) and the phase transition is of first order. Previously this conclusion has been drawn from an analysis of low T structural data [9]. Experimentally one finds that the phase transition is first order [23,26]. The transition temperature is given by

$$T_1 = x^{(2)} \left[J + \frac{B^2}{6C} \right] . \tag{15}$$

With the model of Ref. [19], we obtain J = 4973 K. The quantities $x^{(n)}$ are evaluated numerically with the crystal field coefficients of the same model (Table 1). We find T_1 = 221 K. The discontinuity of the order parameter at the transition is given by

$$\eta_1 = -\frac{B}{2C}. \tag{16}$$

Numerically we find $\eta_1 = -0.021$. The negative value of η_1 follows from the fact that, with the crystal field coefficients of Table 2, $x^{(3)} < 0$. The sign of η is in agreement with an analysis of neutron powder diffraction data [27].

5. Conclusions

Starting from a microscopic model of molecular interactions, we have derived the crystal field in the orientationally disordered phase and studied its dependence on molecular orientations. Rotating the molecule anticlockwise around the [111] axis, the crystal field exhibits minima at angles of 98° and 38°. For the same model we have calculated the discontinuity of the order parameter amplitude at the first order phase transition. We find $\eta_1 < 0$. In order to obtain these results it was essential to consider a model of the molecular structure where the double bonds are described by a distribution of interaction centers along the bond.

References

1. W. Krätschmer, L.D. Lamb, K. Fostiropoulos and D.R. Huffman, Nature 347, 354 (1990).
2. R.M. Fleming, T. Siegrist, P.M. Marsh, B. Hessen, A.R. Kortan, D.W. Murphy, R.C. Haddon, R. Tycko, G. Dabbagh, A.M. Mujsce, M.L. Kaplan and S.M. Zahurak, Mat. Res. Soc. Symp. Proc. 206, 691 (1991).
3. C.S. Yannoni, R.D. Johnson, G. Meijer, D.S. Bethune and J.R. Salem, J. Phys, Chem. 95, 9, (1991); R. Tycko, R.C. Haddon, G. Dabbagh, S.H. Glarum, D.C. Douglass and A.M. Mujsce, J. Phys, Chem. 95, 518 (1991).
4. D.A. Neumann, J.R.D. Copley, R.L. Cappelletti, W.A. Kamitakahara, R.M. Lindstrom, K.M. Creegan, D.M. Cox, W.J. Romanov, N. Coustel, J.P. McCauley, Jr., N.C. Maliszewskyj, J.E. Fisher and A.B. Smith III, Phys. Rev. Lett. 67, 3808 (1991).

5. A. Dworkin, H. Szwarc, S. Leach, J.P. Hare, T.J.S. Dennis, H.W. Kroto, R. Taylor and D.R.M. Walton, C.R. Acad. Sc. Paris, Serie II, 213, 979 (1991).
6. P.A. Heiney, J.E. Fischer, A.R. McGhie, W.J. Romanov, A.M. Denenstein, J.P. McCauley, Jr., A.B. Smith III and D.E. Cox, Phys, Rev. Lett. 66, 2911 (1991).
7. R. Sachidanandam and A.B. Harris, Phys. Rev. Lett. (Comments) 67, 1467 (1991); P.A. Heiney, J.E. Fischer, A.R. McGhie, W.J. Romanov, A.M. Denenstein, J.P. McCauley, Jr., A.B. Smith III, and D.E. Cox, Phys. Rev. Lett. (Reply) 67, 1468 (1991).
8. W.I.F. David, R.M. Ibberson, J.C. Matthewman, K. Prassides, T.J.S. Dennis, J.P. Hare, H.W. Kroto, R.Taylor and D.R.M. Walton, Nature 353, 147 (1991).
9. A.B. Harris and R. Sachidanandam, Phys. Rev. B 46, 4944 (1992).
10. H.W. Kroto, J.R. Heath, S.C. O'Brien, R.F. Curl and R.E. Smalley, Nature 318, 162 (1985).
11. M. Sprik, A. Cheng and M.L. Klein, J. Phys, Chem. 96, 2027 (1992).
12. K.H. Michel, J.R.D. Copley and D.A. Neumann, Phys. Rev. Lett. 68, 2929 (1992).
13. K.H. Michel, J. Chem. Phys. 97, 5155 (1992).
14. H.M. James and T.A. Keenan, J. Chem. Phys. 31, 12 (1959).
15. W. Press and A. Hüller, Acta Cryst. A 29, 252 (1973).
16. M. Yvinec and R.M. Pick, J. Physique 41, 1045 (1980).
17. K.H. Michel and K. Parlinski, Phys. Rev. B 31, 1823 (1985).
18. C.J. Bradley and A.P. Cracknell, The Mathematical Theory of Symmetry in Solids, (Clarendon Press, Oxford, 1972).
19. D. Lamoen and K.H. Michel, Z. Physik B, in press.
20. J.R.D. Copley and K.H. Michel, J. Phys. C, in press.
21. P.C. Chow, X. Jiang, G. Reiter, P. Wochner, S.C. Moss, J.D. Axe, J.C. Hanson, R.K. McMullan, R.L. Meng and C.W. Chu, Phys. Rev. Lett. 69, 2943 (1992).
22. W.I.F. David, R.M. Ibberson and T. Matsuo, Proc. Roy. Soc. A (London), in press.
23. W.I.F. David, R.M. Ibberson, T.J.S. Dennis, J.P. Hare and K. Prassides, Europhysics Lett. 18 , 219 (1992).
24. Ruizhong Hu, T. Egami, Fang Li and J.S. Lannin, Phys. Rev, B 45, Rapid Commun. 9517 (1992).
25. K.H. Michel, Z. Physik B-Cond. Matter 88, 71 (1992).
26. P.A. Heiney, G.B.M. Vaughan, J.E. Fisher, N. Coustel, D.E. Cox, J.R.D. Copley, D.A. Neumann, W.A. Kamitakahara, K.M. Creegan, D.M. Cox, J.P. McCauley Jr. and A.B. Smith III, Phys. Rev. B 45, Rapid Commun., 4544 (1992).
27. W.I.F. David, private communication.

Molecular Dynamics Calculations on C_{60}-Based Crystals

G. Bortel, G. Faigel, G. Oszlányi, S. Pekker, and M. Tegze
Research Institute for Solid State Physics, P.O.B. 49,
H-1525 Budapest, Hungary

Abstract. Molecular–dynamics calulations have been carried out to investigate the phase transitions of pure C_{60} crystals in the solid phase. We employed a constant pressure method which allows changing of the volume and the shape of the unit cell. The interaction between molecules is based on a pairwise, additive, atom–atom Lennard–Jones potential.

The calculations yield a stable fcc structure above transition temperature (T_c). At about T_c the rapid change of the volume and the order parameters versus temperature shows the well known first order phase transition. Some physical parameters have been derived and their values agrees well with measured data. Calulations have been extended to C_{60}-clathrate systems. Results show a somewhat different phase transition than in pure fcc C_{60}.

1. Introduction

In the past two years numcrous theoretical and experimental works were published on the unusual properties of C_{60} related materials. These articles focus on the electronic, magnetic and structural characteristics of the molecule and of different solids based on it. The existence of a structural phase transition in pure C_{60} crystals attracted considerable attention.

To understand the microscopic reasons for the phase transition we concentrated our efforts on temperature dependent x-ray diffraction studies on pure C_{60} and C_{60}-clathrates and on molecular dynamics calculations. First, we developed a model that was able to describe the phase transition of pure C_{60} and then a similar model was applied to the C_{60}-n-pentane clathrate. The results of the calculations are in good agreement with the experimental findings and physically reasonable predictions are expected for more complicated systems.

2. Molecular Dynamics

Molecular dynamics methods follow the motion of some atoms or molecules in a periodically repeated finite simulation box. Macroscopic parameters of the system (energy, temperature, pressure, volume) are calculated with averaging on time and ensemble. We employed Andersen's molecular dynamics method

Springer Series in Solid-State Sciences, Volume 117
Electronic Properties of Fullerenes Eds.: H. Kuzmany, J. Fink, M. Mehring, and S. Roth

extended by Parrinello and Rahman [1]. The main features of the model were the following:

- –constant energy, constant pressure, constant number of molecules
- –fluctuating temperature and volume
- –periodic boundary conditions with fluctuating periodic length
- –classical equations of motion
- –rigid C_{60} and n-pentane molecules
- –atom–atom additive Lennard–Jones potential with cutoff

This method is successfully applicable to the study of solid–solid phase transitions. The details can be found in three articles [2,3,4] dealing with the basics of this method and similar simulations.

3. Results

The first calculations were done for pure C_{60} which has an fcc structure at room temperature. The calculations were performed on a small system containing $2 \times 2 \times 2$ fcc unit cells, 32 molecules. The rapid change of the unit cell volume (fig. 1.) and the standstill of the temperature during the continuous heating or cooling evidently showed a first order phase transition at about 180 K. We also calculated some order parameters introduced by A. Cheng and M. L. Klein [4] to indicate the rotating or frozen state of C_{60} molecules (fig. 2.). However, the four quaternions describing the orientation of a molecule were better indicators of the difference between the two states. Analyzing time evolution of the quaternions one can conclude that the sum of the standard deviations of all quaternions could be used as order parameter.

The calculated physical parameters are the following: isobaric volume thermal–expansion coefficient, isobaric specific heat, adiabatic compressibility and latent heat of the transition:

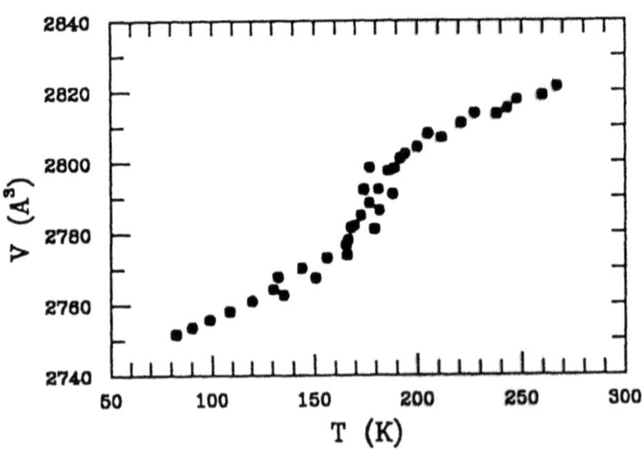

Figure. 1. The rapid change of the volume vs. temperature indicates the first order phase transition of pure C_{60}.

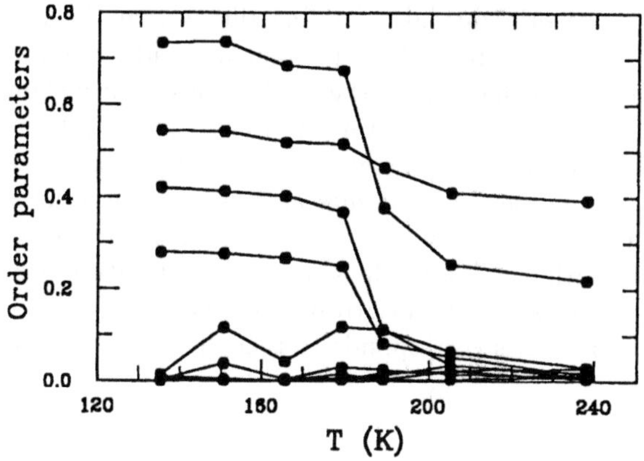

Figure. 2. The $\langle[Q_{6m}]\rangle$ order parameters introduced by A. Cheng and M. L. Klein [4] clearly show the phase transition in pure C_{60}.

	calculated	experimental
α $[K^{-1}]$	7.1×10^{-5}	6.2×10^{-5} [5]
c_p $[Jg^{-1}K^{-1}]$	0.11	0.5 – 1 [6,7]
κ_s $[Pa^{-1}]$	5.5×10^{-11}	4×10^{-11} [8]
L $[Jg^{-1}]$	5.5	6.7 [9]

These calculated values are in good agreement with the experimental data. The agreement can be further improved by use of different interaction potentials or varying the molecule charge densities.

The other system investigated was $2 \times 2 \times 1$ unit cells of C_{60}-n-pentane clathrate [10] containing 16 C_{60} and 16 pentane molecules. The main difference between the pure C_{60} and the C_{60}-n-pentane came from the motion of the pentane molecules which perturbed the free rotation of the C_{60}. The pentanes have two stable orientations as the center of the librational motion and they

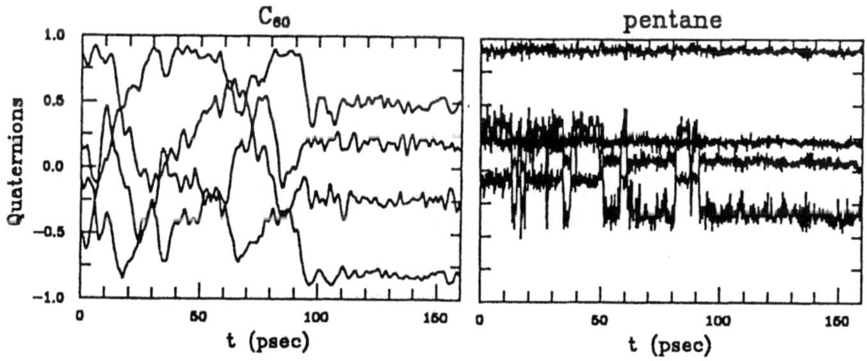

Figure. 3. The quaternions describing the orientation of a molecule during continuous cooling. At about 100 ps (which corresponds to $T_c \approx 120$K) the rotation of C_{60} molecules and the flipping of n-pentanes stops.

sometimes flip between these two orientations. Cooling the system the free rotation of the C_{60} molecules and the flipping of the pentane molecules stops at the same defined temperature. The behaviour of the quaternions clearly show that there is a phase transition (fig. 3.). However, it is not trivial that the flipping of n-pentane and the rotation of C_{60} molecules should freeze in at the same temperature since they have different dynamics at higher temperatures. The understanding of this fact would give a clue to reveal the driving force standing behind this phase transition. This result is in agreement with our x-ray single crystal measurement: the strongly anisotropic thermal motion factor of some atoms of the pentane molecules corresponds to their flip–flop motion.

4. Conclusions

In this paper we reported a molecular dynamics study on two C_{60} related materials. This simple simulation was able to describe the microscopic behaviour of these materials and lead to good qualitative agreement between calculated and measured physical parameters. This gives the hope that it can be employed to model other similar systems with predictive force. Based on the success of our present model we plan to extend our investigations to other related C_{60}-clathrates.

Acknowledgement. This work has been supported by OTKA grants under contract number T4474 and T4222.

References

1. M. Parrinello, A. Rahman: Phys. Rev. Lett. **45**, 1196 (1980)
2. S. Nosé, M. L. Klein: Mol. Phys. **50**, 1055, (1983)
3. D. J. Evans: Mol. Phys. **34**, 317, (1977)
4. A. Cheng, M. L. Klein: Phys. Rev. B **45**, 1889, (1992)
5. P. A. Heiney et al.: Phys. Rev. B **45**, 4544, (1992)
6. T. Atake et al.: Physica C **185-189**, 427, (1991)
7. T. Matsuo et al.: Solid State Commun. **83**, 711, (1992)
8. S. J. Duclos et al.: Nature, **352**, 380, (1991)
9. P. A. Heiney et al.: Phys. Rev. Lett. **66**, 2911, (1991)
10. S. Pekker et al.: Solid State Commun. **83**, 423, (1992)

Orientational Order in A_3C_{60}: Effects on Valence Bands and Infrared Optical Spectra

A.I. Liechtenstein[1], *I.I. Mazin*[1], *O. Gunnarsson*[1], *O.K. Andersen*[1], *S.E. Burkov*[2], *and V.P. Antropov*[1]

[1]Max-Planck-Institut für Festkörperforschung, Heisenbergstr. 1, D-70569 Stuttgart, Germany
[2]McMaster University, IMR, Hamilton, Canada

By analysis of the electronic contribution to the binding in A_3C_{60}, we show that the problem of orientational ordering can be mapped onto the fcc nearest-neighbor antiferromagnetic Ising model with J of the order of 100K. The ground state crystal structure should be therefore antiferromagnetically ordered in two dimensions and disordered in the third. This is not inconsistent with present X-ray powder data. We find that the low-temperature conduction-band structure closely resembles that of the simplest antiferromagnetically ordered (so-called bi-directional) crystal. Calculated optical conductivity can be fitted by a Drude-like contribution and an "interband" Lorenzian peak at 400-500 cm^{-1}.

1 Introduction

After the discovery of superconductivity in solid A_3C_{60} [1], it became important to figure out the low-temperature crystal structure. It was found that the C_{60}–molecules are on an fcc lattice and that the alkali ions fill the octahedral and tetrahedral pores [2]. Each molecule takes one of two possible orientations. In both of them, 6 of the 30 double bonds are parallel with the cubic axes and the two orientations are transformed into each other by any $\pi/2$-rotation around any of the cubic axes. That only these two orientations are possible, was established experimentally [2] and can also be understood from consideration of the space requirements of the alkali ions. The X-ray pattern at room temperature was found consistent with a random occupation of each site by a molecule of either orientation [2]. Such a 3-dimensional, orientational disorder would have tremendous effects on the electronic properties, as demonstrated by a theoretical calculation [3]. The X-ray experiment does not distinguish between static and dynamic disorder, but NMR [4] showed that on a 50μsec time-scale, the disorder in K_3C_{60} is static for temperatures below ≈270K, and dynamic above.]

A theoretical attempt to predict the orientational ordering of fullerenes at T=0, was undertaken by Gunnarsson et al. [5] who estimated structural energies using a tight-binding model with the 60 radial carbon p-orbitals per molecule as basis. For doped C_{60}, the lowest energy was obtained for the so-called bi-directional structure (BDS). This is a crystal in which the orientation is the same for all molecules in each (100)–plane, and alternates between neighboring planes. The BDS thus corresponds to the tetragonal CuAu-structure

Springer Series in Solid-State Sciences, Volume 117
Electronic Properties of Fullerenes Eds.: H. Kuzmany, J. Fink, M. Mehring, and S. Roth

and has space group $P4_2/mnm$. Compared with the uni-directional structure (UDS, $Fm\bar{3}$), where all molecules have the same orientation, the BDS turned out to be energetically favorable by $\sim 10^3$ K/molecule.

In this paper we shall argue, that the low-temperature structure of K_3C_{60} is not 3-dimensionally disordered, but bi-directionally ordered in two dimensions and disordered in the third. The 2-dimensional order is estimated to melt at about 200K in K_3C_{60}. What NMR observes above 270K, is merely that the orientational life-time becomes shorter than the 50 μsec sampled by this technique. We shall see, that with X-ray scattering, the low-temperature 1-dimensionally disordered structure (1DDS) is difficult to distinguish from the high-temperature 3-dimensionally disordered structure (3DDS). Of the infinitely degenerate 1DDS, one coincides with the BDS. Its conduction-band, discussed in Ref.[6], will be shown to be representative of the 1DDS. We shall also show the optical conductivity for the 1DDS.

2 Orientational order

First, we show that the orientational energy of A_3C_{60} can be approximately described by the fcc nearest-neighbor (nn) antiferromagnetic Ising model (AFIM),

$$E = J\frac{1}{N}\sum_{\langle ij\rangle}\sigma_i\sigma_j \equiv JS, \tag{1}$$

where E is the energy per molecule, N is the number of molecules, the sum is over all nn pairs (bonds), and σ_i is ± 1 depending on the orientation of the molecule at site i. For the fcc lattice, the Ising sum S is restricted to the range $-2 \le S \le 6$. To demonstrate the validity of (1), we first observe that the exchange-correlation term is a pair-wise interaction, if we can assume that in each point, the charge density can be considered to be a superposition of, at most, two molecular charges. The Madelung interaction between A^+ and C_{60}^{3-}, and between the A^+–ions, is the same for the two orientations. The Madelung interaction between the C_{60}^{3-} molecules can be decomposed pairwise and, due to the large distance (≈7Å) between the closest atoms on the next nearest-neighbor (nnn) molecules, and to the high multipolarity of the C_{60} molecule, only the interaction between the nn molecules needs to be kept. Hence, it has the form of the AFIM (1). For the one-electron energy sum, we first use the tight-binding (TB) nn-molecule approximation, which excellently reproduces the density-functional energy bands [5, 6]. The contribution from the full bands, plus the one from the *position* of the half-filled t_{1u} band, can be accurately calculated using second-order perturbation theory, and can therefore be decomposed into pair-interactions [5]. We tested this numerically by performing TB-calculations with the basis of 60 radial carbon p-orbitals per molecule (rad-AO basis) [9] for a sc unit cell with all four, with three, or with two of the molecules having the same orientation. Both energy differences yield the *same* contribution (70K) to J when fitted by the AFIM (1).

The contribution from the *broadening* of the t_{1u} conduction band can also be approximated by (1). To show that, we have performed TB band-structure

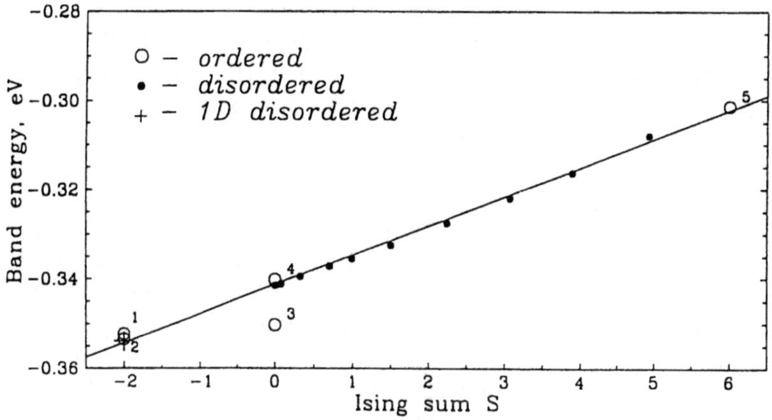

Figure 1: t_{1u} band-broadening energy vs. Ising sum calculated using the single-MO approximation [6] and a sc cell containing 500 molecules with randomly chosen orientations (filled dots). Open dots: Ordered short-period structures, (1) BDS=CuAu=AF (100)-planes ordered like ..*xxxx*.. as described in the text, (2) ...*xyxyxy*.., (3) Cu_3Au, (4) AF MnO=ferromagnetic (111)-planes of alternating orientation, (5) uni-directional=ferromagnetic. Cross: 1DDS.

calculations for a 5 × 5 × 5 sc cell with 500 molecules of randomly chosen orientations, using as basis just the three t_{1u}-molecular orbitals per molecule (single-MO basis) [10]. The calculated band-structure energies are shown by the filled dots in Fig. 1 as a function of the Ising sum. They follow a straight line, i.e. the AFIM (1), fairly well. Even the energies shown by the open dots numbered 1–5 for ordered, short-period configurations are close to the straight line [11]. Comparison between points 1 and 2 renders an estimate of the band-broadening contribution to the nnn interaction, which is seen to be antiferromagnetic and of order 1K.

We have thus demonstrated that the total orientational energy is well represented by the fcc nn AFIM (1). The total value of J, calculated as one eighths the energy-difference between the uni- and bi-directional structures with the TB rad-AO scheme, is 95K [9]. The value of J has also been calculated with the full-potential LMTO-LDA method, which includes effects neglected in the tight-binding calculation (exchange-correlation effects and Madelung interactions). We find that the result, J = 110 K[12], is similar to the tight-binding result.

The fcc nn AFIM has been studied intensively since the beginning of the 60's [13], and it is a classical example of a frustrated anti-ferromagnet. Danielian[14] was the first to describe the ground state: Each (001)–plane has AF–order such that all 4 nn's, at $(\pm\frac{1}{2},\pm\frac{1}{2},0)$, of a given site, (0,0,0), have the opposite spin. The next AF (001)–plane can now be generated from the first, either by the translation $[\frac{1}{2},0,\frac{1}{2}]$, which we shall refer to as x, or by the translation $[0,\frac{1}{2},\frac{1}{2}]$ (y). In both cases, 2 of the 4 additional nn's will have the same spin and 2 will have the opposite spin. For *any* stacking of the planes, e.g. ..*xxxyxyy*.., each site will therefore have 8 nn's with the opposite, and 4 nn's with the same spin. Hence, the ground state is infinitely degenerate with $E = -2J$. While the BDS is *one* of the ground states (*..xxxx..*) [15], the more general ground state is disordered in the z-direction. This is what we call the 1DDS. Next-nearest neighbor interactions will favor an ordered structure (the BDS, in case of ferromagnetic nnn interactions, and the *..xyxy..*-structure, in case of anti-ferromagnetic nnn interactions), but such interactions appear to be only of order 1K (see Fig.1). Since only disorder in 1D is permitted, the entropy per site is zero at $T = 0$ and increases slowly with temperature. For the nn AFIM, the BDS is stabilized over the 1DDS at all $T \neq 0$, but the free-energy difference between the average 1DDS and the BDS remains order $10^{-4}J$ [16], and we therefore expect the low-temperature structure of A_3C_{60} to be a metastable 1DDS, or possibly the ordered*..xyxy..*-structure. At $T_N \approx 1.76J$ ~200K the long-range AF order is lost even within the planes; the "heat of melting" is $\Delta E \approx 0.5J \approx 50$ K≈4 kJ/Mole [16, 17].

A difference between the thermodynamics of the fcc nn AFIM and that of orientation-flips in A_3C_{60}, is that for the latter there is a potential barrier B. Therefore, at temperature T, an orientation has a life-time $\tau(T) \approx \nu^{-1}\exp(B/T)$, where $\nu \approx 10^{12}$/sec is the frequency of the rotational phonons [18]. Since for K_3C_{60}, NMR indicates that $\tau(270\text{K}) \approx 50\mu$sec [4], we estimate that $B \approx 4800$K. This means that the orientational life-time $\tau(T_N)$ at the Neel temperature is about 1sec. In Rb_3C_{60}, NMR shows no dynamical disorder even at room temperature [4] so that for this material, the orientational life-time may be so long that the low-temperature structure depends crucially on the details of the preparation.

As regards the X-ray scattering, one can show that the Bragg reflections are exactly the same for 1DDS and 3DDS. The difference in the diffuse background exists, but can hardly be detected in X-ray scattering, because the relevant wave vectors are too large, $\geq 3\text{Å}^{-1}$, but can be detected in neutron scattering. The published data [19] are more compatible with the 3DDS, but it is not clear how well the sample had been annealed.

3 Valence band structure

How would the 1DDS structure affect the electronic properties? Most published electronic-structure calculations are for the uni-directional structure (UDS) and exhibit a density-of-states peak near the Fermi level [20]. In the 3DDS, this peak is completely smeared out and, as a consequence, $N(\epsilon_F)$ is reduced by about 40% [3]. This would have important implications for the superconductivity. For the BDS, the $N(\epsilon)$-shape is quite different from that of the UDS, but the Fermi level again falls near a peak and $N(\epsilon_F)$ is even a bit larger than in the UDS, although the conduction band is wider [6]. We have now performed TB single-MO calculations for five random stackings of one hundred AF (001)–layers. The resulting density of states for the 1DDS is compared with that of the BDS in Fig. 2, and the similarity is striking. The following

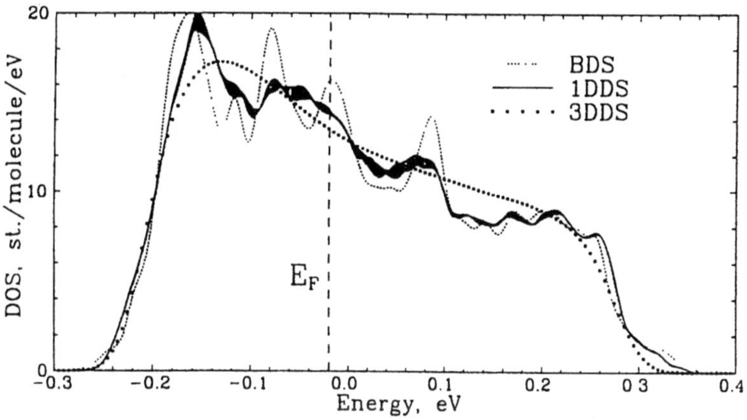

Figure 2: Density of states of the t_{1u}-band in the 1-dimensionally disordered structure (Ising ground state) and in the completely disorder structure, compared to the crystalline bi-directional structure. For the former, the width of the line corresponds to the statistical error. The t_{1u}-TB model was used [6].

facts may provide some understanding of why: Two neighboring (001)–layers always define a BDS–crystal and the probability that this crystal continues l inter-layer spacings upwards, is $1/2^l$. Therefore, the average thickness of the BD-domain surrounding any layer is $2\langle l\rangle = 2\sum_{l\geq 1} l/2^l = 4$ in units of the inter-layer spacing $a/2$, i.e. as large as 29Å. It is also interesting to note that the overall shape of $N(E)$ in the 3DDS model considered by Gelfand and Lu [3] is much closer to the BDS than to the UDS (Fig. 2). Finally we note, that even at $T = 2.8J$ ~300K, the energy of the AFIM is still $-J$ [15], so that about half the molecules have the same local coordination as at *T=0.* Even at room temperature, the electronic structure is therefore intermediate between those of the BDS and the 3DDS.

The electrical resistivity in the normal state might distinguish between the 1DDS and the 3DDS, provided that the intrinsic metallic resistivity is measured. Unfortunately, the experimental situation does not allow firm conclusions about the intrinsic resistivity [19]: Very different temperature dependencies have been reported, and the value of the room-temperature resistivity scatters from 2 to 5 mΩ cm, which corresponds to the mean-free path being less than the fcc lattice parameter, i.e., beyond the minimal metallic conductivity. 3-dimensional orientational disorder is calculated to provide a resistivity of merely 0.3mΩ cm [20] and phonons would, in the high-temperature limit, cause the resistivity $\rho \approx 8\pi^2\hbar\omega_p^{-2}k_BT\lambda$, which is about 3λ $\mu\Omega$ cm /K~1mΩ cm at room temperature, taking $\omega_p = 1.2$ eV.

4 Optical conductivity

An interesting observation comes from the infrared reflectivity [19]. It was found that in Rb_3C_{60} the infrared conductivity can be described as a combination of a Drude peak with extremely low spectral weight (ω_p $\approx$0.013 eV) and a Lorentzian centered at 0.12 eV with the spectral weight $\omega_p \approx 1.66$ eV. We have performed the calculations of the optical conductivity similar to those of Ref.[20], but using, instead of a 3DD cluster, a supercell 30×2, with 30 randomly stacked antiferromagnetic layers (25 k-points in the two-dimensional Brillouin zone). In this case, of course, the conductivity is formally different for the in-plane and the perpendicular polarizations. However, if we add to the in-plane "interband" conductivity the calculated Drude part (with an appropriate relaxation rate), the two conductivity become rather similar (Fig. 3). Interestingly, the calculated conductivity can be reasonably well fitted by a a combination of a Drude peak a Lorentzian, like in the experiment[19]. However, the effective plasma frequency, both for c and a/b polarizations, was about 0.89 eV, nearly 7 times larger than the experimental one. The "interband" peak is situated at 400 cm^{-1} (in-plane) and at 500 cm^{-1} (perpendicular). The corresponding relaxation rates for the Drude parts are 130 cm^{-1} and for the Lorenzians 1100 cm^{-1}.

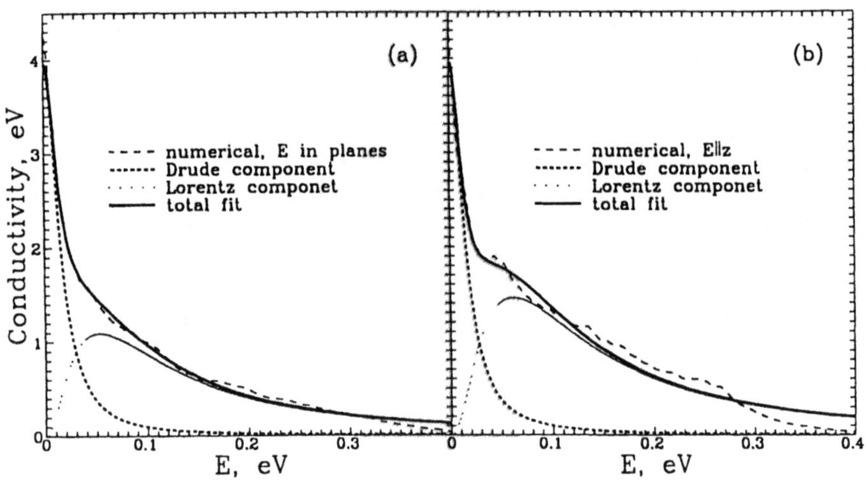

Figure 3: Optical conductivity for (a) in-plane and (b) perpendicular polarizations calculated in the t_{1u}-TB model [6] for the 1DD state and fitted conductivity by a combination of a Drude peak and a Lorentzian.

5 Summary

To summarize, we have shown that the problem of orientational ordering in K_3C_{60} can be mapped onto the fcc nearest-neighbor Ising model with an anti-ferromagnetic coupling constant $J \approx$100K. The ground state is 2-dimensionally anti-ferromagnetically ordered and 1-dimensionally disordered. This structure is difficult to distinguish by powder X-ray scattering from the 3-dimensionally disordered high-temperature structure. The conduction-bands for the two structures are very different. That of the low-temperature structure is well represented by the conduction band of the so-called bi-directional crystal with space group $P4_2/mnm$. The 2-dimensional long-range order vanishes in a first-order phase transition with $T_N \approx 1.76J$. The barrier between orientational flips appears to be large, of order 5000K. Calculated optical conductivity for the 1DD state can be fitted by a Drude-like contribution and an "interband" Lorenzian peak at 400-500 cm^{-1}.

References

[1] A.F. Hebard, M.J.Rosseinsky, R.C.Haddon, D.W.Murphy, S.H.Glarum, T.T.M.Palstra, A.P.Ramirez, and A.R.Kortan, Nature **350**, 600 (1991).

[2] P.W.Stephens, L.Mihaly, P.L.Lee, R.L.Whetten, S.-M.Huang, R.Kaner, F.Diederich, and K.Holczer, Nature, **351**, 632 (1991); P.W.Stephens, L.Mihaly, J.B.Wiley, S.-M.Huang, R.Kaner, F.Diederich, R.L.Whetten, and K.Holczer, Phys.Rev.B., **45**, 543 (1992).

[3] M.P.Gelfand and J.P.Lu, Phys.Rev.Lett., **68**, 1050 (1992).

[4] R.Tycko, G.Dabbagh, M.J.Rosseinsky, D.W.Murphy, A.P.Ramirez, and R.M.Fleming, Phys. Rev. Lett., **68**, 1912 (1992); J.C. Tully and the above, Science **253**, 884 (1991).

[5] O.Gunnarsson, S.Satpathy, O.Jepsen, and O.K.Andersen, Phys.Rev.Lett., **67**, 3002 (1991).

[6] S. Satpathy, V. P. Antropov, O. K. Andersen, O. Jepsen, O. Gunnarsson, and A. I. Liechtenstein, Phys. Rev. **B46**, 1773 (1992).

[7] The TB rad-AO results of the present paper were obtained with a=14.24Åand bond lengths 1.391Å and 1.455Å.. The results were corrected for the structural difference in the repulsive interaction, which is assumed to be atomic and of form A/R^{12}, where R is the inter-atomic distance. A was adjusted so that the correct lattice parameter was obtained for C_{60} in the presence of a van der Waals interaction ($\propto d^{-6}$, where d is the separation between two molecules) of -1.7 eV at equilibrium.

[8] The TB single-MO calculations of the present paper used the *conv.*-values from tables II and III of Ref. [6] for the hopping integrals and the lattice constant 14.10Å.

[9] The largest deviation occurs for the Cu_3Au-structure. Its favourable band-broadening energy is countered by the other components of the total energy.

[10] V. P. Antropov, A. I. Liechtenstein, O.K.Andersen, O.Jepsen, and O.Gunnarsson, to be published.
[11] For a review, see: R. Liebmann, *Statistical mechanics of Periodic Frustrated Ising Systems*, Springer, 1986.
[12] A. Danielian, Phys. Rev., **133**, A1344 (1964).
[13] Note that the *..xxxx..* or *..yyyy..* stackings give the same structure as the AF stacking of *ferromagnetic* (100)- or (010)-planes.
[14] N. D. Mackenzie and A.P. Young, J. Phys. **C14**, 3927 (1981), and Refs. therein.
[15] K. Binder, Z. f. Physik, **B45**, 61 (1981).
[16] K. Prassides, J. Tomkinson, C. Christides, M. J. Rosseinsky, D. W. Murphy, R. C. Haddon, Nature (London), **354**, 462 (1991).
[17] C. Christides, D. A. Neumann, K. Prassides, J. R. D. Copley, J. J. Rush, M. J. Rosseinsky, D. W. Murphy, and R. C. Haddon.
[18] S. Saito and A. Oshiyama, Phys. Rev. Lett. , **66**, 2637 (1991); S. C. Erwin and W. Pickett, Science, **254**, 842 (1991).
[19] L.Degiorgi *et al*, Phys. Rev. B, **46**, 11250 (1992); X. -D. Xiang *at al*, Science, **256**, 1190 (1992); O.Klein *et al*, Phys. Rev. B, **46**, 11247 (1992).
[20] M. P. Gelfand and J. P. Lu, Phys. Rev. B, **46**, 4367 (1992), and Z.Phys.B to be published.

Optical and SEM Studies of Morphology in C_{60} Films

L. Akselrod[1], *H.J. Byrne*[1], *J. Callaghan*[2], *A. Mittelbach*[1], *and S. Roth*[1]

[1]Max-Planck-Institut für Festkörperforschung,
Heisenbergstr. 1, D-70569 Stuttgart, Germany
[2]Physics Department, Trinity College, Dublin 2, Ireland

Abstract. We investigate the variance in the fabrication of C_{60} films to increase crystallinity and grain size within the polycrystalline film structure in sublimed and solution cast films. Ultra Violet/Visible absorption spectroscopy is seen to be strongly dependent on morphology in sublimed films. Strong solvent and post-heat treatment effects on morphology in solution cast films are observed. Further evidence of varying film crystallinity is provided by scanning electron microscopy. Production of large crystalline structures is shown to be most promising in solution cast films.

1. Introduction

Solid state C_{60} [1], can be sublimed under controlled conditions to form vacuum deposited, morphologically different films. This is desirable for optical studies, specifically for uv/vis absorption spectroscopy and the investigation of the HOMO-LUMO transition, the 610nm peak, normally symmetry disallowed, which becomes weakly allowed due to symmetry breaking interactions [2]. The relative strength of this electronic transition can be compared in sublimed films to the next highest transition, at 436nm, which is not present in solution. The intensity and the ratio of these peaks as a function of thickness give insight into the morphological dependence of uv/vis absorption spectra. Scanning electron microscopy allows for effective surface morphology characterization, ideal for discerning the effects of various deposition parameters on sublimed film morphology and the comparison to solvent dependent solution cast films. We intend to show a spectral dependance on morphology and to define relations between deposition conditions and crystallinity, in order to produce large grains for high intensity photoemission.

2. Experimental, Results, and Discussion

The C_{60} was prepared by a modification of the technique of Krätschmer et al. [3], purified by MPLC, and characterized by time of flight mass-spectroscopy to be 99% pure C_{60}. The deposition chamber was vacuumed to 10^{-6}mbars. The powder was outheated at 200°C for one hour and 400°C for one half hour. Deposition rates on unheated quartz substrates were varied by changing the initial amounts of

Springer Series in Solid-State Sciences, Volume 117
Electronic Properties of Fullerenes Eds.: H. Kuzmany, J. Fink, M. Mehring, and S. Roth

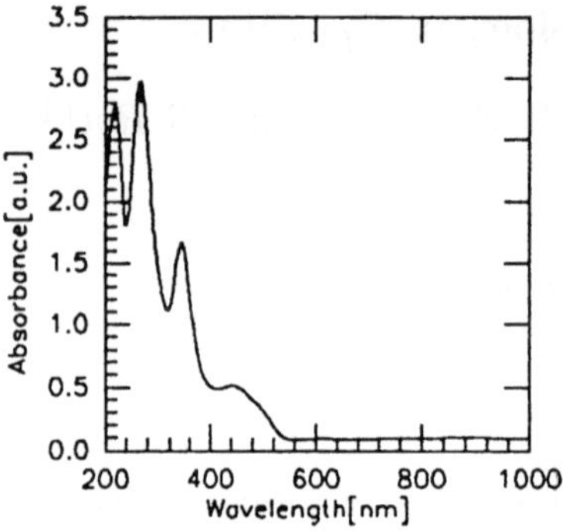

Fig.1:UV/VIS absorption C_{60} film spectrum

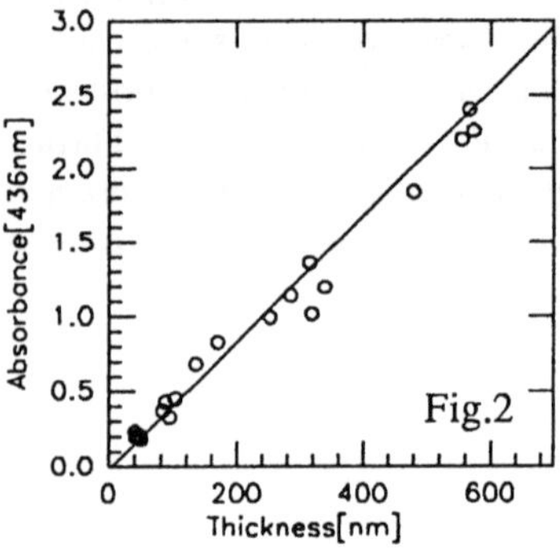

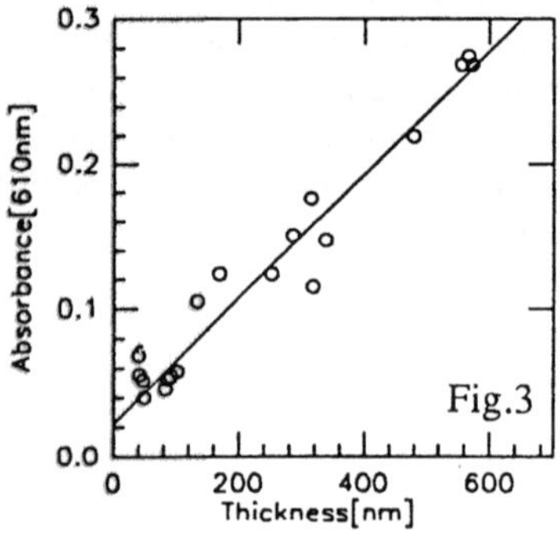

Fig.2,3: Intensity of the 610nm, 436nm peaks versus film thickness

powder or the temperature of deposition. Thickness was controlled by changing deposition times. All samples were kept and measured under argon and the spectra taken in a uv/vis spectrometer. Solution cast films were made by precipitation of C_{60} powder from various solvents and then outheated. SEM provided surface morphology characterization.

Figure 1 shows a uv/vis C_{60} film spectrum. The linear, proportional behavior of the plots (fig.2,3) with some scatter about the linear dependance appears consistent with the Lambert-Beer Law for the structures under investigation. However, the microcrystallinity of films creates questions of its possible affect on optical and electronic properties. High intensity luminescene and transport phenomena [4] indicate that grain size and crystallinity are critical parameters. Morphological affects are clearly observed in the absorption spectra (fig.4). In general, detailed examination of film spectra strongly indicates that they vary with the thickness and the temperature of deposition. This optical variation can be interpreted using a solution and film spectral comparison (fig 5). The 610nm peak, the HOMO-LUMO transition, is seen in both spectra and is thus of a molecular nature, the interactions in C_{60} solution being restricted to solvent-solvent and solvent-solute ones. No 436nm peak is seen in solution and it is consequently the first solid state feature. From figures 5 and 6, the relative contributions of molecular and solid state affects are shown to vary. Figure 6 exhibits a dependence of the ratio on thickness for thin

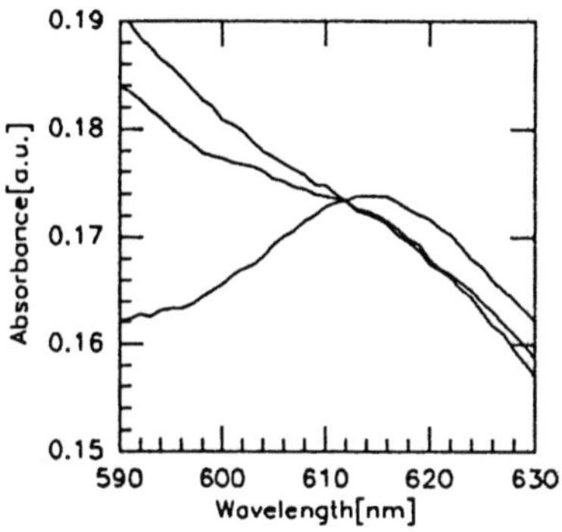

Fig.4: 3 spectra(610nm peak) for films of different rates/thicknesses.

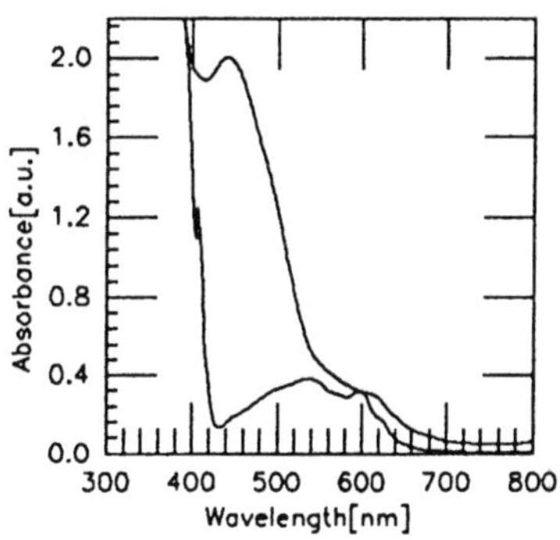

Fig.5: Comparison of solution and film spectra.

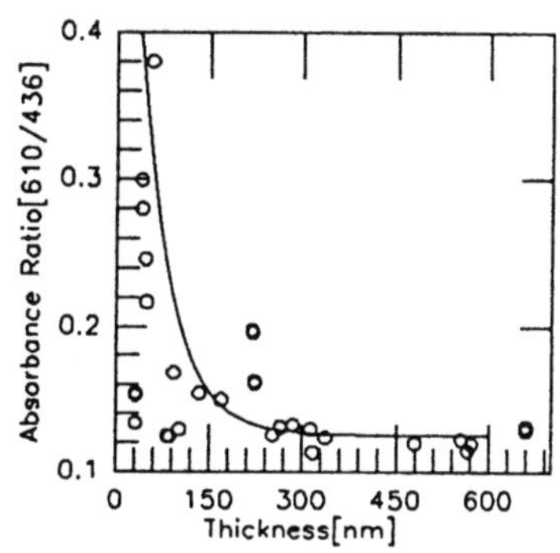

Fig.6: 610/436nm peak absorbance ratio plotted versus thickness

films only. The solid state effect is decreased in the thin films, which lie in the region between the molecular and the solid state.

In screening for desired optical properties, it is thus necessary to screen for desired morphological properties. From SEM photographs1-3, it is seen that variation in the temperature of deposition is an important factor of grain size in sublimed films. Decreased deposition temperatures progressively decrease the number of nucleation sites, contributing to larger crystalline structures, but concurrently to greater overall film inhomogeneity. However, solution cast films contain grain sizes on the order of ten times greater than their sublimed counterparts and are solvent dependent (SEM photographs4-7). The solvent is a factor in the variation of grain size and crystal geometry and solvents with higher saturation points are effective in producing films with somewhat increased overall substrate coverage. Solvent outheating and sample heating after deposition cause a geometrical restructuring of the crystalline structures in the films.

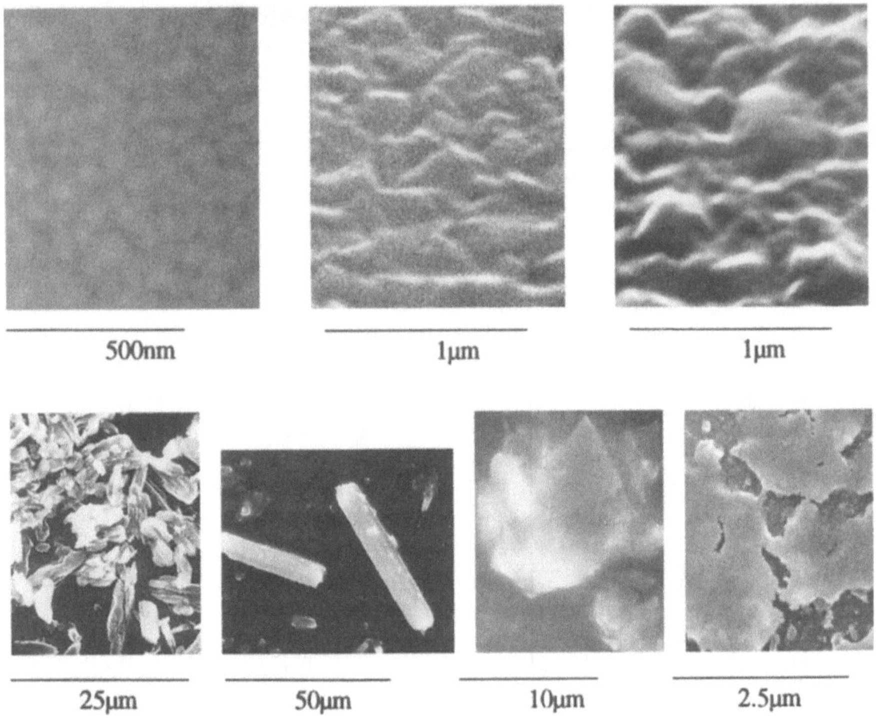

SEM photographs1-3: variation in grain size and homogeneity with deposition temperature(1-400°C, 2-355°C, 3-300°C)
SEM phototographs4-7: Solution cast films precipitated from(4-toluene, 5-hexane,6,7-cumene;4-6-outheated, 7-not outheated)

3. Conclusions

Morphology strongly affects and can be somewhat gleaned from the film's optical properties. Morphological variation is achievable with changes in deposition temperature, rate, and possibly thickness, but with limited control. Solution films offer a new option, having much larger grains, but concomittantly greater inhomogeneity. Work in increasing solution cast film homogeneity and stability holds the greatest promise for creating large grain films for select optical activity.

References

[1] H.W. Kroto et. al., Nature, **318** (1985) 162.
[2] S. Leach et. al., Chemical Physics, **160** (1992) 451-466.
[3] W. Krätschmer, L. Lamb, K. Fostiropolus and D. Huffman, **347** (1990) 345.
[4] H.J. Byrne, et. al., these proceedings.

Characterization of a Pure C_{60} Powder Using Transmission Electron Microscopy

A. Loiseau[1], *G. Van Tendeloo*[2], *and P. Bernier*

[1]Laboratoire de Physique du Solide and LEM/CNRS ONERA, B.P. 72, F-92322 Châtillon Cédex, France
[2]EMAT, University of Antwerp (RUCA), Groenenborgerlaan 171, B-2020 Antwerp, Belgium
[3]Laboratoire de Résonance Magnétique de Matériaux Solides, GDPC-USTL, case 026, F-34095 Montpellier Cédex 05, France

1. Introduction

Complementary to neutron and X ray diffraction, electron microscopy (TEM) is a very powerful technique for characterizing crystalline materials at microscopic and atomic scales. This technique has been rapidly used to study C60 and C70 solids [1, 2] and has allowed in particular to identify various lattice defects present in these materials and to determine different phase transitions through in situ experiments [1-5].

In this paper, we present a preliminary investigation of pure C60 powder under different imaging conditions. The observed peculiarities are compared with previous studies .

2. Experimental details

The C60 powder has been prepared at the GDPC Montpellier according to the procedure described in [6]. Samples for electron microscopy studies were prepared by smoothly crushing the fullerite powder and fragments were then slightly glued on a copper grid [3]. The crushing allows to obtain fragments transparent to the electrons from the core of the powder crystallites, which is free from oxydation. It has been shown indeed that the surface is rapidly contaminated by oxygen which easily penetrates and locates in interstitial sites, modifying the molecule librations [7]. Furthermore the crushing, because of the weak van der Waals bonds of the material, will only break the crystallites and do not produce lattice defects. Microscopy experiments were performed with a Philips CM20 analytical microscope and with a Jeol 4000EX at room temperature.

3 Characterization of the FCC structure

The electron diffraction patterns shown in Fig 1 correspond to the simple zone axes of the fcc structure, which is now well known to be the stable phase at room temperature. As explained in detail in [8], h00 reflections present in (001) and (110) zone axes are only an effect of double reflection and disappear when other reflections are not excited.

Springer Series in Solid-State Sciences, Volume 117
Electronic Properties of Fullerenes Eds.: H. Kuzmany, J. Fink, M. Mehring, and S. Roth

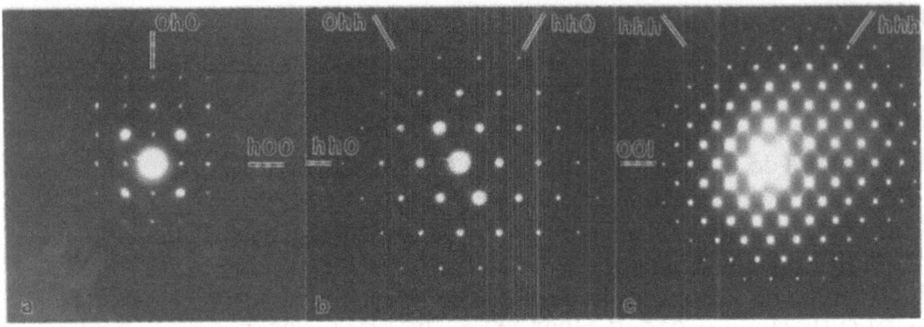

Fig 1: Diffraction patterns of the (002), (111) and (110) zone axes.

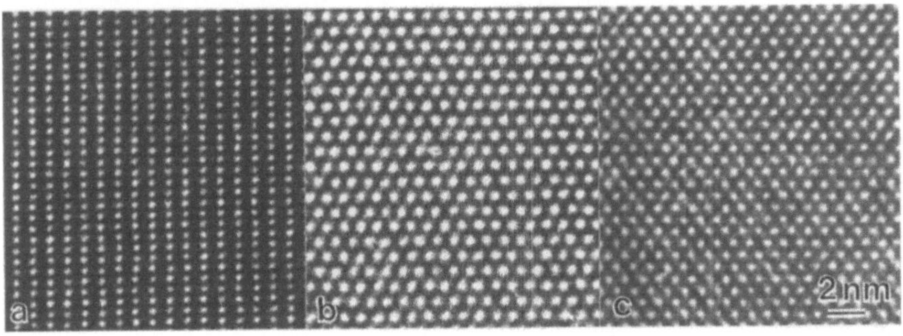

Fig.2: High resolution images of the structure projected along [112] in a), [111] in b) and [110] in c) (CM20, objective aperture = 10 nm^{-1}, defocus near Scherzer defocus).

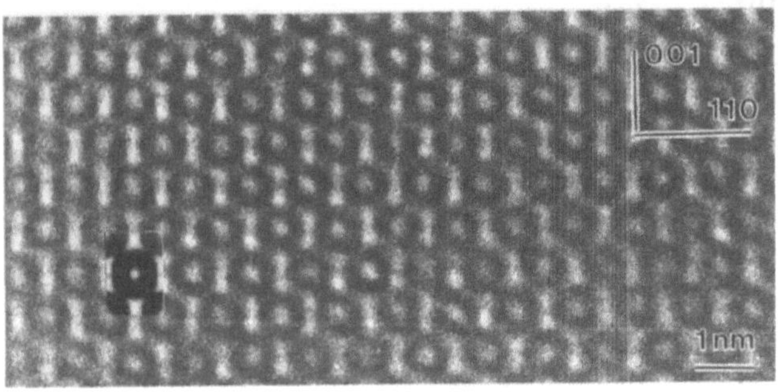

Fig.3: High resolution image of the structure projected along [110] and its simulation (Jeol 4000EX at 400kv, obj. aperture = 40 nm^{-1}, Δf= -50 nm, thickness = 5 - 7nm).

The structure is also well characterized from the high resolution images of Fig 2,3,4 which reveal different typical projections of an fcc structure. For the interpretation of the micrographs, simulations have been performed with the EMS code [9] assuming the model structure due to Fleming [10] of static molecules in a fcc lattice with the Fm3 symmetry group. No significant difference is found with calculations assuming a spherical atomic distribution [4].

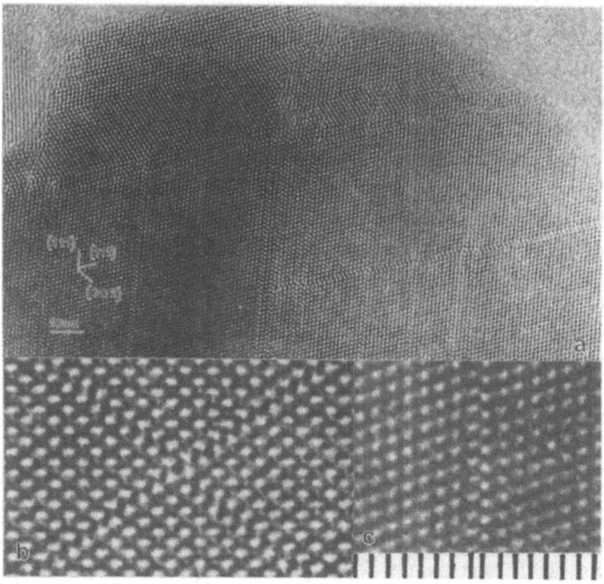

Fig.4: High resolution image along [110] of a defect area in a) with enlargements showing a stacking fault intersection in b) and a interspacing relaxation of the (111) planes in c).

These simulations show, that different operating conditions allow to image, either the fcc lattice only, with in that case molecules represented by white dots located in their center, or structural details of the atomic distribution. The first condition imaging is well suited for studying the lattice and its defects: typical arrays of the fcc lattice are shown in Fig 2,4. On the other hand, as shown in [3,4,11], in detailed images of the molecules, the atomic distribution yields an annular ring which can either be white or dark with a dot in the centre of opposite colour. This is typically what is observed in Fig 3 where the molecules are seen as dark rings with a white dot in their center separated by elongated white bars.

4. Lattice defects.

A high crystalline quality is revealed from the high resolution images. Very wide areas are defect free. However, (110) diffraction patterns can reveal for certain areas, as shown in Fig 1c, diffuse intensity along $\langle hhh \rangle^*$ directions.

The diffuse intensity corresponds to defects in the stacking of the dense (111) type planes, which is for the perfect fcc lattice ABCABC. This disorder can be: i) twinning or microtwinning, a twin being characterized by the sequence ..ABCBA.., ii) thin hcp ribbons ABAB..., iii) disordered distribution of stacking faults, either intrinsic ..ABCACABC.. or extrinsic ..ABCACBCABC.., a stacking fault being the limited case of twin or hcp lamella reduced to two planes.

The nature of the defects is clearly identified when imaging the lattice projected along a <110> direction, which reveals directly the stacking of two kinds of dense planes seen edge on (Fig 4). Crystallites are found to contain a more or less high density of intrinsic stacking faults and of microtwins but no hcp lamellae could be identified. Stacking faults and twinning are very frequent in C60 materials [1-5].

Furthermore, a relaxation in the interspacing of the dense planes adjacent to the stacking fault or to the twin interface can be observed in certain cases as shown in Fig 4b. Finally, Fig 4 reveals a remarkable feature of intersecting stacking faults lying in different (111) planes. Simple geometric considerations [12] indicate that such intersections can result from stacking faults produced during the crystal growth but not from a deformation process. This observation confirms the fact, already noticed in view of the other kinds of lattice defects observed [3], that the C60 phase has a very low stacking fault energy.

5. Surface observation: faceting and relaxation

Fig. 5 shows a sample surface observed immediately after the sample preparation and which is not yet covered by the usual strongly disordered layer.

The surface viewed along a [110] direction is microfaceted with more or less extended steps in (111) and (100) type planes.These planes are precisely the facets developed during the growth of large crystallites. This indicates that the surface is not contaminated at all by oxydation and not pertubed by any kind of irradiation. Different reconstructed surfaces have been indeed observed for surfaces covered by oxygen [3], with in particular a relaxation of the planes interspacing in the surface neighborhood. In the present case, more refined observations should be done to show if any relaxation also occurs.

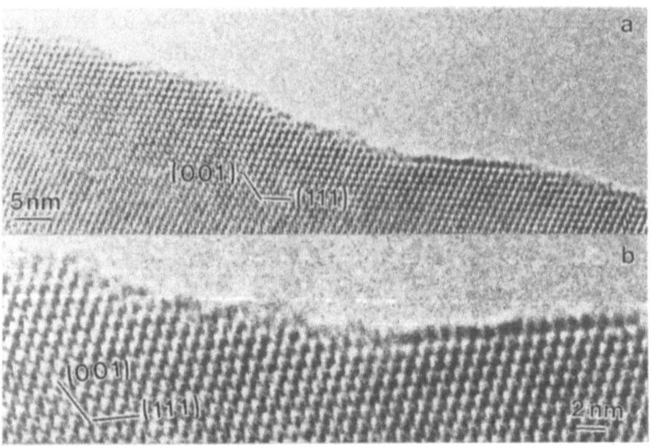

Fig.5: High resolution image along [110] of a sample surface in a) and enlargement in b).

6. Conclusion

The crystal and purity quality of a C60 powder have been characterized at an atomic level by TEM. Besides large defect free areas, growth defect configurations typical of a fcc crystal with a very low stacking fault energy have been identified, and clean sample surfaces could be observed. Owing to these peculiarities, the material is propitious to a refined study of the orientational order and in situ cooling experiments will be realized in the near future to study the ordering transitions.

Acknowledgments

S. Amelinckx is gratefully acknowledged for useful scientific discussions. Image simulations have been performed at Laboratoire des Solides Irradiés, C E A - Ecole Polytechnique (Palaiseau).

References

[1] S. Wang and P.R. Buseck, Chem. Phys. Lett., **182**, 1 (1991).
[2] G. Van Tendeloo, M. Op de Beeck, S. Amelinckx, J. Bohr and W. Krätschmer, Europhys. Lett., **15**, 295 (1991).
[3] S. Muto, G. Van Tendeloo and S. Amelinckx, Philos. Mag., in press (1993)
[4] F. Banhart, M. Forster, W. Kratschmer and H. Schaefer, Philos. Mag., **65**, 283 (1992)
[5] G. Van Tendeloo, S. Amelinckx, S. Muto, M. Verheijen, P. Van Loosdrecht and G. Meyer, Ultramicroscopy, in press (1993).
[6] A. Zahab, J. Sauvajol, L. Firley, R. Aznar and P. Bernier, J.Phys.I , **2**, 7 (1992).
[7] M. Wohlers, H. Werner, R. Schlögl, this issue (1993).
[8] S. Amelinckx, C. Van Heurck, D. Van Dyck, and G. Van Tendeloo, Phys. Stat. Sol.(a), **131**, 589 (1992).
[9] P. Stadelmann, Ultramicroscopy, **21**, 131 (1987).
[10] R. Fleming et al , Mat. Res. Soc. Symp. Proc., **206**, 691 (1991).
[11] G. Van Tendeloo, C. Van Heurk, J. Van Landuyt, S. Amelinckx, M. Verheijen, P. van Loosdrecht and G. Meijer, J. Phys. Chem., **96**, 7424 (1992).
[12] S. Amelinckx, communication privée (1992)

Surface Imaging by Scanning Tunneling Microscopy of $C_{60/70}$ Thin Films on Au(111) with Different C_{60}/C_{70} Ratios

H.P. Lang, V. Thommen-Geiser, R. Hofer, and H.-J. Güntherodt

Institut für Physik der Universität Basel,
Klingelbergstr. 82, CH-4056 Basel, Switzerland

Abstract. Fullerene thin films have been prepared on Au(111) substrates by sublimation from fullerene powders with different C_{60} / C_{70} ratios. Scanning tunneling microscopy (STM) images of the film surfaces show a hexagonal lattice built-up of two types of ball-shaped molecules (differing in diameter and corrugation). The ratio of smaller to larger balls compares well with the C_{60} / C_{70} ratio determined by high pressure liquid chromatography (HPLC) on hexane solutions of the respective fullerene powders. STM images show defects such as vacancies, interstitials and boundaries between different stacking domains (ABC and CBA). Dynamic displacement of C_{70} molecules and vacancies is observed.

1. Introduction and Experimental

Scanning tunneling microscopy (STM) has proven to be a useful method to image surfaces of fullerene thin films on metals or semiconductors in air or in ultrahigh vacuum [1-4] and has confirmed the room-temperature structure of solid C_{60} (fcc with a nearest-neighbour-distance of 1 nm). In this contribution, a local value for the C_{60} / C_{70} ratio will be deduced from STM images of $C_{60/70}$ thin films. Different types of lattice defects will be discussed.

Fullerene films of a thickness of only a few monolayers have been prepared by sublimation of fullerene powder at 350-370 °C in a vacuum of 10^{-3} Pa onto a mica-supported, flame-annealed Au(111) thin film substrate equilibrated at a temperature of 220-230 °C. STM was performed in air at ambient temperature and pressure using a mechanically prepared $Pt_{90}Ir_{10}$ tip. A bias voltage between 0.1 and 0.3 V and a tunneling current between 0.5 and 2 nA was applied. All images are recorded in the constant current mode and therefore give topographic height information which is encoded to a grey-scale. High pressure liquid chromatography (HPLC) was performed on hexane solutions of the fullerene powders to determine the C_{60} / C_{70} ratio (for details see [5]).

2. Results and Discussion

$C_{60/70}$ thin film samples exhibit by STM crystalline $C_{60/70}$(111) domains that are extended over the size of the Au(111) substrate terraces (several hundred

Springer Series in Solid-State Sciences, Volume 117
Electronic Properties of Fullerenes Eds.: H. Kuzmany, J. Fink, M. Mehring, and S. Roth

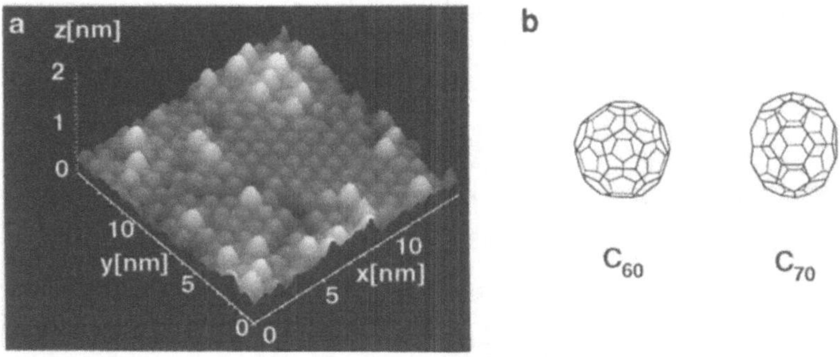

Figure 1: (a) Three-dimensional representation of an area of 15 × 15 nm^2 of the surface of a $C_{60/70}$ film on Au(111). The molecules appearing brighter are C_{70} molecules, the darker ones are C_{60} molecules. (b) Wire-models of C_{60} and C_{70}.

Table 1. C_{70} content of different fullerene mixtures.

	Mixture 1	Mixture 2	Mixture 3
HPLC (corrected values)	19.4 %	18.2 %	23.6 %
STM (averaged values)	15.3±2.1 %	16.6±3.7 %	20.4±3.9 %

nanometer in diameter). The fullerene molecules form a hexagonal lattice with a nearest-neighbour-distance of 1 nm. Two different kinds of molecules are observed in films prepared from a mixture of C_{60} and C_{70} powder. A three-dimensional representation of a $C_{60/70}$ film surface is shown in figure 1a. The two molecular species (cf. wire-models in figure 1b) differ in size (apparent diameters: 0.7 and 0.8 nm) and corrugation (peak-to-peak values: 0.3 and 0.5 nm). The speculation is obvious that the molecules appearing larger are to be identified with C_{70} molecules, whereas the smaller ones are assigned to C_{60} molecules. By counting the two species of molecules, a local C_{60} / C_{70} ratio can be inferred from the STM images. We have performed this analysis on $C_{60/70}$ films prepared under the same conditions (temperature, pressure, duration of sublimation) from three different $C_{60/70}$ powder sources (with different C_{60} / C_{70} ratios). For each $C_{60/70}$ source, STM images of different regions on the film surface were evaluated, the ratios averaged and the standard deviation calculated. Since the vapour pressures of C_{60} and C_{70} are dissimilar [6], we do not expect the C_{60} / C_{70} ratio in the powder to be accurately transferred to the sublimed film. A HPLC determination of the C_{60} / C_{70} ratio in hexane solutions of the corresponding fullerene sources yields in all three mixtures a slightly higher value for the C_{70} content than observed in the STM images. The HPLC values determined from peak integration have been corrected for molar absorptivities of C_{70} and C_{60}, $\epsilon(C_{70})/\epsilon(C_{60}) = 1.2$ at 280 nm [7]. The results for the C_{60} / C_{70} ratios are compiled in table 1.

Surfaces of fullerene thin films exhibit numerous defects such as vacancies, interstitials and domain boundaries. Figure 2 shows two types of molecular C_{60}

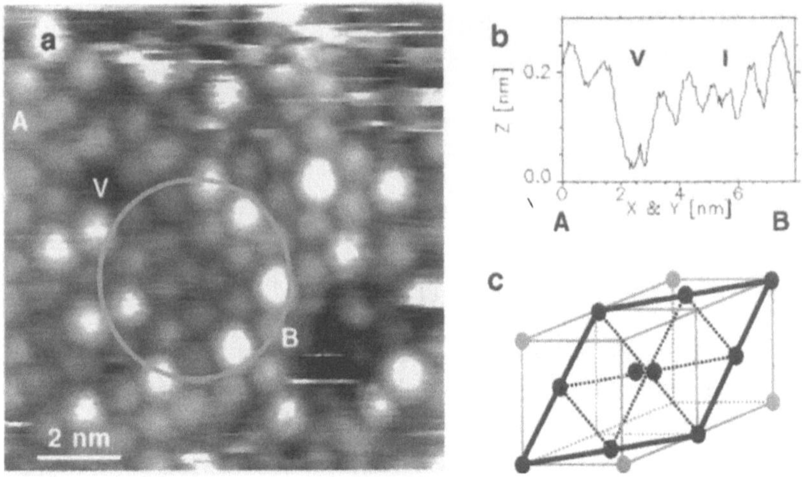

Figure 2: (a) STM image of lattice point defects in a $C_{60/70}$ film: a vacancy (**V**) and a split interstitial **I** (center of the circle). (b) Line section from **A** to **B**. (c) Model of a split interstitial in a fcc lattice. The (111) plane is highlighted.

point defects: a vacancy (marked by **V**) and an interstitial (situated close to the center of the circle in the STM image in figure 2a). Both defects can easily be found in the line section (figure 2b) between the points marked with **A** and **B** in figure 2a. Interstitials in a fcc lattice are in most cases *split interstitials*, i.e. two molecules occupy a single regular lattice site. The long axis of a split interstitial is parallel to <100>. Figure 2c shows a schematic view of a split interstitial in a fcc lattice (the (111) face is highlighted).

Successive STM images reveal dynamic rearrangement of the film surface. This is demonstrated by the three STM images shown in figures 3a-c which show the same area imaged at time intervals of 15 s. Close to the position marked by **1**, a C_{70} molecule moves to a vacancy site. Near **2** a C_{70} molecule disappears and a vacancy is left. Possibly, this C_{70} molecule has moved to the site marked by **3** in figure 3b, and vanishes in figure 3c.

Another type of lattice defect can be observed in figure 3a: a stacking order boundary (marked by **SB**). Such boundaries occur when domains of different layer stacking (ABC and CBA) meet. Figure 3d schematically shows a model for such a boundary (the A layer is shown in black, the B layer in dark grey and the C layer in light grey). The existence of stacking order boundaries is evidence for epitaxial growth of $C_{60/70}$ on Au(111) implying a well-defined growth relation between Au(111) and $C_{60/70}$(111).

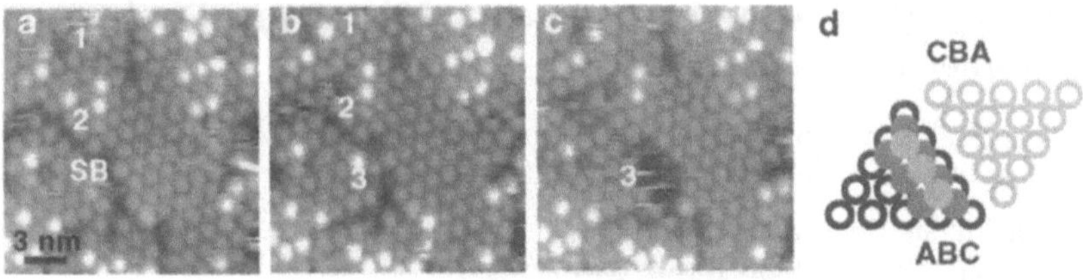

Figure 3: (a)-(c) Series of STM images recorded at time intervals of 15 s showing rearrangement of the film surface. A C_{70} molecule moves to a vacancy site (marked by **1**), another C_{70} molecule is removed (**2**) and shifted to a different lattice site (**3**), and then finally disappears from the imaged area. A stacking domain boundary between domains of ABC and CBA stacking is marked by **SB**. (d) Model for a stacking domain boundary (A-layer: black, B-layer: dark grey, C-layer: light grey).

Acknowledgements. We thank the research groups from Frankfurt (H. Werner and R. Schlögl) and Montpellier (A. Zahab and P. Bernier) for generously supplying us with fullerene powder. We are grateful to C. Bolm and M. Felder (Institute of Inorganic Chemistry, University of Basel) for the HPLC data. The Swiss National Science Foundation is acknowledged for financial support.

References

[1] R.J. Wilson, G. Meijer, D.S. Bethune, R.D. Johnson, D.D. Chambliss, M.S. de Vries, H.E. Hunziker and H.R. Wendt, Nature **348**, 621 (1990).

[2] J.L. Wragg, J.E. Chamberlain, H.W. White, W. Krätschmer and D.R. Huffman, Nature **348**, 623 (1990).

[3] H.P. Lang, V. Thommen-Geiser, J. Frommer, A. Zahab, P. Bernier and H.-J. Güntherodt, Europhys. Lett. **18**, 29 (1992).

[4] Y.Z. Li, J.C. Patrin, M. Chander, J.H. Weaver, L.P.F. Chibante and R.E. Smalley, Science **252**, 547 (1991), *idem* Science **253**, 429 (1991).

[5] H.P. Lang, V. Thommen-Geiser, C. Bolm, M. Felder, J. Frommer, R. Wiesendanger, H. Werner, R. Schlögl,A. Zahab, P. Bernier, G. Gerth, D. Anselmetti and H.-J. Güntherodt, Appl. Phys. A **56**, 197 (1993).

[6] J. Abrefah, D.R. Olander, M. Balooch and W.J. Siekhaus, Appl. Phys. Lett. **60**, 1313 (1992).

[7] J.M. Hawkins, T.A. Lewis, S.D. Loren, A. Meyer, J.R. Heath, Y. Shibato and R.J. Saykally, J. Org. Chem. **55**, 6250 (1990).

Low-Temperature Scanning Tunneling Microscopy Study of C_{60} Fullerite

S. Behler, H.P. Lang, S.H. Pan, V. Thommen-Geiser, R. Hofer, M. Bernasconi, and H.-J. Güntherodt

Institut für Physik der Universität Basel,
Klingelbergstr. 82, CH-4056 Basel, Switzerland

Abstract. We have used a scanning tunneling microscope (STM) to study monolayer coverages of C_{60} on a Au(111) substrate at 4.5 K. The STM images clearly show internal structure on all C_{60} molecules. The patterns can be related to a structural model of the C_{60} molecule assuming the corrugation maxima correspond to the centers of the carbon ring structures.

1. Introduction and Experimental

Scanning tunneling microscopy (STM) has proven to be a useful tool for the study of organic materials because of its ability to "see" single molecules [1]. C_{60} thin films on metal substrates have been investigated by STM at ambient temperature. Due to the fast rotational motion (10^{10} Hz) of the molecules at this temperature [2], intramolecular contrast is observed only occasionally [3, 4]. In this paper we present a STM image of a C_{60} thin film obtained at 4.5 K which clearly shows intramolecular structure on all molecules. This result provides strong evidence for the freezing of the rotational motion of C_{60} molecules at low temperatures, as indicated by NMR and neutron diffraction [2, 5].

Since C_{60} fullerite itself is not intrinsically conductive, we have prepared C_{60} thin films by sublimation in vacuum onto a mica-supported Au(111) substrate heated to 225 °C. The duration of sublimation was chosen to give a coverage of a few monolayers. This ensures the conductivity required for STM investigations.

Mechanically sharpened $Pt_{90}Ir_{10}$ tips were used in all STM measurements. After imaging in vacuum (2×10^{-5} mbar) at ambient temperature, the sample was cooled down to 4.5 K with the STM over one hour. Helium exchange gas was introduced to a total pressure of about 5×10^{-3} mbar to keep the temperature stable during the low temperature measurements.

2. Results and Discussion

Figure 1a shows a STM image of the C_{60} film obtained at ambient temperature. The molecules appear as smooth spheres due to the fast rotational motion. Some regions exhibit close packed molecules (C_{60}(111) face), whereas

Springer Series in Solid-State Sciences, Volume 117
Electronic Properties of Fullerenes Eds.: H. Kuzmany, J. Fink, M. Mehring, and S. Roth

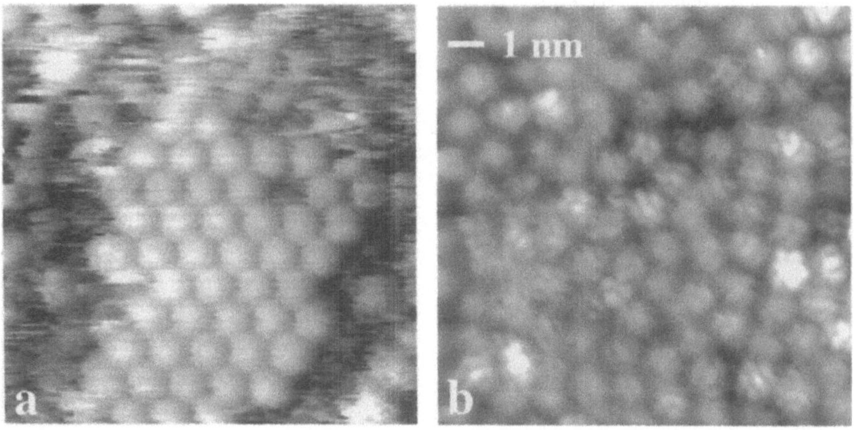

Fig. 1. STM images of C_{60} molecules on a Au(111) substrate obtained (a) at 295 K and (b) at 4.5 K. The high frequency rotational motion of the molecules at 295 K presumably prevents the intramolecular structure from being recorded (tunneling current I_t = 300 pA, sample bias voltage U_s = 100 mV). At 4.5 K intramolecular contrast is clearly observed within the C_{60} molecules (I_t = 300 pA, U_s = -20 mV).

in other regions the scanning process seemed to change the arrangement of the fullerenes, when the same area was imaged repeatedly. The distance between neighbouring molecules is 10 Å, which is in good agreement with X-ray diffraction results (10.07 Å) [6].

A typical image obtained at 4.5 K is shown in Figure 1b. The smallest distance between two molecules is still 10 Å. In contrast to the room temperature images, we can also resolve internal structure on nearly all molecules which indicates the freezing of the rotational motion. The corrugations of these features are measured to be about 0.3 Å. Typical distances between neighbouring corrugation maxima are about 3 to 4 Å. No significant dependence of these patterns on the scanning parameters such as scanning speed, scanning direction, sample bias voltage, polarity or tunneling current was observed. Enlarged images of four distinctive patterns are shown in Figure 2a-d.

The three-dimensional structure of the C_{60} molecule and the unknown tip geometry complicate the interpretation of the STM image. The symmetry and size of the observed patterns cannot be explained if one assumes that carbon atoms correspond to the corrugation maxima. A comparison of the patterns with a structural model of the C_{60} molecule suggests that the corrugation maxima are situated at the centers of the carbon ring structures. For example, the four patterns in Figure 2a-d are correlated with the corresponding schematics of the C_{60} molecule in the four different orientations shown in the same Figure.

Electron transport through the C_{60} molecule is crucial for the STM imaging process. The mechanism of tip-to-substrate electron tunneling through adsorbed organic molecules is not yet fully understood [1]. Images obtained by STM reflect the structure of the adsorbed molecules, although they do not always reflect all the atom sites. The intramolecular features in our STM image

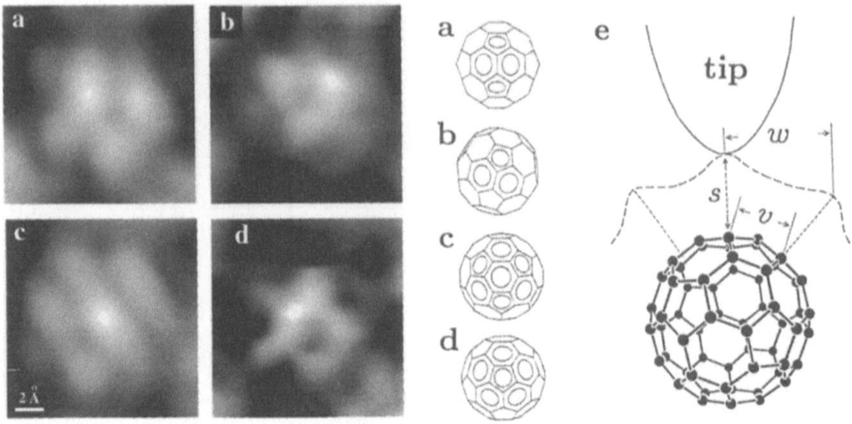

Fig. 2. (a)-(d) Enlarged STM images of distinctive patterns in Fig. 1b and schematic views of corresponding carbon ring structures of the C_{60} molecule. (e) Structural model of the C_{60} molecule. The scanning profile shows that due to the three-dimensionality of the C_{60} molecule and the tip – molecule separation s a larger distance between corrugation maxima w (compared to the theoretical distance between ring centers v) is observed.

mainly show that the C_{60} molecules can be found in different orientations at 4.5 K.

The theoretical distance v between ring centers on the C_{60} molecule is about 2.2 Å, but the STM images show a larger distance w of 3 to 4 Å between corrugation maxima (Fig. 2). This may be explained by taking into account the tip – molecule separation s and the three-dimensionality of the C_{60} molecule. Figure 2e shows a model of the C_{60} molecule and the contours imaged by the STM tip assuming the corrugation maxima are situated at the centers of the carbon rings. The tip – molecule separation causes a larger apparent distance between neighbouring corrugation maxima. By calculating the tip – molecule separation s from the observed distance between corrugation maxima w, a value of about 2 Å is obtained. This is a reasonable tip – sample separation in STM.

In summary, we have observed intramolecular features of C_{60} molecules by STM at 4.5 K providing strong evidence for the freezing of rotational motion at low temperature. Correlation of various patterns of these features to C_{60} molecules in different orientations has been suggested. We preliminary assign the observed corrugation maxima to the centers of the carbon ring structures.

Acknowledgements

We would like to thank H. Werner and R. Schlögl (University of Frankfurt, Germany) for generously providing us with pure C_{60} material. This work was financially supported by the Swiss National Science Foundation.

References

[1] J. Frommer. *Angew. Chem., Int. Ed. Engl.* **31**, 1298 (1992).

[2] C. S. Yannoni, R. D. Johnson, G Meijer, D. S. Bethune, and J. R. Salem. *J. Phys. Chem.* **95**, 9 (1991).

[3] T. Chen, S. Howells, M. Gallagher, L. Yi, D. Sarid, D. L. Lichtenberger, K. W. Nebesny, and C. D. Ray. *J. Vac. Sci. Technol. B* **9**, 2461 (1991).

[4] Y. Zhang, X. Gao, and M. J. Weaver. *J. Phys. Chem.* **96**, 510 (1992).

[5] W. I. F. David, R. M. Ibberson, T. J. S. Dennis, J. P. Hare, and K. Prassides. *Europhys. Lett.* **18**, 219 (1992).

[6] P. A. Heiney, J. E. Fischer, A. R. McGhie, W. J. Romanow, A. M. Denenstein, Jr. J. P. McCauley, III A. B. Smith, and D. E. Cox. *Phys. Rev. Lett.* **66**, 2911 (1991).

Layer Disorder in C_{60}-Ether Clathrates

M. Tegze, G. Oszlányi, G. Bortel, G. Faigel, and S. Pekker
Research Institute for Solid State Physics, P.O.B. 49,
H-1525 Budapest, Hungary

Abstract. We report structural studies on new C_{60} clathrates formed with ether as a guest molecule. Single crystal x-ray diffraction shows that in C_{60}-ether two different pseudo tetragonal unit cells exist, which correspond to different compositions. The unit cells are based on the same relative displacements of the square–planar layers as found in the C_{60}-n-pentane clathrate [1,2]. However, the stacking sequence of the layers is different and reciprocal space maps reveal a significant stacking disorder. Model calculations suggest that the disorder comes mainly from the random sequence of the relative displacements.

1. Introduction

The succesfull production of macroscopic quantities of C_{60} [3] opened a new field in chemistry. The new materials based on these cagelike carbon molecules show a wide variety of interesting structural and physical properties.

While the crystallization of C_{60} by sublimation produces pure crystals with fcc structure, the precipitation from organic solvents often results in solvent containing crystals. These compounds are clathrates, a fact that little attention has been paid to in the search for other C_{60} compounds.

This tendency of C_{60} to form clathrate type compounds has been systematically utilized for the production of C_{60} clathrates containing small linear molecules [4].

2. Experimental

The preparation of C_{60}-ether clathrate is described elsewhere [4,5]. Briefly, platelet shaped single crystals up to the size of $0.5 \times 1.0\ mm^2$ were precipitated from a saturated toluene solution of C_{60} due to the effect of diethyl ether, introduced to the solution by isotherm distillation [4].

X-ray crystal diffraction measurements were performed on several single crystals using a four circle diffractometer. Copper K_α radiation was separated by a pyrolitic graphite monochromator.

Two different types of crystals were found. Both of them showed effects of disorder: the diffraction peaks were broadened in one direction of the reciprocal space.

Springer Series in Solid-State Sciences, Volume 117
Electronic Properties of Fullerenes Eds.: H. Kuzmany, J. Fink, M. Mehring, and S. Roth

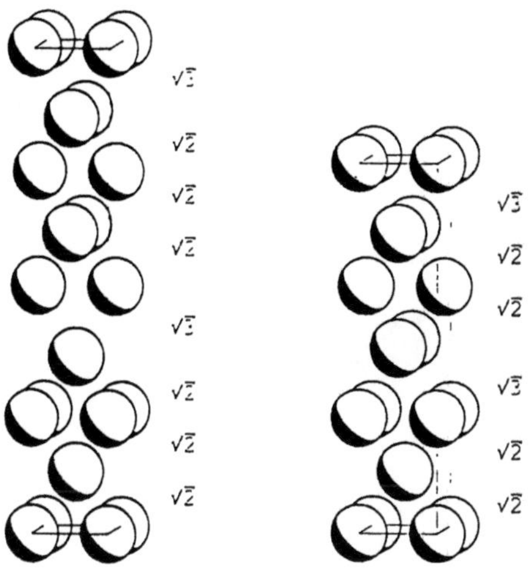

Figure 1. The two different structures of C_{60}-ether. Only the C_{60} molecules are shown.

The strongest peaks were indexed by an A-centered pseudo tetragonal unit cell with lattice parameters:

I.	a = 10.05(1) Å,	b = 10.05(1) Å,	c = 45.85(4) Å
II.	a = 10.04(1) Å,	b = 10.04(1) Å,	c = 59.82(4) Å

Most of the weak (and broad) peaks can be indexed by releasing the centering condition.

The cell dimensions suggest that the basis of the cell is the same square planar layer of C_{60} molecules as in the case of the C_{60}-n-pentane [4]. The relative shifts of consecutive layers can be $(1, 1, \sqrt{2})R$ (for short: $\sqrt{2}$) or $(1, 0, \sqrt{3})R$ (for short: $\sqrt{3}$), where $R = 5.03$ Å is the van der Waals radius of the C_{60} molecule. The stacking sequences in the two C_{60}-ether structures are $\sqrt{2}\sqrt{2}\sqrt{3}$ and $\sqrt{2}\sqrt{2}\sqrt{2}\sqrt{3}$, respectively (Fig. 1). The ether molecules are situated in the channels of the $\sqrt{3}$ layers.

To clarify the details of the disorder we measured 2-dimensional reciprocal space maps. Figure 2a–c shows selected reciprocal planes with tetragonal indices. The most obvious feature of these maps is the reflection broadening in the l direction while reflections in the h and k directions are perfectly sharp.

These experimental results suggest that while the C_{60} molecules constitute perfect square layers in the *a*–*b* plane, the stacking of these layers in the *c* direction is not perfectly regular.

3. Model Calculations

We have performed model calculations for the diffraction pattern of one of the C_{60}-ether structures ($\sqrt{2}\sqrt{2}\sqrt{3}$). Good agreement with the experimental results was obtained using the following assumptions: 1. The C_{60} molecules are freely rotating. 2. The contribution of ether molecules is neglected. 3. The

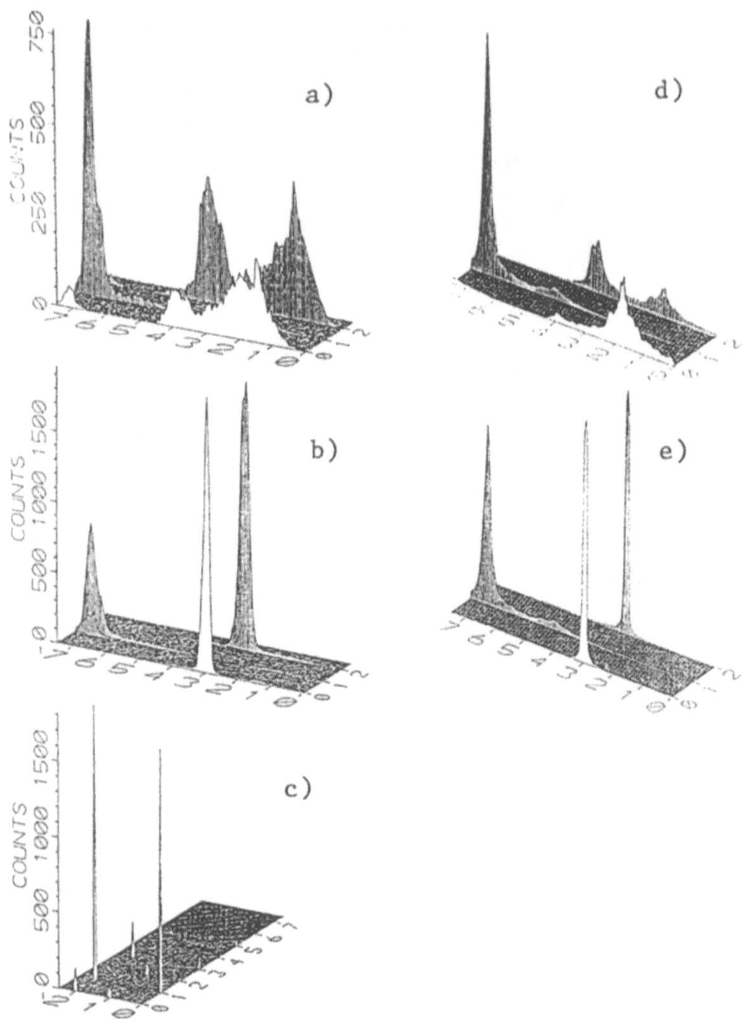

Figure 2. Intensity maps of the a,d: (h,1,l), b,e: (1,k,l), c: (h,k,3) reciprocal lattice planes of the $\sqrt{2}\sqrt{2}\sqrt{3}$ structure. a,b,c: measured, d,e: calculated.

stacking of C_{60} layers with $\sqrt{2}$ and $\sqrt{3}$ shifts is random with the restrictions: 4. The ratio of the number of $\sqrt{2}$ and $\sqrt{3}$ shifts is 2:1. 5. The possibility of two adjacent $\sqrt{3}$ shifts is excluded.

The calculated intensity maps are shown on Fig. 2d–e.

4. Conclusion

The two different structures found in C_{60}-ether show striking similarities to C_{60}-n-pentane. All three structures are based on square-planar layers of C_{60}

molecules which form channels containing guest molecules. However, in the case of C_{60}-ether there is a significant stacking disorder which comes mainly from a random sequence of $\sqrt{2}$ and $\sqrt{3}$ layer shifts. The condition that no $\sqrt{3}$ shifts can be adjacent means that apart from the apparent disorder, C_{60}-n-pentane and the two C_{60}-ether structures are members of the same $(C_{60})_{n+1}X_2$ homologous series (with n = 1, 2, and 3, respectively) described in Ref. 4.

Acknowledgement

This work has been supported by OTKA grants under contract number T4474 and T4222.

References

1. R. M. Fleming et al.: Phys. Rev. B, **44**, 888 (1991)
2. S. Pekker, G. Faigel, K. Fodor-Csorba, L. Gránásy, E. Jakab, M. Tegze: Solid State Comm. **83**, 423 (1992)
3. W. Krätschmer, L. D. Lamb, K. Fostiropoulos, D. R. Huffman: Nature, **347**, 354, (1990)
4. S. Pekker, G. Faigel, G. Oszlányi, M. Tegze, T. Kemeny, E. Jakab: *Proc. of ICSM'92*, Göteborg, 1992; Synthetic Metals (to be published)
5. G. Oszlányi, G. Bortel, G. Faigel, S. Pekker, M. Tegze: J. Phys. CM (in press)

Part IV

Infrared and Raman Spectroscopy, Luminescence, Transient Optics, and NLO

Vibrational Spectroscopic Studies of Solid C_{60} Films

P.C. Eklund[1,2], Z.-H. Dong[1,2], Y. Wang[1,2], M.J. Holden[1], K.A. Wang[1,2], P. Zhou[1,2], A.M. Rao[1,2], G. Dresselhaus[3], and M.S. Dresselhaus[4]

[1]Department of Physics and Astronomy, University of Kentucky, Lexington, KY 40506, USA
[2]Center for Applied Energy Research, University of Kentucky, Lexington, KY 40511, USA
[3]Francis Bitter National Magnet Laboratory, M.I.T., Cambridge, MA 02139, USA
[4]Department of Physics, M.I.T., Cambridge, MA 02139, USA

Abstract. Experimental studies of the vibrational modes of pristine and photopolymerized solid C_{60} films are presented. A modest flux of visible or ultraviolet light is shown in the temperature range $250 < T < 373$ K to transform C_{60} into a second solid phase which is identified with a polymeric form. A photochemical "2+2 cycloaddition" reaction mechanism is proposed to form the C_{60} polymer.

Solid C_{60} crystallizes with the molecules centered on positions of a face centered cubic (fcc) lattice (d=14.2 Å)[1]. Consistent with the weak van der Waals bonds between molecules in the solid state, an orientational ordering transition is observed at $T_o \sim 260$ K[2, 3]. Above T_o, the molecules are spinning about their fcc lattice positions, and below T_o this motion freezes out exponentially, and the system finally settles in a merohedrally disordered phase[1, 4] in which the electron rich double bonds on one molecule are either opposite the electron deficient pentagonal face (83%) or hexagonal face (17%) on an adjacent molecule[5].

Perhaps the best evidence for the nearly ideal molecular behavior of solid C_{60} comes from vibrational spectroscopy[6]. Both infrared (IR) and Raman spectroscopies[6, 7] reveal a vibrational spectrum with strong first-order-allowed intramolecular modes whose frequencies are in good agreement with both phenomenological[8] and *ab initio* [9] calculations for a free molecule. Numerous weaker features in the IR[10] and Raman[11] spectra can be observed, however, which can be attributed to either 2nd and higher order processes, or to the isotopic activation (e.g. $^{13}C^{12}C_{59}$) of otherwise optically silent modes. Unlike conventional solids, where the 2nd order IR or Raman spectra reveal broad features whose widths stem from the dispersion in the optic and acoustic phonon branches, these higher order features in solid C_{60} are very narrow, consistent with the properties of a molecular solid. Due to the icosahedral symmetry of a C_{60} molecule, group theory predicts [12,13] the existence of 46 distinct vibrational mode frequencies ($2A_g+8H_g$ {Raman-active}; $4F_{1u}$ {IR-active}; A_u, $5F_{2u}$, $6G_u$, $7H_u$, $3F_{1g}$, $4F_{2g}$, $6G_g$ {silent}). In Fig. 1, we display Raman spectra taken at temperatures T=20 and 523 K on ~7000 Å films of C_{60} which were vacuum deposited onto room temperature substrates. As can be seen, the spectra are quite rich, exhibiting numerous, sharp, second-order (or higher-order) features. The low T spectrum is in good agreement with the 40 K spectrum obtained by van Loosdrecht below ~ 1600 cm^{-1} on single crystal C_{60}[14]. We have taken spectra at 20 and 523 K to compare the properties of C_{60} in the low T, orientationally ordered phase to those of the unique high T phase in which the molecules spin about their fcc lattice positions. In this way, we can study the effect of the low T phase crystal field symmetry on the spectrum.

We consider first the region in the spectrum (Fig.1) up to and including the $H_g(8)$ line at 1577 cm^{-1}. In this region, 44 lines are observed experimentally, 10 of which are the Raman-allowed 1st order lines. Two possible identifications are possible for the 34 additional lines: they could be assigned either to *silent modes* isotopically activated in 1st order, or to *higher-order combination or overtone scattering*. It should be realized that ^{13}C is 1.1 % abundant, and therefore there is a ~60 % probability that a C_{60} molecule will actually be $^{13}C^{12}C_{59}$ rather than $^{12}C_{60}$. Thus, these two molecular types appear in almost equal numbers in our films which were originated from natural

Springer Series in Solid-State Sciences, Volume 117
Electronic Properties of Fullerenes Eds.: H. Kuzmany, J. Fink, M. Mehring, and S. Roth

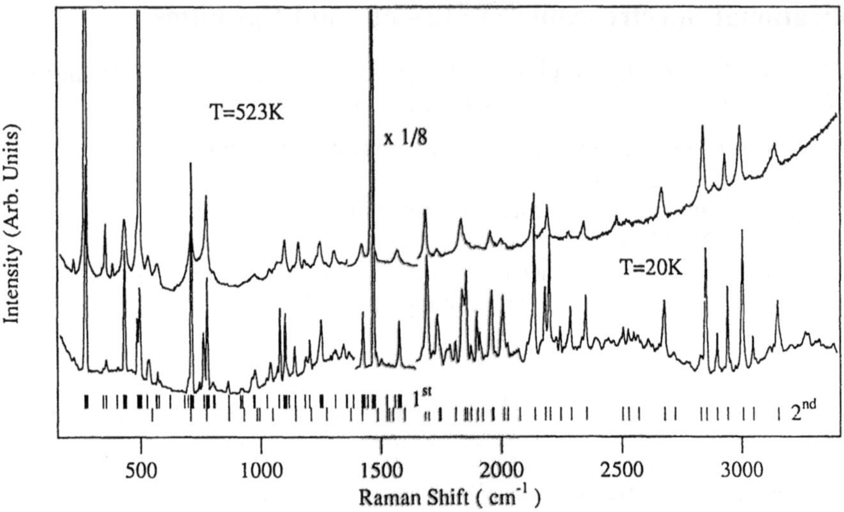

Fig.1 Raman spectra for solid C_{60} films (~7000Å) taken at T=523 K and 20 K using 4880Å Ar laser radiation. The mode frequencies shown schematically in the figure are: 46 fundamental frequencies (upper, or "1st"), where the Raman-allowed modes are indicated by the thicker ticks; 2nd order modes $(A_g+H_g)\otimes(A_g+H_g)$ (lower, or "2nd"). Some of the calculated frequencies cannot be resolved in the schematic representation at the bottom of the figure.

graphite rods. Isotopic substitution breaks the icosahedral symmetry and activates (in principal) all the Raman-silent modes. The experimental possibility of observing these isotopically activated modes is well documented, for example, in benzene[15]. For C_{60}, we tentatively assign[11] the existence of the following lines to isotopic activation: $H_u(1)/343$; $F_{2u}(1)/355$; $G_u(1)/399$; $G_g(1)/486$; $F_{1u}(1)/526$; $F_{2g}(1)/566$; $F_{1u}(2)/576$; $F_{1u}(3)/1183$, where the number following the symmetry label corresponds to the experimental frequency in cm^{-1} which are found in reasonable agreement with other experiments (neutron scattering [16] and photoluminescence[17]) and calculations [18]. Thirteen other lines in the region below the $H_g(8)$ line may also be due to isotopically-activated modes, or alternatively, they could be assigned to a 2nd order Raman-allowed line[11]. Positions of calculated modes are shown schematically at the bottom of Fig. 1: upper row (marked "1st"), the thick and thin lines, respectively, indicate 1st order allowed Raman modes and the other 36 fundamental intramolecular mode frequencies; bottom row (marked "2nd") indicate the calculated positions for a subset of 2nd order modes (discussed below) obtained from A_g and H_g symmetry vibrations. As is evident in the figure, virtually all the lines observed are present in both the orientationally disordered phase (523 K) as well as in the ordered phase (20 K). Therefore, we cannot assign any of the lines observed in the 20 K spectrum to intramolecular modes activated (or split) by the low T crystal field, as has been suggested by van Loosdrecht et al.[14]. Significant thermal broadening is apparent, however, even on the large frequency scale of the figure.

From group theoretical considerations, a total of 812 Raman-active 2nd order $(\omega_1 + \omega_2)$ modes are expected with A_g (151) and H_g (661) symmetry. Above ~ 1600 cm^{-1}, we find that most of the strongest higher order lines in Fig.1 can be assigned to a subset of these modes involving combinations or overtones of A_g and H_g modes, in agreement with the conclusion of Denisov et al.[19]. This subset is identified by taking the direct product $(A_g + H_g)\otimes(A_g + H_g)$, and their calculated positions are indicated in the bottom row of Fig. 1. Many other experimental lines can be assigned to combination mode scattering involving an H_g (or A_g) mode and a second, Raman-silent mode [11]. Overall, quite a few of the 2nd order assignments are still tentative, however, because other Raman-allowed second-order lines also have nearby frequencies. More work is still needed to sort out the rich structure in Fig. 1.

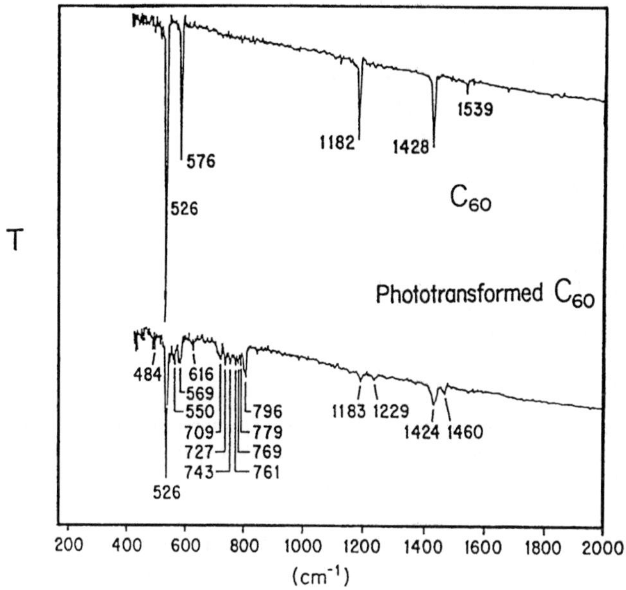

Fig.2 Fourier transform infrared transmission spectra of pristine C_{60} and phototransformed C_{60}. Film samples were vacuum deposited on KBr substrates.

Considerable confusion has persisted in the literature about the experimental frequency for the $A_g(2)$ mode in solid C_{60} [20]. In large part, this has stemmed from numerous studies on samples studied either in air or exposed to too high a laser flux[21]. According to group theory, the $A_g(2)$ line should be 100% polarized, and a line reported at 1469 cm^{-1} (T=300K) has this property[6, 21]. A second value for this frequency has been reported at ~1458 cm^{-1}[20, 22-24]. We have pointed out, however, that this 1458 cm^{-1} line is unpolarized and is the result of a phototransformation [21] of solid C_{60} which can occur with the sample maintained either in a vacuum or in an inert atmosphere. Phototransformation can be achieved with a modest flux of light with a photon energy greater than the bandgap (~1.7 eV[25]). We have identified the ~1458 cm^{-1} line with the photodimerization of adjacent C_{60} molecules, leading eventually to a photopolymerization of the lattice[26]. The subsequent, light-induced covalent coupling of C_{60} molecules lowers the molecular symmetry thereby activating numerous otherwise optically silent intramolecular modes. This activation can be appreciated by comparing, for example, the T=300 FTIR transmission spectra (Fig. 2) for a pristine C_{60} film and that of a photopolymerized film. In the spectrum for pristine C_{60}, the strong absorption lines are identified with the 4 F_{1u} modes, in agreement with early work[7]. In the case of phototransformed C_{60}, numerous new lines, activated by the broken molecular symmetry associated with C-C bonds crosslinking adjacent molecules, are observed. Direct evidence for this cross linking is shown in the laser desorption mass spectrum of a polymerized film (Fig. 3) showing a series of peaks associated with integer multiples of the C_{60} mass (720 amu) out to a mass equivalent to 20 C_{60} molecules linked together[26, 27]. Evidence for the strength of these crosslinking bonds has been provided by the observation of a new, low frequency Raman line at ~ 118 cm^{-1} for the photopolymer which has been identified with the stretching of these intermolecular bonds [26, 28]. This frequency lies in the gap between the lattice modes and the intramolecular modes of solid C_{60}.

In Fig. 4 we demonstrate the T-dependence of the photopolymerization process using Raman scattering to monitor changes in the region surrounding the $A_g(2)$ mode in pristine C_{60}[29]. A 3000 Å C_{60} film deposited on a Cu substrate (for thermal conduction) was studied at the various

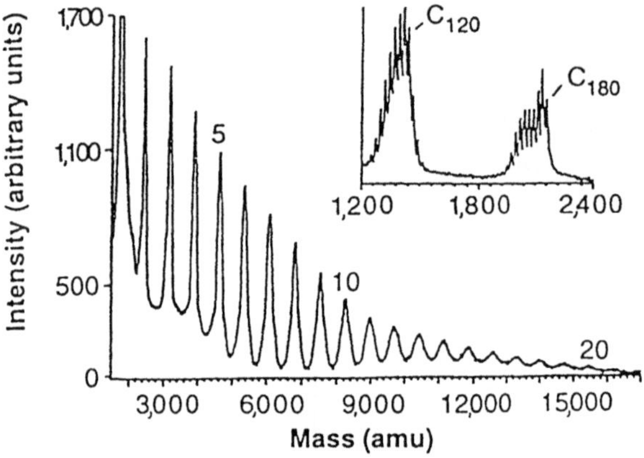

Fig.3 Laser desorption mass spectrum (LDMS) of a C_{60} film (~2000Å) phototransformed with UV-visible radiation from a 300-watt Hg arc lamp for 2 days. Inset: LDMS spectrum shown on an expanded scale in the region of the dimer (C_{120}) and trimer (C_{180}).

temperatures indicated. A fresh spot on the sample was used at each T and exposed first to 488 nm radiation (450 mW/mm^2) for 20 min. before the Raman spectra shown in the figure were acquired rapidly (10 sec) at this T. Between 20 K and 230 K very little effect of the irradiation is noticed; the film appears insensitive to the incident light flux, and the spectrum of pristine C_{60} with a strong $A_g(2)$ line near ~ 1469 cm^{-1} is obtained. However, for T > 230 K the effects of the irradiation first become evident; the intensity in the intrinsic 1469 cm^{-1} line decreases at the expense of the growth of a new, broader ~1458 cm^{-1} line identified with the photopolymer. At T=260 K, the data in Fig.4 show that the conversion to the photopolymer is complete during this 20 min. exposure to 488 nm radiation. The ratio R of the integrated intensities of the 1458 and 1469 cm^{-1} lines resulting from the irradiation of the film is plotted vs. temperature in Fig. 5, where a clear threshold for the phototransformation is observed at a temperature very near the orientational order transition T_o ~ 255-260K. This observation has interesting implications regarding the photodimerization mechanism, as we discuss next.

"2 + 2 cycloaddition" is a well known photochemical reaction mechanism which can join (via C-C bonds) otherwise weakly coupled hydrocarbon molecules[30]. The mechanism has been reported to require reactive double bonds (C=C) on adjacent molecules which are nearly parallel and separated by less than 4.2Å[30]. The "2+2 cycloaddition" reaction results in a rearrangement of the carbon bonds in these double bonds to form a four-membered ring as the bridge between the molecules (i.e.,)(+)(→ □). In many cases, which might involve either aromatic or chain-like hydrocarbon molecules, the incident photon is used to place one of these molecules in the first excited triplet state (T_1). This excited molecule then reacts with the adjacent molecule in its ground state. In C_{60}, several experiments indicate that significant population of T_1 is possible by optical pumping, followed by a ~100% efficient[31, 32] intersystem crossing. Furthermore, the T_1 state is long lived (~ 40 μsec[31, 32]). Another observation consistent with the importance of the T_1 state in the photopolymerization of C_{60} is the "hardening"[21] of the film against phototransformation by the incorporation of O_2 into the film (O_2 is known to quench the T_1 state[33]). The observed threshold for the photopolymerizability of solid C_{60} (Fig. 5) is indeed consistent with our proposed photochemical "2+2 cycloaddition" mechanism[29], since the topochemical requirement of parallel C=C bonds on adjacent C_{60} molecules cannot be met in the low T phase, for reasons discussed in the first paragraph of the text. However, above T_o, the molecules first begin to spin rapidly about their lattice positions and it is then possible for one of the 30 reactive double bonds on one C_{60} to be aligned nearly parallel with one of 30 such bonds on an adjacent C_{60}. Of course, other experiments, such as NMR will be needed to confirm the presence of the four-membered ring that we propose to exist between molecules in photopolymerized C_{60}.

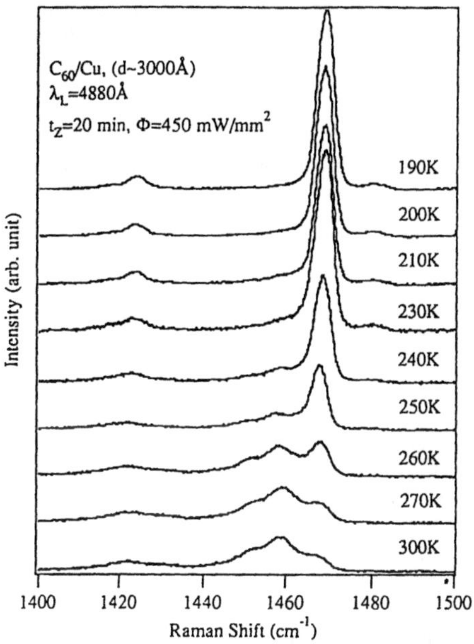

Fig.4 Raman spectra of a C_{60} film (d~3000Å) on a Cu substrate as a function of temperature (T), and collected after 20 min pre-exposure to laser irradiation at the same T.

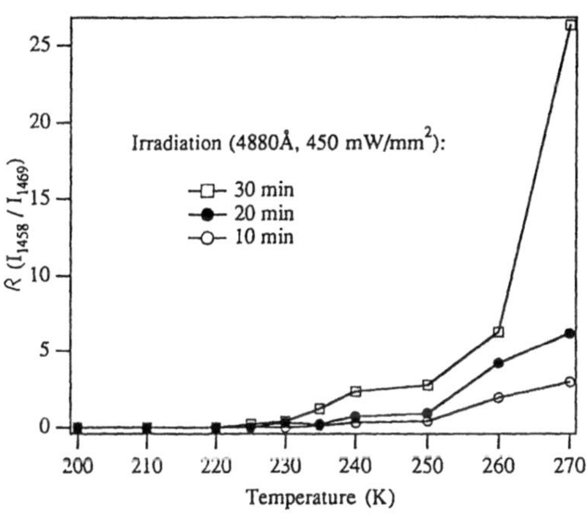

Fig.5 Ratio (R = I_{1458}/I_{1469}) of the integrated intensity of the *unpolarized* 1458 cm^{-1} line in phototransformed C_{60} to that of the *polarized* 1469 cm^{-1} "pentagonal pinch" mode of pristine C_{60}.

Finally, we should mention that it is also possible to use Raman scattering to measure the thermal decomposition of the C_{60} polymer[28]. By monitoring the region between 1400 and 1500 cm^{-1}, we have tracked the disappearance of the 1458 cm^{-1} mode associated with the photo-induced polymer with increasing T. We find that photopolymerized films begin to transform back to pristine solid C_{60} at ~ 100 °C . At this temperature, the intensity under the 1458 cm^{-1} line was found to decrease, as the intensity under the ~1469 cm^{-1} line first begins to grow. The transformation back to pristine C_{60} is complete at T~ 170 °C. Furthermore, the T-dependence of the ~118 cm^{-1} mode was also studied and found to mirror the behavior of the ~1458 cm^{-1} line, beginning to lose intensity at 100 ° C and extinguishing at ~170 °C.

Acknowledgements: This work was supported, in part, by grants from the United States National Science Foundation #EHR-91-08764 (Univ. of Kentucky) and #DMR-92-01878 (M.I.T.).

References

1. Heiney, P., *J. Phys. Chem. Solids* **53**, 1333 (1992).
2. Heiney, P.A., *et al.*, *Phys. Rev. Lett.* **66**, 2911 (1991).
3. David, W.I.F., *et al.*, *Nature* **353**, (1991).
4. David, W.I.F., Ibberson, *et al.*, *Europhys. Lett.* **18**, 219 (1992).
5. For a general review of Fullerenes, see Dresselhaus, M.S., *et al.*, *J. Mat. Res.* in press (1993).
6. Eklund, P.C.,*et al.*, *J. Phys. Chem. Solids* **53**, 1391 (1992).
7. Meijer, G. & Bethune, D.S., *Chem. Phys. Lett* **175**, 1 (1990).
8. Jishi, R.A., *et al.*, *Phys. Rev. B.* **45**, 13685 (1992).
9. Negri, F., *et al.*, *J. Am. Chem. Soc.* **113**, 6037-6040 (1991).
10. Wang, K.A., *et al.*, M.S., *Phys. Rev. B* ,unpublished (1993).
11. Dong, Z.H., *et al.*, *Phys. Rev.*, in press (1993).
12. Weeks, D.E. & Harter, W.G., *J. Chem. Phys.* **90**, 4744 (1989).
13. Dresselhaus, M.S., *et al.*, *Phys. Rev. B* **45**, 6923 (1991).
14. van Loosdrecht, P.H.M., *et al.*, *Chem. Phys. Lett.* **198**, 587 (1992).
15. Painter, P.C. & Koenig, J.L., *Spectrochimica Acta.* **33A**, 1003 (1977).
16. Coulombeau, C., *et al.*, *J. Phys. Chem.* **96**, 22 (1992).
17. Nissen, M.K., *et al.*, *Phs. Rev. Lett.* **69**, 2423 (1992).
18. Jishi, R.A., *et al.*, *Phys. Rev. B*, in press (1993).
19. Denisov, V.N. , *et al.*, *Soviet Phys. JETP* **75**, 158 (1992).
20. Duclos, S.J., *et al.*, *Solid State Comm.* **80**, 481 (1991).
21. Zhou, P., *et al.*, *Appl. Phys. Lett.* **60**, 2871 (1992).
22. Tolbert, S.H., *et al.*, *Chem. Phys. Lett.* **188**, 163 (1992).
23. Matus, M. & Kuzmany, H., *Appl. Phys.* **A56**, 241 (1993).
24. Akers, K., *et al.*, *Science* **259**, 1152 (1993).
25. Rao, A.M., *et al.*, *Phys. Rev. B.* unpublished (1993).
26. Rao, A.M., *et al.*, *Science* **259**, 955 (1993).
27. Cornett, D.S., *et al.*, *J. Phys. Chem.* to be published (1993).
28. Wang, Y., *et al.*, *Chem. Phys. Lett.*, unpublished (1993).
29. Zhou, P., *et al.*, *Nature* unpublished (1993).
30. Venkatesan, K. & Ramamurthy, V., *Photochemistry in Organized and Constrained media* p133 (VCH, New York, 1991).
31. Arbogast, J.W., *et al.*, *J. Phys. Chem* **95**, 11-12 (1991).
32. Haufler, R.E., *et al.*, *Chem. Phys. Lett.* **179**, 449 (1991).
33. Hung, R.R. & Grabowski, J.J., *J. Physical Chem.* **95**, 6073-6074 (1991).

Crystal Field Effects and Electronic Excitations in Single-Crystal C_{60}

P.H.M. van Loosdrecht

Research Institute for Materials, University of Nijmegen, Toernooiveld, 6525 ED Nijmegen, The Netherlands

Abstract. The symmetry lowering by the crystal field in solid C_{60} leads to Davydov splittings and activation of molecular modes, as observed by Raman spectroscopy on single crystal C_{60}. The influence of 514 nm irradiance on the Raman spectrum of C_{60} is discussed in terms of the long living excitonic triplet state of C_{60}. At high irradiance the measured spectrum shows the vibrational properties of the excited state, rather than those of the ground state. The model used also explains the influence of the presence of oxygen on the Raman spectrum.

1 Introduction

Solid Buckminsterfullerene (C_{60}) is a van der Waals bonded crystal with a large ratio between the intra- and intermolecular forces. The physical properties of this solid are therefore strongly influenced by the molecular properties of C_{60}. Nevertheless, solid state interactions lead to a number of interesting phenomena in this material such as the orientational ordering transition at 260 K, the presence of delocalised one dimensional excitons, and superconductivity in alkali-doped material.

Raman spectroscopy forms a convenient tool to study solid state interactions, since the vibrational spectrum of a molecule, as well as optical scattering processes involving the vibrations, are highly sensitive to symmetry. The first part of this paper discusses some of the solid state effects as observed by Raman spectroscopy on single crystal C_{60}. In studying the Raman spectrum of solid C_{60} it has been found that the vibrational properties are highly sensitive to the laser power incident on the samples, as well as to the presence of oxygen. The origin of these effects is briefly discussed in the last part of this paper. Before turning to the Raman experiments we first discuss the crystal preparation and the structural aspects of solid C_{60}.

2 Crystal preparation

Fullerene rich soot is produced from a DC arc discharge between two high purity graphite electrodes in a 0.2 atm He environment. C_{60}, C_{70}, and higher fullerenes are separated using soxhlet extraction on this soot in boiling decaline.

Springer Series in Solid-State Sciences, Volume 117
Electronic Properties of Fullerenes Eds.: H. Kuzmany, J. Fink, M. Mehring, and S. Roth

Pure C_{60} (> 99.5 %) is extracted from the obtained fullerene mixture using liquid column chromatography. Further purification is performed by washing the C_{60} powder in cyclohexane in order to remove any residual poly-aromatic hydrocarbons. The solvent still present in the obtained powder is removed by a drying procedure at 200 °C under 10^{-5} Torr for several hours. The purity (> 99.9 %) of the C_{60} powder is verified using quantitative mass-spectrometry and nuclear magnetic resonance.

Large C_{60} single crystals (< 8 mm^3) are grown using a vapour transport method.[1] In this method about 10 mg of C_{60} powder is placed in the first compartment of a quartz tube (ø1 cm×50 cm) containing several equidistant (1 cm) necks. After the drying procedure the C_{60} is sublimed at 500 °C under dynamic pumping. The C_{60} solidifies again in the next compartment of the tube which kept at a temperature of about 400 °C. Surprisingly, not all material present in the first compartment is sublimed. A small fraction of the material stays behind, and will not even sublime at 800 °C. Both mass-spectrometry and Raman spectrometry indicate, however, that a considerable fraction of the residual material is C_{60}. The residual material is removed by melting off the first compartment, and is not used in subsequent steps of the crystal growth. Sublimation can serve as an efficient final purification step. Therefore, the material of the second compartment is additionally sublimed and solidified for two or three times in the same manner. During these steps all the material sublimes and is vapour transported to the subsequent compartment. In the final step, the actual growth step, the high purity C_{60} is kept in a single sealed off compartment, heated to 500 °C for half an hour, and then slowly cooled down to room temperature in several hours. In this way crystals are obtained having typical dimensions .5×.5×.5 mm^3, which are often macroscopically untwinned and show the characteristic {100} and {111} growth forms of the beautiful morphology of the fcc phase. The high purity of the crystals is once more checked using mass-spectroscopy and Raman spectroscopy on the crystals themselves.

3 Crystal structure

The structure of the crystals has been studied using x-ray diffraction[2], electron diffraction and high resolution electron microscopy[3]. At room temperature the crystals are in the well established fcc phase[4], which arises by virtue of the orientational disorder of the molecules. Below the orientational ordering transition at $T = 260$ K the crystals show the simple cubic phase[5] with four orientationally inequivalent molecules per primitive cell. In electron diffraction experiments on defect free regions of C_{60} crystals crushed at liquid nitrogen temperature van Tendeloo *et al.* found strong evidence for the existence of a superstructure with doubled axes. Strong support for the existence of this superstructure has been found independently by triplet electron paramagnetic resonance experiments.[6] In the superstructure two groups of molecules can be distinguished, differing in orientation by a rotation over 60° around the <111> directions. The exact temperature of the phase transition to the superstructure is at present unknown. Because usually only a small fraction of a macroscopic crystal is truly defect free it seems doubtful that one can show the existence of the superstructure using x-ray diffraction, indeed experiments in this direction have not found evidence for it yet.[7] Finally, evidence has been found for a glass-like transition near 90 K in which the reorientational motions completely freeze out.[8]

4 Crystal field effects

The high I_h symmetry of the C_{60} molecule has a strong influence on the number of energetically distinguishable phonon modes and on the selection rules in optical scattering experiments. For infrared absorption only four active modes are expected for the free molecule, whereas in Raman scattering one expects 10 active modes. Although the properties of solid C_{60} are to a large extend determined by the intra-molecular forces, it has been shown that one can not neglect the crystal field which effectively lower the symmetry of the molecules.[9, 10, 11]
Figure 1 depicts the correlations between the molecular modes and the modes in the fcc and sc phases. Since all modes belonging to the gerade irreducible representations of the T_h group are Raman active it is clear from figure 1 that all gerade molecular silent modes become active in the solid state. Furthermore, the molecular modes split up. In the fcc phase this occurs for the four- and fivefold degenerate modes, whereas in the sc phase all modes, including the A_g modes, are expected to split up.
This activation of the silent modes, as well as the splitting is illustrated in figure 2 for several modes. Apart from this, most of the modes show a strongly decreasing linewidth, and shift towards higher frequency upon decreasing temperature. These changes occur predominantly at the orientational phase tran-

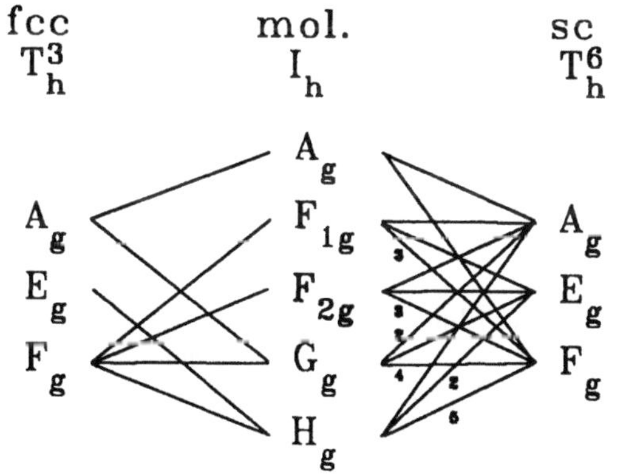

Figure 1: Correlation diagram of the gerade molecular modes (I_h-symmetry) to the modes in the fcc (T_h^3) and the sc (T_h^6) phase. The multiplicity of the correlation lines is one unless indicated otherwise. The correlations for the ungerade modes are analogous.

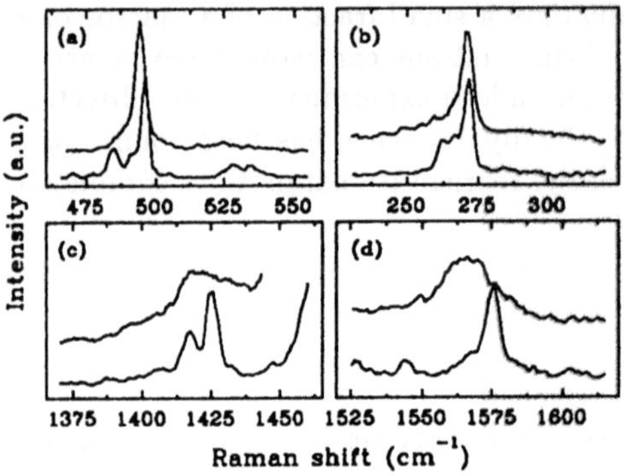

Figure 2: Parts of the unpolarised Raman spectra of single crystal C_{60} at T = 280 K (upper curves) and T = 40 K (lower curves) showing splittings, broadenings, lineshifts, and activation of new lines in the vicinity of the (a) 496 cm^{-1}, (b) 273 cm^{-1}, (c) 1425 cm^{-1}, and (d) 1575 cm^{-1} lines.

sition. It has been argued that the decreasing linewidth results from a decreased vibrational-rotational coupling in the orientationally ordered phase.[9] The frequency changes result mainly from differences in the population of electronically excited states above and below the orientational phase transition. Electronic excitation of the molecules can be avoided by using a excitation energy smaller than the S_0-T_0 energy gap. In this case lineshifts of the order of 2 cm^{-1} are observed, induced by the lattice contraction at the orientational phase transition.[10]

5 Electronic excitations

The Raman spectrum of C_{60} strongly depends on the irradiance and the presence of oxygen.[12, 13] The influence of the irradiance on the pentagonal pinch mode is shown in figure 3. At low irradiance a single mode is observed at 1468 cm^{-1}. As the laser power increases a new mode appears in the spectrum, at a somewhat lower frequency, with a simultaneous decrease of the intensity of the original mode. At high enough irradiance the original mode has vanished and only the power induced mode is observed. It should be noted that the observed effects are not due to laser heating, and are reversible in the sense that immediately after (within seconds) the laser power has been reduced the original spectrum reappears. Only above 500 W/cm^2 irradiance the process becomes irreversible. It is most likely that using these high irradiances the C_{60} is polymerised in the same manner as has been found by Eklund et al.[14]

The above described reversible behaviour has been attributed to a progressing triplet exciton population upon increasing irradiance, and can be understood in terms of a simple four level model for C_{60} where the 1468 cm^{-1} mode is the ground state mode, and the induced mode is due to electronically excited C_{60}.[12] Essential in this model is the long lifetime (0.4 ms) of the triplet state. As is well known, this lifetime is drastically reduced in the presence of oxygen. This explains the confusion in literature concerning the position of the pentagonal pinch mode. In the presence of oxygen the triplet state is effectively quenched and Raman spectroscopy measures the ground state properties of

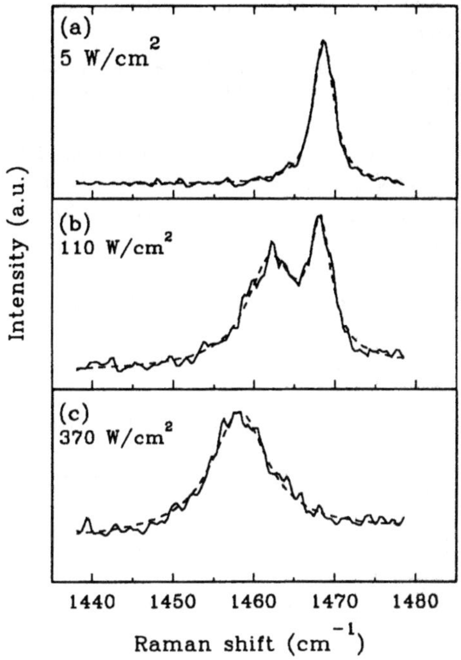

Figure 3: Irradiance dependence of the 514 nm Raman spectrum ($T =$ 40 K) in the vicinity of the pentagonal pinch mode. Clearly a new mode appears in the spectrum for higher irradiances, at the cost of the intensity of the original mode.

C_{60}, yielding 1468 cm^{-1} for the highly polarised pentagonal pinch mode. In the absence of oxygen the molecules are easily excited into the triplet manifold and "pentagonal pinch" mode is found at a lower frequency. Moreover, this mode is not purely symmetrical anymore and partly depolarised. It is surprising that at high irradiance the original mode completely vanishes, indicating a nearly full depopulation of the ground state. If the induced mode is indeed due to triplet state C_{60}, one would expect that decay processes such as triplet-triplet annihilation limits the depopulation of the ground state. Obviously this is not the case here. It is very well possible that the excited states are a precursor for the polymerisation process. In this view the excited molecules form metastable clusters, which eventually lead to polymerisation.

Acknowledgments. Collaboration with M.A. Verheijen, P.J.M. van Bentum, G. Meijer, and others are gratefully acknowledged. Part of this work has been supported by the Dutch Foundation for Fundamental Research of Matter (FOM).

References

[1] M.A. Verheijen, G. Meijer, H. Meekes, E. Raas, and P. Bennema, Chem. Phys. Lett. **191**, 339 (1992).

[2] J. de Boer, University of Groningen, unpublished.

[3] G. van Tendeloo, S. Amelinckx, M.A. Verheijen, P.H.M. van Loosdrecht, and G. Meijer, Phys. Rev. Lett. **96**, 7424 (1992).

[4] R.M. Flemming, T. Siegrist, P.M. March, B. Hessen, A.R. Kortan, D.W. Murphy, R.C. Haddon, R. Tycko, G. Dabbagh, A.M. Mujsce, M.L. Kaplan, and S.M. Zahurak, Mat. Res. Soc. Proc. **206**, 691 (1991).

[5] P.A. Heiney, J.E. Fisher, A.R. McGhie, W.J. Romanow, A.M. Denenstein, J.P. McCauly Jr., A.B. Smith III, and D.E. Cox, Phys. Rev. Lett. **66**, 2911 (1991).

[6] E.J.J. Groenen, O.G. Poluektov, M.Matsushita, J. Schmidt, J.H. van der Waals, and G. Meijer, Chem. Phys. Lett. **197**, 314 (1992).

[7] J.E. Fisher et al., preprint.

[8] X.D. Shi, A.R. Kortan, J.M. Williams, A.M. Kini, B.M. Savall, and P.M. Chaikin, Phys. Rev. Lett. **68** (1992), 827.

[9] P.H.M. van Loosdrecht, P.J.M. van Bentum, and G. Meijer, Phys. Rev. Lett. **68**, 1176 (1992).

[10] P.H.M. van Loosdrecht, P.J.M. van Bentum, M.A. Verheijen, and G. Meijer, Chem. Phys. Lett. **198**, 587 (1992).

[11] L.R. Narasimhan, D.N. Stoneback, A.F. Hebard, R.C. Haddon, and C.K.N. Patel, Phys. Rev. B **46**, 2591 (1992).

[12] P.H.M. van Loosdrecht, P.J.M. van Bentum, and G. Meijer, Chem. Phys. Lett. **205**, 191 (1993).

[13] S.J. Duclos, R.C. Haddon, S.H. Glarum, A.F. Hebard, and K.B. Lyons, Sol. St. Comm. **80**, 481 (1991).

[14] P.C. Eklund et al., this volume.

Stability of Single-Crystal and Thin-Film Raman Spectra in C_{60}

M. Matus, J. Winter, and H. Kuzmany

Institut für Festkörperphysik, Universität Wien, A-1090 Wien, Austria

Abstract. We have analysed the unusual light induced changes in the Raman spectra of C_{60} single crystals and thin films. This changes revealed a strong dependence on the exposure of the material to ambient conditions. The changes are suppressed for temperatures above 80°C and for temperatures moderately below room temperature. Excitation to a triplet state and interaction with oxygen are considered as the main sources for the unusual behavior of the spectra.

1. Introduction

Since the early days of fullerene research, Raman spectroscopy proved to be a valuable tool for the investigation of this new allotropic modification of carbon. The existence of only 10 Raman lines [1] together with only 4 IR lines was the first evidence for the exceptionally high symmetry of the C_{60} molecule. On the other hand, Raman spectra of the undoped material were observed to be unstable if taken at room temperature. Duclos et al. [2] found a change of the spectra under the influence of light and oxygen. Similar results, and a characteristic influence of oxygen on the luminescence were reported by us in a preceding paper [3].

On the other hand, C_{60} can be converted to a highly reactive triplet state by irradiation with light [4]. Recently this effect has been observed in low temperature Raman spectra [5]. In a very recent work a photo-induced polymerisation of C_{60} [6] was found which could finally be correlated with the instability of the Raman spectra [7].

We report the development of the Raman spectra of solid C_{60} after different exposures to ambient conditions and for excitation at different temperatures. As an indicator for the stability we used the intensity of the $A_g(2)$ Raman mode at $1468\,\mathrm{cm}^{-1}$.

2. Experimental

Measurements were performed on thin films and single crystals, respectively. Fullerene films were prepared by evaporation of the purified material on silicon and quartz substrates. Single crystals of pure C_{60} with typical dimensions of 0.5–2 mm were grown using the vapor phase technique described in detail in Ref.8. The crystals were exposed to ambient conditions for varying periods of time to study the effect of oxygen on the stability. The thin film measurements were performed in vacuum in a temperature range of 200 K–420 K.

The Raman measurements were performed on a Dilor XY spectrometer with a liquid N_2 cooled CCD-detector. For the excitation the 514 nm line of an Ar^+ laser with intensity of the order of $100\,\mu$W was used. This yields a power density at the sample of about $14\,\mathrm{W/cm}^2$.

Springer Series in Solid-State Sciences, Volume 117
Electronic Properties of Fullerenes Eds.: H. Kuzmany, J. Fink, M. Mehring, and S. Roth

3. Results

The instability of the Raman spectra appears as a decrease of the Raman line at 1468 cm^{-1}, and as an increase of several broader lines next to it. This is shown in Fig.1 for a C_{60} single crystal. The bleaching effect was observed for laser intensities as low as 1 W/cm^2.

The dynamics of the bleaching process is strongly sample dependent. Figure 2 shows the relative intensity of the 1468 cm^{-1} mode as a function of laser light exposure for a C_{60} crystal under three different conditions (all at room temperature). First the crystal was measured in the sealed evacuated glass tube in which it was prepared (curve a). Then the spectrum was taken in air, immediately after removal of the crystal from the tube (curve b). The last experiment was performed after keeping the sample for six days in air (curve c). It is obvious that storage in air stabilizes the material against the photo transformation. Thin evaporated films which were exposed to air for an extended period of time were found to be even more stable [9]. Interestingly, there is no observable frequency shift of the pinch mode, even though its intensity changes dramatically with time.

The temperature at the irradiation process turns out to be crucial for the stability of the spectra. Whereas from previous measurements [9] it was found that the spectra are stable for low-temperature excitation, the present investigation proved a stability also for high-temperature excitation. Figure 3 shows the time dependence of the $A_g(2)$ mode intensity for four different sample temperatures. We measured a thin C_{60} film at 145°C and found no indication of a line bleaching during a one hour experiment. In stepwise lowering the sample temperature to 120°C and to 100°C, the material remained stable and the line intensity even increased with lowering the temperature. The bleaching of the 1468 cm^{-1} mode started only at temperatures as low as 80°C and showed a similar behavior as the one for the sample in Fig.2

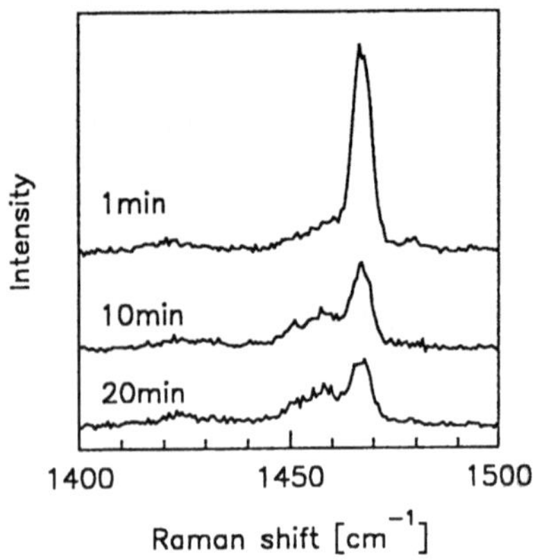

Fig.1: Raman spectra of a C_{60} single crystal near the pinch mode after different exposure times to laser light.

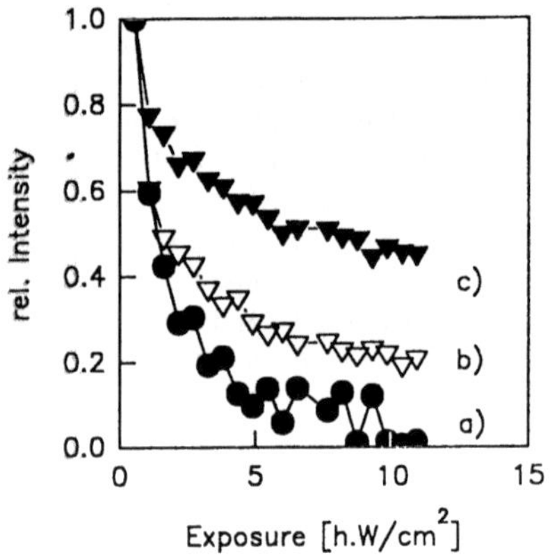

Fig.2: Relative intensity of the $A_g(2)$ mode at 1468 cm^{-1} as a function of laser light exposure (514 nm, 19 W/cm^2). a) C_{60} crystal in vacuum; b) after removal from glass tube; c) after exposure for 6 days in air.

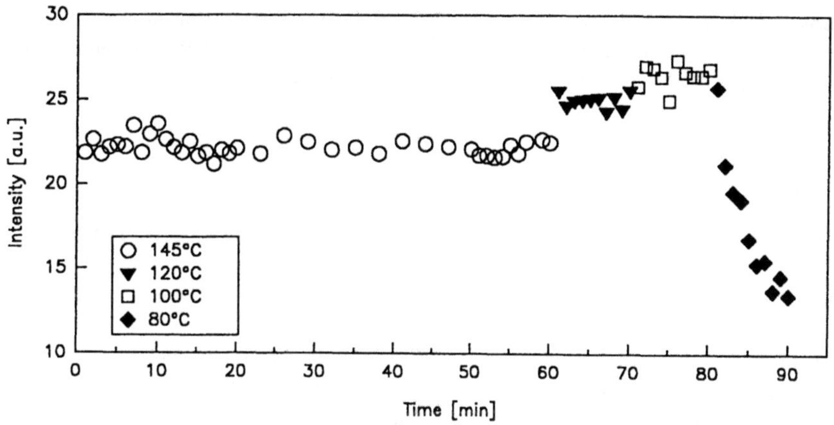

Fig.3: Intensity of the $1468\,cm^{-1}$ Raman mode of a C_{60} thin film (Laser: 514 nm, $14\,W/cm^2$). The measurement was started at a sample temperature of 145°C, then continued at 120°C, 100°C, and 80°C, respectively.

4. Discussion

To understand the observed effects, we propose the following qualitative model. Light with energy above the S_0–T_0 energy gap [4] produces a chemically highly reactive triplet C_{60} molecule. This excited molecule can react with neighbour C_{60} molecules to form a polymer like compound, as discussed by Eklund et al. [6]. This compound is probably similar to, but less stable than, the material described recently by Haluška et al. [8] as a cocoon which is left over after the sublimation of light exposed single crystals. The Raman spectrum of the new product might be different from the spectrum of pure C_{60}, resulting in an intensity decrease of intrinsic lines. The resonance condition for the new compound may be different from pure C_{60}, so that it appears as a much weaker scatterer. The fact that the Raman spectrum is stable at temperatures above 80°C indicates that either the lifetime of the triplet state is strongly reduced at higher temperatures, or the generated compound is not thermally stable. The latter fact agrees again with the observed instability of the light-induced product described by Eklund et al. [6]. The increase in intensity of the $1468\,cm^{-1}$ Raman mode with decreasing temperature can be explained by a change in the resonance behavior which we discussed in more detail in Ref.9.

It is well known that oxygen diffuses readily into solid fullerenes [10], which is also evident from Fig.2 b. In the presence of oxygen, the triplet state of C_{60} is efficiently quenched [11]. In this case, the rate for a chemical modification of C_{60} is strongly reduced. Thus the observed Raman spectrum originates from two types of materials. Some of the molecules are in a clean environment, but others have accumulated oxygen in their neighborhood. In both cases the pinch mode frequency is the same.

The fact that the photo reaction of C_{60} at room temperature occurs even for laser power densities of less then $1\,W/cm^2$ suggests that it is more or less impossible to record an intrinsic C_{60} spectrum using the conventional Ar^+ laser as an excitation source. In using light with energy below the S_0–T_0 gap one can avoid this effect, as it has recently been shown [5].

Acknowledgment. We acknowledge M. Haluška for the preparation of the C_{60} single crystals and valuable discussions with P.H.M. van Loosdrecht and P.C. Eklund. This work was supported by the FFWF in Austria, grant P8172.

References

[1] D.S. Bethune, G. Meijer, W.C. Tang, H.J. Rosen, Chem. Phys. Lett. **174** (1990), 219.

[2] S.J. Duclos, R.C. Haddon, S.H. Glarum, A.F. Hebard, K.B. Lyons, Solid State Commun. **80** (1991), 481.

[3] M. Matus, H. Kuzmany, E. Sohmen, Phys. Rev. Lett. **68** (1992), 2822.

[4] S. Leach, M. Vervloet, A. Desprès, E. Bréheret, J.P. Hare, T.J. Dennis, H.W. Kroto, R. Taylor, D.R.M. Walton, Chem. Phys. **160** (1992), 451.

[5] P.H.M. van Loosdrecht, P.J.M. van Bentum, G. Meijer, Chem. Phys. Lett. **205** (1993), 191.

[6] A.M. Rao, P. Zhou, K.-A. Wang, G.T. Hager, J.M. Holden, Y. Wang, W.-T. Lee, X.-X. Bi, P.C. Eklund, D.S. Cornett, M.A. Duncan, I.J. Amster, Science **259** (1993), 955.

[7] P.C. Eklund, (1993), this volume.

[8] M. Haluška, H. Kuzmany, M. Vybornov, P. Rogl und P. Fejdi, Appl. Phys. **A 56** (1993), 161.

[9] M. Matus, H. Kuzmany, Appl. Phys. **A 56** (1993), 241.

[10] R. Schlögl, (1993), this volume.

[11] J.W. Arbogast, A.P. Darmanyan, C.S. Foote, Y. Rubin, F. N. Diederich, M.M. Alvarez, S.J. Anz, R.L. Whetten, J. Phys. Chem. **95** (1991), 11.

Resonant Raman Scattering of the Normal and Superconducting State of A_3C_{60} Films (A = Rb,K)

C. Taliani[1], *V.N. Denisov*[2], *A.A. Zakhidov*[3], *G. Stanghellini*[1], *G. Ruani*[1], *and R. Zamboni*[1]

[1]Istituto di Spettroscopia Molecolare, CNR, Via de'Castagnoli 1, I-40126 Bologna, Italy

[2]Institute of Spectroscopy, Russian Academy of Sciences, Troitsk, 142092 Moscow, Russia

[3]Department of Thermophysics, Uzbek Academy of Sciences, Katartal 28, 700135 Tashkent, Uzbekistan

Abstract. Significant spectral changes upon crossing superconducting T_c have been observed by resonant Raman scattering (RRS) (excited at 1.16 eV) in Rb_3C_{60} and K_3C_{60} thin films. In RRS spectra $H_g(1)$ and $H_g(2)$ modes clearly show Fano-lineshapes due to resonance with electronic continuum. The electronic Raman scattering is clearly observed as a broad background (fwhm = 450 cm^{-1}) at low energy. The large linewidth observed in $H_g(2)$ (fwhm = 60 cm^{-1}) indicates a sizable electron-phonon coupling. Below T_c only $H_g(2)$ mode changes dramatically its asymmetric lineshape becoming more symmetric in the superconducting phase. The fitting of experimentally observed Hg(2) lineshapes shows the abrupt change of the Fano parameters at T_c. Interband t_{1u} - t_{1u}^* transitions are suggested as a possible origin for low energy electronic excitations.

1. Introduction

Since the discovery of superconductivity in alkali metal doped fullerene A_xC_{60} [1] a number of papers appeared [2-8] studying the changes of Raman spectra with variation of x, with the aim to clarify which phonons (either intermolecular [9], intramolecular [10-14] or librational [15])are important (if any) in the superconducting (SC) pairing mechanism. In most of those studies visible excitation by Ar-laser is used [3-6] at λ =545 nm which is not resonant in x=3 case as e.g. mentioned in [6]. The main result for x=3 (SC composition) was the disappearence of all higher energy Hg modes [3-6] which was interpreted as due to overdamping caused by strong interaction of namely those higher energy H_g vibrations with electrons, as predicted theoretically [11,12].Estimates based on a simple metallic picture of intraband el-ph scattering for A_3C_{60} attributed this el-ph interaction as responsible for the SC pairing [11,12],and Raman results [3-6] are supposed to support this picture.

However, it has been realized recently that only interband electronic transitions can contribute to el-ph interaction in A_3C_{60} [17], making estimates of Hg modes overdamping somewhat suspicious. It should be noted also that the disappearence of higher Hg modes should not be necessarily connected with strong el-ph interaction, but may be caused by some other reasons e.g. decrease

Springer Series in Solid-State Sciences, Volume 117
Electronic Properties of Fullerenes Eds.: H. Kuzmany, J. Fink, M. Mehring, and S. Roth

of their intensity due to different resonance conditions in x=3 compared to x=0 or 6. Thus more studies of Raman scattering are required, to clarify the role of el-ph interaction and hopefully to choose among phonon-mediated [10-15] and electronic [18,19] mechanisms of superconductivity.

The advantages of IR excited Raman (at 1.16 ev) in studying A_3C_{60} is its resonance character,due to the new t_{1u} - t_{1g} electronic transition appearing upon doping [20]. We have shown previously [2,7,8] that resonance Raman scattering (RRS) allows to observe the low-energy Hg and Ag modes with additional features,compared to visible excited Raman. Higher modes also can not be observed in this case, but e.g. Ag(2) mode at 1447 cm^{-1},(which is the most intense in VIS-Raman) disappears in RRS, but due to the effect of destructive interference [21]. We have found that in SC the lower Hg modes are broadened, (compared to insulating x=0, and x=6 phases) suggesting their strong interaction with electrons, consistently with recent neutron scattering [22] and Raman [5] data which also indicate stronger el-ph interaction for lower energy Hg modes. It was shown that Hg(1) and Hg(2) modes demonstrate strong Fano-type behaviour, due to resonance with some low-energy electronic excitations (LEEE) [2,7,8] of the similar h_g symmetry.

In the present paper we make an attempt to clarify the origin of the Fano-resonance, the source of the large broadening of $H_g(2)$ and the nature of the electronic continuum, by going to low temperatures [8] ,and studying the changes of RRS uppon crossing the superconducting Tc.

2. Experimental

C_{60} has been prepared and purified by conventional arc-discharge method, while film deposition and doping of thin films by K and Rb has been performed either in vapour phase (as described earlier [23] or from K- and Rb dispensers). Superconducting properties of A_3C_{60} were investigated by low field signal (LFS) of microwave absorption and ESR [23]. Raman spectra at excitation λ_i = 1064 nm were analyzed by means of a Fourier spectrometer (Bruker IFS 88). At excitation λ_i = 514.5 nm Raman spectra were detected by a multichannel spectrometer. The laser power density did not exceed 100 W/cm^2. The spectral resolution was about 4 cm^{-1}. The low temperature Raman scattering measurement have been performed in a continuous flow liquid He cryostat with a temperature controller that allows to vary the temperature from RT to 7 K.

3. Results

Typical RRS spectra of A_3C_{60} films are presented at Fig.1.

Comparing first the room temperature Raman spectra of Rb_3C_{60} excited at two different laser wavelengths (shown at Fig. 1) one may point out the following features:

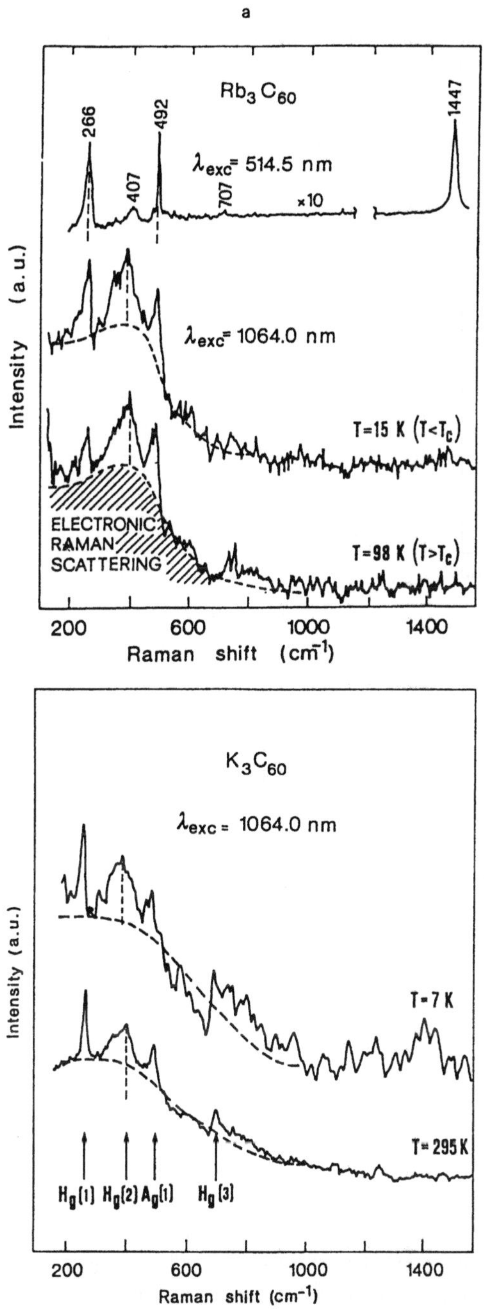

Fig. 1. a) Room temperature Raman spectra at visible (2.41 eV) and infrared excitation (1.16 eV) in Rb_3C_{60} film compared to low temperature one (below Tc) at 1.16 eV. b) RRS (at 1.16 eV) of K_3C_{60} thin film at 295 K and at 7 K.

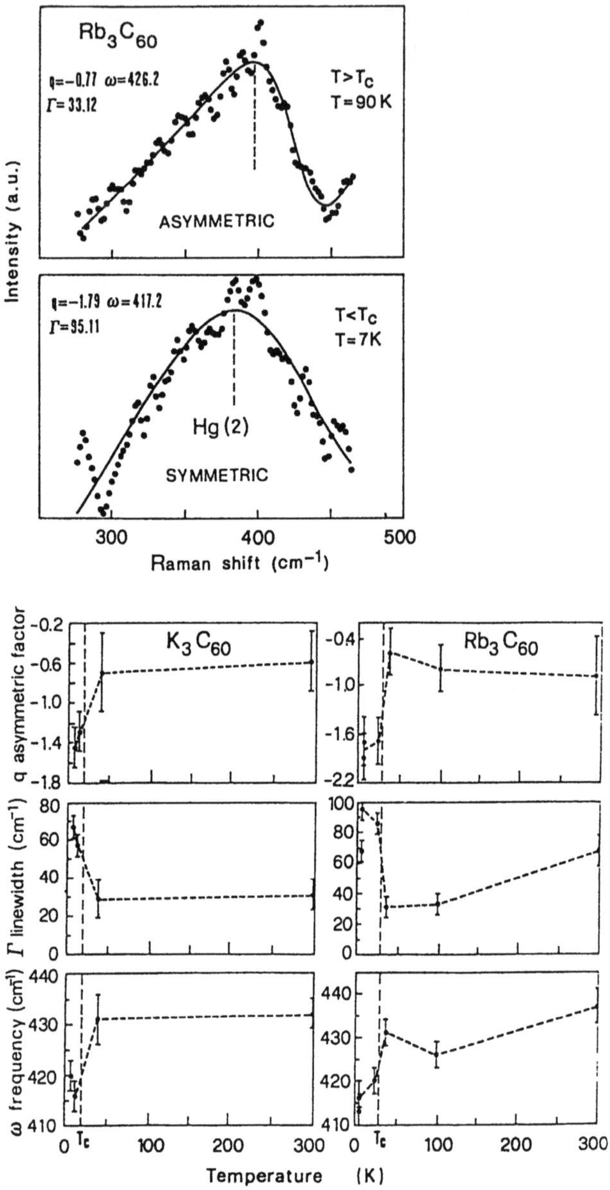

Fig.2. a) The fitting of experimental $H_g(2)$ vibron line (dots) above and below T_c in Rb_3C_{60} by Fano-type lineshape (solid line), b) Temperature dependence of Fano parameters: center frequency (ω_o), linewidth (Γ) and the asymmetry (q). For details of Fano-lineshape see Eq (1) and The dashed lines are guides for eyes.

a) The most clear difference is the existence of the broad background at low energy upon IR excitation, which we have assigned to scattering by low energy electronic excitations (LEEE) in our first report on K_3C_{60} powders [2].

b) The lower $H_g(1)$ mode at 266 cm^{-1} appears quite similar at both excitations: it is broadened, and red shifted compared to pristine C_{60} (271 cm^{-1}) and shows a Fano-type asymmetrical shape [7,8]

c) The next $H_g(2)$ mode is at 407 cm^{-1} if excited at 2.41 eV. It shows a larger broadening (fwhm = 20 cm^{-1}) and much smaller intensity compared to $H_g(1)$ in agreement with recent data [5] of Raman scattering in ultrathin films. However this band is much broader (fwhm = 60 cm^{-1}), significantly more intense relatively to $H_g(1)$, and 7-9 cm^{-1} red shifted upon excitation at 1.16 eV. Moreover it shows clearly the asymmetry and Fano antiresonance lineshape both for K and Rb cases [7,8].

d) A_g modes are observed as narrow lines but with different intensities at two excitations: $A_g(1)$ mode at 492 cm^{-1} is significantly weaker at 1.16 eV excitation; $A_g(2)$ at 1447 cm^{-1}, which is the strongest line at 2.41 eV [3-6], vanished when excited at 1,16 eV due to destructive interference but not due to overdamping (see details in our earlier [21]).

Let now . turn to T-dependence of RRS: the low temperature RRS measurements, exciting at 1.16 eV, of both the Rb_3C_{60} and K_3C_{60} thin films have shown no significant change in the spectrum untill the superconducting critical temperature. At T_c the spectrum shows an abrupt change of the lineshape of only $H_g(2)$ mode: it becomes broader more symmetric and the peak shifts towards lower energy (Fig. 1.). This behaviour is shown in more detail in Fig.2, with the fitting by Fano lineshape, as discussed below. Note that the LEEE background is not changing its intensity or shape with temperature, neither are the Hg(1) and Ag(1) modes.

4. Discussion

The main questions arising from the previous results, which need understanding, are the following:

- Why electronic Raman scattering background can be observed only upon 1.16 ev excitation, and what is the nature of the LEEE?
- What is the origin for enormous broadening and Fano-lineshape of Hg(2) and why it changes below Tc and, is it connected with the mechanism of SC pairing.

To give some reasonable explanations we make two natural assumptions: 1.It is reasonable to assume that the t_{1u}-t_{1g} allowed transition, which appears upon doping, has in metallic state about the same energy of 1.2 eV, as observed in the insulating x=6 phase [20]. 2.We suppose that the lowest t_{1u} (3-fold degenerate)-derived conduction band splits into subbands, due to J-T type or other interactions, so that t_{1u}-t_{1u}^* interband transitions can be excited by Raman scattering process.The energy of this inter t_{1u}-band intermolecular charge excitations is expected to be very low (0.1 eV) from band structure calculations [24],meaning comparable to that of Hg(2) energy. Then one may give answers, at least partially, to the above questions. Our IR excitation at 1.16 eV is in resonance with optically allowed t_{1u}-t_{1g} transition which means that electronic Raman scattering cross section should be enhanced due to contribution of resonant second order diagrams (with $H_{int} \infty$ (Ap)), while at VIS excitation only usual nonresonant first order diagrams ($H_{int} \infty A^2$) [25] are important. Interference of those electronic scattering processes with usual Raman phonons, according to the charged phonon effect [16-17] (i.e. coupling of $H_g(2)$ to t_{1u}-t_{1u}^* transition), may give the observed Fano-lineshapes. The corresponding calculations are currently in progress.

Turning to the question of linewidths of H_g modes one should note that $H_g(1)$, $H_g(2)$ and some other low energy Hg modes have been shown [10,13.14] to be involved significantly into dynamic J-T distortions which should be reflected in the broadening of Raman lines due to J-T electron-phonon coupling. Since x = 6 phases of C_{60} films exhibit much narrower lines, it may be suggested that namely electron-phonon interaction with free electrons is the origin of the broadening. However as pointed recently by M.Rice et.al.[16], due to negligible dispersion of intramolecular vibrations, their interaction with free electrons should be described not in terms of intraband electronic transitions [11,12], but in terms of interband electronic transitions. In this case the intramolecular vibron linewidth should be determined not by $N(E_f)$ (the density of states at the Fermi level), but by the density of interband electronic transitions: $N(t_{1u}$-$t_{1u}^*)$ [16] (maximum at x=3 and zero at x=6).Since Hg(2) is nearly in exact resonance with the t_{1u}-t_{1u}^* separation, one may suppose that the $N(t_{1u}$-$t_{1u}^*)$ should be largest at the energy of Hg(2) i.e. around 400 cm^{-1}, and thus linewidth of Hg(2) should also be the largest, among other Hg modes. Moreover, due to resonance conditions at 1.16 eV excitation, the intensity of Hg(2) should be much increased compared to 2.41 eV excitation, and thus its larger linewidth can be better resolved experimentally.

To clarify the T-dependence of Hg(2) mode we have fitted the lineshape shown in Fig.2,a) with the phenomenological Breit-Wigner-Fano formula:

$$I(\omega) = (1 + \varepsilon / q\Gamma)^2 / (1 + (\varepsilon / \Gamma)^2), \qquad \varepsilon = \omega - \omega_0 \qquad (1)$$

with ω and ω_0 being the Raman shift and the center frequency, Γ the linewidth, and q the inverse asymmetry parameter, which in the limit of $q \rightarrow \infty$, gives a symmetric Lorentzian line. The temperature dependences of ω_0, Γ and q, plotted in Fig.2,b) show abrupt changes of all the three parameters at Tc in both K and Rb doped A_3C_{60} superconductors. It should be noted that this T-dependence is quite similar to that observed for B_{1g}= 335 cm^{-1} phonon behaviour in YBaCuO high Tc superconductor below T_c [25,26]. However in YBaCuO the energy of B_{1g} is close to superconducting gap 2Δ, which allowed to explain the dramatic evolution of the B_{1g} phonon at least qualitatively [25,26]. On the contrary fulleride superconductor, with a gap 2Δ of 48 cm^{-1} and 60 cm^{-1} respectively for K and Rb doped C_{60} [27], one should not expect, at least in a classic BCS picture, any change of lineshape in the Raman scattering of phonons with energy very different from 2Δ. On the other hand the temperature dependence of the

Raman scattering in Rb_3C_{60} and K_3C_{60} is evidently correlated with the superconducting T_c.The origin of such a behaviour is not clear at the moment. One may interpret this behaviour as due to the change of the interaction between the $H_g(2)$ mode and LEEE below T_c due to the change of $N(t_{1u}\text{-}t_{1u}^*)$ at the opening of the gap in the t_{1u} band, in conditions of sharp resonance of Hg(2) with LEEE. Hopefully RRS studies of alkali earth metals (Ca,Ba) doped fulleride superconductors, which should have different LEEE spectrum (due to the different nature of the conduction band (t_{1g} vs t_{1u}) may clarify this question.

In conclusion we have shown that at room temperature two lowest energy H_g modes are strongly interacting with electrons and show Fano-lineshapes in A_3C_{60} (A = K, Rb) single phase films. This observations indicate the possible role of low-energy radial H_g modes in strong electron-phonon coupling and hence probably in SC mechanism as recently predicted [10,13,14]. Further studies are required to clarify the origin of the abrupt lineshape change of Hg(2) below Tc and the role of LEEE in SC-ty of fullerides. RRS studies of Ca_5C_{60} and Ba_6C_{60} superconductors are expected to be helpfull for this purpose.

5.Acknowledgments

We wish to thank Prof.M.Rice for stimulating and extremely valuable discussions on the origin of Hg modes broadening and nature of LEEE, observed in RRS.

6.References

1. Haddon R.C. et al., Nature,350, 320 (1991)
2. Danieli R. et.al., Solid. State Commun. 81, 257 (1992)
3. Duclos S.J. et.al.,Science 254, 1625 (1991)
4. Zhou P. et.al.,Phys.Rev.B. (in print)
5..Mitch M.G. et.al.,Phys.Rev.Lett. 68, 883 (1992),
6.P.C.Eklund,P.Zhou,K.A.Wang.G.Dresselhaus, M.S.Dresselhaus, J.Phys.Chem.Sol.53,1391 (1992)
7. V.N.Denisov, A.A.Zakhidov, R.Danieli,G.Ruani, R.Zamboni and C.Taliani, Int.J.Modern.Phys. B6/23-26,4019 (1993)
8. G.Ruani,V.N.Denisov, A.A.Zakhidov, R.Zamboni and C.Taliani, in "Laser deposition of Advanced Materials", eds M.Allegrini et.al,1992 Edizioni ETS,Pisa,p.122
9. Zhang F.C., Ogata M., Rice T.M., Phys. Rev. Lett. 67, 3452 (1991)
10. Johnson K.H., McHenry M.E. and Clougherty D.P., Physica C 183, 319 (1991)
11. Varma C.M., Zaanen J. and Raghavachari K., Science 254, 989 (1991)
12. Schluter M.A. et.al., Phys. Rev. Lett. 68, 526 (1992)
13.Jishi R.A. and Dresselhaus M.S., Phys. Rev. B45, 2597 (1992)
14. Asai Y. and Kawaguchi Y., Int.J.Mod.Phys.B6/23-26,(1993)
15. Dolgov O. and Mazin I.,Solid. State Comm. 81, 935(1992)

6. M.J.Rice and H.-Y.Choi, this Proceedings of IWEP'93

17. M.J.Rice, Phys.Rev.Lett.,37,36 (1976), and Proc.IWEP'91,

18.. Baskaran G. and Tossati E., Curr. Sci. 61, 3452 (1991), see also this Proceedings

19. Chakravarty S and Kivelson S., Europhys. Lett. 16, 751 (1991)

20.K.-J.Fu, W.L.Karny, et.al, Phys.Rev. B46, 1937 (1992)

21. V.N.Denisov,B.N.Mavrin, G.Ruani,R.Zamboni and C.Taliani,Sov.Phys. JETP,75,158 (1992)

22. Prassides K. et.al., Nature 354, 462 (1991), Europhys. Lett.,

23.A.A.Zakhidov,A.Ugawa,K.Imaeda,K.Yakushi,H.Inokuchi, et.al, Sol.St.Comm,79 939, (1991)

24. Y.-N.Xu,M.-Z.Huang, and W.Y.Ching, Phys.Rev.B44,13171 (1992)

25.M.Cardona, Phys.C 185-189,65 (1991)

26.R.Feile,P.Leiderer,J.Kowalewski, W.Assmus, J.Schubert and U.Poppe, Z.Phys.B.73, 155 (1988)

27. L. Degiorgi, et al., Phys. Rev. B, 69, 2987 (1992).

Raman Scattering and Lattice Dynamics of Fullerides M_6C_{60} (M = K, Rb, Cs)

V.N. Denisov[1], *A.S. Lipin*[1], *B.N. Mavrin*[1], *A.A. Zakhidov*[2,*], *G. Ruani*[2], *R. Zamboni*[2], *and C. Taliani*[2]

[1]Institute of Spectroscopy, Russian Academy of Sciences, Troitsk, 142092 Moscow, Russia

[2]Istituto di Spettroscopia Molecolare, CNR, Via de'Castagnoli 1, I-40126 Bologna, Italy

*Permanent address: Department of Thermophysics, Uzbek Academy of Sciences, Katartal 28, 700135 Tashkent, Uzbekistan

Abstract. Raman spectra of fullerides M_6C_{60} (M = K, Rb, Cs) in the low energy region (20-300 cm^{-1}) at excitation by λ = 514.5 nm and 647.1 nm are studied. One of the bands (33 cm^{-1} in K_6C_{60}, 45 cm^{-1} in Rb_6C_{60} and Cs_6C_{60}) is assigned to the librational mode of C_{60} molecules while other bands are due to modes corresponding to translations of predominantly metal atoms. An analysis of the lattice dynamics of fullerides in the assumption of intermolecular potential consisting of the sum of Lennard-Jones potential (for C-C intermolecular interactions), the Born-Mayer potential (for M-C interactions) and the of the Coulomb potential has been carried out.

1. Introduction

An interest to low-frequency vibrational spectra of fullerides is motivated by two reasons. First, these spectra are necessary to construct the correct intermolecular potential defining molecular dynamics, elastic constants and thermodynamic parameters of these compounds. Seconddly different mechanisms of superconductivity in fullerides are broadly discussed emphasising role of different vibrations in this mechanism. Models involving low-frequency phonons are still under way [1-3].

Previously Raman spectra of M_6C_{60} have been already investigated, but these investigations are restricted to the region of intramolecular vibrations (ω > 200 cm^{-1}) [4]. We have recently reported low-frequency Raman studies of K_6C_{60} films [5] and here we report results of both experimental Raman spectra and calculations of low-frequency vibrational spectra in insulating fullerides M_6C_{60} (M = K, Rb, Cs)

Intermolecular spectra of fullerides have been studied by various techniques. In neutron scattering spectra of C_{60} [6,7] and Rb_6C_{60} [8] only one band, near 20 and 48 cm^{-1} respectively, was observed at low temperatures which was assigned to librations of the C_{60} molecules. In infrared transmission spectra of C_{60} instead of two expected absorption bands 8 bands were observed [9] of which bands at 26.6 and 53.6 cm^{-1} were assigned to infrared allowed modes.

Lattice dynamics of C_{60} was analyzed in detail [10,11]: in particular, the mode frequencies at k = 0, forms of vibrations, phonon dispersion and sound velocities

Springer Series in Solid-State Sciences, Volume 117
Electronic Properties of Fullerenes Eds.: H. Kuzmany, J. Fink, M. Mehring, and S. Roth

were calculated. Such a detailed analysis is desirable also for M_6C_{60} to interpret vibrational spectra and to predict physical properties of fullerides.

2. Experimental

The C_{60} films of thickness of about 1 mm were deposited on substrates of crystalline silicon and glass [12] and doped in sealed quarts tubes by a conventional technique [13]. Raman spectra were excited by the argon-krypton laser radiation at grazing incidence of laser beam and were detected by a multichannel spectrometer with triple monochromatization [14] and resolution of about 4 cm^{-1}. The laser power did not exceed 10 mW and the density of exciting radiation was about 100 W/cm^2.

3. Discussion of spectra

According to group-theoretical analysis [5] intermolecular vibrations are classified as follows:

$$Ag + Eg + 3Fg + 4Fu \qquad (1)$$

All even modes (Ag + Eg + 3Fg) are Raman-active. The librations of the C_{60} molecules are assigned to the Fg mode. Among the Fu modes one is acoustical and three Fu modes are infrared-active.Raman spectra of M_6C_{60} were obtained at laser excitations λ = 514.5 nm and 647.1 nm (Fig.1). From previous Raman study [6] it is known that the lowest intramolecular band of M_6C_{60} is located near 270 cm^{-1}. Therefore all bands below this frequency can be considered as possible intermolecular bands. The intramolecular band near 270 cm^{-1} is split into doublet Eg + Fg [6,7] and this splitting decreases in the series K-Rb-Cs (Fig.1). The intensity of the low-frequency component of this doublet increases in going from χ = 647.1 nm to 514.5 nm (Fig. 1) giving evidence of a strong resonance character of this component. Such a resonant behaviour of bands is inherent to Eg modes [6,7]. Therefore the low-frequency component of the doublet near 270 cm^{-1} may be assigned to the Eg mode.In experimental spectra (Fig.1) bands below 110 cm^{-1} are observed. However a number of these bands is one less than expected from group-theoretical analysis (1). The bands at 33, 48,~75 and 105 cm^{-1} in K_6C_{60} spectra, at 45, 55,~60,77 and 110 cm^{-1} in Rb_6C_{60} spectra and at 45,~53 and 60 cm^{-1} in Cs_6C_{60} are found. Their intensities are not large in comparison with intramolecular bands. The Raman intensity of modes related to the metal atoms are small due to the ionic character of the M-C bonding and the intensity of librational modes is not large because of the high symmetry of C_{60} molecules.

Previously the band at 33 cm^{-1} that is most intense in K_6C_{60} spectra was assigned by us [5] to C_{60} librations and it is consistent with neutron scattering data [7,8]. The librational frequency at 33 cm^{-1} in K_6C_{60} is essentially higher than the measured one in C_{60} (~20 cm^{-1} [6-8]). An increase of this frequency is likely due

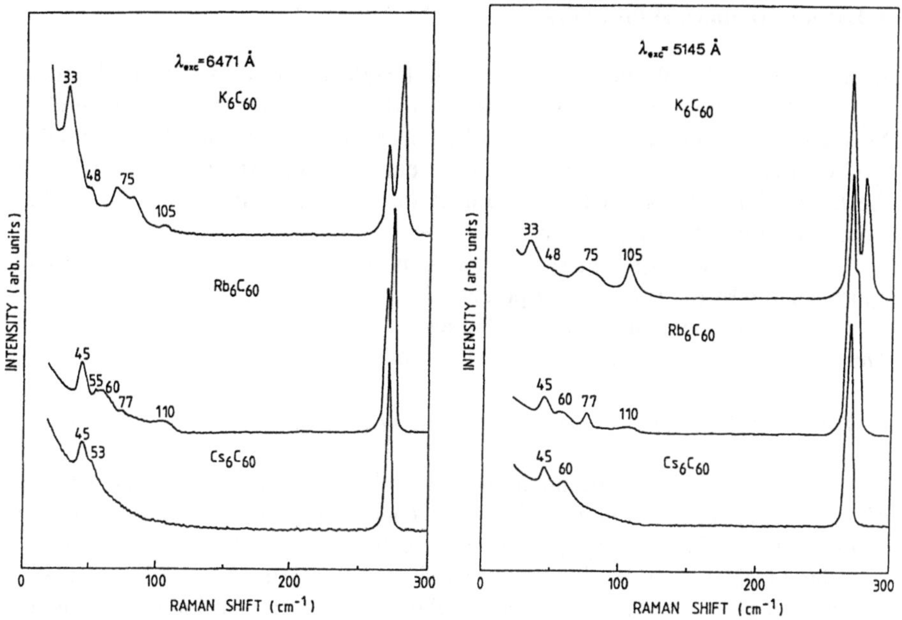

Fig.1. Intermolecular Raman spectra of M_6C_{60} fullerides upon excitation by: a) λ = 647.1 nm, and b) λ = 514.5 nm.

to an interaction between alkali atoms and C_{60} molecules. According to neutron data [7,8] the band at 45 cm^{-1} in Rb_6C_{60} spectra should be assigned also to the C_{60} librations. The high frequency shift of the librational mode in Rb_6C_{60} in comparison with K_6C_{60} can be due to an increase of the ionic radius of the metal ion that gives rise to a strengthening of M-C interactions and, hence, to an increase of the librational force constant. We suppose that the intense broad band at 45 cm^{-1} in Cs_6C_{60} spectra should be assigned also to librational motions of C_{60}.

According to group-theoretical analysis (1) all other Raman bands should be related to translations of the metal ions whose frequencies decrease with a growth of the ion mass. Therefore one should expect a decrease of these frequencies in a series K->Rb->Cs. One should pay attention to changes in translational spectra in going from λ = 647.1 nm to 514.5 nm. From Fig. 1 one can see that the band intensities at 105 cm^{-1} (K_6C_{60}), 77 cm^{-1} (Rb_6C_{60}) and 60 cm^{-1} (Cs_6C_{60}) are higher. This is be due to resonant properties of Raman spectra. Therefore we expect that these bands, having the similar resonant character, are assigned to the same symmetry. Moreover, we assigned these bands to Eg modes, since the intensity of Eg modes (that are components of doublets of intramolecular modes [4,5]) have a resonant character [5].

4. Lattice dynamics of fullerides

A model of rigid molecular ions with bond lengths taken from structural data of M_6C_{60} [15] is used. We have suggested that the ion charges does not depend on M. We have taken into account 3 contributions into the intermolecular potential:
1) C-C interaction of carbon atoms of different molecules which was described by the Lennard-Jones potential with parameters e = 31.5 K and s = 3.36 A taken from calculations for C_{60} crystal;
2) a short-range repulsion between the carbon atoms and the metal ions approximated by the Born-Mayer potential;
3) a Coulomb interaction of charges.
Then the intermolecular potential has the form:

$$F(r)=4e((s/r)^{12}-(s/r)^6)+b\exp(-r/r_0)+qQ/r \qquad (2)$$

In (2) b and r_0 parameters were found in the following way. At first the $b(r_0)$ dependence was obtained from the condition of the pressure absence in model lattice. Then the r_0 parameter was found by means of adjusting the calculated frequency of the librational mode to the measured one. The calculated frequencies of intermolecular modes in the centre of the Brillouin zone of the M_6C_{60} crystals are represented in Table 1 and compared with the experimental ones.

We divide the Raman-active modes of M_6C_{60} into two groups: the low-frequency one containing 3 modes (Ag + 2Fg) including the Fg librational mode and the high-frequency one (Fg + Eg). Between these groups of frequencies the Raman-forbidden Fg modes are located. In the low-frequency group the Raman intensity of the totally symmetric mode Ag is small and it was not seen in Raman spectra. The position of this mode relative to the librational mode should change

Table 1. The calculated and measured mode frequencies (in cm^{-1}) in M_6C_{60}.

	K_6C_{60}		Rb_6C_{60}		Cs_6C	
	Calc.	Exp.	Calc.	Exp.	Calc.	Exp.
Fg^{lib}	33.8	33	45.1	45	45.5	45
Ag	43.3	-	44.4	-	36.4	-
Fg1	47.2	48	55.3	55	44.5	45
Fu1(TO)	56.4 |		57.9 |		48.6 |	
Fu1(LO)	67.2 |	70	65.8 |	60	52.9 |	53
Fu2(TO)	76.1 |		66.1 |		52.9 |	
Fu2(LO)	85.3 |		67.3 |		55.0 |	
Fu3(TO)	91.6		92.9		82.1	
Fg2	91.7	-	84.5	-	70.1	-
Eg	104.9	105	77.0	77	61.2	60
Fu3(LO)	122.1		112.7		100.1	

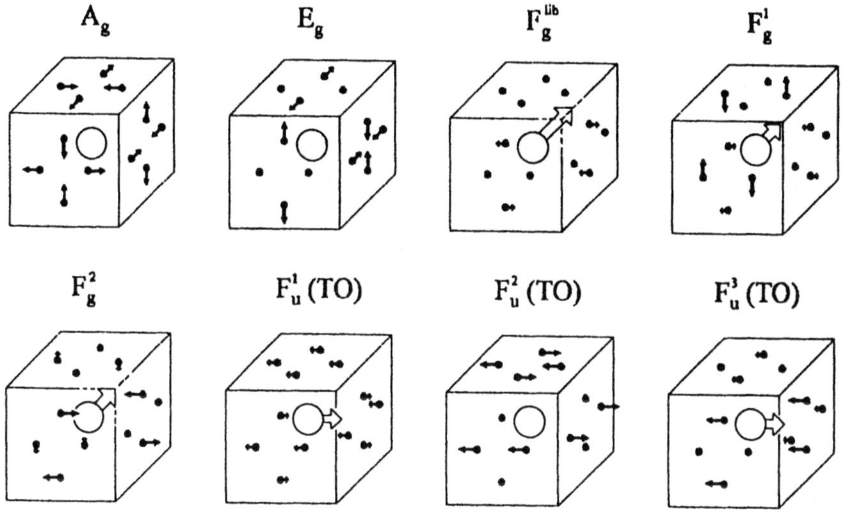

Fig.2. Forms of intermolecular normal modes in M_6C_{60}. Metal atoms (black circles) are located on all faces of cell and shown only on 3 faces. The C_{60} molecules that occupy corners and a centre of cell are shown only in a centre of unit cell. Black arrows correspond to the displacement vectors of translational motion of M atoms and C_{60} molecules and open arrows indicate axial rotation vectors for a librational (rotational) motion of C_{60} molecules.

in the series K->Rb->Cs_6(Table 1). In K_6C_{60} its frequency is higher than Fg , but in Rb_6C_{60} and Cs_6C_{60} it becomes lower. We could assign only one band to the high-frequency group. It was exhibited clearly in K_6C_{60} (105 cm^{-1}), Rb_6C_{60} (77 cm^{-1}) and Cs_6C_{60} (60 cm^{-1}) spectra, especially at excitation λ = 514.5 nm (Fig.1,b)). According to calculations (Table 1) and resonant property, this band should be assigned to the Eg mode with very good correspondence of the calculated and measured frequencies of this Eg modes.

Beside Raman-allowed bands that we have discussed above we have observed peculiarities in spectra which are more than likely due to a violation of selection rules. For example, the broad band near 75 cm^{-1} in K_6C_{60}, the weak bands near 60 and 110 cm^{-1} in Rb_6C_{60} and the weak band at 53 cm^{-1} in Cs_6C_{60} are assigned to these peculiarities.

Let us consider now forms of intermolecular vibrations of M_6C_{60} shown at Fig.2. One can see that Ag and Eg translational vibrations have a simple form with displacements of the metal atoms only. But the Fg and Fu modes have more complex forms because of mixing of vibrations of the same symmetry. Nevertheless, the form of the Fg^{lib} mode corresponds in the main to the librational motion of C_{60} molecules and the contribution of librations to other Fg modes is small.

References

1. F.C.Zhang, M.Ogata and T.M.Rice, Phys.Rev.Lett. 67, 3452(1991).
2. O.V.Dolgov and I.I.Mazin, Solid State Commun. 81, 935(1992).
3. I.I.Mazin,O.V.Dolgov, A.Golubov,S.V.Shulga, Phys.Rev.B 47, 538 (1993)
4. P.Zhou, K.-A.Wang,P.C.Eklund, M.S.Dresselhaus, G.Dresselhaus and R.A.Jishi, Phys.Rev.B 46, 2595(1992) and references therein.
5. V.N.Denisov, B.N.Mavrin, G.Ruani, R.Zamboni and C.Taliani, Sov. Phys.JETP 75, 158(1992).
6. D.A.Neumann, J.R.D.Copley, W.A.Kamitakahara, J.J.Rush, R.L.Cappelletti, N.Coustel, J.P.McCauley, J.E.Fisher, A.B.Smith, K.M.Cregan and D.M.Cox, J.Chem.Phys. 96, 8631(1992).
7. C.Christides, D.A.Neumann, K.Prassides, J.R.D.Copley, J.J.Rush,M.J.Rosseinsky, D.W.Murphy and R.C.Haddon, Phys.Rev. B, in press.
8. L.Pintschovius, B.Renker, F.Gompf, R.Heid, S.L.Chaplot, M.Haluska and H.Kuzmany, Phys.Rev.Lett. 69, 2662 (1992)
9. S.Huant, J.B.Robert, G.Chouteau, P.Bernier, C.Fabre and A.Rassat, Phys.Rev.Lett. 69, 2666(1992).
10. T.Yildirim and A.B.Harris, Phys.Rev. B 46,7878(1992).
11. A.S.Lipin and B.N.Mavrin, Phys.Stat.Sol., in press.
12. W.Kraetschmer, K.Fostiropoulos and D.R.Huffman, Chem.Phys.Lett. 170, 167(1990).
13. A.A.Zakhidov, A.Ugawa, K.Imaeda,K.Yakushi, H.Inokuchi, et al., Sol.St.Com.,79 939 (1991)
14. A.F.Goncharov, V.N.Denisov. B.N.Mavrin and V.B.Podobedov, Zh.Exp.Teor.Fiz. 94, 321(1988).
15. O.Zhou, J.E.Fisher,et al., Nature 351, 462(1991).

Stability of the Fulleride Phase K_1C_{60} as Observed from Raman Spectroscopy

J. Winter and H. Kuzmany

Institut für Festkörperphysik, Universität Wien, A-1090 Wien, Austria

Abstract. Potassium doping of C_{60} to K_xC_{60} with $0 \leq x \leq 6$ was studied by Raman spectroscopy at temperatures slightly above room temperature. A new phase with x = 1 was found. By careful preparation of the samples, single – phase K_1C_{60} could be obtained. The new phase decays at a temperature of about 370 K into C_{60} and K_3C_{60}. Quenching the samples showed that the new phase with x = 1 can be supercooled to low temperatures. The x = 0, 1, 3, and 6 structures revealed a nearly linear relation between charge on the C_{60} and the position of the Raman line for the pinch mode.

1. Introduction

Since the discovery of experimental techniques to produce large quantities of fullerenes by Krätschmer et al. [1], and in particular since the report by Hebard et al. [2] on metallic conductivity and superconductivity after doping, these materials became very attractive objects for various fields of science. Three dominating phases for the system M_xC_{60}, where M is an alkali metal, have been observed : x = 0, 3, and 6 with an fcc(I), an fcc(II) and a bcc Bravais lattice, respectively [3]. By a special sample treatment an intermediate bct phase with x = 4 was reported for powdered samples with M = K, Rb, Cs [4].
Raman spectroscopy turned out to be a very useful tool to watch the doping process [5 - 9]. Even in the first *in-situ* experiment by Haddon et al. [5], a strict phase separation into C_{60}, K_3C_{60}, and K_6C_{60} phases was observed at room temperature. One of the most remarkable differences in the Raman spectra between the three phases is the downshift of the A_g pinch mode of approximately 6 cm^{-1} per elementary charge on the C_{60} molecule. Performing the doping process at elevated temperatures, another phase was observed with the A_g pinch mode at a position at 1459 cm^{-1} in the Raman spectra. By following the quasilinear downshift of the pentagonal pinch mode, this additional phase was interpreted as a phase K_1C_{60} [9].

2. Experimental

The samples were prepared by vacuum deposition of purified C_{60} at 420° C on quartz slides for several hours. The thickness of the investigated films was between 0.3 μm and 0.8 μm with a diameter of about 8 mm. Four contacts for a van der Pauw resistivity measurement were fixed on the films. The doping process of the samples was performed in a modified Haddon-type cell which allowed the heating and cooling of the films. The doping was performed in several steps with the potassium at 390 K and the C_{60} sample at 415 K. During the experiments the vacuum in the cell was better than $6 \cdot 10^{-7}$ mbar. Raman spectra were excited with a conventional argon laser. Typical intensities for excitation were 20 W/cm^2. The scattered light was analysed with a Dilor XY spectrometer at

514.5 nm excitation wavelength in a backscattering geometry, and detected by a nitrogen cooled CCD – multichannel detector.

3. Results

Fig. 1 shows a typical behavior of the Raman response for the pinch mode during doping the films with potassium at a temperature of 415 K. The figure shows clearly the decrease of the pinch mode of the undoped material at 1467 cm^{-1} and the increase of a new line at 1459 cm^{-1}. Since in Raman measurements with a laser power of 20 W/cm^2 at 415 K no bleaching of the pinch mode could be observed, this new line is not a result of a photo – transformation, but rather originates from the new phase K_1C_{60}. After 180 minutes the sample was completely doped and the homogeneity of the K_1C_{60} was proved by measuring at positions on the front side and on the back side of the film. Cooling the sample shows a decay of the K_1C_{60} phase in the two well known room temperature phases C_{60} and K_3C_{60}. Fig. 2 illustrates this behavior for different temperatures.
Heating the films showed the opposite behavior. The C_{60} and the K_3C_{60} phase disappeared, and at 415 K the K_1C_{60} phase was again established. The cooling and heating process was completely reversible, and a small hysteresis was observed, which can be taken as an indication for a first – order phase transition. Rapid cooling from 445 K to 155 K shows that at least a part of the K_1C_{60} phase could be retained at this low temperature. However, on heating the sample to 250 K the K_1C_{60} phase disappeared rapidly.

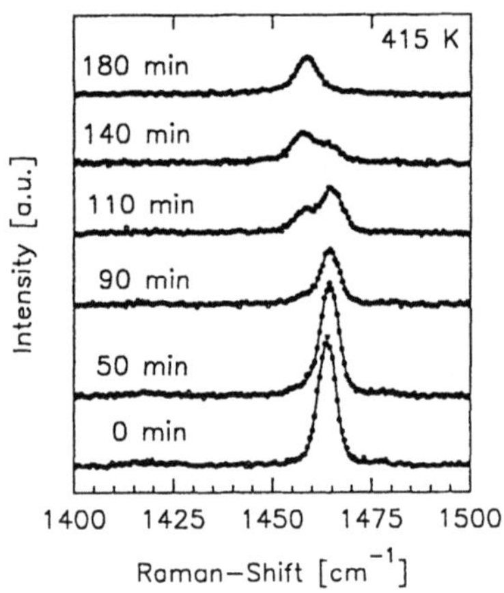

Fig. 1: Raman response for the pinch mode of C_{60} at 415 K after doping for 0, 50, 90, 110, 140, and 180 minutes with potassium.

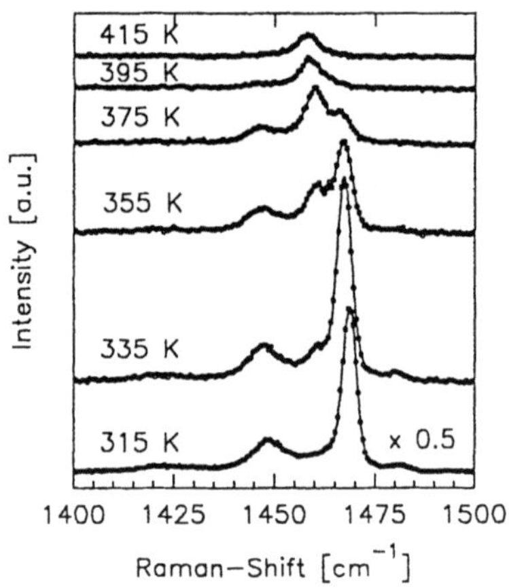

Fig. 2: Cooling process of the K_1C_{60}. The spectra show the decay of the K_1C_{60} phase into the phases K_3C_{60} (1449 cm^{-1}) and C_{60} (1469 cm^{-1})

4. Discussion

Fig. 3a shows the position of the A_g pinch mode from the doped phases versus the valence state at a temperature of 415 K. The data were obtained from a film with a thickness of about 0.3 μm. The thickness of the film was observed to influence the line position

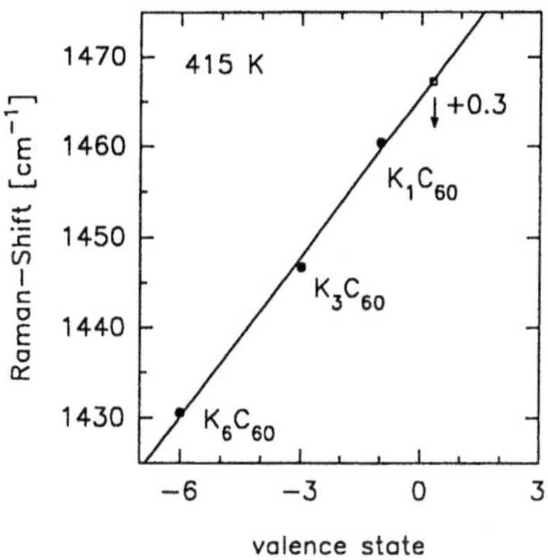

Fig. 3a: Calibration curve for the frequency of the pinch mode and the valence state in solid C_{60} at 415 K.

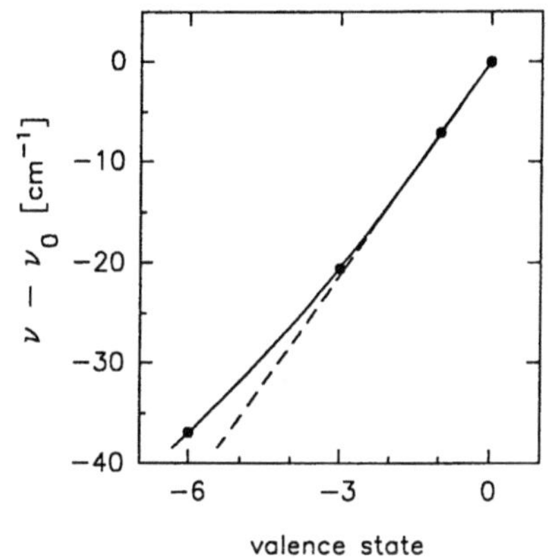

Fig. 3b: Frequency shift (ν - ν_0) of the pinch mode versus valence state of C_{60} and the deviation from the linear relation (dashed line).

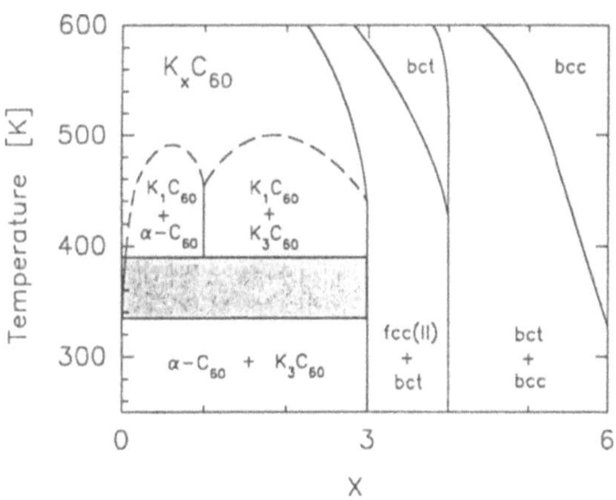

Fig. 4: A suggested phase diagram for the K – C_{60} system.

of the pinch mode for all phases. Thin films have an 1.5 wavenumbers upshifted line, as compared to thick films. The accuracy of the line position is 0.5 wavenumbers. Following a linear behavior for the frequency – charge relation for the three doped phases x = 6, 3, and 1 the position of the measured A_g mode of the undoped material is not at zero charge. Extending the linear behavior obtained for the charged molecules to zero charge yields a predicted frequency of 1465.6 cm^{-1} for the intrinsic pinch mode, which is about 1.6 cm^{-1} lower than the measured value. This result was interpreted in a previous paper [9] as an influence of oxygen on the pinch mode. Very recent work showed that the presence of oxygen modifies only the stability of the pinch mode against photo – transformation but not its frequency [10 - 12]. Thus, the overall frequency – charge relation should not be considered as linear, but rather as parabolic. Fig. 3b shows the results obtained from a parabolic fit and the deviation from the linear relation.

Summarising the data obtained from Raman spectra, we suggest a phase diagram for the K_xC_{60} – system as shown in Fig. 4. The diagram is an extension of a phase diagram proposed previously by Zhu et al. for the main phases of A_xC_{60} [14]. The extension refers in particular to the part $0 \leq x \leq 3$ at elevated temperatures. The solid lines in this special part were obtained from Raman measurements in the temperature range from 445 K to room temperature. The dashed lines are hypothetical. In the temperature range of the grey region all three phases (C_{60}, K_1C_{60} and K_3C_{60}) were observed during cooling or heating the sample. However, within the experimental error, the decay of the single K_1C_{60} – phase starts always at the same temperature, independent of the stoichiometry, which is in good agreement with results from photoelectron spectroscopy by Poirier et al. [13]. For $3 \leq x \leq 6$ we copied the phase diagram of Zhu et al. [14]. A remarkable difference between our measurements and the measurements of other groups [13, 15] is, however, the slightly lower temperature for the stability range of K_1C_{60} in our case.

Acknowledgment. We acknowledge valuable discussions with M. Matus, J.H. Weaver and J.E. Fischer. This work was supported by the FFWF, project P8172.

References

[1] W. Krätschmer, L.D. Lamb, K. Fostiropoulos, and D.R. Huffman, Nature **347**, 354 (1990).
[2] A.F. Hebard, M.J. Rosseinsky, R.C. Haddon, D.W. Murphy, S.H. Glarum, T.T.M Palstra, A.P. Ramirez, and A.R. Kortan, Nature **350**, 600 (1991).
[3] O. Zhou, J.E. Fischer, N. Coustel, S. Kycia, Q. Zhu, A.R. McGhie, W.J. Romanow, J.P. McCauley, A.B. Smith, and D.E. Cox, Nature **351**, 462 (1991).
[4] R.M. Fleming, M.J. Rosseinsky, A.P. Ramirez, D.W. Murphy, J.C. Tully, R.C. Haddon, T. Siegrist, R. Tycko, S.H. Glarum, P. Marsh, G. Dabbagh, S.M. Zahurak, A.V. Makhija, and C. Hampton, Nature **352**, 701 (1991).
[5] R.C. Haddon, A.F. Hebard, M.J. Rosseinsky, D.W. Murphy, S.J. Duclos, K.B. Lyons, B. Miller, J.M. Rosamilia, R.M. Fleming, A.R. Kortan, S.H. Glarum, A.V. Makhija, A.J. Muller, R.H. Eick, S.M. Zahurak, R. Tycko, G. Dabbagh, and F.A. Thiel, Nature **350**, 320 (1991).
[6] S.J. Duclos, R.C. Haddon, S. Glarum, A.F. Hebard, and K.B. Lyons, Sience **254**, 1625 (1991).
[7] T. Pichler, M. Matus, J. Kürti, and H. Kuzmany, Phys. Rev. **B 45**, 13841, (1992).
[8] K.A. Wang, Y. Wang, P. Zhou, J.M. Holden, S. Ren, G.T. Hager, H.F. Ni, P.C. Eklund, G. Dresselhaus, and M.S. Dresselhaus, Phys. Rev. **B 45**, 1955, (1992).
[9] J. Winter and H. Kuzmany, Solid State Communication **84**, 935 (1992).
[10] P.H.M. v. Loosdrecht, this volume.
[11] P. C. Eklund, this volume.
[12] M. Matus and H. Kuzmany, Applied Physics **A 56**, 241 (1993).
[13] D.M. Poirier and J.H. Weaver, submitted to Phys.Rev **B-15**, Rapid Communication.
[14] Q.Zhu, O. Zhou, N. Coustel, G.B.M. Vaughan, J.P. McCauley, W.J. Romanow, J.E. Fischer, and A.B. Smith, Science **254**, 545 (1991).
[15] Q. Zhu, O. Zhou, N. Bykovetz, J.E. Fischer, A.R. McGhie, W.J. Romanow, C.L. Lin, R.M. Strongin, M.A. Cichy, and A.B. Smith, Phys.Rev **B-15** (1993), in print.

Raman Spectra of ^{13}C-Substituted and Li-Intercalated Fullerenes

S. Lefrant[1], J.P. Buisson[1], P. Bernier[2], J.M. Lambert[2], A. Zahab[2], and C. Mathis[3]

[1]Laboratoire de Physique Cristalline, Institut des Matériaux, Université de Nantes, F-44072 Nantes Cédex 03, France
[2]G.D.P.C., Université de Montpellier, F-34095 Montpellier Cédex 05, France
[3]Institut Ch. Sadron, 6, rue Boussingault, F-67083 Strasbourg Cédex, France

Abstract. Raman spectra of both C_{60} and C_{70} fullerenes are at the present time well known and understood on both experimental and theoretical points of view. We have extended these studies to ^{13}C substituted samples (82% substitution). As expected, an overall downshift of the Raman frequencies is observed, in good agreement with the values predicted from the total mass change of the molecules. In parallel studies carried out on the same samples intercalated with K at the superconducting level (K_3C_{60}), a significant shift in the T_c transition temperature has been measured. The isotope effect is therefore discussed in the framework of the BCS theory of superconductivity and compared to other published results.

We have also investigated the Raman spectra of C_{60} doped with Li by a chemical method, according to a procedure which involves the use of organo-alkali compounds. Results are consistent with a charge transfer from Li atoms to fullerene molecules. Upon annealing, up to 250°C, these compounds present a desorption or a diffusion of Li atoms, a part of the system going back to neutral crystalline C_{60}, as evidenced by Raman, X-rays and NMR experiments.

1. Introduction

Since the discovery of the new stable form of carbon molecules [1], the fullerenes, and later on, their production in significant quantities by using an electric arc [2], an important number of papers have been published in solid state physics and chemistry, and numerous possible applications have been reported. Also, the discovery that fullerenes become superconducting after doping with alkali metals [3] has broadened the field of applications and interest for scientists. All kinds of techniques have been extensively used to investigate the properties of these fascinating compounds, and among them, Raman scattering, to investigate the vibrational properties of both pristine C_{60} and metal-doped C_{60} in both superconducting and insulating phases.

As compared to recent studies [4] carried out on β-(BEDT-TTF)$_2I_3$, in which central atoms were substituted with ^{13}C and in which a downshift of 0.3-0.4 K of the T_c temperature was measured, putting clearly in evidence, in view of the modifications observed in the C=C stretching vibrations, the role of the intramolecular vibrations in the electron pairing process, we extended the well-known Raman studies [5,6] on C_{60} to ^{13}C substituted C_{60} (82% substitution), to determine in which way the substitution affects the vibrations in this high symmetrical molecule. In a parallel investigation, different groups have measured the T_c temperature in metal-doped substituted C_{60} [7-12] and it turns out that results are rather different from one group to the other. In our samples, the T_c downshift was found to be of the order of 1.8 K, leading to an unexpected high value for the α coefficient of the BCS theory of superconductivity.

Also, we present Raman data obtained on C_{70} also substituted with 82% of ^{13}C atoms, as well as preliminary measurements of Li-intercalated C_{60} samples, in which intercalation has been achieved by using a chemical method involving the use of organo-alkali compounds.

Electronic Properties of Fullerenes Eds.: H. Kuzmany, J. Fink, M. Mehring, and S. Roth

2. Experimental Results

2.1 Sample preparation

The preparation of the fullerene powder used in our study follows the standard procedure of Krätschmer et al.[2]. An electric arc between two graphite rods fixed on water-cooled electrodes connected to a high power dc current source was used. A special fabrication of carbon rods was carried out for enriched samples, made from 10 g of ^{13}C enriched amorphous carbon powder and 5g of non-enriched bakelite binder. These two components were grinded and a high temperature treatment under a high pressure resulted in fragile but free standing rods. A dc current of 100 A and a pressure of 200 Torr of pure He were the conditions of evaporation of these rods. The fullerene molecules were then extracted by using the standard Soxhlet procedure, and 99.9% pure C_{60} was in a final step obtained by column chromatography. Before doping the material for superconductivity measurements, a time of flight spectroscopy analysis [13] determined the concentration of ^{13}C atoms and its distribution in the C_{60} molecules. It was found that the distribution of C_{60} masses was centered on m/z=770, corresponding to an average enrichment of 82% and moreover, no sign of contamination with either C_{70} or $^{12}C_{60}$ molecules was detected.

In the case of Li-intercalated compound, the starting product consisted in a C_{60}/C_{70} mixture with approximately 80% of C_{60}. A THF solution of the lithium salt of benzophenone (Bz) dianion was added under vacuum to a steered suspension of C_{60} in THF. The electron transfert to the C_{60} is demonstrated by the conversion of the purple benzophenone dianion to the blue radical ion. This reaction proceeds until the redox potential of the reduced C_{60} becomes close to the one of the benzophenone dianion (≈ -2.3 V).

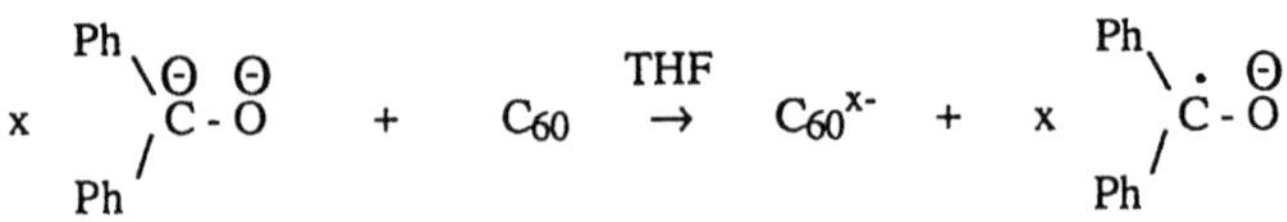

The doping level at the thermodynamic equilibrium could be estimated from the weight of the C_{60} and from the volume and concentration of the solution of $Bz^{2-}(Li^+)_2$ added at the point where the dianion stops to react. It was found to be close to 4 (x≈4). The solubility of $C_{60}^{4-}(Li^+)_4$ in THF being very low, it was possible to remove the excess of lithium salts from the reduced C_{60} by internal distillation of pure THF without significant loss of matter. The reduced C_{60} was then dried by pumping under vacuum. Notice that all experiments were done in glass apparatus sealed under high vacuum using the break-seal technique.

2.2 Experimental data

Figure 1 shows Raman spectra of $^{13}C_{60}$ recorded with a Jobin-Yvon T64000 multichannel spectrometer equipped with a CCD detector. Samples were investigated under a microscope with a power of ≈ 20 μW onto ≈ 15 μm^2 at $\lambda_{exc.}$= 514.5 nm.

Taking into account the amount of substitution (82%) as well as the random statistical distribution of ^{13}C in the fullerene molecules, one may expect theoretically an overall downshift of the frequencies of the $^{13}C_{60}$ compared to those of $^{12}C_{60}$ (frequency ratio: 0.967). This is exactly what is found experimentally. On Table 1, we have collected the frequencies for both C_{60} and C_{70} enriched samples.

Since fullerenes have a strong molecular behaviour and therefore, the interactions in the solid state are rather weak, it appears reasonable that electron-phonon couplings involved in the pairing process will implicate mainly intramolecular vibrational modes, although Raman data alone do not allow to draw such conclusions. The superconducting transition measured in $K_3{}^{13}C_{60}$ by an ac susceptibility technique led to a value of T_c of 17.8 ± 0.3K. This represents a downshift of ≈ 1.8K compared to K_3C_{60} doped in the same conditions. This high value of ΔT_c was corroborated by NMR measurements of the temperature dependence of the ^{13}C nuclear spin relaxation times T_1 [13].

The Raman spectrum presented in Fig.2 for Li-intercalated samples is weak and of rather poor quality. Numerous weak features are observed and among them, parasite lines coming from the microscope objective. Nevertheless, from the main band peaked at 1457 cm^{-1}, one can

Table 1. Experimental Raman frequencies of $^{13}C_{60}$ and $^{13}C_{70}$.

$^{13}C_{60}$	257	415	476	---	745	1059	1200	1380	1419	1526
$^{13}C_{70}$	244	392	436	485	546	675	710	741	915	1023
	1140	1185	1211	1287	1321	1395	1418	1465	1505	

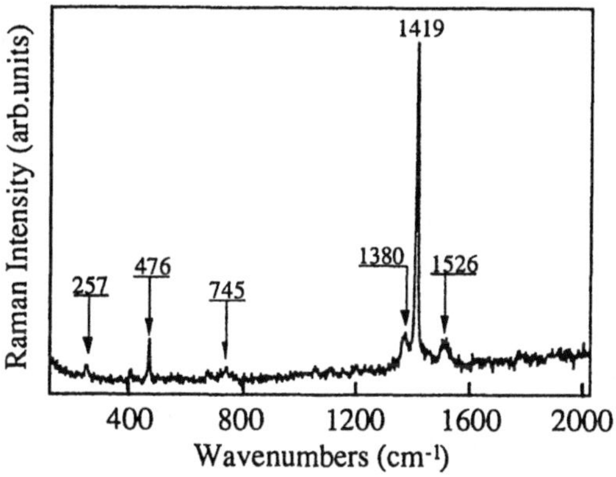

Fig.1 Raman spectrum of $^{13}C_{60}$ recorded at T=20°C with $\lambda_{exc.}$=514.5 nm

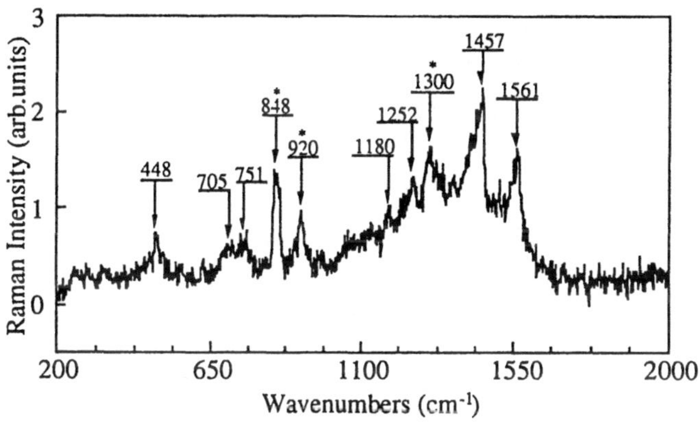

Fig.2 Raman spectrum of Li-intercalated C_{60} recorded at T=20°C with $\lambda_{exc.}$=514.5 nm
(* lines coming from the microscope objective)

confirm an electron transfer from the dopant to the molecule. If we take into account previous results carried out on metal-doped C_{60} at different doping levels, the Raman shift of this so-called "pinch" mode is ≈6 cm^{-1}/electron transferred to the molecule, one would achieve in this case an electron transfer of ≈3 electrons. This slightly differs from what is expected from the chemical reaction and one may suppose different phases in the doped material. Upon thermal annealing at 230°C, a partial recovery of the Raman spectrum of the undoped C_{60} is observed, with also remaining lines of presumably different origins such as the chemical products which participate to the reactions. These preliminary results need obviously to be confirmed on other samples.

3. Discussion and Conclusion

On the basis of these results obtained in sustituted C_{60} samples, the applicability of the BCS theory of superconductivity may be questionned. Different groups have measured for substituted samples at different levels either a low value for the α factor (0.2-0.4) [8,9,10] or a much higher value (1.2-1.8) [7,11,13]. On one hand, most samples partially substituted present in the mass spectrum different peaks characteristics of contamination with $^{12}C_{60}$ or even $^{13}C_{70}$. On the other hand, only one sample has been reported [14] to be substituted with 100% of ^{13}C atoms. Therefore, it is not clear at the present time which parameter is the most important to influence the superconducting temperature of the M_3C_{60} sample (M= alkali metal), being sure as demonstrated by Fleming et al. [15] that the observed shift in T_c in substituted samples is not due to a small change in the lattice parameter.

References

1. H.W. Kroto, J.R. Heath,S.C. O'Brien, R.F. Curl and R.E. Smalley, Nature **318**, 162 (1985)
2. W. Krätschmer, L.D. Lamb, K. Fostiropoulos and D.R. Huffman, Nature **347**, 354 (1990)
3. R.C. Haddon, Nature **350**, 320 (1991)
4. P. Auban-Senzier, C. Bourbonnais, D. Jérome, C. Lenoir, P. Batail, E. Canadell, J.P. Buisson and S. Lefrant, J. Phys. **I 3**, 871 (1993)
5. D.S. Bethune, G. Meijer, W.C. Tang and H.J. Rosen, Chem. Phys. Lett. **174**, 219 (1990)
6. T.J. Dennis, J.P. Hare, H.W. Kroto, R. Taylor and D.M. Walton Spectrochimica Acta, **47A**, 1289 (1991)
7. T.W. Ebbesen, J.S. Tsai, K. Tanigaki, J. Tabuchi, Y. Shimakawa, Y. Kubo, I. hirosawa and J. Mizuki, Nature **355**, 620 (1992)
8. A.P. Ramirez, A.R. Kortan, M.J. Rosseinsky, S.J. Duclos, A.M. Mujsce, R.C. Haddon, D.W. Murphy, A.V. Makhija, S.M. Zahurak and K.B. Lyons, Phys. Rev. Lett. **68**, 1058 (1992)
9. D.S. Bethune, W.Y. Lee, M.S. de Vries, J.R. Salem, W.C. Tang, H.J. rosen, C.A. Brown and H.C. Dorn, to be published in Phys. Rev. B
10. Chia-Chun Chen and C.M. Lieber, J. Am. Chem. Soc. 114 (1992)
11. A.A. Zakhidov, K. Imaeda, D.M. Petty, K. Yakushi and H. Inokuchi, Phys. Lett. A **164**, 355 (1992)
12. S. Lefrant, E. Faulques, C. Godon, J.P. Buisson, P. Auban-Senzier, D. Jérome, C. Fabre, A. Rassat, A Zahab, J.M. Lambert and P. Bernier, Synth. Met. **55**, 3044 (1993)
13. P. Auban-Senzier, G. Quirion, D. Jérome, J.M. Lambert, A Zahab, P. Bernier, S. Della-Negra, C. Fabre, A. Rassat, J.P. Buisson and S. Lefrant (to be published)
14. Chia-Chun Chen and C.M. Lieber, MRS-Fall Meeting abstract, (1992)
15. R.M. Fleming, MRS-Fall Meeting abstract, (1992)

Charge Transfer and Electron-Vibrational Mode-Coupling in Alkali-Metal-Doped C_{60}

T. Pichler and H. Kuzmany

Institut für Festkörperphysik, Universität Wien, A-1090 Wien, Austria
and Ludwig Boltzmann-Institut für Festkörperphysik, A-1060 Wien, Austria

Abstract. *In situ* IR-reflectivity measurements are reported for K and Rb doped C_{60}. The analysis for the phases A_xC_{60} (x=0,1,3,4,6) revealed the line positions for the T_{1u} modes. A detailed study of the reflectivity of K_3C_{60} showed in addition structures which were assigned to 6 distorted Raman-active H_g modes and one moderately distorted Raman-active A_g mode. The IR activity of the IR forbidden gerade modes can be explained by a coupling of the phonons to virtual $t_{1u} \rightarrow t_{1u}^*$ transitions. Using this model the total electron molecular vibration coupling constant for A_3C_{60} is found to be 0.89. This compares to 0.77 found previously from a conventional model. The consequences for superconductivity are discussed.

1. Introduction

Doping C_{60} with alkali metals leads to an accommodation of the latter in octahedral and tetrahedral lattice sites. For A_3C_{60} a metallic phase with an fcc lattice [1], and further doping, semiconducting phases A_4C_{60} with a bct, and A_6C_{60} with a bcc lattice [2] are formed. Very recently an A_1C_{60} phase was found by Raman, X-ray and PES experiments [3–5]. Measurements of the optical properties for the $x = 0, 3$, and 6 phase by different spectroscopic techniques, such as optical absorption, resonance Raman, photoemission, inverse photoemission, and electron energy-loss spectroscopy, revealed characteristic and doping-induced transitions at less than 0.5 eV and at 1.2 eV, which were assigned to the intraband transition in the t_{1u}-derived band and to the $t_{1u} \rightarrow t_{1g}$ transition, respectively [6,7]. Measurements of the optical conductivity of Rb_3C_{60} and K_3C_{60} between 14 cm^{-1} and 50000 cm^{-1} on pressed pellets showed that the $x = 3$ material has a relatively small Drude component and a broad intraband transition [8]. No IR active T_{1u} modes were observed [8]. Well-expressed structures between 1300 cm^{-1} and 500 cm^{-1} in the optical spectra of thin films have been interpreted as originating from activation of gerade modes (H_g and A_g) by coupling to free carriers [9].

We performed detailed measurements of the IR-reflectivity in the range between 400 and 5000 cm^{-1} for A_xC_{60} (A=K, Rb) in a continuous doping range $0 \leq x \leq 6$. From this the lineshift of the four T_{1u} modes with the valence state of the C_{60} molecule is reported. In addition, the optical conductivity of the metallic A_3C_{60} phase was analysed in more detail.

2. Experimental

Smooth films from purified C_{60} with a thickness of 1.2 μ were sublimed on a Si wafer with very high resistivity. Four electrical contacts were provided on the film for a van der Pauw conductivity measurement during the doping process. The minimum of the resistivity was

Springer Series in Solid-State Sciences, Volume 117
Electronic Properties of Fullerenes Eds.: H. Kuzmany, J. Fink, M. Mehring, and S. Roth

used to identify the metallic state. Doping was performed *in situ* in a home-made Haddon type IR-cell inserted in a Bruker IFS 66V spectrometer by evaporating the alkali metal in a dynamical vacuum of $1 \cdot 10^{-4}$ Pa. The temperature of the sample was increased up to 415 K in order to equilibrate the doped material. The reflectivity spectra of the C_{60} coated wafer was measured with a resolution of $0.5\,cm^{-1}$ between 5000 and $400\,cm^{-1}$, where the four T_{1u} modes and the $t_{1u} \rightarrow t_{1u}^{*}$ transitions occur.

3. Results

The line positions for the different stable phases were identified after a careful equilibration process. In the case of A_1C_{60} we used Rb as the alkali metal, as only the Rb_1C_{60} phase is stable at room temperature. The reflectivity spectra of the stable phases C_{60}, Rb_1C_{60}, K_3C_{60}, K_4C_{60} and K_6C_{60} in the range of the 4 T_{1u} modes at 300 K are shown in Fig. 1. In the case of K_3C_{60} and K_4C_{60}, two of the T_{1u} modes could not be observed.

In Fig. 2 the line positions of the four T_{1u} modes, revealed from the reflectivity spectra, are plotted vs the valence state of the C_{60} molecule. For all frequences a quasilinear relation is obtained.

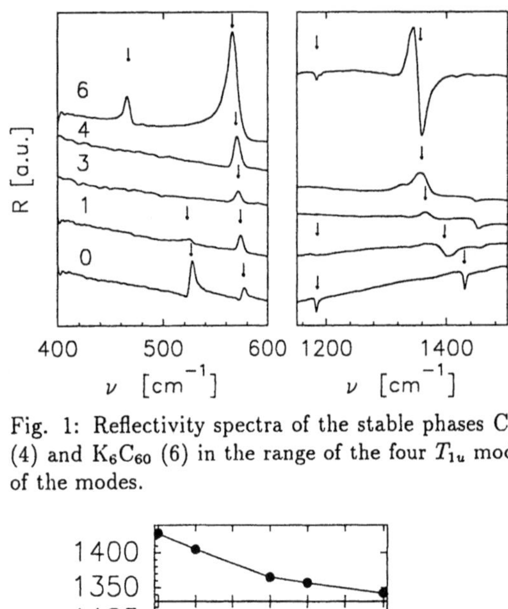

Fig. 1: Reflectivity spectra of the stable phases C_{60} (0), Rb_1C_{60} (1), K_3C_{60} (3), K_4C_{60} (4) and K_6C_{60} (6) in the range of the four T_{1u} modes. The arrows indicate the position of the modes.

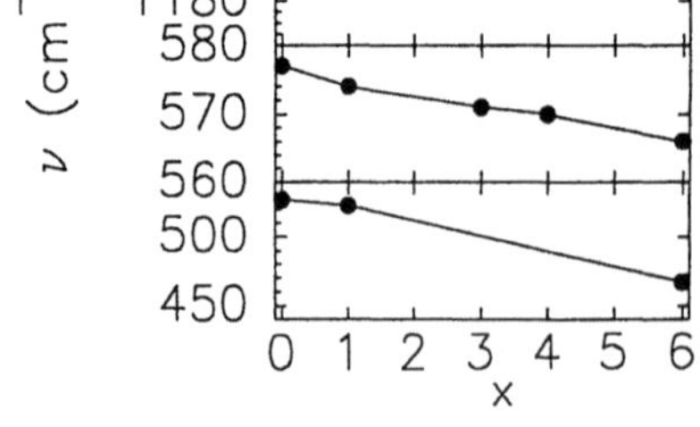

Fig. 2: Lineshift for the four T_{1u} modes vs valence state of C_{60}.

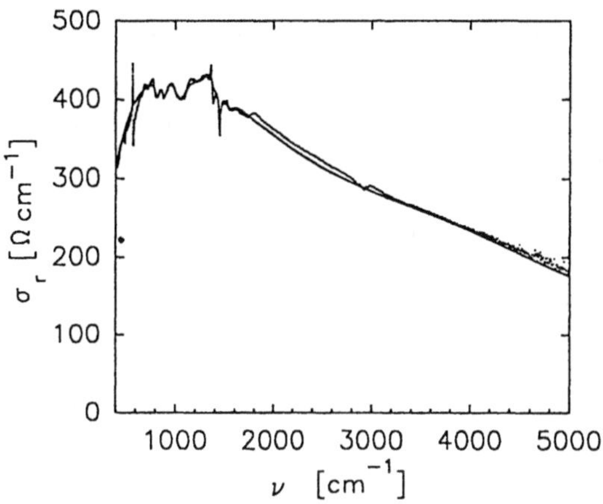

Fig. 3: Real part of the optical conductivity for K_3C_{60}. The smooth curve is a fit according to the charged phonon model of Rice et al. [10].

In addition we analysed the metallic K_3C_{60} compound in detail. Fig. 3 shows the real part of the optical conductivity of K_3C_{60} as evaluated by a Kramers-Kronig transformation. The smooth curve is a fit using the charged phonon model of Ref. 10, as discussed below.

4. Discussion

The lineshift and the enhancement of the T_{1u} modes with increasing doping (Fig. 1) was recently described by a charged phonon effect due to the coupling of the T_{1u} phonons to the doping-induced electronic $t_{1u} \rightarrow t_{1g}$ transition [11,12]. This theory predicts a linear lineshift and a quadratic increase of the integrated oscillator strength of the T_{1u} modes with the charge state of the C_{60} molecule. The nearly linear shift for the modes shown in Fig. 2 is in good agreement with this model. Also, the strong increase of the oscillator strength follows the theoretical predictions. A quantitative comparison of the experimental oscillator strengths with the theory is, however, not yet possible, because of experimental difficulties.

The electronic backround of the optical conductivity in Fig. 3 was described by a very small Drude part and two electronic interband $t_{1u} \rightarrow t_{1u}^{*}$ contributions, as reported previously for pressed pellets [8]. The plasma frequency, line positions, linewidths, and oscillator strengths of the two electronic transitions are $\omega_{PL} = 3500\,\text{cm}^{-1}$, $\omega_{L1} = 925\,\text{cm}^{-1}$, $\tau_{L1} = 2825\,\text{cm}^{-1}$, $f = 0.795$, $\omega_{L2} = 4000\,\text{cm}^{-1}$, $\tau_{L2} = 4000\,\text{cm}^{-1}$, $f = 0.205$. The observed T_{1u} modes were at $\nu_2 = 570\,\text{cm}^{-1}$ and $\nu_4 = 1362\,\text{cm}^{-1}$.

The additional structures in the spectrum, basically between the two large T_{1u}-modes, respectively, have been described by gerade modes which, in I_h-symmetry, have H_g- and A_g-character [9]. Two different models can be applied for this description.

One uses a classical Fano theory to characterize the activated gerade modes. As was shown in Ref. 9, a very good agreement between fitted lines and the observed optical conductivity is obtained, whence the line positions and the line widths of the gerade modes could be evaluated. Using a classical metal band theory [13] the electron-phonon coupling constants λ_i of the gerade modes were evaluated. A total electron phonon coupling constant $\lambda = 0.77$ was obtained for a density of orbital states at the Fermi level of $N(0) = 10$ eV^{-1}.

The other model is more sophisticated and uses the underlying charged phonon interaction [10,14]. In this case the intra-t_{1u}-band excitations do not contribute to the electron-phonon coupling, because the usual adiabatic condition $\omega_{phonon}/q_F v_F \ll 1$, where q_F and v_F are the Fermi wave vector and the Fermi velocity, respectively, is not satisfied for K_3C_{60}. However, a coupling of the phonons to virtual inter-t_{1u}-band excitations is possible, and induces a dipole moment in orientally disordered K_3C_{60}, but not in a highly ordered compound. The latter fact explains why the gerade modes, in our case, generate only small structures, as compared to the strong IR activity observed for the A_g modes in some semiconducting charge-transfer salts [15]. In this case a similar coupling to free carriers occurs and leads to a dipole moment even for the ordered material. According to the charged phonon model, the optical conductivity is in a first approximation described by:

$$\sigma(\omega) = \sigma_{IB}(\omega)\left[1 + bx\frac{A(\omega)}{1 - xA(\omega)}\right]$$

where x is the charge on the molecule, σ_{IB} is the optical conductivity for the contribution of the electronic $t_{1u} \rightarrow t^*_{1u}$ transitions, and

$$A(\omega) = \chi(\omega)\sum_i \lambda_i D^0_i(\omega)$$

describes the coupling to the phonons (χ denotes the normalized $t_{1u} \rightarrow t^*_{1u}$ polarizability, λ_i the corresponding electron phonon coupling constants, and D^0_i the phonon propagators). For a disorder parameter $b = \langle\mu\rangle^2/\langle\mu^2\rangle$ of 0.15 ($\langle\mu\rangle$ stands for an average dipole moment) we have obtained a satisfactory evaluation of our data, shown as the smooth line in Fig. 3. According to this model, the magnitudes of the Fano indentations strongly depend on the amount of disorder. The parameters of our fit, together with the observed Raman lines of C_{60} and K_3C_{60} and the calculated interband electron phonon coupling constants, are shown in table 1.

The effective BCS coupling constant is evaluated from

$$\Lambda = N_F\frac{\omega_T}{2}\frac{A(0)}{1 - xA(0)}$$

and $\Lambda = 0.89$ is obtained, when the density of states at the Fermi level N_F is taken to be 10 eV^{-1} and the disorder parameter is 0.15. This is similar to the value found by the other model, and strongly supports the description of K_3C_{60} as a superconductor in the medium strong coupling limit as given in Ref. 1.

Table 1: Parameters for the fit of the observed optical conductivity with the charged phonon model as shown in Fig.3 together with the observed Raman lines of C_{60} and K_3C_{60}. All frequencies and damping constants γ are given in cm^{-1}; λ is dimensionless.

	$H_g(1)$	$H_g(2)$	$H_g(3)$	$H_g(4)$	$H_g(5)$	$H_g(6)$	$H_g(7)$	$H_g(8)$	$A_g(1)$	$A_g(2)$
Raman										
x=0	273	430	710	774	1099	1244	1428	1575	498	1469
x=3	269	405	670	709	–	–	–	–	496	1446
IR x=3										
$\omega_{T,i}$	–	–	670	724	815	890	1055	1545	–	1447
γ_i	–	–	10	35	40	52	102	60	–	9
$\Delta\omega_i$	4	25	36	50	284	354	373	30	2	23
λ_i	–	–	0.001	0.004	0.025	0.040	0.099	0.01	–	0.034

Concluding, we found the position of the T_{1u} modes for A_xC_{60} and the value of the total electron vibrational mode coupling constant. Our analysis gives strong support for the explanation of superconductivity in the medium to strong coupling limit for A_3C_{60}.

Acknowledgments. Valuable discussions with M. J. Rice and H. J. Choi are greatly acknowledged. Work supported by the FFWF, project P8172.

References

[1] P.W. Stephens, L. Mihály, P.L. Lee, R.L. Whetten, S.M. Huang, R. Kaner, F. Deiderich, and K. Holczer, Nature **351**, 632 (1991).

[2] O. Zhou, J.E. Fisher, N. Coustel, S. Kycia, Q. Zhu, A.R. McGhie, W.J. Romanow, J.P. McCauley Jr, A.B. Smith III and D.E. Cox, Nature **351**, 462 (1991).

[3] J. Winter and H. Kuzmany in this volume and Solid State Commun. **84**, 935 (1993).

[4] Q. Zhu, O. Zhou, N. Bykovetz, J.E Fischer, A.R. McGhie, W.J. Romanow, C.L. Lin, R.M. Strongin, M.A. Cichy and A.B. Smith, Phys. Rev **B 15**, submitted.

[5] D.M Poirier and J.H Weaver, Phys. Rev **B 15**, submitted.

[6] T. Pichler, J. Kürti, M. Matus, and H. Kuzmany, Solid State Commun. **81**, 859 (1992).

[7] J. Fink and E. Sohmen, Physikalische Blätter, Vol **48**, No. 1, 11 (1992).

[8] L. Degiorgi in this volume and L. Degiorgi, P. Wachter, G. Grüner, S.M. Huang, J. Wiley, and R.B. Kaner, Phys. Rev. **B 46**, 11250 (1992).

[9] T. Pichler, M. Matus, and H. Kuzmany, Solid State Commun. **86**, 221 (1993).

[10] M.J. Rice and H.J. Choi in this volume.

[11] T. Pichler, J. Kürti, and H. Kuzmany, Condensed Matter and Materials Commun., (to be published)

[12] M.J. Rice and H.Y. Choi, Phys. Rev. **B45**, 10173 (1992).

[13] C. M. Varma, J. Zaanen, and K. Raghavachari, Science **254**, 989 (1991).

[14] M.J. Rice, Phys. Rev. Lett. **37**, 36 (1976).

[15] M.J. Rice, L. Pietronero, and P. Brüesch, Solid State Commun. **21**, 757 (1977).

Fullerene-Doped Conducting Polymers: Effects of Enhanced Photoconductivity and Quenched Photoluminescence

K. Yoshino[1], *X.H. Yin*[1], *K. Muro*[1], *S. Kiyomatsu*[1], *S. Morita*[1], *A.A. Zakhidov*[2,*], *T. Noguchi*[3], *and T. Ohnishi*[3]

[1]Department of Electronic Engineering, Faculty of Engineering, Osaka University, 2-1 Yamada-Oka, Suita, Osaka 565, Japan
[2]Department of Thermophysics, Uzbek Academy of Sciences, Katartal 28, Chilanzar, 700135 Tashkent, Uzbekistan
[3]Tsukuba Research Laboratory, Sumitomo Chemical Co., Ltd., 6 Kitahara, Tsukuba, Ibaraki 300-32, Japan
*Present address: Istituto di Spettroscopia Molecolare, CNR, Via de'Castagnoli 1, I-40126 Bologna, Italy

Abstract. It is found that fullerenes (C_{60},C_{70}),due to their strong electron accepting abilities can be hole generators in conducting polymers sensitizing photoinduced charge transfer.Here we report that photoconductivity of poly(2,5-dialkoxy-p-phenylene-vinylene) OO-PPV is found to be remarkably enhanced by several orders of magnitude upon introduction of several mol % of C_{60}. Positive polarons (P^+) photogenerated with increased efficiency due to autoionization of excitons and/or photopumping from fullerene are considered to be responsible for enhanced photoconductivity. Photoluminescence of polymer is strongly quenched upon C_{60} doping due to dissociation of excitons accompanied by electron transfer to fullerene.

1. Introduction

C_{60} is usualy considered to be a host for dopants:various donors have been intercallated into C_{60} lattice, making it superconducting [1], or ferromagnetic [2].

In contrast, we have recently demonstrated that C_{60} and C_{70} can be doped in conducting polymers (CP) e.g. poly(3-alkylthiophene) (PAT), whose main chains are composed of a highly extended conjugate electron system [3-7]. Two types of characteristics were found in C_{60}-doped PAT depending on alkyl chain length: (1) marked change in the absorption spectrum, quenching of photoluminescence, slight enhancement of conductivity and slight change in ESR spectrum were observed in poly(3-hexyl thiophene), suggesting charge-transfer doping even in the ground state [3,4]. (2) Small changes in the absorption spectrum, ESR and conductivity, but marked quenching of photoluminescence and enhancement of photoconductivity were observed in poly(3-octadecylthiophene), suggesting ineffective ground state charge transfer, while the occurrence of the photoinduced charge-transfer effect, i.e. photodoping [5].

In present paper, characteristics of C_{60}-doped poly(2,5-dialkoxy-p-phenylene vinylene), (RO-PPV) whose molecular structure is shown in inset of Fig.1, are studied and discussed in terms of its electronic energy structure relative to that of C_{60}. We also discuss some general advantages of C_{60} intercallation into CP mattrix.

Springer Series in Solid-State Sciences, Volume 117
Electronic Properties of Fullerenes Eds.: H. Kuzmany, J. Fink, M. Mehring, and S. Roth

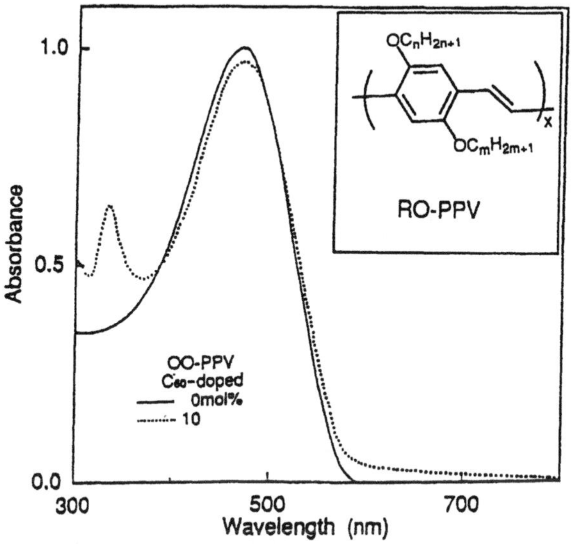

Fig.1. Absorption spectra of (OO-PPV + C_{60}) as a function of C_{60} concentration, inset shows molecular structure of RO-PPV: OO-PPV corresponds to n = m = 8, MOO-PPV: to n = 1, m = 8.

2. Experimental

RO-PPV was prepared by dehydrohalogenation by treating 2,5-dialkyl-p-xylene dibromide with t-C_4H_9OK in mixed solvent (t-C_4H_9OH-Xylene). Among various RO-PPVs with different alkoxy chain lengths, poly(2,5-dioctyloxy-p-phenylene vinylene) (OO-PPV) and poly(2-metoxy5-octyloxy-p-phenylene vinelene) (MOO-PPV) were used in present experiment.Both C_{60} (provided by Science Laboratories Co.,Ltd.) and RO-PPV of appropriate molar ratios were dissolved in toluene. Then a thin film was prepared by casting the solution on an ITO (indium-tin oxide) coated quartz plate. Gold deposited by evaporation on the film served as a second electrode. Details of preparations and measurements can be found in our [7,8]

3. Results and Discussion

As shown in Fig.1, the absorption spectrum of OO-PPV is not changing much upon C_{60} doping; only the absorption peak at about 330nm due to C_{60} appeared for a higher concentration of C_{60}. Electrical conductivity of OO-PPV is also not enhanced upon C_{60} doping contrary to the case of doping with conventional dopants such as iodine and BF_4^-, in which electrical conductivity is usually enhanced by more than ten orders of magnitude up to 10^{-1}S/cm. This also indicates that C_{60} is not transfering charge to OO-PPV in the ground state.

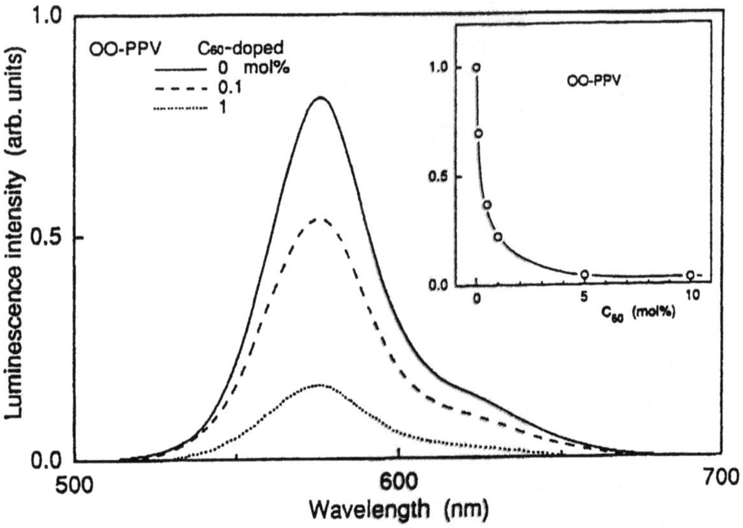

Fig.2. Photoluminescence spectra of OO-PPV $+C_{60}$. Inset shows PL intensity as a function of C_{60} concentration.

On the other hand, photoluminescence of OO-PPV changes markedly upon C_{60} doping as shown in Fig.2. Though the wavelength (580nm) of the emission peak do not change, its intensity decreased drastically with increasing C_{60} concentration as also evident in inset of Fig.2. Quenching of photoluminescence has also been observed in C_{60}-doped PAT and is interpreted in terms of dissociation of exciton-polarons (Ex-P) upon collision with C_{60} molecules.[3-5]

As shown in Fig.3, photoconductivity is also enhanced markedly. The photoconduction response at the photon energy corresponding to the band gap energy of MOO-PPV was especially enhanced by the introduction of a small amount of C_{60} as shown in inset of Fig.3. Moreover,two new photoconduction response bands have appeared at about 1.8 eV and 3.5 eV at higher doping levels.This bands correspond to optical absorption of C60, though the band at 1.8 eV is usually a forbiden one and hence rather weak.Recently, we have reported the similar enhancement of photoconductivity upon C60 doping in poly(3-octadecylthiophene) [5]. We have interpreted this enhancement in terms of photoinduced charge transfer.

The present observation of enhancement of photoconductivity in OO-PPV and MOO-PPV upon C_{60} doping can be explained by the similar concept of photoinduced charge transfer which then suggests that the top of the valence band of RO-PPV is lower in energy than LUMO of C_{60},so that C_{60} is a typical "weak CT-dopant" [9].

With the aid of the electronic energy diagram of Fig.4, the enhancement of photoconductivity is explained as follows: Electron transfer from RO-PPV to C_{60} is unfavourable in the ground state which is consistent with the small changes in the absorption spectrum and electrical conductivity in RO-PPV by C_{60} doping. Upon irradiation of light with photon energy exceeding the band gap energy,

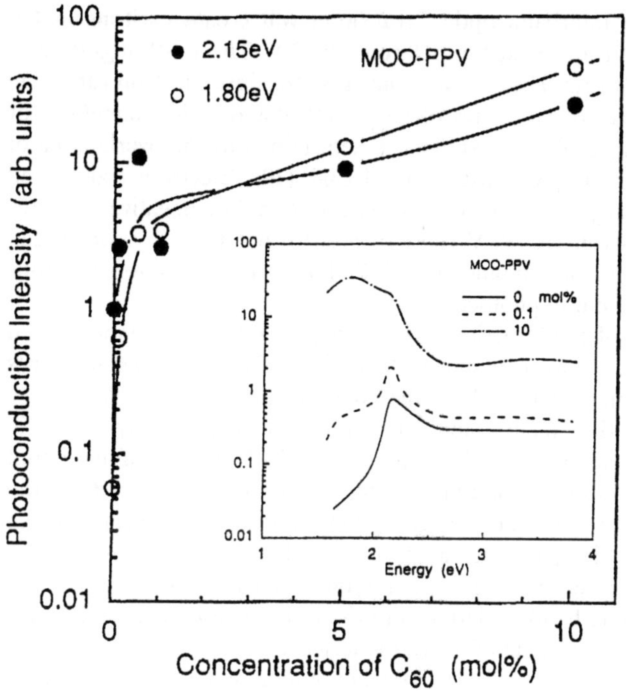

Fig.3. Dependence of the photoconduction intensity of MOO-PPV on C_{60} concentration. Inset shows photoconduction spectra at three different C_{60} concentrations.

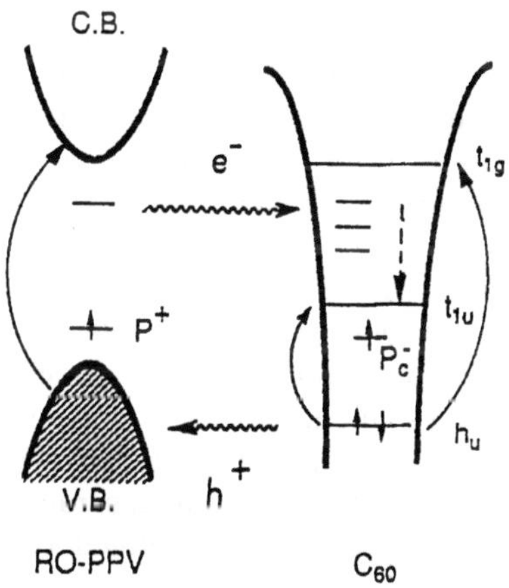

Fig.4. Energy diagram of OO-PPV and C_{60} with and schematic explanation of Ex-P dissociation by e⁻(electron) transfer and h^+ (hole) injection from directly excited C_{60} molecule.

2.2eV, of RO-PPV, the interband optical transition will create excitons in CP chains which quickly self trap to exciton-polarons (Ex-P). The Ex-P migrating to the vicinity of neutral C_{60} reaches an autoionizing state, since electron tunneling from the Ex-P level to t_{1g} or t_{1u} states of C_{60} is facilitated. This tunneling (or phonon-assisted hopping) probability should be determined by the overlapping of π orbitals of RO-PPV and C_{60} and extension of Ex-P [10] Electrons transferred by this photoinduced charge transfer to C_{60} relax to immobile negative polarons Pc- in C_{60} [11] but holes in OO-PPV relax to positive polarons P^+, which can become mobile escaping Coulomb binding with P^- during thermalisation, and thus resulting in the enhancement of photoconductivity.

For the enhancement of photocarrier generation probability, the photoexcited Ex-P must encounter a C_{60} molecule during its lifetime. One C_{60} molecule exists for 50-100 monomer units of OO-PPV at the concentration of 1 mol% of C_{60}. Hence, the finding that the enhancement of photoconductivity mainly occurs in the concentration range below 1 mol%, suggests that Ex-P migrates along about 50-100 monomer units during its lifetime. At higher concentrations of C_{60}, the lifetime of Ex-P may become shorter but further enhancement of photoconductivity will no longer occur, as observed. This interpretation can also explain the concentration dependence of photoluminescence quenching. That is, shortening of Ex-P lifetime by dissociation with the introduction of several mol% of C_{60} also results in the quenching of photoluminescence.

At high concentration of C_{60} its direct optical excitation at 1.8 eV band (which apparenty becomes partially allowed, due to interaction with CP chains, as evident from Fig.3) and 3.5 eV band results in hole injection into valence band of RO-PPV followed by relaxation to P^+, also contributing to PC enhancement, but not PL quenching.

Recently, Sariciftci et al.[13] observed photovoltaic phenomena in the poly(2-methoxy, 5-(2'-ethyl-hexyloxy)-p-phenylene vinylene)-C_{60} junction and also photo-induced ESR, which can also be interpreted in terms of photoinduced electron transfer from polymer to C_{60} and is consistent with our observation of enhancement of photoconductivity and quenching of photoluminescence in RO-PPV upon C_{60} doping

It should be noted that effect of photoconductivity enhancement reported here is quite different from the one known in molecularly doped conventional polymers, like polycarbonate [14] in which the insulating polymer matrix plays the passive role, while transport of charges occures via hoppings among molecular dopants. However to clarify quantitatively the role of polaron mobility in the CP network, one has to separate it from the photogeneration enhancement. To study separately the effect of increased photogeneration we have found recently that photoinduced absorption in IR active vibration
modes and electronic bands of Ps in PAT is increased 9.5 times upon C60 doping, proving that concentration of P^+ (or bipolarons BP^{2+}) is increased [15] To check the effect of C60 on the mobility one should use the xerografic discharge technique,by which holes have been found recently as main photocarriers in PVK doped by C_{60} [16].

4. Conclusions

The results of our study can be summarized as follows. Photoconductivity is enhanced by several orders of magnitude and photoluminescence is markedly quenched upon C_{60} doping of several mol% in RO-PPV, suggesting that C_{60} becomes an effective dopant upon light irradiation with photon energy exceeding the band gap energy of RO-PPV or absorption bands of C_{60}.

Concluding we would like to note that intercallaton of fullerenes into various mattrices is a very promising field,not only because the properties of the host can be changed in a usefull way (e.g. recently the C_{60} sensitization of photorefractive polymer [17], C_{70}- mediated enhancement of electron transport [18] and other interesting phenomena has been reported).What is more exciting and intriguing is the possibility to modify and hopefully improve the unusual properties of fullerene molecules itself and cooperative phenomena of C_{60} solid complexes by making composites with electronically and optically active mattrices like conducting polymers.

References

1) A.F. Herbard, M.J. Rosseinsky,et al, Nature 350 (1991) 600.

2) P.-M. Allemand, K.C. Khemani,et al, Science 253 (1991) 301.

3) S. Morita, A.A. Zakhidov and K.Yoshino: Solid State Commun. 82 (1992) 249.

4) S. Morita, A.A. Zakhidov, T. Kawai, H. Araki and K. Yoshino: Jpn. J. Appl. Phys. 31 (1992) L890.

5) K. Yoshino, S. Morita, T. Kawai, H. Araki, X.H. Yin and A.A. Zakhidov: ICSM'92 Proc., to be published in Synth. Met.

6) K.Yoshino,X.H.Yin,S.Morita and A.A.Zakhidov, Jpn.J.Appl.Phys.,32,L140, (1993)

7) K.Yoshino, X.Xong Yin, K.Muro, S.Kiyomatsu, S.Morita,A.A.Zakhidov, T.Noguchi, T.Ohnishi. Jap.J.Appl. Phys.32,L357, (1993)

8) S.Morita, S.Kiyomatsu,X.H.Yin, A.A.Zakhidov
T.Noguchi,T.Ohnishi and K.Yoshino, J.Appl.Phys. (submitted)

9) A.A.Zakhidov, Synth.Met.,41-43,3393 (1991), see also Proc.ELORMA'87, p.216, Moscow, 1987.

10) A.A.Zakhidov, Synth.Met., 41-43,3649 (1991)

11) I. Murase, T. Ohnishi, T. Noguchi and M. Hirooka: Polym. Commun. 28 (1987) 229.

12) K.Harigaya. J.Phys.Soc.Jap.,60,4001 (1991)

13) N.S. Sariciftci, L. Smilowitz, D. Braun, G. Srdanov, V. Srdanov, F. Wudl and A.J. Hegeer: Synth. Met.,Proc. ICSM'92 (to be published)

14)see e.g.in L.B.Schein," Electrophotography & Development Physics",(Springer,Berln,1988)

15).S.Luzzati, C.Botta, A.A.Zakhidov, K.Yoshino, S.Morita, (to be published)

16) Y.Wang,Nature,356,585,(1992)

17)S.M.Silence,C.A.Walsh,J.C.Scott,W.E.Moerner, Appl.Phys.Lett., 61, 2967, (1992)

18) K.C.Hwang and D.Mauzerall, Nature,361, 138, (1993)

Femtosecond Time-Resolved Optical Studies of Photoexcited States in C_{60}

T.N. Thomas[1], *R.A. Taylor*[1], *J.F. Ryan*[1], *D. Mihailovic*[2], *and R. Zamboni*[3]

[1]Department of Physics, University of Oxford, Parks Road, Oxford, UK
[2]Institut Josef Stefan, 39 Jamova, 61111 Ljubljana, Slovenia
[3]Istituto di Spettroscopia Molecolare, CNR, Via de'Castagnoli 1, I-40126 Bologna, Italy

Abstract. We have measured the relaxation of photoexcited states in C_{60} using femtosecond excitation at energies corresponding to the $h_u \rightarrow t_{1u}$ bandgap. Induced absorption with a fast 2 ps relaxation is observed, which has bimolecular kinetics. The induced absorption is spectrally-broad, with two peaks: one at 1.92 eV which may arise from a photo-induced transition due to distortion of the C_{60} icosahedral symmetry; the other, at 2.25 eV, is assigned to excited state absorption from the t_{1u} level to a higher-lying h_g excited state at ~4.0 eV.

1 Introduction

In its ground-state, molecular C_{60} has the symmetry (I_h) of a truncated icosahedron [1] approximately 7 Å in diameter. The solid form is a semiconducting molecular crystal with an fcc lattice where the inter-molecular interactions (Van der Waals) are relatively weak. The energy levels of the isolated molecule are only weakly modified in the solid: e.g. the $h_u \rightarrow t_{1u}$ energy gap is seen to shift down by only ~124 meV (from 1.91 to 1.79 eV [2]). These energies are those of excitons which are related to the free carrier gap, determined by other methods [3] to be 2.3 eV. The lowest singlet and triplet exciton energies have been indicated, by high-resolution electron energy loss spectroscopy, to be at 1.72 and 1.55 eV respectively [4]. Two-photon absorption resonances at ~0.9 eV seem to confirm the lowest singlet energy value to be around 1.8 eV [5]. Although one-photon optical transitions in this region are nominally forbidden the t_{1u} state is mixed, by specific odd parity (Herzberg-Teller active) vibrational modes, with states of even parity, giving rise to weak absorption [2].

2 Experimental

The techniques for generating femtosecond laser pulses which are used in this work have been described in detail elsewhere [6]. 100 fs, 610 nm (2.03 eV) pulses from a colliding-pulse modelocked dye laser are amplified to 4.0 µJ. A spectrally-broad femtosecond continuum is generated by focusing some of the amplified beam into an ethylene glycol jet. The 2.03 eV laser pulses and the continuum were used as pump and probe respectively in a standard time-resolved pump-probe transmission experiment. The pump beam was brought to a 70 µm focus at the sample, the probe spot size being slightly smaller. The induced transmission ΔT/T and reflection ΔR/R

Springer Series in Solid-State Sciences, Volume 117
Electronic Properties of Fullerenes Eds.: H. Kuzmany, J. Fink, M. Mehring, and S. Roth

changes were measured by a differential lock-in detection technique. Fractional changes as small as 10^{-4} were detected. To take into account the different time delays at different wavelengths in the continuum (chirp), we used two-photon absorption measurement in ZnO.

Samples, in the form of thin films ~200 nm in thickness, were prepared by sublimation of chromatographically separated C_{60} material on to a cooled quartz substrate at an ambient pressure of 6×10^{-6} torr. The films were subsequently exposed to air, and the experiments were also carried out in the presence of oxygen. All our measurements were made at a temperature of 296K. The excitation densities were estimated by measuring the absorbed pump power and assuming a uniform excitation across the spot and through the sample.

3 Results

After excitation by 2.03 eV photons we find a spectrally-broad increased absorption with two peaks at 1.92 eV and 2.25 eV (see figure 1). The data were obtained at excitation densities of $\sim10^{19}$ cm^{-3}. At all energies near the gap we also find a relatively fast (~2 ps) and a slow (≥1 ns) component in the recovery of the induced absorption.

The temporal response of the induced absorption seen at 2.03 eV (see figure 2) is characterised by a rise which follows the laser pulse, and a subsequent ~2 ps relaxation to a long lived plateau (see inset to figure 2). The density dependence of the induced absorption was measured, the results of which are summarised in figure 2. The peak heights shown in figure 2 have been normalised for the purposes of comparison. A distinct density dependence of the relaxation time is observed,

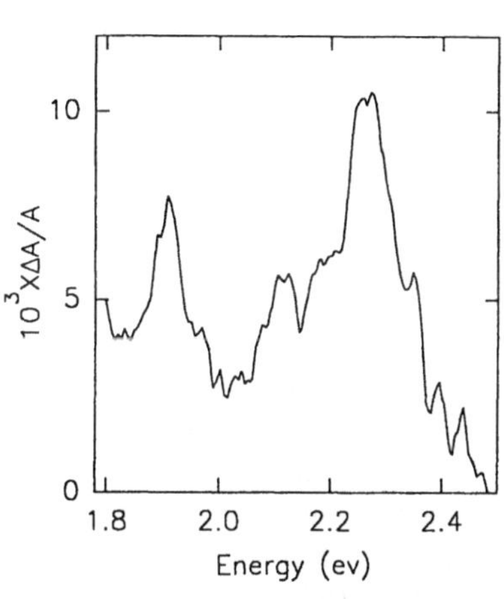

Figure 1 The induced absorption spectrum obtained on photo-excitation. There is a positive chirp of 2.05 ps across the entire spectrum. The delay at 2.48 eV is 3.3 ps.

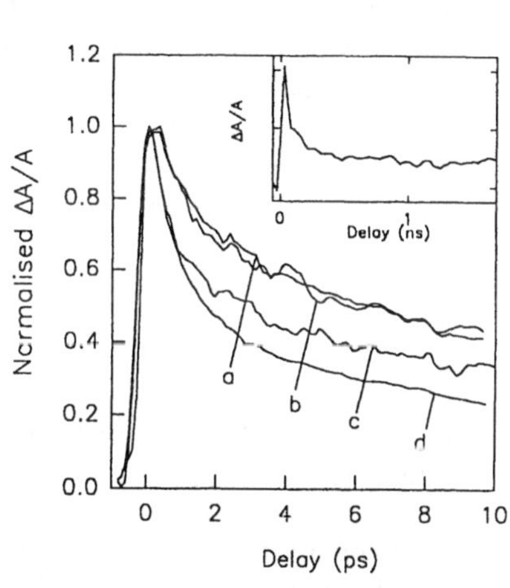

Figure 2 Normalised induced absorption vs. delay at four excitation densities (a, b, c and d: 1, 1.5, 3, 6 $\times10^{19}$ cm^{-3} respectively). The inset shows the long time behaviour.

showing faster decay rates at higher densities. The peak value of the ΔA/A shows a linear increase with carrier density up to 4.5%, at 5×10^{19} cm^{-3}, where saturation begins.

4 Discussion

In order to explain these results we first discuss the electronic structure of the material and possible electron-hole excitations which might give rise to the two induced absorption peaks at 1.92 eV and 2.25 eV. The latter feature is attributed to excited state transitions between t_{1u} and h_g singlet excitons, i.e. excited state absorption. Transmission measurements of solutions of C_{60}^- [8] and ellipsometry of M_6C_{60} [9] also detect additional absorption (when compared to neutral C_{60}) at ~2.5 eV, which in the latter case was attributed to transitions between the t_{1u} and the h_g states.

Apart from the HOMO-LUMO gap, there are no obvious transitions which could give rise to an absorption at 1.92 eV. For this reason we believe that the latter, and the surrounding background induced absorption, is due to a photo-induced changes in the nominally forbidden $h_u \rightarrow t_{1u}$ transition. Calculations [10, 11] suggest that occupation of the t_{1u} state reduces the symmetry from icosahedral to D_{5h} (rugby-ball shape) by slightly relaxing the equatorial molecular conjugation. This may give rise to a new, polaron-like state 0.1 eV below the conduction band (with a similar state 0.1 eV above the valence band) thus changing the density of states and the effective oscillator strength. However, excited state absorption cannot be ruled out since Jahn-Teller shifts of the energy levels are expected.

The measured 2 ps decay time is consistent with measurements made at similar excitation densities by Dexheimer *et al.* [13] and Brorson *et al.* [14]. Also, a 43 ps relaxation has been measured at much lower densities [15]. The fast decay process observed here may involve polaron-like distortion of the photo-excited t_{1u} states. However, on the basis of the assignment to excited state absorption, we suggest that the density dependent relaxation time corresponds to a depopulation of the singlet excitons by exciton-exciton annihilation. At high densities, the photo-generated excited excitons are expected to exchange energy via a dipole-dipole interaction [16] resulting in free carriers, i.e.

$$S^* + S^* \Rightarrow S + e + h$$

where S^* and S denote excited and relaxed singlet excitons respectively. The relaxation kinetics of such an Auger-like process at early times (a few ps) are predominantly bimolecular at high densities, with a long-time contribution from exciton recombination and singlet-to-triplet inter-crossing (both times > ns). The data have been fitted by such a model (see figure 3) as represented by:

$$\frac{dn}{dt} = -\beta n - \gamma n^2$$

where n represents the S^* density. The model assumes a linear relation between the

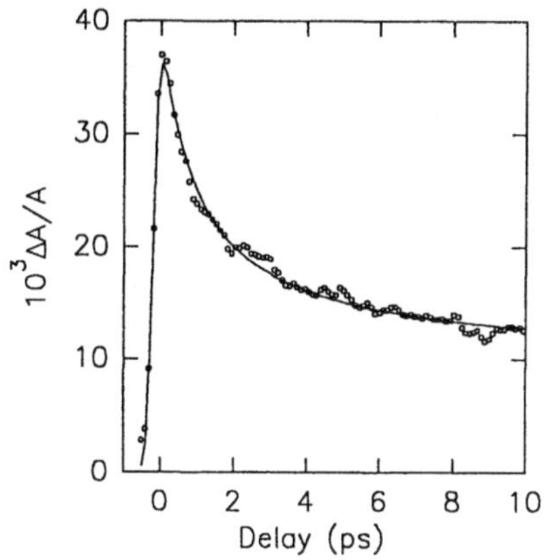

Figure 3 Induced absorption (hollow circles) measured at an excitation density of $3\times10^{19}cm^{-3}$. The fit (solid line) to the data is described in the text.

induced absorption seen and the density *n*. Also, an offset is added to the fit to account for the long-lived absorption possibly due to relaxed singlet excitons generated by annihilation of the excited excitons. β and γ were found to be $3.5\times 10^{10}s^{-1}$ and $1.3\times10^{-7}cm^3s^{-1}$ respectively for the data presented in figure 3, with a large uncertainty (>50%) arising from the estimation of exciton density.

This interpretation of the relaxation is consistent with an exciton having more oscillator strength than the free electron and hole, so that the decrease in exciton population results in a reduced excited state absorption. The relaxed exciton population produced is not expected to generate free carriers, thus contributing to the observed long-lived excited state. The free carriers produced may be observable in a study of the density dependence (up to $10^{20}cm^{-3}$) of photo-conduction whilst photo-exciting below the free carrier threshold (i.e. below 2.3 eV).

5 References

1. H.W. Kroto, J.R. Heath, S.C. O'Brien, R.F. Curl, R.E. Smalley: Nature (London) **318**, 162-163 (1985)
2. F. Negri, G. Orlandi, and F. Zerbetto: J. Chem. Phys., **97**, 6496-6503 (1992)
3. R.W. Lof, M.A. van Veenendaal, B. Koopmans, H. T. Jonkman, and G.A. Sawatzky: Phys. Rev. Lett. **68**, 3924-2927 (1992)
4. A. Lucas, G. Gensterblum, J.J. Pireaux, P.A. Thiry, R. Caudano, J.P. Vigneron, Ph. Lambin, W. Krätschmer: Phys. Rev. B **45**, 13694-13702 (1992)
5. C. Taliani, CNR, Bologna, private communication.
6. W.H. Knox, M.C. Downer, R.L. Fork, and C.V. Shank: Opt. Lett. **9**, 552-554 (1984)
7. M. K. Kelly, P. Etchegoin, D. Fuchs, W. Krätschmer, K. Fostiropoulos: Phys. Rev. B, **46**, 4963-4968 (1992)
8. M.A. Greaney, S. M. Gorun: J. Phys. Chem. **95**, 7142-7144 (1991)
9. Y. Wang, J.M. Holden, A.M. Rao, W.T. Lee, X.X. Bi, S.L. Ren, G.W. Lehman, G.T. Hager, P.C. Eklund: Phys. Rev. B, **45**, 14396-14399 (1992)

10. Kikuo Harigaya: Phys. Rev. B **45**, 13676-13684 (1992)
11. Barry Friedman: Phys. Rev. B **45**, 1454-1457 (1992)
12. Y. Kajii, T. Nakagawa, S. Suzuki, Y. Achiba, K. Obi, K. Shibuya: Chem. Phys. Lett. **181** , 100-104 (1991)
13. S.L. Dexelheimer: International Conference on Quantum Electronics Technical Diest Series **9**, 544-546 (1992)
14. S. D. Brorson, M.K. Kelly, U. Wenschuh, R. Buhleier, J. Kuhl: Phys. Rev. B **46**, 7329-7332 (1992)
15. R.A. Cheville N.J. Halas: Phys. Rev. B **45**, 4548-4550 (1991)
16. A. Bergman, M. Levine, J. Jortner: Phys. Rev. Lett. **18**, 593-596 (1967)

Broadband Electroluminescence in Fullerene Crystals

A.T. Werner, J. Anders, H.J. Byrne, W.K. Maser, M. Kaiser, A. Mittelbach, and S. Roth

Max-Planck-Institut für Festkörperforschung,
Heisenbergstr. 1, D-70569 Stuttgart, Germany

Abstract. The observation of electroluminescence from crystalline fullerenes is described. A broad band emission spectrum, extending from 400nm to 1100nm is observed. The spectrum has a primary maximum at 920nm and a weaker feature centered on 420nm. The spectral characteristics are independent of the applied field and the longer wavelength region is identical to that measured in the high excitation density photoluminescence spectrum. In addition, the electroluminescence intensity increases with the cube of the injection current, strengthening the association to the nonlinear phenomena observed in the highly excited state of fullerenes.

1. Introduction

Luminescence studies in fullerenes have shown that a highly nonlinear emission process is affected at high excitation densities [1] and transport studies support the association of this phenomenon with a transition from a localised molecular-like to a banded or extended state behaviour, i.e. an optically driven Mott-like transition [2]. The resultant nonlinear emission process is characterised by a cubic dependence of the output on the input intensity accompanied by a strong red shift of the emission spectrum. The unique nature of this process prompts an investigation of its applications potential, in particular in the form of an electroluminescent device.

In this report, electroluminescence from crystalline samples is described, which has a spectral distribution attributable to the C_{60} fullerene. Furthermore, the nonlinear dependence of the emission intensity on the injection current strongly associates it with the previously reported nonlinear phenomena in fullerenes.

2. Sample Preparation and Experimental Setup

C_{60} microcrystalline powders were generated from chromatographed solutions of starting material prepared by the method of Mittelbach et al. [3]. Residual solvent was removed by annealing under dynamic vacuum. Mass spectroscopic and rest gas analyses indicate a purity >99.9%. Crystallites were generated by sublimation of the powder under vacuum, and had size in the order of ~100-500μm. The sample employed was a crystalline platelet of dimensions ~ 300 x 300 x 200 μm^3. X-ray diffraction measurements showed a poly- rather than single crystalline

Springer Series in Solid-State Sciences, Volume 117
Electronic Properties of Fullerenes Eds.: H. Kuzmany, J. Fink, M. Mehring, and S. Roth

structure. Gold and aluminium electrodes, of thicknesses 200nm and 15nm respectively, were deposited on the large facets of the crystal under a vacuum of ~10^{-6} mbar. Care was taken to minimise exposure to oxygen and, after contacting with gold wire and silver paste under argon atmosphere, the sample was transferred to an argon-filled glass cell. A Keithley 617 electrometer was employed as the power source and upon application of a voltage, the emitted light was monitored by a photodiode array at the exit of a Chromex 250IS imaging spectrometer with a 300 line/mm grating blazed at 500nm.

3. Results and Discussion

Below ~5V no emission was detectable. Increasing the voltage above this value, however, resulted in the rapid emergence of a broad emission which was clearly visible to the naked eye. The emission originated from the bulk of the crystal although it was not spatially uniform and was observed perpendicular to the field direction. The spectrum extends from the blue to the near infrared and, as is shown in figure 1, has a broad principle maximum at ~920nm with a weaker feature at ~420nm. In the short wavelength region the photoluminescence has not been monitored and the origin of the observed feature is therefore unclear. The spectral profile in the long wavelength regime was observed to be independent of applied field, the magnitude increasing with increasing bias. In this region, the spectrum is notably similar to that of the measured high excitation density photoluminescence spectrum in terms of both spectral positioning and width (figure 1) [2,4]. In order to further examine this similarity, the dependence of the emission intensity on the injected current was monitored. A strongly nonlinear dependence was observed and a logarithmic representation, shown in figure 2, indicates a cubic increase in the emission intensity upon increasing the current above a threshold value, J_0, of ~1mA, corresponding to the application of ~5V. The current densities achieved are indicative of a sample conductivity far in excess of those previously reported for fullerenes [5], but the measurements concerned were performed on polycrystalline

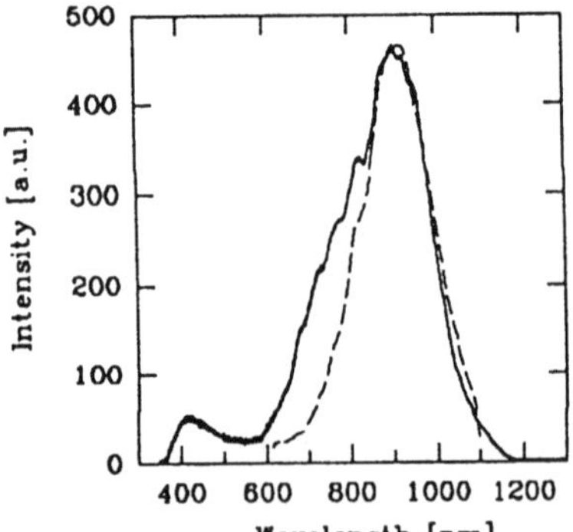

Figure 1; Spectral distribution of the electroluminescence (solid line) and high intensity photoluminescence (dashed line) of the fullerene crystallite.

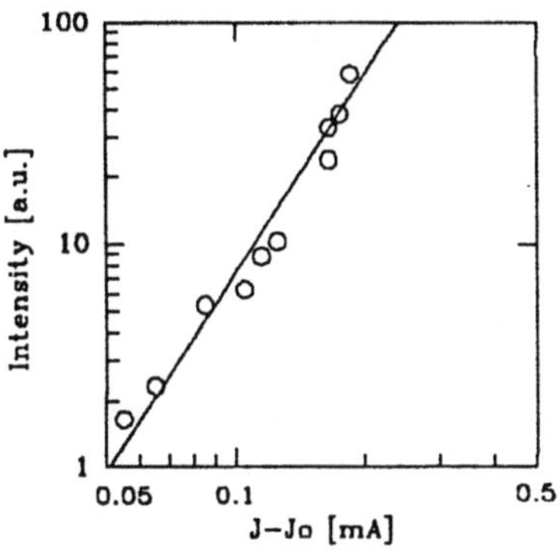

Figure 2; Intensity at 920nm vs the current (J-Jo), above the threshold current, Jo. The solid line shows a slope of three.

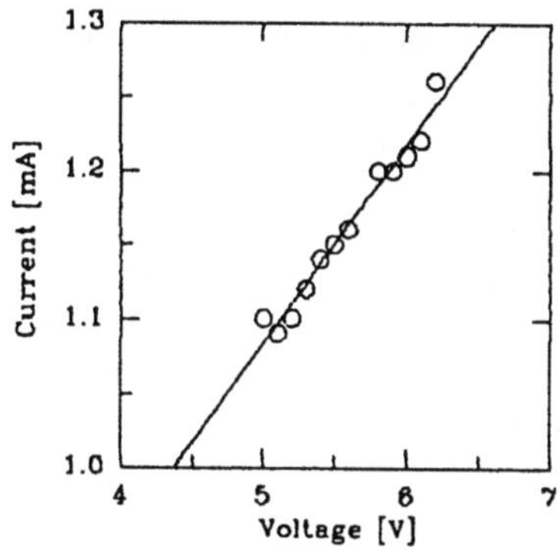

Figure 3; J-V Characteristics of the aluminium-fullerene-gold structure.

films which had been exposed to oxygen. Recent reports indicate that the removal of oxygen from single crystal fullerene results in an increase in conductivity by many orders of magnitude [6]. The high conductivities of the oxygen-free crystal described here are therefore not unexpected, and furthermore, similar values have been observed in other crystals subsequent to the measurements reported here.

The similarity of the electroluminescence described in figure 1 and the photoluminescence spectrum of fullerenes in the highly excited state clearly identifies it as characteristic of the fullerene molecule [1]. The observed cubic increase of the electroluminescence is reminiscent of the nonlinear behaviour of the photoluminescence as well as the photoconductivity in the highly excited state of fullerenes [1,2], and may therefore be associated with a similar phenomenon.

The nonlinear dependence of the electroluminescence on the current is clearly indicative of an interaction process which depends on the carrier density. The nature of the device performance is, however, unclear. Figure 3 shows the current-voltage characteristics of the structure in the region of the onset of emission. A roughly linear dependence is observed and the characteristics are largely independent of voltage polarity, as, indeed, is the emission. Clearly the device may not be considered a light emitting diode. Consideration of the energy levels of the combined aluminium-fullerene-gold system confirms this. The ground state of fullerenes is reported to lie 6.8eV below the vacuum level [7], whereas the Fermi levels of aluminium and gold lie 4.2eV and 5.2eV below respectively. While electron injection into the upper fullerene levels may occur from either electrode, the mechanism for complimentary injection of holes into the ground state, required for radiative recombination, is not apparent.

Transport in fullerene has been shown to be largely dominated by intermolecular hopping processes and defect states [6]. Carriers injected into molecular-like states hop through the crystal under the influence of the field. Such processes prevail below the emission threshold. Above threshold, the spectral profile of the emission as well as the nonlinear dependence of the emission on injection current associate it with the nonlinear phenomena observed in the highly optically excited state of fullerenes [1,2]. The implications of this observation are that the emission is not from molecular states but rather has as origin a state of high carrier density. Such a build-up of high carrier densities may occur at crystalline domain boundaries which act as bottlenecks to the conduction process. This build-up process may result in the turn on of a luminescence process analogous to that observed in the highly excited state of fullerenes, which, as is shown in the luminescence/ current characteristic, depends on the cube of the current. The contribution of such an accumulation process is consistent with the observed spatial inhomogeneity of the emission. Improved device performance should be affected by a more appropriate choice of electrode materials.

4. Conclusions

The device performance is not yet ideal and can be optimised by better choice of electrode materials and use of single crystalline material. Nevertheless, an electro-luminescence which is truely characteristic of the material properties is reported for fullerenes for the first time. The primary emission covers the range of ~700nm to 1100nm, promising broad band operation of an optimised device in the near infrared. A weaker emission is observable in the region of 400nm. Both the nonlinear nature and spectral characteristics of the emission clearly associate it with the previously reported high excitation density photoluminescence and photoconductive phenomena, providing further indications that this is indeed a phenomenon inherent to fullerene, and highlighting its potential applications.

References

[1] H.J. Byrne et al., Appl. Phys., A56 (1993) 235.
[2] H.J. Byrne et al., these proceedings
[3] A. Mittelbach et al., Angew. Chem., 104 (1992) 1681.
[4] J. Feldmann et al., Europhys. Lett., 20 (1992) 553.
[5] J. Mort et al., Phys.Lett., 186 (1991) 284.
[6] T. Aria et al., to be published in the proceedings of the International Conference of Synthetic Metals, Sweden (1992)
[7] G.K. Werteim et al., Science, 252 (1991) 1419.

Nonlinear Phenomena in the Highly Excited State of C_{60}

H.J. Byrne, W.K. Maser, M. Kaiser, L. Akselrod, J. Anders, W.W. Rühle, X.-Q. Zhou, A. Mittelbach, and S. Roth

Max-Planck-Institut für Festkörperforschung,
Heisenbergstr. 1, D-70569 Stuttgart, Germany

Abstract. Under high intensity illumination, the optical and electronic properties of fullerenes are seen to undergo dramatic, nonlinear changes. The photoluminescence emission is seen to increase with approximately the third power of the input intensity above an apparent threshold intensity. Associated with this nonlinear increase is the emergence of a long lifetime emission component and a redshifting of the emission spectrum. Above the threshold intensity the photoconductive response increases with approximately the cube of the input power. In the highly excited state, the photoconductive response becomes relatively temperature independent compared to the thermally activated behaviour observed at low intensities. The characteristics of the temperature dependence are associated with a metallic-like phase in the highly excited state and therefore an optically driven insulator to metal transition is proposed as a description of the observed phenomena.

1. Introduction

In this paper, the influence of highly populating the excited state of C_{60} on the optical and electronic properties are reported. The photoluminescence and photoconductive properties are monitored in the low and high intensity regimes, and dramatic, nonlinear changes are observed. The characteristics of the changes exclude a purely intramolecular interpretation of the phenomenon, and the observed nonlinear changes in the photo-induced transport properties support an intramolecular to intermolecular phase transition as the origin of the behaviour.

2. Sample Preparation and Experimental Technique

C_{60} microcrystalline powders were generated from chromatographed solutions of starting material prepared by the method of Mittelbach et al. [1]. Residual solvent was removed by annealing under dynamic vacuum. Mass spectroscopic and rest gas analyses indicate a purity >99.9% with no oxygen content. Scanning electron microscopy shows these powders to be polycrystalline with crystallites of sizes ranging from microns to tens of microns. Single crystallites were extracted from a supersaturated hexane solution, or by sublimation of the powder under vacuum, and had size in the order of ~100-500μm. Samples were fixed on a cold finger in a helium cryostat which could be cooled to ~10K. All measurements were performed under vacuum and exposure to oxygen was minimised at all times.

For the purpose of photoluminescence studies, excitation was performed with a tuneable Rhodamine 6G cavity dumped dye laser, operating at 585nm with pulsewidth ~20psec at a repetition rate of 400 kHz, pumped by a mode-locked Ar^+ - ion laser with a repetition rate of 80 MHz. The spotsize on the sample was ~10μm radius. The luminescence from the samples was dispersed by a 0.32 m focal length monochromator and detected by a two dimensional synchroscan streak camera with an S-20 photocathode and a time resolution of 20psec. The synchroscan is driven by an electrical 40 MHz trigger from the modelocker of the source laser.

The photoconductive response was measured under illumination with 532nm radiation, derived from a frequency doubled, actively modelocked Nd^{3+}:YAG laser of pulsewidth 100psec operating at 80MHz. Given the long lifetime of the excited species in fullerenes [2], the source may be considered to be continuous. In addition the laser was chopped with a 1:1 duty cycle at a frequency of 75Hz. The sample employed was an oxygen-free C_{60} crystal and was mounted across a single gap of 200μm spacing by Cu/Be- pressure contacts and the input powers were varied in the range of 1 to 200mW. The spot size was ~300μm diameter.

3. Results and Discussion

Under low illumination conditions a weak luminescence output is observable at room temperature from all samples. The spectrum exhibits a considerable amount of structure which may be attributed to the molecular vibrational modes [3]. The temporal decay of the luminescence fits well to a single exponential of lifetime 1.2nsec [4], consistent with the depopulation of the first excited singlet state by intersystem crossing, with efficiency close to one, to the triplet state [5]. As in C_{60} solutions, no long lifetime, phosphorescence component is observable and the triplet lives as long as 30msec [2]. On increasing the input intensity, a dramatic change in the emission spectrum is observable. Figure 1 shows the evolution of the spectrum over the range 0.1mW to 3.0mW. The shift is accompanied by a broadening of the spectrum and a dramatic increase in the luminescence output. Investigations of the temporal decay of the emission indicate that the spectral changes may be associated with the emergence of a long lived emission component. The spectrally integrated temporal streak as a function of intensity is shown in figure 2. Excitation occurs at t~900psec, whereupon the fast rise and subsequent decay of the 1.2nsec component is observable. The temporal range from 900-0psec corresponds to the backsweep of the streak camera, a time window of 6.25 to 7.15nsec after excitation, during which a strongly intensity dependent, longlived component is observable. The magnitude of this component, as measured from the reverse swept component, increases with the cube of the average input

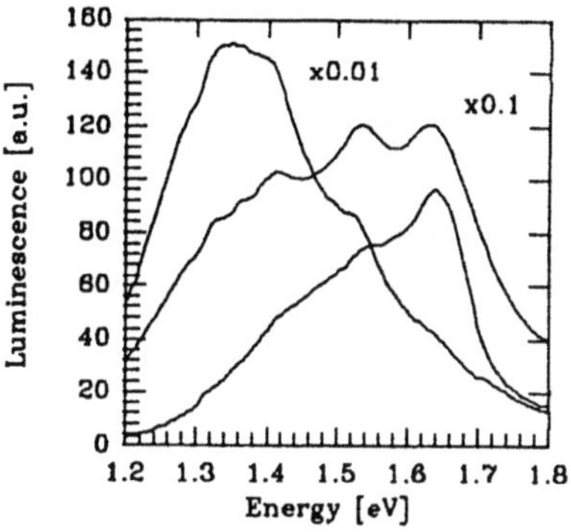

Figure 1: Spectra of C_{60} powder at room temperature illuminated with 0.1 (—), 0.5 (----) and 3.0mW (···).

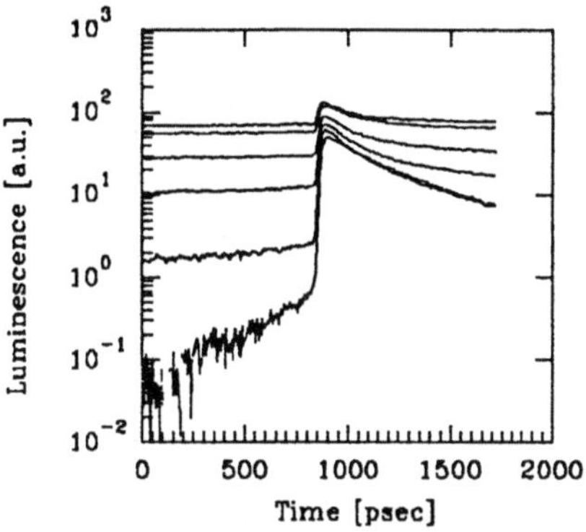

Figure 2: Dependence of the temporal decay on increasing illumination from 0.5 to 3.0 mW, in steps of 0.5 mW.

power as shown in figure 3. (The discrepancy in thresholds between figures 1 and 2 and figure 3 is due to difference in samples). Furthermore, within the time window of the reverse sweep, the decay time of the luminescence, extracted from a fit to a single exponential decay, is seen to increase with intensity, with a dependence of at least squared [4]. The intensity onset of this behaviour appears to be a threshold value and varies with position on the sample by a factor of the order of two. This phenomenon, which is fully reversible, is exhibited by powders and the single crystallite samples prepared from both solution and vapour phase, although not in thin films of crystallite dimensions ~200nm [4]. Its occurrence appears to be largely independent of oxygen content. It may therefore be considered an intrinsic solid state phenomenon although the "threshold" intensities can vary considerably from sample to sample. Furthermore C_{70} has been observed to exhibit similar behaviour although it has not as yet been fully characterised.

The features of the observed behaviour, including the nonlinear evolution of the emission spectrum, a continuous, nonlinear increase in the luminescence efficiency as well as lifetime, may not easily be accounted for by a model system based on static energy levels. The apparent existence of a threshold intensity for this behaviour is indicative of the requirement of a critical excited state density. This is a strong departure from the intramolecular processes observable in solution, and, in order to further investigate this phenomenon, the transport properties of the system under low and high level illumination were investigated.

Under low intensity illumination, the photoconductive response of fullerene crystals is linear in intensity [6] and the temperature dependence shows the transport process to be thermally activated, characteristic of a molecular insulator or semiconductor. The activation energy varies from sample to sample, and, for the example shown in figure 4(a), has a value ~110meV. The observation of this molecular insulator-like behaviour is consistent with the assignment of the luminescence emission to intramolecular transitions. At high intensities, the intensity dependence of the photoconductive response initially reduces to a square

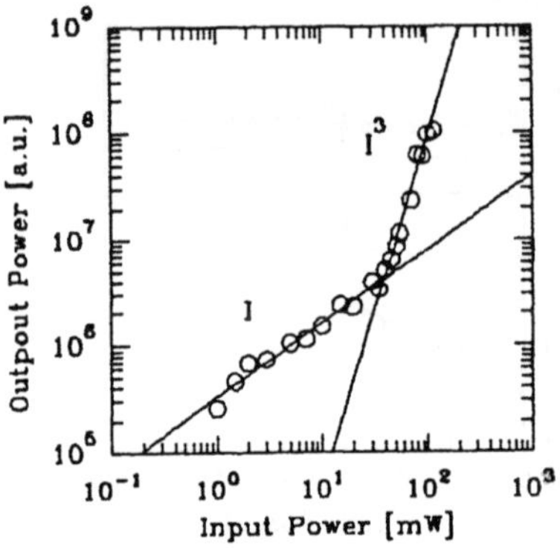

Figure 3: Dependence of the long lived luminescence component on intensity.

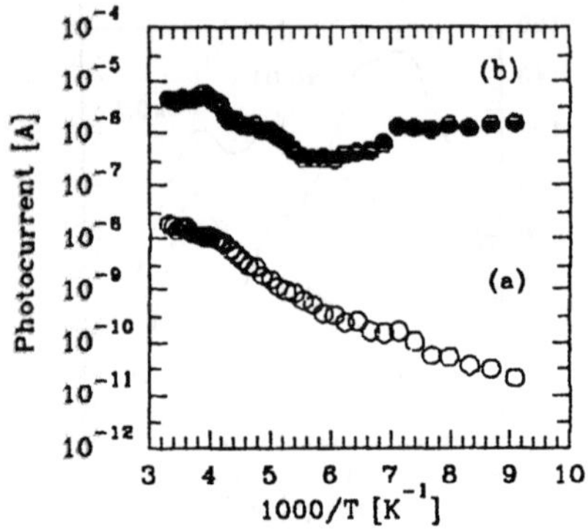

Figure 4: Temperature dependence of the photoconductive response. (a) low and (b) high illumination conditions.

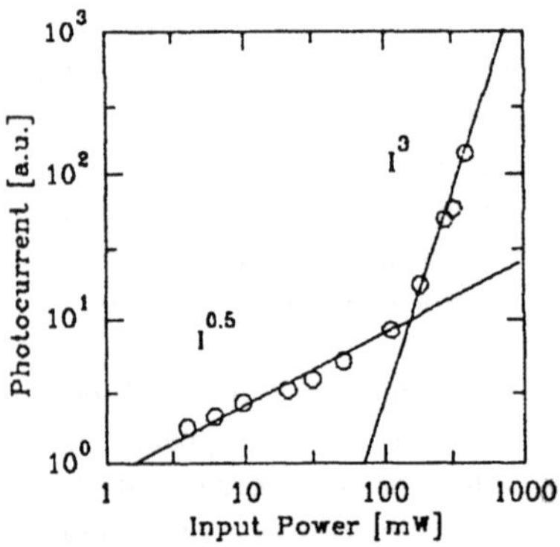

Figure 5: Intensity dependence of the photocurrent in fullerene crystal.

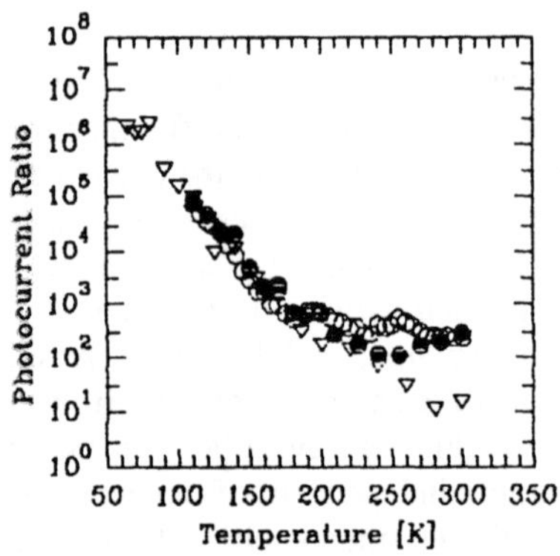

Figure 6: Temperature dependence of the ratio of the high to low intensity response for three samples.

root dependence due to the contribution of bimolecular recombination processes [7]. As is shown in figure 5, however, this dependence evolves into a cubic dependence upon further increase in intensity. The onset of the cubic dependence of the photoconductive response coincides with the visible onset of the nonlinear photoemission and the nonlinear behaviour of the transport properties may, therefore be associated with the nonlinearity in the optical properties.

The increase of the photoconductive response over two orders of magnitude is indicative of a similar increase in the mobility of the photo-generated charge carriers. In figure 4(b), the temperature dependence of a crystal illuminated with 100mW 532nm radiation is compared to that of the same crystal illuminated with 1.0mW (figure 4(a)). The measurement was performed by sweeping the temperature and monitoring the high and low intensity response at each temperature. At each point, adjustment of the laser spot on the sample was performed to optimise the response. Whereas the low intensity photoconductive response decreases by three orders of magnitude over the temperature range measured, the high intensity response remains constant within an order of magnitude. Such a temperature independence is characteristic of a metallic-like behaviour. The observed variations as a function of temperature may be partly the result of the point to point realignment, but may also arise from the fact that the intensities used are below the saturation of the observed nonlinear increase in the photocurrent and so the material is a mixed phase material. The temperature dependence is a combination of that of the weakly excited and fully excited material and therefore a clearer illustration of the temperature dependence of the relative increase of the photoconductive response of fullerenes in the highly excited state may be obtained by viewing the ratio of the high intensity to the low intensity response. This ratio, shown in figure 6, normalises for variations in experimental response, and the reproducibility of the material behaviour may be seen by comparing the ratios obtained from three different samples. The curves have not been normalised to overlap at any point and the good reproducibility indicates that the observed "insulator to metal" transition is intrinsic to the highly excited state of fullerenes.

Under high intensity illumination, both the optical and electronic properties of the solid state of fullerenes are seen to undergo nonlinear changes. The temperature dependence of the photoconductive response of the highly excited state clearly indicates that these changes are the result of the emergence of a metallic phase at high excitation densities. The observed behaviour may be interpreted as a transition from a material in which the localised molecular orbitals are largely unperturbed in the ground or weakly excited state, to one in which, under high level illumination, next nearest neighbour excited states interact to form an extended, highly conductive state. Further evidence of such a reversible transition may be seen in the Raman measurements of Loodsrecht et al. [8]. In this extended state, optical relaxation is no longer restricted by the symmetry selection rules applicable to the molecular states, and so a dramatic increase in the luminescence emission is observed. Nonradiative relaxation from this state requires coupling of the electronic wavefunction to the weak intermolecular phonons and the increased restriction of the intramolecular nonradiative route may account for the observed nonlinear increase in the radiative lifetime. The mechanism of the interaction has, as yet, to be elucidated and although it may have origin in a critical overlap of the excited states, the considerable shift in the emission maximum is indicative of an energy stabilisation interaction which may originate in spin or electron correlation interactions. Similar processes give rise to condensation of the exciton gas to an electron hole plasma, manifest as an optically driven Mott-like insulator to metal transitions in indirect semiconductors [9].

4. Conclusions

Investigations of the optical and electronic properties of fullerene indicate that the molecular nature of the material is largely preserved in the solid state. The luminescence emission is weak and the spectrum possesses vibronic structure which is attributable to molecular vibrations. The temperature dependence of the photoconductive response is characteristic of a molecular insulator. Under high illumination conditions, however, both the optical and transport properties undergo significant nonlinear changes. The photoemission and photocurrent increase with the third power of the input intensity and the temperature independence of the photoconductive response at high intensities indicates that these changes are the result of an optically induced insulator to metal transition in the highly excited state of fullerenes. The implication of the observed behaviour is that fullerenes in the solid state are finely balanced between a molecular crystal and an extended state material, and a transition between these two phases may be optically induced.

5. References

[1] A. Mittelbach, W. Hönle, H.G. von Schnering, J. Carlsen, R. Janiak and H. Quast, Angew. Chem., in press.

[2] S.P. Sibley, S.M. Argentine, A.H. Francis, Chem. Phys. Lett., 188 (1992) 187.

[3] C. Reber et al., J. Phys. Chem., 95 (1991) 2127.

[4] H.J. Byrne, W.K. Maser, W.W. Rühle, A. Mittelbach and S. Roth, Appl. Phys. A56, (1993) 235

[5] T.W. Ebbesen, K. Tanigaki, S. Kuroshima, Chem. Phys. Lett., 181 (1991) 501.

[6] M. Kaiser, J. Reichenbach, H.J. Byrne, J. Anders, W. Maser, S. Roth, A. Zahab and P. Bernier, Solid State Commun., 81 (1992) 261.

[7] H. Yonehara and C. Pac, Appl. Phys. Lett., 61 (1992) 575.

[8] P.H.M. van Loosdrecht, et al., these proceedings

[9] M. Combescot and P. Nozières, J. Phys. C: Solid State Phys., 5 (1972) 2369

Linear and Nonlinear Optical Properties of Fullerenes and Some Metal Derivatives

J. Callaghan[1], *D.N. Weldon*[2], *F.Z. Henari*[1], *W. Blau*[1], *and D.J. Cardin*[3]

[1]Department of Pure and Applied Physics, Trinity College, Dublin 2, Ireland
[2]Department of Chemistry, Trinity College, Dublin 2, Ireland
[3]Department of Chemistry, University of Reading, Reading RG11 6AD, UK

Abstract. Fullerene C_{60} and C_{70} have been prepared using the contact arc method, purified and used to synthesise metal derivatives with the metals iridium (Ir), platinum (Pt), and palladium (Pd). Non-linear optical studies of these materials in the sub nanosecond regime have been carried out employing the technique of Saturation Spectroscopy permitting the magnitudes of the imaginary component of the third order non-linear susceptibility to be obtained.

1. Introduction

The non-linear optical properties of fullerenes have been likened to those of conjugated organic polymers, where the behaviour is attributed to extensive π electron delocalization, resulting in a highly polarizable material. Recently Tutt and Kost [1] have reported optical limiting in solutions of C_{60} as have also Henari et al. [2]. This behaviour has been explained in terms of excited state absorption, and has been observed in a number of materials including semiconductors [3], dyes and organometallic materials [4].

Here we report the synthesis of metal derivatives of C_{60} and C_{70} and a comparative study of their intensity dependant absorption with their parent molecule. The metals used were the transition metals iridium (Ir), platinum (Pt), and palladium (Pd), each metal being co-ordinated with two carbon atoms of the fullerene cage.

2. Synthesis of (η^2-$C_{60/70}$M) Derivatives and Experimental Proceedure.

Fullerenes were prepared in the usual manner using the contact arc method [5] For a typical preparation of the fullerene derivatives, 3.10^{-5} moles of C_{60} or C_{70} powder was dissolved in 25ml of dry degassed toluene. The metal reactants were dissolved in a similar amount of dry degassed toluene and added dropwise over ten minutes to a stirring solution of the fullerene using a cannula. Work-up of the products involved removal of the toluene on a vacuum line followed by washing the product in a Schlenk tube with dry degassed ethanol. The powders were then dried on a vacuum line for 2 hours. All manipulations of the products were carried out in a Millar-Howe drybox with O_2 and N_2 levels of less than 1ppm. The products were

Springer Series in Solid-State Sciences, Volume 117
Electronic Properties of Fullerenes Eds.: H. Kuzmany, J. Fink, M. Mehring, and S. Roth

η²-coordination in metallofullerenes

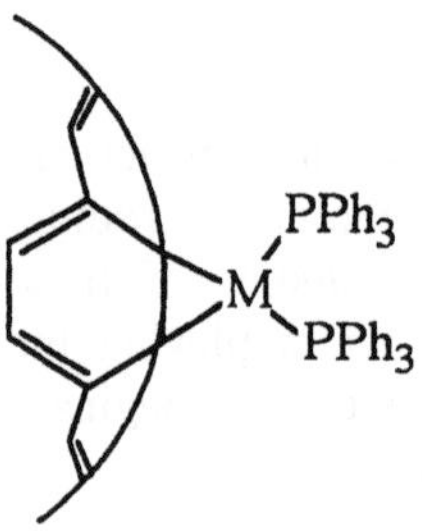

Fig. 1 Structural representation of C_{60}/C_{70} metal derivative. M=Ir, Pd, or Pt

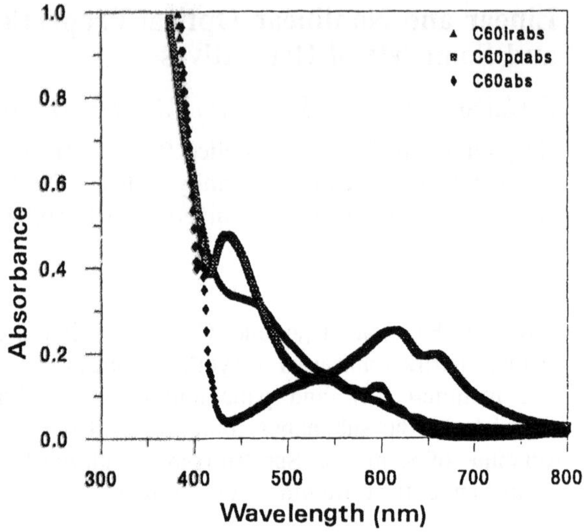

Fig. 2 Absorption spectra of C_{60}, C_{60}Pd and C_{60}Ir

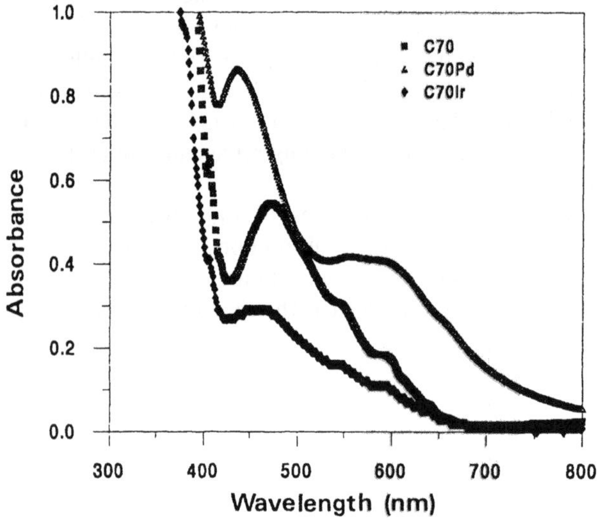

Fig. 3 Absorption spectra of C_{70}, C_{70}Pd and C_{70}Ir

analysed and structures assigned using a combination of IR, UV/VIS/NIR, NMR and FABMS. Fig.1 shows a schematic representation of η^2- bonding to the metal.

Solutions of the metal derivatives were made up in dry degassed toluene, each with a concentration of 2g/L. The absorption spectra for C_{60}, C_{60}Ir, C_{60}Pd, and the corresponding derivatives of C_{70} are shown in Figs 2 and 3, respectively.

The non-linear absorption experiments were performed using a PRA LN100 Nitrogen pumped dye laser with Rhodamine 6G as the lasing dye. This laser has a

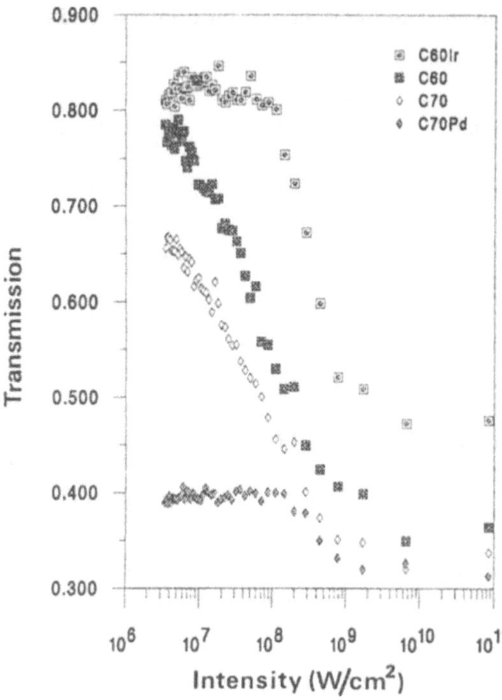

Fig. 4 Plot of intensity dependant transmission for C_{60}, C_{70} C_{60}Ir and C_{70}Pd

pulsewidth of 500ps and lasing at 588nm gave a typical energy output of 40μJ/pulse. Samples under investigation were contained in 1mm quartz cuvettes and placed on a linear translation stage. This permits the sample to be moved through the focal plane of a 5cm lens. Variations in the transmitted fluence with incident intensity are monitored using wide area photodiodes, and a digital storage oscilloscope interfaced to a PC.

The maximum incident intensity was 100GW/cm^2, occurring at focus. Data points were averaged over 20 laser shots with a repetition rate of 1Hz. A plot of the transmission as a function of incident intensity is shown in Fig. 4 for C_{60}, C_{70}, C_{60}Ir and C_{70}Pd.

3. Results and Discussion

Addition of the metal sidegroups has greatly changed the absorption spectra compared to those of C_{60} and C_{70}. In the case of C_{60}Pd the λ_{max} is now around 750nm, 100nm further in the red than that of C_{60}. Also, though it cannot be seen from Fig.2, the large C_{60} peak at 350nm has been shifted well below 300nm. The palladium and platinum (not shown) derivatives both have a sharp peak near 450nm, but in the iridium compound this is absent. Indeed the visible section of the

Table 1 List of calculated absorption cross sections and $\chi^{(3)}$ for C60, C70 and their metal derivatives

Compound	σ_0 (cm^2)	σ_{ex}/σ_0	$\chi^{(3)}$ (m2/V2)
C60	$1.45 \cdot 10^{-18}$	4.16	$1.78 \cdot 10^{-16}$
$(\eta^2-C_{60})Pd(PPh_3)_2$	$5.61 \cdot 10^{-18}$	1.62	$1.55 \cdot 10^{-17}$
$(\eta^2-C_{60})Pt(PPh_3)_2$	$1.14 \cdot 10^{-17}$	1.44	$1.32 \cdot 10^{-16}$
$(\eta^2-C_{60})IrCOI(PPh_2Me)_2$	$2.56 \cdot 10^{-18}$	2.78	$4.14 \cdot 10^{-17}$
C70	$2.94 \cdot 10^{-18}$	2.74	$7.55 \cdot 10^{-17}$
$(\eta^2-C_{70})Pd(PPh_3)_2$	$1.16 \cdot 10^{-17}$	1.22	$5.78 \cdot 10^{-17}$
$(\eta^2-C_{70})IrCOI(PPh_2Me)_2$	$3.46 \cdot 10^{-18}$	2.58	$2.64 \cdot 10^{-17}$

iridium absorption spectra has no distinguishing features, being characterised solely by a gradual increase in the absorption from 700nm down to 350nm with small very weak transitions observable near 600nm, 550nm and 475nm

The palladium and platinum derivatives of C_{70} show qualitatively the same behaviour as those of C_{60}. We see absorption starting at 800nm for C_{70}Pd and the large peaks at 350nm and 480nm being shifted into the blue. For the C_{70}Ir species the same blueshift is observed but the λ_{max} remains the same as that of C_{70}.

C_{60} has been shown to possess a long lived first excited triplet state with a lifetime between 40µs [6] and 400µs [7]. Initial excitation of the molecule results in populating the excited singlet levels S_1, S_2,..S_n. These rapidly relax to the first excited singlet and, via a fast intersystem crossing, populate the first excited triplet. This level thus acts as an accumulation site from which excited state absorption may result. Thus to a large extent there are two levels involved in any excitation process, as the population of the S_1 state is much smaller than that of S_0 or T_1. For a two level system the following equation may be derived for the change in absorption ($\Delta\alpha$), in terms of the saturation intensity I_{sat} and the ground and excited state absorption cross sections (σ_g,σ_{ex}) [2].

$$\mathrm{Im}(\chi(3)) = \frac{\varepsilon_0 c n \lambda N_A C(\sigma_{ex} - \sigma_0)}{16\pi I_{sat}}$$

where N_A is Avogadro's constant,n the linear refractive index, λ the laser wavelength and C the molar concentration (Table 1).

For the fullerene derivatives we see that the Im $\chi^{(3)}$ values are smaller than those of the parent C_{60} or C_{70}. and consequently show a smaller degree of optical limiting behaviour. This may be explained in terms of a reduction of the conjugation in the molecule and a decrease in the level of electron delocalization, brought about by the bonding of two of the carbon atoms in the fullerene cage to the metal sidegroup, but differences in the photodynamics may also contribute to this.

In conclusion new metal derivatives of fullerenes have been prepared and optical limiting behaviour observed in them. This phenomenon has been analysed using the same rate model that has been shown to be relevant to the parent fullerenes (but the authors acknowledge that this may not be entirely valid) and the magnitudes of their Im $\chi^{(3)}$ calculated. Further experiments are underway to obtain the size of the real component and will be reported shortly.

References

[1] L. W. Tutt and A. Kost, Nature, 356 (1992) 225

[2] F. Z. Henari, J. Callaghan. H. Stiel, W. Blau and D. J. Cardin, Chem. Phys. Lett 199 (1992) 144.

[3] H. J. Eichler. F. Massmann and C. H. Zaki, Opt. Commun. 40 (1982) 302

[4] W. Blau, H. Byrne, W. M. Dennis, J. M. Kelly, Opt. Commun. 56 (1985) 25

[5] W.Kratschmer. L. D. Lamb, K. Fostiropoulos and D. R. Huffman, Nature 347

[6] J. W. Arbogast, A. P. Darmanyan, C.S. Foote, Y. Rubin, F. N. Diederich, M. M. Alvarez. S. J. Anz and R. L. Whetten. J. Phys. Chem. 95 (1991) 11.

[7] M. R. Wasielewski, M. P. O'Neill, K.R. Lykke, M. J. Pelin and D. M. Gruen, J. Am. Chem. Soc. 113 (1991) 2774

Optical Properties of C_{60}-Diethyl Ether Clathrate Single Crystals

K. Kamarás[1,2], *V.G. Hadjiev*[1,3], *C. Thomsen*[1], *M.K. Kelly*[1], *M. Stutzmann*[1], *S. Pekker*[2], *K. Fodor-Csorba*[2], *G. Faigel*[2], *and M. Tegze*[2]

[1]Max-Planck-Institut für Festkörperforschung, Heisenbergstr. 1, D-70569 Stuttgart, Germany
[2]Research Institute for Solid State Physics, Hungarian Academy of Sciences, P.O. Box 49, H-1525 Budapest, Hungary
[3]Faculty of Physics, Sofia University, BG-1126 Sofia, Bulgaria

Abstract. We characterized single crystals of C_{60} clathrate with diethyl ether by infrared and Raman microspectrometry, photothermal deflection spectroscopy and ellipsometry. Our results are consistent with a picture of noninteracting host and guest molecules.

Clathrates of C_{60} constitute a growing family of solids based on fullerene molecules. Thin plate-like crystals of excellent morphology suitable for optical spectroscopy can be grown from solutions of C_{60} in certain solvents [1, 2]. In this paper we present optical properties of the C_{60}·diethyl ether clathrate.

Details of sample preparation are given in Ref. [1]. We investigated single crystals of lateral dimensions 200x200 μm^2 and 3-5 μm thickness by microspectrometric methods, described in Refs. [3] and [4], and spectroscopic ellipsometry [5]. X-ray diffraction shows slight variations in crystal structure of different batches and structural disorder along the c-axis [2], but the spectra were reproducible from sample to sample and on different spots of the same crystal.

The infrared and Raman spectra of the crystals are shown in Fig. 1 and 2, respectively. In Fig. 1 we also show the spectrum of a pristine C_{60} crystal for comparison. The two IR active lines of C_{60} which fall into the measurement range (1183 and 1429 cm^{-1}), as well as the ten Raman-active lines [6] show minimal change upon clathrate formation. Most of the other bands in the infrared spectrum can be attributed to diethyl ether vibrations (marked in Fig. 1.). The ones we cannot account for are suspiciously close to Raman active (717, 1100 cm^{-1}) and silent (729,1480 cm^{-1}) C_{60} modes. The crystal structure of the material is not known at present, but the appearance of these lines indicates the loss of center of inversion. In the Raman spectra, the corresponding features are very weak (577 cm^{-1} IR active; 726, 755, 1480 cm^{-1} silent) and are hardly distinguishable from the lines emerging as a result of phototransformation due to illumination by the exciting laser [7].

Ether bands cannot be observed in the Raman spectra, because of conjugation effects increasing the Raman cross section of C_{60} by orders of magnitude compared to simple aliphatic systems. (This observation implies a practical

Springer Series in Solid-State Sciences, Volume 117
Electronic Properties of Fullerenes Eds.: H. Kuzmany, J. Fink, M. Mehring, and S. Roth

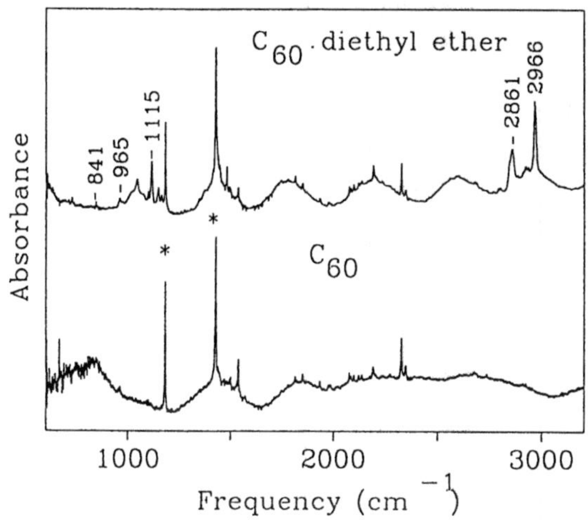

Figure 1: The infrared spectrum of a C_{60} and a $(C_{60})_2(C_4H_{10}O)$ single crystal. C_{60} vibrations are marked by asterisks, ether bands by their frequency.

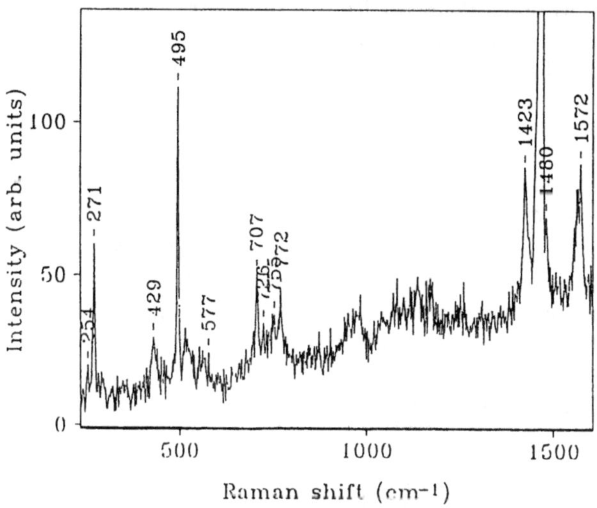

Figure 2: The Raman spectrum of $(C_{60})_2(C_4H_{10}O)$

concern: organic impurities can be present in C_{60} in substantial amounts without being noticed in Raman spectra. Raman spectroscopy, on the other hand, is extremely sensitive to higher fullerenes whose infrared lines are very close to the C_{60} ones. A careful purity control should therefore involve both methods.)

The high-energy spectrum, shown in Fig. 3, does not show drastic changes compared to pristine C_{60} either, proving the lack of electronic interaction between the constituents. The absorption spectrum derived from ellipsome-

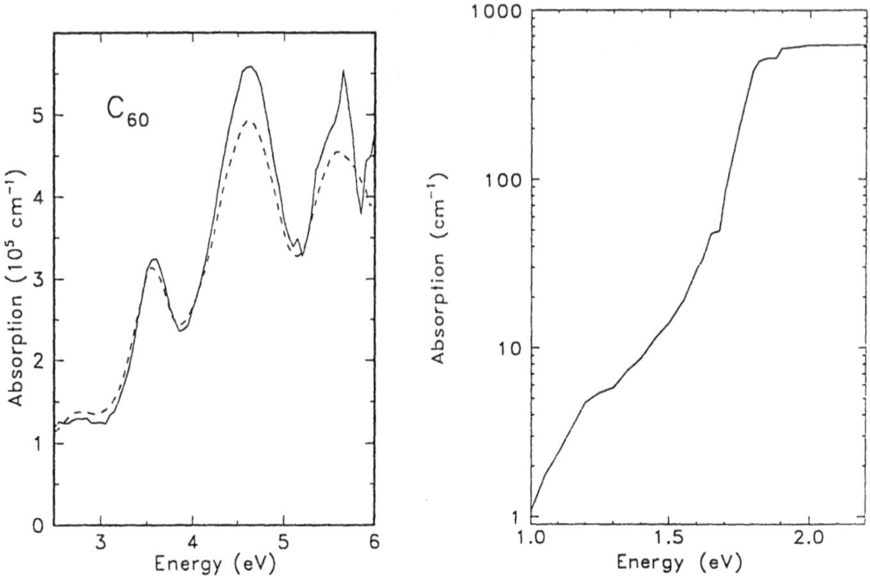

Figure 3: Left side: Absorption calculated from ellipsometric measurements for $(C_{60})_2(C_4H_{10}O)$ (solid line) and a pristine C_{60} film (Ref. [5], dashed line). Right side: PDS spectrum of C_{60}·diethyl ether.

tric measurements reveals the same peaks at 2.6, 3.5, 4.4 and 5.6 eV, as in a C_{60} film with 10 per cent C_{70} content [5]. This agreement shows that C_{70} in this concentration has very little influence on the visible/UV part of the spectrum. The region around the absorption threshold can be studied in detail by photothermal deflection spectroscopy (PDS). The PDS spectrum of the $(C_{60})_2(C_4H_{10}O)$ crystal is very similar to that of the C_{60} film studied by Skumanich [8], with a gap value about 0.1 eV higher (1.6 eV, determined from the change of slope). Another remarkable similarity is the exponential character of the absorption edge between 1.6 - 1.8 eV. Skumanich assigned this feature to the "Urbach edge" caused by disorder in the film; in our crystals, it is probably the signature of the structural disorder mentioned above.

In conclusion, we characterized $(C_{60})_2(C_4H_{10}O)$ single crystals with optical methods from the infrared to the UV region. As the number of clathrate systems continues growing, we may gain more insight into the properties of crystalline C_{60} as well.

We are indebted to H. Byrne for fruitful discussions. Research in Hungary was funded by the Hungarian National Foundation for Scientific Research under contract numbers 2950 and T4222. One of us (K.K.) thanks the Alexander-von-Humboldt-Foundation for support.

References

[1] S. Pekker, G. Faigel, K. Fodor-Csorba, L. Gránásy, E. Jakab and M. Tegze, Solid State Comm. **83**, 423 (1992)

[2] S. Pekker, G. Faigel, M. Tegze, T. Kemény and E. Jakab, Synthetic Metals **55-57**, 3021 (1993)

[3] K. Kamarás, V. G. Hadjiev, C. Thomsen, S. Pekker, K. Fodor-Csorba, G. Faigel and M. Tegze, Chem. Phys. Lett. **202**, 325 (1993)

[4] K. Kamarás, A. Breitschwerdt, S. Pekker, K. Fodor-Csorba, G. Faigel and M. Tegze, Appl. Phys. A **56**, 231 (1993)

[5] M.K. Kelly, P. Etchegoin, D. Fuchs, W. Krätschmer and M. Fostiropoulos, Phys. Rev. B **46**, 4903 (1992)

[6] P. H. M. van Loosdrecht, P. J. M. van Bentum, M. A. Verheijen and G. Meijer, Chem. Phys. Lett. **198**, 587 (1992)

[7] P. C. Eklund, Z.-H. Dong, Y. Wang, M. J. Holden, K. A. Wang, P. Zhou, A. M. Rao, G. Dresselhaus and M. S. Dresselhaus, this volume.

[8] A. Skumanich, Chem. Phys. Lett. **182**, 486 (1991)

Optical Properties of Some New Fulleroids and Fullerene Polymers

A. Hassanien[1], *T. Mrzel*[1], *P. Venturini*[1,4], *F. Wudl*[4], *D. Mihailovic*[1], *J. Gasperic*[1], *B. Kralj*[2], *D. Zigon*[2], *S. Milicev*[3], *and A. Demsar*[5]

[1]Department of Solid State Physics, Josef Stefan Institute, University of Ljubljana, 61111 Ljubljana, Slovenia
[2]Department of Physical Chemistry, Josef Stefan Institute, University of Ljubljana, 61111 Ljubljana, Slovenia
[3]Department of Fluorine Chemistry, Josef Stefan Institute, University of Ljubljana, 61111 Ljubljana, Slovenia
[4]Institute for Polymers and Organic Solids, University of California, Santa Barbara, CA, USA
[5]Iskra Elektrooptika, Stegne 6, 61111 Ljubljana, Slovenia

Abstract. Optical absorption and luminescence spectroscopy are performed on a number of new fulleroids with the aim of investigating the change in electronic properties as various functional groups are bonded to the molecule. The compounds studied include phenyl-, p-methylphenyl- and p-chlorophenyl fulleroids, as well as compounds produced by chemical and photo-polymerization. The bonding of additional groups appears to give rise to allowed transitions in the visible and UV. Chemical or photo "polymerization" of C_{60} gives rise to an insoluble insulator with new absorption peaks and luminescence in the visible and UV, indicative of breaking of icosahedral symmetry.

1. Introduction

The aim of the present preliminary study is to investigate using optical spectroscopy how large the electronic structure of C_{60} changes are with the addition of different numbers of different functional groups to the C_{60} molecule. The electronic structure of the undistorted C_{60} molecule is fairly well characterized using optical and vibrational spectroscopy[1,2,3,4,5]. Using optical absorption together with luminescence and Raman and infrared absorption spectroscopies for materials characterization we attempt here to give a preliminary characterization of the electronic structure of a number of new Fulleroids and C_{60} polymers or conglomerates in view of determining to what extent the molecule is distorted with the addition of a) different groups and b) different numbers of added functional groups.

2. C_{60} "Polymer"

2.1. Experimental

A thin film of purified C_{60} (Texas Fullerenes) was evaporated at 400° C and 10^{-4} Torr on a quartz substrate [6,7]. During thin film deposition the substrate was exposed to light from a quartz halogen bulb (200 W, 9V). A control C_{60} film was prepared without illumination keeping all other conditions the same. The films were characterized using optical absorption, luminescence, Raman and infrared spectroscopy to observe differences in electronic structure between the illuminated and non-illuminated C_{60} films.

2.2. Results and discussion

Luminescence and absorption spectra of the described films are shown in Fig. 1. Absorption spectra of the illuminated C_{60} film shows a new feature at 5.4 eV. Observed are also shifts of h_u-t_{1g}, g_g, h_g-t_{1u} and h_u-h_g transitions, as well as different ratio of intensities for these absorptions.

Springer Series in Solid-State Sciences, Volume 117
Electronic Properties of Fullerenes Eds.: H. Kuzmany, J. Fink, M. Mehring, and S. Roth

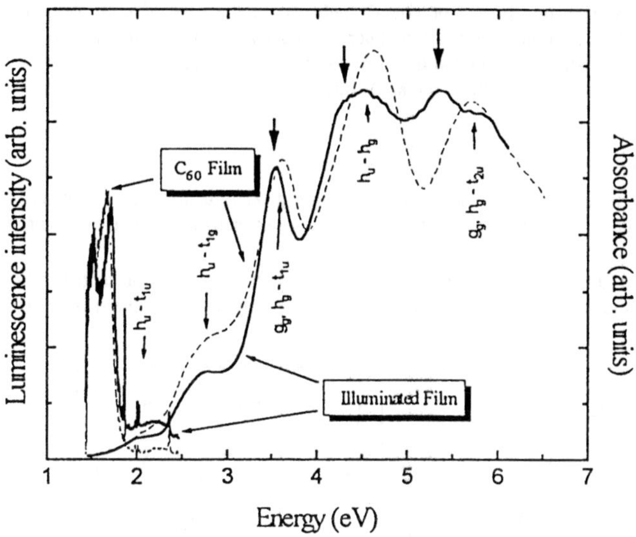

Fig 1. Comparison of luminescence and absorption spectra of "photopolymerized" C_{60} (solid) and pristine C_{60} films (dashed).

Like absorption spectra also luminescence spectra show shifted positions and changed luminescence intensities when comparing non-illuminated and illuminated C_{60} films. However, the changes due to illumination are not very large and occurs primarily in the 1.8-2 eV region, where originally in the pristine films there is no C_{60} luminescence from the LUMO-HOMO transition. It suggests that some symmetry breaking is taking place, but the basic electronic structure on the C_{60} molecule is not changed significantly. The insolubility of the illuminated C_{60} sample is suggesting that significant photo-polymerization of C_{60} might indeed have occurred.

3. C_{60} derivatives

3.1. Synthesis and characterization

The C_{60} derivatives were made by performing a.) radical reactions ($C_{60}Ph_1$, $C_{60}Ph_n$, $C_{60}(PhMe)_n$ $C_{60}Cl_1$, $C_{60}(OCOPh)_n$, $C_{60}(OCOPhOMe)_n$, b.) dipolar addition (4,4'-Bis(dibenzylamino)-2,2'-biphenylene Fulleroid: $BDABC_{61}$) [8,9] and c.) ozonation ($C_{60}O_n$) [10].

Radicals were generated thermally and photoliticaly. Phenyl radicals for synthesis Ph_1C_{60}, Ph_nC_{60} and $C_{60}(PhMe)_n$ were obtained thermally from peroxide precursors. Radicals for preparation of $C_{60}Cl_1$, $C_{60}(OCOPh)_n$ and $C_{60}(OCOPhOMe)_n$ were generated photoliticaly from appropriate iodobenzen precursors. Products of radicals reactions were sometimes precipitated from solution and were easy to separate. In other cases extraction and other separation methods were performed, but unfortunately not always giving rise to single products (for this series). Dipolar addition of diazo compound with C_{60} gives C_{61} and higher fulleroids. To obtain $BDABC_{61}$ mixture of 9-Diazo-2,7-bis(dibenzylamino)fluorene was stirred with C_{60} in toluene overnight. The product was purified by flash chromatography. For ozonation C_{60} was dissolved in toluene and exposed to ozone at -80°C. After heating to room temperature a precipitate was observed. The material was washed in hexane and dried. The materials were characterized first by mass spectroscopy. Electronic ionization mass spectroscopy was possible only for the phenylated C_{60} derivatives. In other cases we were not able to record mass spectra because of fragmentation due to heating. For BDAB C_{61} proton NMR was taken. Raman and infrared spectra were recorded to detect the presence of the appended units as well as the presence of the C_{60} cage. In the case of single appended units, we were fairly certain that only one functional group was bonded (i.e. one phenyl or one Cl radical). In the case of multiple group additions, we were unable to determine the exact number of appended groups, although for the $C_{60}Ph_n$ compound, from mass spectra we find that n>7. For observing changes in electronic structure also luminescence and absorption spectra of the compounds was taken.

3.2. Results and discussion

Luminescence spectra of prepared C_{60} derivatives are shown in Fig. 2. With increasing number of groups on the C_{60} molecule, both luminescence efficiency and peak energy of the luminescence energy apparently increases.

A rough correlation between luminescence intensity and position of luminescence is presented in Fig. 3. (note that the precise number of groups has not been determined, only a general trend is shown.)

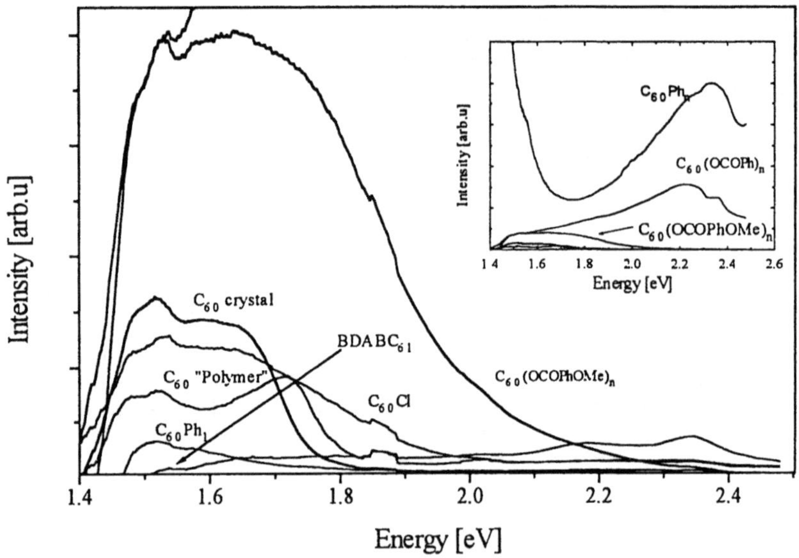

Fig. 2 Luminescence spectra of various Fulleroids. The insert shows the more intensely luminescing materials, of which $C_{60}Ph_n$ has extremely strong luminescence.

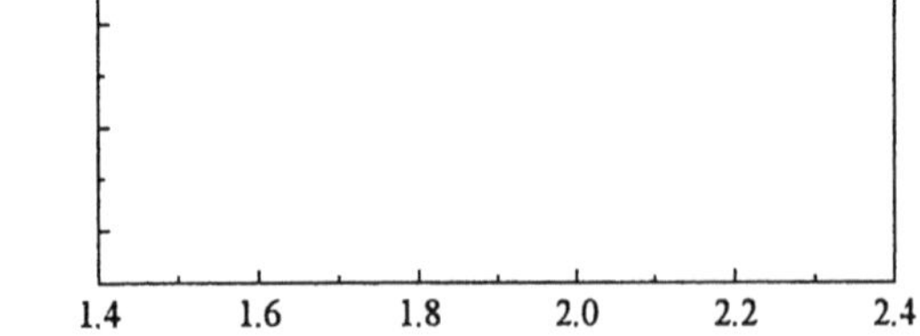

Fig. 3. Luminescence intensity and peak position of luminescence for a number of different fulleroids.

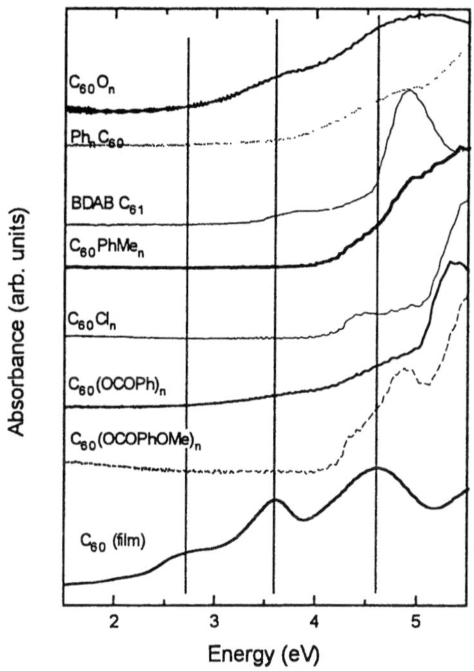

Fig. 4 Absorption spectra of some fulleroides and C_{60}

Luminescence intensity is not very different from C_{60} if only one group is bonded on the C_{60} molecule. More groups bonded to C_{60} seem to increase fluorescence intensity dramatically. Very strongly fluorescent is $C_{60}Ph_n$ (where n>7), which is found to be a soluble orange strongly flourescing dye, comparable to laser dyes. To confirm this hypothesis, it will be necessay to perform a separation of the various products according to the number of added functional groups, which for the present investigation was not attempted for $n>1$.

Absorption spectra of some C_{60} derivatives shows a systematically increasing band-gap with different functional groups attached to C_{60}. On the other hand charge-transfer salts of C_{60}; especially the alkali-doped compounds give rise to black materials with absorption in the visible and infrared [11]. The increase in band-gap is consistent with the observation of increasing luminescence energy shown in Figures 2 and 3. The large difference in the absorption spectra of C_{60} and BDAB C_{61} is rather surprising, especially since this compound has been purified and the pure single functional group material isolated and well characterized. All the transitions below 3.5 eV have virtually dissappeared, with only a weak absorption remaining at 3.7 eV and the first strong absorption as high as 5 eV.

4. Conclusion

The absorption spectra of photo-"polymerized" C_{60} films are presented for the first time, showing a new transition at 5.3 eV, and a shift of some other optical transition energies in the range up to 5 eV. Consistent with symmetry breaking of the C_{60} molecule in the "polymerization" process, the material shows an increase in luminescence intensity in the visible. Addition of functional groups to the C_{60} molecule appears to result in a significant change of the molecular structure, witnessed by a consistently increasing band-gap and increasing luminescence intensity as more groups are bonded to the C_{60} molecule. Further studies of the purified derivative materials are presently underway.

5. References

[1] H.W.Kroto, A.W.Allaf, S.P.Balm, Chem.Rev., **91**, 1213 (1991)

[2] J.H.Hare, H.W.Kroto, R.Taylor, Chem.Phys.Lett. , **177**, 394, (1991)

[3] C.reber, l.Yee J.I.McKiernan, J.I.Zink, R.S.Williams, N.W.Tong, D.A.A.Ohlberg, R.L.Whetten, F.N.Diederich, J.Phys.Chem. **95**, 2127 (1991)

[4] C.I.Frum, R.Engleman, H.G.Hedderich, P.F.Bernath, L.D.Lamb, D.R.Huffman, Chem.Phys.Lett. **176**, 504 (1991)

[5] W.Kraetschmer, L.D.Lamb, F.Fostiropoulos, D.R.Huffman, Nature **347**, 354 (1990)

[6] Y.Saito, Y.Ishikawa, A.Oshita, H.Shinohara, H.Nagashima, Phys.Rev.B **46** 1846 (1992)

[7] D.M.Poirier, T.R.Ohno, H.G.Kroll, Y.Chen, B.J. Benning, J.H.Weaaver, L.P.F.Chibante, R.E.Smalley, Science **253**, 646 (1991)

[8] J.R.Morton, K.F.Preston, P.J.Krusic, S.A.Hill, E.Wassermann, J.Am.Chem.Soc. **114**, 5454 (1992)

[9] J.R.Morton, K.F.Preston, P.J.Krusic, E.Wasserman, J.Chem.Soc.Trans. **2**, 1425 (1992)

[10] T.Suzuki, Q.Li, K.C. Khemani, F.Wudl, Oe. Almarsson, Science **254**, 118 (1991)

[11] T.Pichler, M.Matus, J.Kurti and H.Kuzmany, Sol.Stat.Comm. **81**, 859 (1992)

Part V

Magnetic Resonance and Magnetic Ordering

^{13}C Nuclear Relaxation in the Normal State of the K_3C_{60} Superconductor

K. Holczer

Department of Physics, University of California, Los Angeles, CA 90024, USA

Abstract ^{13}C NMR results on K_3C_{60} showing that nuclear relaxation is nonexponential below 250 K. The nonexponential nuclear relaxation at low temperatures is an intrinsic property of the frozen molecular motion state and is associated with intramolecular variations of the electron spin density distribution. Dipolar coupling to the electron spins at the Fermi level of the carbon $p\pi$ - t_{1u} band is identified as the dominant source of ^{13}C relaxation. The measured mean value of $(T_1T)^{-1}$ is determined not only by the *total* density-of-states at the Fermi level but also by the intramolecular electron spin density distribution.

Not long after macroscopic quantities of C_{60} buckminsterfullerene became available [1], more than a dozen superconductors described by the formula A_3C_{60} (where A = alkali metal) were discovered [2-4], and there emerged an empirical correlation of T_c with the lattice parameter [5-7], i.e. with the nearest neighbor C_{60} - C_{60} distance. Supported by band-structure calculations [8], this has led to a near-consensus in the literature, that a BCS-like picture can account for the superconductivity in these materials. In this picture, the attractive interaction between the electrons is of intramolecular nature, i.e. it is not influenced by the C_{60} - C_{60} distance, and the variation of the Fermi-level density-of-states $n(e_F)$ of the narrow intermolecular band is responsible primarily for variations in T_c.

^{13}C spin - lattice relaxation rates $(1/T_1)$ for K_3C_{60} and Rb_3C_{60} were reported and interpreted by Tycko et al. [9,10] in terms of normal metal relaxation (Korringa law) giving a density-of-states of 20 - 25 state/molecule/eV in K_3C_{60} and "about 40 percent higher in Rb_3C_{60}", in agreement with the expectations of the simple density of state argument. Their analysis is based on the assumption, supported by calculations [11], that the relaxation is dominated by contact (scalar) hyperfine coupling to the conduction electrons, although the measured Knight-shift values were notedly inconsistent with this interpretation [9,10,12]. Furthermore, it was noted that the estimated value of the scalar hyperfine coupling is rather uncertain and that dipolar and orbital contributions to the electron - nuclear coupling may be non-negligible. The absolute values derived for $n(e_F)$ were therefore taken with caution, but the relative values found for the two materials were claimed to provide *quantitative* support for the $n(e_F)$ variation as the principal source of the different T_c - s. This conclusion was reached in spite of the observed nonexponential relaxation, implying a relaxation rate distribution of unidentified origin [9]. The shape of the magnetization recovery curve is not a simple exponential decay, as expected for spin-1/2 nuclei like ^{13}C. Figure 1 illustrates this for a representative set of data on K_3C_{60}, where the curves are normalized in amplitude and in time scale to the initial

Springer Series in Solid-State Sciences, Volume 117
Electronic Properties of Fullerenes Eds.: H. Kuzmany, J. Fink, M. Mehring, and S. Roth

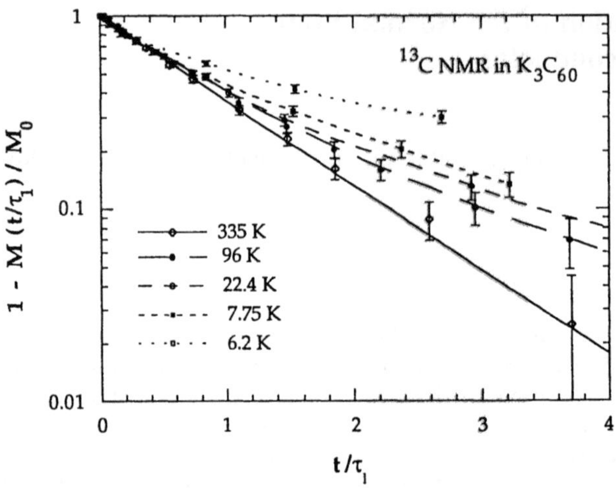

Figure 1. Semilog plot of the nuclear magnetization recovery vs. reduced time, t/t_1, where t_1 is the initial slope of the recovery curve, at the various temperatures indicated.

slope, τ_1, of each recovery curve. An exponential relaxation is observed only in the high temperature range, T > ~ 250 K, where the spectra are narrowed by the molecular motion.

Four temperature ranges can be qualitatively distinguished:

(I). 250 - 300 K where a single narrow line provides a simple exponential relaxation,

(II). from about 100 to 250 K where both the shape of the spectra and the shape of recovery curves are changing with temperature,

(III). from roughly 8 to 100 K where the spectral shape does not change significantly and the shape of magnetization recovery, though non-exponential is only slightly temperature dependent,

(IV). far below T_c (T < 8 K), where the recovery shape varies dramatically with temperature.

In the range (I), the evaluation of T_1 is straightforward. In (III), as the shape of the relaxation curve varies little (the scaled plots at 22.4 K and 96 K on Figure 1 are nearly superimposable), evaluating a kind of mean relaxation rate and considering its temperature dependence is justified. Tycko et al. [9] used a stretched-exponential fit — $M_0\{1-\exp[-(t/t)^{\beta}]\}$ — with a temperature-independent β to recover a characteristic relaxation time, without clearly identifying a mechanism to give such a behavior.

There are many other possibilities to define an average T_1 like the 1/e point of the relaxation curve or a single exponential fit [12], provided that in a reduced timescale the distribution of the fitted data points are kept closely similar. We have found that the temperature dependence one can arrive at, using any of these definitions, are very similar, with a change in numerical values of about 5 %.This is illustrated on Figure 2, where all possible fits are presented on a typical data set (T = 36 K) and the usually defined "T_1" -s are about the same.

In Figure 3 we show the overall temperature dependence of $1/(T_1T)$, but as the meaning of T_1 differs in the above defined temperature ranges they are distinguished

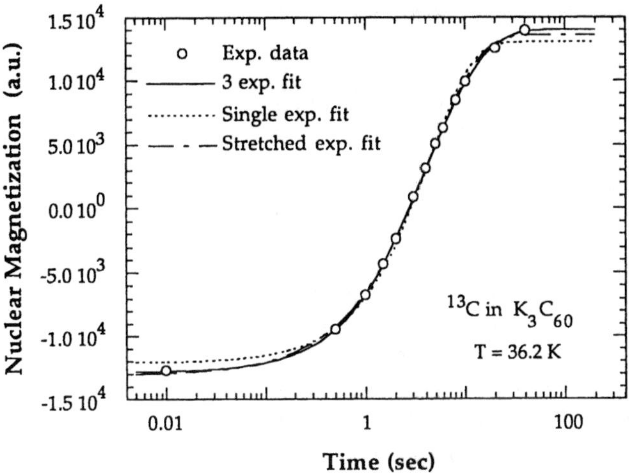

Figure 2. Nuclear magnetization recovery curves for K_3C_{60}. The points are experimental data, the solid line is a three-exponential ($< T_1 >$ = 3.7 sec, see text), the dashed is a single exponential (T_1 = 4.33 sec), and the dashed - dotted is a stretched exponential fit ($\beta = 0.085$ and τ = 4.40 sec). Note that the 1/e point of the relaxation curve is at 4.15 sec.

by different symbols. The different partial data sets reported in the literature by others [9,10,12] are in overall agreement with the data represented on Fig. 3, in the overlapping temperature ranges.

The quasi-constant nature of T_1T in ranges (I) and (III) suggests that one is faced with the well known Korringa relaxation of a metal. In the classical case the conduction electron spins are predominantly coupled to the nuclear spins through the (Fermi) contact interaction, proportional to the square of the *s*-wave component of the wavefunction at the nuclear site. In that case the paramagnetic (Knight) shift, K, of the resonance frequency scales with T_1T, as expressed by the Korringa relation:

$$T_1T\,K^2 = \frac{h}{8\pi^2 k_B}\left(\frac{\gamma_e}{\gamma_n}\right)^2 \qquad (1)$$

where γ_e and γ_n are the gyromagnetic ratios of the electron and the nuclei respectively. For ^{13}C nuclei, the value of the right hand side of eq. (1) is 4.34 x 10^{-6} K-s. In the high temperature range (I) where all anisotropic contributions to the shift are averaged out by the molecular motion, a positive shift of ~ 45 ppm is observed compared to the insulating C_{60} solid. Using this as a reference, one obtains T_1TK^2 = 1.8x10^{-7} K-s, which is ~25 times smaller than the Korringa constant, as was noticed earlier [9]. To fulfill eq. (1) a Knight shift of ~ 200 ppm would be required. Arguments have been given that this discrepancy could result from a chemical shift contribution originating from the three extra charges transferred to the C_{60} molecular orbitals. However, the ^{13}C resonance of the insulating A_6C_{60} and A_4C_{60} do not show a substantial chemical shift compared to the pure C_{60} (< 50 ppm); therefore it is unlikely that a negative chemical shift of 200 ppm would result from the three

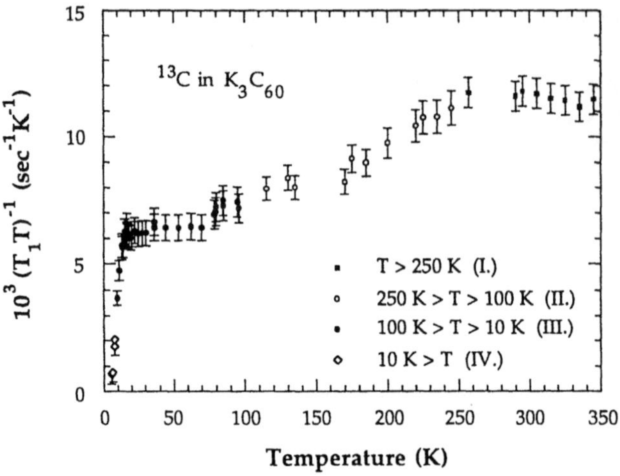

Figure 3. The temperature-dependence of the average T_1 plotted as $(T_1T)^{-1}$ vs. temperature. Different symbols are used in the four temperature ranges where the relaxation curve is of different shape as described in the text (see also Fig 2.).

electrons transferred in K_3C_{60}. On the other hand if T_1 is dominated by dipolar or orbital couplings, it is natural that eq. (1), i.e. the Korringa law, does not apply.

In the low temperature range (III), the relaxation is nonexponential, while the temperature-independence of the spectra indicates that the molecular motion is already essentially frozen. As the probability of finding two ^{13}C in the same C_{60} ball is about 10 % (the ^{13}C abundance is about 1 %) nuclear spin diffusion is expected to be negligible and insufficient to average out the microscopic T_1 differences. Therefore the spin-lattice relaxation may be multi-exponential due to the locally different electron densities or hyperfine couplings, resulting in a temperature independent relaxation shape. This spectral inhomogeneity could be situated at three different levels:

A. A structural disorder, due to the random occupancy of the two crystallographically equivalent C_{60} orientations, results in a distribution of local electron densities on the different molecular sites in the crystal, and hence on their various carbon atoms.
B. The angular dependence of the couplings (dipolar and orbital) is such that in a powder average a nonexponential relaxation may naturally emerge. (Note that the local symmetry of the carbon sites is very low although the lattice is cubic.)
C. An intramolecular differentiation of the spin (charge) density of the three structurally distinct carbon sites [16] of the nonrotating molecules, which would be present even in the perfectly ordered solid.

The first possibility would lead qualitatively to the same kind of results for the ^{39}K NMR as well. The observed ^{39}K relaxations in the same sample (to be reported elsewhere [17]) are found to be good single exponentials in this temperature range making this possibility unlikely. An independent argument is that we find no observable correlation between the recovery shape or the value of T_1 and the position of different selected spectral ranges [14]. Therefore the origin of nonexponential relaxation is either B. or C., which are intramolecular in nature. It is difficult to distinguish solely from experiments between these two cases. For case B., for example, even in a single-crystal sample we expect identical results to that of the present powder samples, because the 60 different sites in the unit cell will result in a nearly perfect angular distribution of the pπ orbitals (and therefore of T_1) with respect to the applied field. Similarly, in case C., unresolved spectra are expected also in single crystals as a consequence of the anisotropic chemical and spin shift of the large number of sites in the unit cell. In experiments using ^{13}C enriched samples, both effects are expected to be washed out by nuclear spin diffusion in a similar way [22].

Support for case C. has been provided by Yannoni et al. [18], who have resolved distinct shifts for the three carbon sites in Rb_3C_{60} at low-temperatures using Magic-Angle Spinning ^{13}C-NMR (MAS), and have also found similar indications in the case of K_3C_{60}. A fit to a sum of three exponentials (also included on Fig. 2.) with the respective weights of 1:2:2, corresponding to the 12:24:24 occupancy of the distinct carbon sites is numerically satisfactory. As for the numerical values, fitting with three exponentials yields two distinct possible sets for the T_1 values: either (case *a*) 1.7, 3.6 and 9.2 s with corresponding intensity ratios of 1:2:2; or (case *b*) 2.3, 5.1 and 15.5 s with 2:2:1 weighting. Case *a* is the order of Yannoni et al. [18], i.e. the *s*-electron density goes like the *p*-electron density, but case *b* gives a somewhat better fit. Case *a* is also counter-intuitive, because the carbon-atom sites closest to the cavity of the octahedral K^+ ion are electron-poor. In other words, the largest positive-shift sites observed by MAS seems to be associated with the longest-T_1, because of the dominance of the dipolar coupling. The weighted average of $1/T_1$ gives respectively 3.7 and 3.78 s for average relaxation time, $T_1^{av} = 1/\langle T_1^{-1} \rangle$. These values are 15 - 20 % shorter than values extracted directly from other fits (which are equivalent within 5 % as discussed above, also see caption of Fig. 2).

The unusual character of the metallic fullerides is that the conduction electrons are residing on the pπ-t_{1u} orbitals of C_{60}, which are of p-wave character. Whereas Haddon et al. [11] estimated the curvature in the C_{60} sphere results in a sufficient sp^3-hybridization to give a significant scalar coupling, recent band structure calculations show [15] that the s electron density of the conduction electrons is negligibly small. The LDA calculations of Antropov et al. [15] can also explain quantitatively the observed nuclear relaxation rate, with a dominant dipolar coupling to the electrons from pπ orbitals with a minor (10 %) contribution from orbital coupling. While T_1 values calculated by Antropov et al. [15] agree with the experimentally observed average T_1 values, we infer from the experiment a more important variation of the local density of states for the different carbon sites than the LDA results obtained on a nonrelaxed structure. Calculations of Andreoni et. al. [19] indicate a significant deformation of the C_{60} molecule in K_3C_{60} which could relieve this discrepancy.

As a consequence of this analysis the $(T_1^{av}T)^{-1}$ is about 7. 5 x 10^{-3} $(Ks)^{-1}$ in the range (III), and results in a smaller drop of $(T_1T)^{-1}$ from the range (I) to (III) than that represented on Fig. 3. The remaining 35 % variation is significant. The occurrence of molecular motion at high temperature introduces an extra relaxation channel for the nuclear spins through a modulation of the NMR shift similar to that seen in the pure C_{60}. To assign the temperature variation to molecular motion [20] is also consistent with recent NMR experiment under pressure [21] showing that the temperature dependence of $1/T_1T$ disappears under moderate pressure, i.e. the molecular motion is hindered efficiently by lattice contraction. A detailed account of these effects will be published elsewhere [14].

We conclude on the basis of the above considerations, that ^{13}C relaxation in K_3C_{60} is governed by the dipolar coupling between the nuclear spins and the conduction electrons situated on the C_{60} molecular pπ orbitals. The non exponential relaxation observed at low temperature is a consequence of the different (local) spin densities on the symmetry inequivalent carbon sites of the molecule, i. e. in band structure language it reflects the distribution of the local density of states at the Fermi level within the molecule. The average $(T_1T)^{-1}$ depends on the repartition of the spin density on the three carbon sites and not only on the total susceptibility therefore the ratio of the $(T_1T)^{-1}$ values for K_3C_{60} and Rb_3C_{60} is proportional to the ratio of $n^2(e_F)$ only if the real space projected local density of states distributions on the molecules in the two materials are the same. Our preliminary results on Rb_3C_{60} indicate this not to be the case, a wider distribution of T_1 is observed in Rb_3C_{60}. In the frame of the above philosophy the higher average $(T_1T)^{-1}$ can result in from a different distribution of the local density of states within the C_{60} molecule as well. Further experimental effort to assure that this difference is intrinsic and not related to material quality problems is currently under progress.

Acknowledgement I'm indebted to R. L. Whetten for our continuous partnership ever since I work in this field. These results were born out of a long standing collaboration with many colleagues in different laboratories, in particular the samples were prepared in the laboratory of R. B. Kaner at UCLA and the NMR experiments performed in Orsay in collaboration with H. Alloul, Y. Yoshinari and O. Klein. I would like to thank useful discussions to W. Andreoni, O. K. Andersen and I. I. Mazin and express my gratitude to the CNRS and the Laboratoire des Physique des Solides for providing a warm working environment which made my stay during the course of this work memorable. I'm also indebted to G. Kriza for useful discussions and for sending preprint of his work.

References

1. W. Krätschmer, L. D. Lamb, K. Fostiropoulos, and D. R. Huffman; *Nature,* **347**, 354-358 (1990).
2. A. F. Hebard, M. J. Rosseinsky, R. C. Haddon, D. W. Murphy, S. H. Glarum, T. T. M. Palstra, A. P. Ramirez, and A. R. Kortan;*Nature,* **350**, 600 (1991); M. J.
3. ^{13}C NMR experiments were performed in a 7 Tesla superconducting magnet using samples prepared as described previously (see Rosseinsky, et al.,; *Phys. Rev. Lett.* **66**, 2830 (1991); K. Holczer, O. Klein, S-M. Huang, R. B. Kaner, K-J. Fu, R. L. Whetten, and F. Diederich; *Science,* **252**, 1154 (1991).) and sealed in a Pyrex tube under He atmosphere. About 200 mg K_3C_{60} sample (^{13}C in natural abundance) was collected from the several hundred syntheses made at the UCLA (typical quantity of about 15 mg) and selected as "good" samples showing high diamagnetic shielding fraction (in excess of 40%). Analysis of the

^{13}C NMR spectra with very low repetition rate shows a presence of about 15 % of pure C_{60} in the sample tube. Apart from the pure C_{60} phase, all other possible phases are below the detection sensitivity, i.e. less than 5%.
4. K. Tanigaki, I. Hirosawa, T. W. Ebbesen, J. Mizuki, Y. Shimakawa, Y. Kubo, J. S. Tsai, S. Kuroshima; *Nature,* **356**, 419 (1992).
5. G. Sparn, J. D. Thompson, R. L. Whetten, S.-M. Huang, R. B. Kaner, F. Diederich, G. Grüner, and K. Holczer; *Phys. Rev. Lett.* **68**, 1228 (1992).
6. R. M. Fleming et al.; *Nature* **352**, 787 (1991).
7. O. Zhou, G. B. M. Vaughan, Q. Zhu, J. E. Fischer, P. A. Heiney, N. Coustel, J. P. McCauley, A. B. Smith; *Science,* **255**, 833 (1992).
8. For a very detailed description of these bands, see: S. Satpathy, V. P. Antropov, O. K. Andersen, O. Jepsen, O. Gunnarsson and A. I. Lichtenstein; "Conduction-Band Structure of Alkali-Doped C_{60}," *Phys. Rev B.*, **46**, 1773, (1992).
9. R. Tycko, G. Dabbagh, M. J. Rosseinsky, D. W. Murphy, A. P. Ramirez, and R. M. Fleming; *Phys. Rev. Lett.* **68**, 1912, (1992).
10. R. Tycko, et al. *Science* **253**, 884 (1991).
11. R. C. Haddon; *J. Am. Chem. Soc.*, **109**, 1676, (1991), and K. Raghavachari; unpublished results.
12. W. H. Wong, M. Hanson, W. G. Clark, G. Grüner, J. D. Thompson, R. L. Whetten, S.-M. Huang, R. B. Kaner, F. Diederich, P. Petit, J.-J. André, K. Holczer; *Europhys. Lett.* **18**, 79 (1992).
13. K. Holczer et. al.; to be published
14. Y. Yoshinari et. al.; to be published.
15. V. P. Antropov, I. I. Mazin, O. K. Andersen, A. I. Liechtenstein and O. Jepsen; "Dominance of Spin-Dipolar NMR Relaxation Mechanism in Fullerene Superconductors", Preprint, to be published
16. P. W. Stephens, L. Mihály, P. L. Lee, R. L. Whetten, S.-M. Huang, R. B. Kaner, F. Diederich, and K. Holczer; *Nature*, **351**, 632 (1991).
17. K. Holczer et.al.; to be published
18. C. S. Yannoni et. al.;*Synthetic Metals*, in press
19. W. Andreoni, F. Gygi, M. Parinello; *Phys. Rew. Lett.* **68**, 823, (1992), and W. Andreoni , P. Giannozzi, M. Parinello; to be published.
21. G. Kriza preprint.
21. G. Quirion, C. Bourbonnais, E. Barthel, P. Auban, D. Jérome, J. M. Lambert, A. Zahab, P. Bernier, C. Fabre and A. Rassat; *Europhys. Lett.*, **21**, 233-235, (1993)

Is the ^{13}C Nuclear Spin-Lattice Relaxation in A_3C_{60} (A = Rb, K) Unconventional?

M. Mehring

2. Physikalisches Institut, Universität Stuttgart,
Postfach 80 11 40, D-70511 Stuttgart, Germany

Abstract: I want to point out that the simple Korringa relation does not hold for A_3C_{60}, i.e. density of states information cannot be obtained from T_1 data. A modified Korringa relation must be applied instead by including a Korringa scaling factor S_K, which is obtained as $S_K = 2.5$ (Rb_3C_{60}) and 2.7 (K_3C_{60}). The deviation from the Korringa value ($S_K = 1.0$) points at appreciable electron - electron correlation in A_3C_{60}.

1. Knight shift and Korringa relation

The isotropic Knight shift of NMR lines in metals can be expresed as[1, 2]

$$K_z = \hbar a_{zz}\left(\frac{\gamma_e}{\gamma_n}\right)\chi(0) \tag{1a}$$

where a_{zz} (in rad/s) is the component of the hyperfine tensor parallel to the external field, γ_e and γ_n are the electron and nuclear gyromagnetic ratios, and the normalized susceptibility $\chi(o)$ can be expressed for an uncorrelated Fermi liquid by

$$\chi(0) = \frac{1}{2}N(E_F) = \frac{1}{4}D(E_F) \tag{1b}$$

both in terms of the density of states $N(E_F)$ and in term of the density of states per spin $D(E_F)$, in order to avoid confusion. Stoner enhancement etc. can readily be included. In cubic metals a_{zz} is replaced by the isotropic hyperfine interaction a leading to an isotropic Knight shift K. It will be demonstrated that this is not legitimate for A_3C_{60}. The nuclear spin lattice relaxation in case of both isotropic as well as dipolar hyperfine interaction can be expressed for a powder as[3, 4, 5]

$$\frac{1}{T_1T} = \hbar k a^2(1+\tfrac{1}{2}\varepsilon)I_K \text{ with } I_K = \sum_q \frac{\operatorname{Im}\chi^{+-}(q,\omega_0)}{\hbar\omega_0} \tag{2}$$

where k is the Boltzmann factor, $\varepsilon = d^2/a^2$ is the square of the ratio of the dipolar and isotropic hyperfine interaction and where $\chi^{+-}(q,\omega_0)$ is the dynamic susceptibility at scattering wavevector q and the Larmor frequency ω_0. I would like to remark that Eq.(2) represents a powder average over the orientationial dependence of the relaxation rate. The magnetization recovery is therefore given by a sum of exponentials with different T_1 values, i.e. it appears as a non-exponential recovery curve. This can be calculated readily and will be discussed elsewhere. A short time expansion, i.e. evaluating the inital slope of the recovery curve, results in the relaxation rate given by Eq.(2). The expression for T_1 given in Eq.(2) is very general and is not even restricted to metals. It can be reformulated in terms of a sum rule[3, 5]

$$T_1TK^2(1+\tfrac{1}{2}\varepsilon)C_0S_K = 1 \tag{3a}$$

with

Springer Series in Solid-State Sciences, Volume 117
Electronic Properties of Fullerenes Eds.: H. Kuzmany, J. Fink, M. Mehring, and S. Roth

$$C_0 = 4\pi\frac{k}{\hbar}\left(\frac{\gamma_n}{\gamma_e}\right)^2 \quad \text{and} \quad S_K = \frac{I_K}{4\pi\chi(0)^2} \tag{3b}$$

The classical Korringa relation[1, 2]

$$\frac{1}{T_1 T} = C_0 K^2 = \pi k \hbar a^2 N(E_F)^2 \tag{4}$$

is only recovered in cubic metals ($\varepsilon = 0$) and if the value $S_K = 1.0$ holds which is true only for a 3D free electron model. Only in this case it is legitimate to derive the density of states from Eq.(3a). Neither is $\varepsilon = 0$ nor is $S_K = 1.0$ in case of the A_3C_{60} material as will be shown below.

2. Results on Knight shift

The isotropic hyperfine interaction a and the dipolar hyperfine interaction d have been determined for ^{13}C in Rb_3C_{60} by G. Zimmer et. al. [6] recently from a carefull analysis of the temperature dependence of both the susceptibility and the Knight shift. Separation of the Knight shift tensor from the chemical (orbital) shift tensor in Rb_3C_{60} allowed also to determine the anisotropic hyperfine interaction. The following average values for the three non-equivalent carbon atoms were obtained[6]:

$$a = 2\pi\, 2.0\; 10^6\; s^{-1} \qquad \text{and} \qquad d = 2\pi\, 9.2\; 10^6\; s^{-1}$$

This corresponds to the local fields $a/\gamma_n = 0.19$ T and $d/\gamma_n = 0.74$ T. These values are on the same order of magnitude as in many organic conductors[7, 4]. The dipolar contribution is usually dominant because of the spin density in the p_z-orbital giving rise to an axially symmetric hyperfine tensor as is well established in solid state magnetic resonance. The isotropic hyperfine interaction does not vanish as would be expected on first sight, but is always appreciable due to core polarization effects. The values for the hyperfine interaction are therefore standard and by no means "unconventional". Although the recently calculated dipolar interaction using local density theory (LDA)[8] seems to come rather close to our value for d, the experimentally determined isotropic value is, however, much larger than the calculated one (which was negligible in [8]). By applying Eq.(1) it is rather straightforward to obtain the density of states from the measured Knight shift. The following values are obtained from magic angle spinning (MAS) ^{13}C-NMR([9] and F. Rachdi, private communication): K(Rb_3C_{60} at 300K) =47 ppm and K(K_3C_{60} at 300K) =37 ppm. Both values were obtained by taking the difference between the measured isotropic shift and the isotropic shift of pure C_{60}. In addition 6 ppm were subtracted in order to account for the charged state.
The following density of state values are obtained:

Rb_3C_{60}(300K): $D(E_F) = 8.6$ (eV C_{60} spin)$^{-1}$ and K_3C_{60}(300K): 6.8 (eV C_{60} spin)$^{-1}$

This is quite close to the ones calculated from local density theory. The corresponding total susceptibility amounts to $\chi_S = 2.8\; 10^{-4}$ emu/mole for Rb_3C_{60} which aggrees with recent ESR data[10], but is in disaggreement with static susceptibility measurements[11]. These are known, however, to suffer from an unknown off-set. In summary the Knight shift and the density of states are in aggrement with local density bandstructure theory and do not hint at "unconventional" behaviour of A_3C_{60}.

3. Results on spin lattice relaxation

By using the sum rule (Eq.(3)) the dimensionless parameter S_K can be determined directly from the known values of T, T_1 and K. No theoretical parameter nor any assumption other than the validity of Eq.(3) is needed. Note that Eq.(3) is very general and does not infer the Korringa relation, which, it can

Table: Experimentally determined S_K values for some classical metals compared with A_3C_{60} data.

classical metals				A_3C_{60}	
^7Li	^{23}Na	^{27}Al	^{63}Cu	Rb_3C_{60}	K_3C_{60}
0.6	0.59	0.83	0.53	2.5	2.7

be recovered, however, by setting $S_K = 1.0$. In fact an S_K value less than 1.0 is expected for an ordinary metals (see the Table). By using the experimental parameters for Rb_3C_{60}(300K): $1/T_1T = 0.015\ K^{-1}\ s^{-1}$ [12] and for K_3C_{60}(300K): $1/T_1T = 0.010\ K^{-1}\ s^{-1}$[13], the following values are obtained:

$$Rb_3C_{60}(300K){:}\ S_K = 2.5 \qquad \text{and} \qquad K_3C_{60}(300K){:}\ S_K = 2.7$$

When ignoring the dipolar hyperfine interaction [12], these values would be about a factor of ten larger and even more so in contradiction with the classical Korringa relation as is also pointed out by Holczer et. al.[14]. In the Table these values are compared with the ones for some classical metals. Note that indeed the S_K values for A_3C_{60} are appreciably larger than expected for an ordinary metal. Such large S_K values (and even larger) have been reported before in other organic conductors[3, 4] and in high-T_c superconductors[5]. They point at large fluctuations in q-space and could be caused antiferromagnetic correlations. These data could therefore provide evidence for the proposed large electron-electron correlation in these materials[15]. I would like to point out that S_K is designed in such a way (see Eq.(3b)) that density of state variations cancel, i.e. the temperature variation of S_K (S_K decreases with decreasing temperature as is evident from the data in [12, 13]) cannot be accounted for by a temperature dependent change in the density of states[13]. Instead it could be explained in terms of decreasing antiferromagnetic correlations (with decreasing temperature), possibly due to an increasing bandwidth caused by lattice contraction. Further investigations have to clarify this conjecture.

This contribution is partly based on a more detailed publication[16].

Acknowledgements: I would like to acknowledge discussions with F. Rachdi, P. Bernier, G.Zimmer and M. Helmle. A financial support was given by the Fonds der Chemischen Industrie.

References.

[1] A. Abragam, The Principles of Nuclear Magnetism. Clarendon Press,Oxford (1961)

[2] C.P. Slichter, Principles of Magnetic Resonance (Springer, Berlin, 1989), 3rd ed.

[3] M. Mehring, in Low-Dimensional Conductors and Superconductors, eds. D. Jerome and L.S. Caron, NATO ASI Series, New York: Plenum Publishing Co., 185 (1987)

[4] M. Helmle and M. Mehring, in Nuclear Spectroscopy on Charge Density Wave Systems, ed. T.Butz, Kluwer Academic Publishers 1992

[5] M. Mehring, Appl. Magn. Res. 3, 383 (1992)

[6] G. Zimmer, M. Helmle, M. Mehring, F. Rachdi, J. Reichenbach, L. Firlej, and P. Bernier, Europhysics Letters (submitted) and this meeting.

[7] D. Köngeter and M. Mehring, Phys. Rev. **B39**, 6361 (1989)

[8] V.P. Antropov, L.I. Mazin, O.K. Andersen, A.I. Liechtenstein, and O. Jepsen, Phys. Rev. B. (in press)

[9] C.S.Yannoni, H.R.Wendt, R.L.Siemens, J.R.Salem, D.Nguyen, J.Lyerla, R.D.Johnson, M.Hoinkis, M.S.deVries, M.S.Crowder, C.A.Brown, D.S.Bethune, L.Taylor, P.Jedrzejewski, and H.C.Dorn, Synth. Metals (in press)

[10] K. Holczer (private communication at this meeting)
[11] A.P.Ramirez, M.J.Rosseinsky, D.W.Murphy, and R.C.Haddon, Phys. Rev.Lett. **69**, 1687(1992)
[12] R.Tycko, G.Dabbagh, M.J.Rosseinsky, D.W.Murphy, A.P.Ramirez, and R.M.Fleming, Phys. Rev.Lett. **68**, 1912 (1992)
[13] G. Quirion, C. Bourbonnais, E. Barthel, P. Auban, D. Jerome, J.M. Lambert, A. Zahab, P. Bernier, C. Fabre and A. Rassat, Europhys. Lett. **21**, 233 (1993)
[14] K. Holczer, O. Klein, H. Alloul, Y. Yoshinari, F. Hippert, S.M. Huang, R.B. Kaner, and R.L. Whetten, Europhys. Lett. (to be submitted)
[15] (a) G. Sawatzky (this meeting)
(b) S. Chakravarty and S. Kivelson, Europhys. Lett. **16**, 751 (1991)
[16] M. Mehring (to be published)

NMR Investigation of K_3C_{60} Under Pressure

G. Quirion[1], *C. Bourbonnais*[2], *R. Kerkoud*[1], *E. Barthel*[1], *P. Auban*[1], *D. Jérome*[1], *J.M. Lambert*[3], *A. Zahab*[3], *P. Bernier*[3], *C. Fabre*[4], and *A. Rassat*[4]

[1]Laboratoire de Physique des Solides (Associé au CNRS), Université de Paris-Sud, F-91405 Orsay, France

[2]Centre de Recherche en Physique du Solide, Université de Sherbrooke, Sherbrooke, Québec J1K 2R1, Canada

[3]Groupe de Dynamique des Phases Condensées, Université des Sciences et Techniques du Languedoc, F-34060 Montpellier, France

[4]Laboratoire d'Activation Moléculaire, Ecole Normale Supérieure, F-75231 Paris, France

Abstract. We report the pressure dependence of ^{13}C nuclear spin-lattice relaxation time T_1 in the normal state of K_3C_{60}. From the $1/T_1T$ pressure profile, a Fermi liquid theory is used to investigate the variation of the density of states and the amplitude of electron-electron interaction. Finally, within a phonon-mediated pairing model for superconductivity, the combined pressure dependence of T_c and $1/T_1T$ are shown to give strong support to a pairing mechanism via high energy modes (> 1200 K).

1. INTRODUCTION

The discovery of superconductivity in alkali fullerides A_3C_{60} (A=K, Rb, Cs) [1-3] has stimulated enormous interest in the electronic properties of these molecular crystals. Unfortunately, the air sensitivity of these compounds has substantially slowed down the elucidation of the related properties. In spite of this, the observed relation between T_c and the lattice constant corroborates the fact that the density of states is effectively a relevant parameter for superconductivity [4]. Moreover, based on band structure calculations, some authors have shown that a BCS-like picture can probably account for all known observations [5]. However, owing to the uncertainty in the density of states, the energy of the vibrational modes and the electron-phonon coupling strength leading to superconductivity in these materials have yet to be determined.

One of the peculiar advantages of the alkali-doped fullerides lies in their high compressibility [6-8]. In this respect, we show how ^{13}C nuclear spin-lattice relaxation measurements under pressure can be used as a tool to probe the induced variation of the density of states in these organic conductors. These results can then be used in order to obtain the first purely experimental functional dependence between T_c and the density of states and, in turn, the nature of the exchanged phonon in the pairing mechanism.

Springer Series in Solid-State Sciences, Volume 117
Electronic Properties of Fullerenes Eds.: H. Kuzmany, J. Fink, M. Mehring, and S. Roth

2. Results and discussion

NMR data presented here have been obtained using an isotopically enriched sample (10% ^{13}C) whose preparation has been described elsewhere [9]. Before measurements, the composition of the sample has been checked using Magic-Angle spinning ^{13}C NMR. We found that the K_3C_{60} phase constitutes the main part of the system with a small residue of pure C_{60} and no other phases present. By an AC susceptibility measurement, the superconducting transition temperature is found to occur at 19.6 K. We have also measured the temperature dependence of the ^{13}C spin-lattice relaxation time T_1 at atmospheric pressure. As given elsewhere [9], the results obtained at 10 Tesla agree with those reported by Tycko et al. [10] and Holczer et al. [11].

In the previous reports on K_3C_{60} [9-11], the analysis of the ^{13}C spin-lattice relaxation rates $1/T_1$ in the normal state has been based on the assumption that the nuclear relaxation rate is dominated by the Fermi-contact hyperfine term which leads to the well Known Korringa law. However, this interpretation is incompatible with the measured Knight-shift since the conduction band is mainly formed by p-orbitals. Antropov et al. [12] have examined the possibility for other relaxation mechanisms to contribute. They found that the relaxation rate in these compounds is rather governed by the dipole interaction between the electronic and the nuclear spins. In this case, the expression for the relaxation rate then reads

$$\frac{1}{T_1T} \propto \left(\frac{4}{5}\langle r^{-3}\rangle_{pp}\, N(E_F)\right)^2, \qquad (1)$$

where $[N(E_F)]$ is the density of states and $<1/r^3>$ is the average of the dipolar interaction.

In Figure 1, we present the pressure dependence of $1/T_1T$ obtained at 20 K (10 Tesla), 80 K and 300 K (3 Tesla) for K_3C_{60} up to 12 Kbar. At room temperature, we observe that $1/T_1T$ diminishes rapidly when we increase the pressure from 1 bar to 5 Kbar and then it displays the same pressure dependence as observed at 80 K and 20 K. Let us first consider the magnitude of the observed variation of $1/T_1T$ at room temperature compared to what should be expected from the variation of the density of states. According to the band structure calculations of Fleming et al. [13], the measured contraction of the lattice parameter at 10 Kbar [8] should produce a reduction of about 11% in the electronic density of states. This in turn should lead to a reduction of 22% in $1/T_1T$, a value smaller than the observed 35% variation (see fig.1). To account for the enhanced variation of $1/T_1T$, we can therefore consider the possibility that electron-electron correlations play a significant role in these compounds. Adopting a Fermi liquid picture, we can rewrite the electronic part of the relaxation rate as

$$\frac{1}{T_1T} \propto [N(E_F)]^2\ (1+F_o^a)^{-1}\ [1+F_o^a(1-\frac{2}{3\pi^2})]^{-1} \qquad (2)$$

where $F_o^a = N(E_F)\, f_o^a$ is the Fermi liquid interaction parameter involved in the enhancement of the spin susceptibility $\chi_s \propto N(E_F)\,(1+F_o^a)^{-1}$. Then, within this model we get

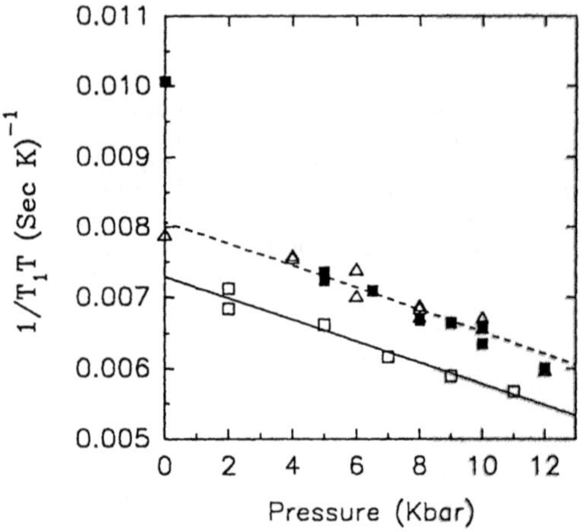

Fig. 1 Pressure dependence of the ^{13}C spin-lattice relaxation time T_1 in K_3C_{60} plotted as $(T_1T)^{-1}$ vs P at 300 K (fill squares), at 80 K (triangles) and at 20 K (open squares). The continuous line is a linear regression fit of the data obtained at 80 K and 20 K.

$$\frac{\delta(T_1T)^{-1}}{(T_1T)^{-1}} = \frac{2}{[1+F_o^a]}\,\frac{\delta N(E_F)}{N(E_F)} \tag{3}$$

for the relative variation of $1/T_1T$. If we use now the measured variation of $1/T_1T$ at room temperature (35%) and the expected reduction of the density of states (11%), we obtained from eq.(3) that $F_o^a \approx -0.37$. Thus, if the band calculations are meaningful, we have to conclude that correlation effects play a significant role at room temperature. This conclusion is also supported by the temperature variation of $1/T_1T$. Indeed, the thermal contraction of the lattice constant (0.7%) [14] should produce a variation of 7% in the density of states [13]. In this case F_o^a must be approximately equal to -0.5 in order to explain the observed variation of $1/T_1T$ (30%) which is close to the value obtained above. Therefore, within this picture, the observed temperature and pressure variation of $1/T_1T$ can be entirely understood in terms of lattice contraction effects on the density of states, these effects being enhanced by correlations. It is worthwhile to note that a similar situation exists in the context of the Bechgaard salts [15]. According to eq.(2), we should also observe a linear scaling relation between the observed enhancements in T_1^{-1} *and* $T\chi_s^2(T)$. Actually, if we carefully examine the temperature dependence of the magnetic susceptibility [16] obtained by ESR (the most reliable technique in order to get the electronic contribution), we effectively observe a variation of ≈ 13% which is consistent with the observed temperature dependence of $1/T_1T$ for K_3C_{60}. A similar decrease for the Knight-shift should be also expected; however, such a variation is likely to be difficult to measure as the Knight-shift is rather small (43 ppm) [10]. Finally, the interpretation presented here is also supported by recent specific heat and magnetic susceptibility measurements realized by Ramirez et al. [17]. Their estimation of the interaction parameter ($F_o^a \approx -0.46$) at ambient pressure is also consistent with our analysis. It is also worthwhile to note that the NMR relaxation rate calculation realized by Antropov et al. [12] is also consistent with our estimation of the strength of correlations.

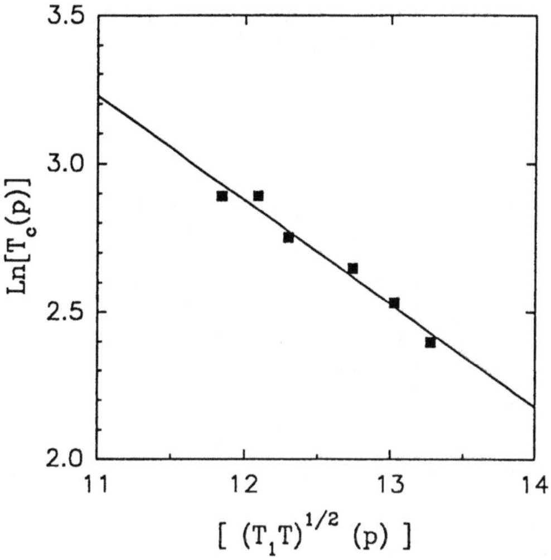

Fig. 2 Plot of the natural logarithm of the superconducting transition T_c against $\sqrt{T_1T(p)}$ obtained at different pressure. Here, $\sqrt{T_1T(p)}$ is proportional to the inverse of the density of states $N(E_F)$. The continuous line is a linear regression fit of the data.

Finally, one can also use the pressure dependence of $1/T_1T$ in order to estimate which pairing mechanism is involved for superconductivity in A_3C_{60} compounds. For that, we will assume that superconductivity in K_3C_{60} can be described by the weak coupling result $T_c \propto \Omega_{ph} \exp^{-1/N(E_F)V}$ where Ω_{ph} is the energy of the phonon and V is the electron-phonon coupling strength. If V is assumed to be unaffected by pressure, one can combined the pressure dependence of T_c [6,7] and $1/T_1T$ using $1/T_1T \propto [N(E_F)]^2$, in order to estimate Ω_{ph}. In fig. 2, we present a plot of $Ln(T_c(p))$ vs $\sqrt{T_1T(p)}$ obtained just above the superconducting phase transition (20 K). As shown by this plot, we effectively observe a linear behaviour which intercept the Y axis at $\Omega_{ph} \approx 1250$ K. The effect of correlations will increase Ω_{ph} even further. Thus within this simple weak coupling picture, our results show that high frequency phonons are likely to be involved in superconductivity. This is consistent with Raman spectroscopy [18] and inelastic neutron scattering [19] studies which indicate that phonons around 1500 K are effectively coupled to electronic states.

In conclusion, we believe that the results presented here have illustrated two main points. First of all, the possibility that electron-electron interactions play a role on the properties of these compounds is indicated self-consistently by the temperature and pressure dependences of $1/T_1T$. Finally, within a BCS framework for the expression of T_c, the reduction of the electronic density under pressure is found to favour a pairing mechanism that would involve high frequency intramolecular vibrations of the C_{60} molecules.

Acknowledgements

One of us (G.Q.) would like to thank the Natural Sciences and Engineering Research Council of Canada (NSERC), the Canadian Institute of Advanced Research (CIAR), le Groupe de Recherche C_{60} (GDR) and the French Ministry of Foreign Affairs for financial support. We are also grateful to J.C. Ameline for building the pressure cell and his help for the experiments.

References

[1] A.F. Hebard et al., Nature (London) 350, 600 (1991).
[2] M.I. Rosseinsky et al., Phys. Rev. Lett. 66, 2830 (1991).
[3] K. Holczer et al., Science 252, 1154 (1991).
[4] R.M. Fleming et al., Nature (London) 352, 787 (1991).
[5] M. Schlüter et al., J. Phys. Chem. Solids 53, 1473 (1992).
[6] E. Schirber et al., Physica C 178, 137 (1991).
[7] G. Sparn et al., Science 252, 1829 (1991).
[8] O. Zhou et al., Science 255, 833 (1992).
[9] G. Quirion et al., Europhysics Letters 21, 233 (1993).
[10] R. Tycko et al., Phys. Rev. Lett. 68, 1912 (1992).
[11] K. Holczer and R.L. Whetten, prepared for special issue of Carbon, edited by H. Kroto.
[12] V.P. Antropov et al., Preprint, to be published.
[13] R.M. Fleming et al., Nature (London) 352, 787 (1991).
[14] K. Tanigaki et al., Nature 356, 419 (1992).
[15] P. Wzietek et al., Synthetic Metals 41-43, 2435 (1991); J. Physique (France) January 1993.
[16] W.H. Wong et al., Europhys. Lett. 18, 79 (1992).
[17] A.P. Ramirez et al., Phys. Rev. Lett. 69, 1687 (1992).
[18] S.J. Duclos et al., Science 254, 1625 (1991).
[19] K. Prassides et al., Nature (London) 354, 462 (1991).

NMR on Alkali-Doped C_{60}

G. Zimmer[1], M. Helmle[1], M. Mehring[1], F. Rachdi[2], J. Reichenbach[2], L. Firlej[2], and P. Bernier[2]

[1]2. Physikalisches Institut, Universität Stuttgart, D-70569 Stuttgart, Germany
[2]Groupe de Dynamique des Phases Condensées, U.S.T.L., F-4060 Montpellier, France

Abstract: In order to unravel the electronic and structural properties of Rb_3C_{60} and K_4C_{60} ^{13}C and ^{87}Rb pulsed NMR spectroscopy was applied. The following results were obtained: The ^{87}Rb NMR line shift is found to be temperature dependent for all three lines observed and can be interpreted in terms of a small hyperfine coupling of the conduction electrons to the Rb nuclear spins. It is also shown that ^{87}Rb NMR is a sensitive probe to structural changes. From ^{13}C NMR we obtained information on the dynamics of C_{60} and on the hyperfine coupling.

1 Introduction

Charge transfer and the local symmetry of the alkali sites are important parameters which determine the physical properties of alkali doped C_{60}. We found the ^{87}Rb NMR to be an adequate method to address these problems. Different phenomena were observed while varying the temperature: The high temperature range ($T > 300\,K$) is characterized by the onset of C_{60} rotation on the ms-μs timescale. In the temperature range $80\,K < T < 300\,K$ the hyperfine interactions were extracted from the comparison of the different Knight shifts with the susceptibility. Just above the superconducting transition we observe a phase transition in Rb_3C_{60}. From ^{13}C NMR on K_4C_{60} we have obtained informations about the Knight shift and hyperfine relaxation [1].

2 Results on Rb_3C_{60}

Temperature dependent ^{87}Rb measurements were performed at 108.9 MHz and ^{13}C measurements at 45.6 MHz. The T_1 relaxation time was measured by an inversion recovery pulse sequence, all the spectra except the ^{13}C MAS spectra were recorded by a Hahn echo with phase alternation.

^{87}Rb NMR from octahedral and tetrahedral sites

In the whole temperature range we observe three clearly distinguishable lines. Two of them, marked by O and T in fig. 1, correspond to the octahedral and tetrahedral sites in Rb_3C_{60}, respectively. The third line T' is rather unexpected for a pure Rb_3C_{60} phase, and two possibilities must be considered: The extra line could be (i) due to interstitials in the Rb_3C_{60} phase, which seems to be supported by a recent SEDOR experiment [2], (ii) or alternatively T' belongs to the Rb_4C_{60} phase, which we found to be present in the sample as a small admixture as seen in the ^{13}C MAS (Fig. 1). The observed temperature dependence (Fig. 2) of the line shift and the resulting hyperfine coupling constant is con-

Springer Series in Solid-State Sciences, Volume 117
Electronic Properties of Fullerenes Eds.: H. Kuzmany, J. Fink, M. Mehring, and S. Roth

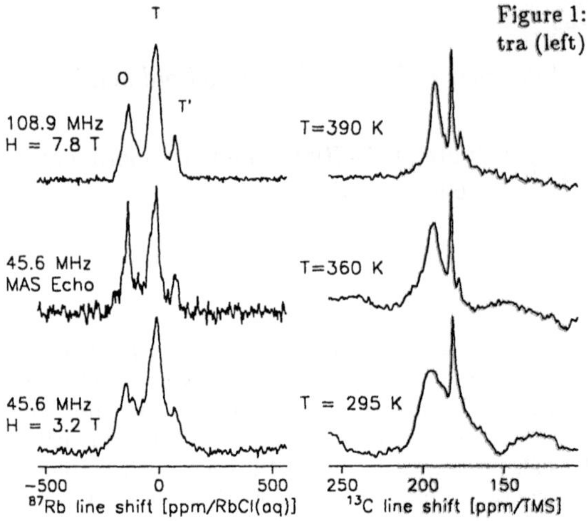

Figure 1: ^{87}Rb NMR room temperature spectra (left) and ^{13}C MAS spectra(right).

sistent with the first possibility, it is, however, inconsistent with the T_1^{-1} relaxation rates, which are much larger for the *T'* line than for the others.
Another question concerns the quadrupole interaction of ^{87}Rb which is a 3/2 spin system and exhibits therefore three transitions, one sharp central transition ((-1/2) - (1/2)) and two quadrupolar broadened satellite lines ((-3/2) - (-1/2)) and ((1/2) - (3/2)) in a low symmetry site. In fact, we observe signature of the satellite transitions in low field spectra, where the line width is considerably smaller. We observe a scaling of the spectra with the magnetic field as can be seen in fig. 1, where the spectra are plotted on the ppm scale (normalized magnetic field) indicating that the main part of the interactions is magnetic. In the static spectrum the satellites are visible and show a separation of about 3 kHz for the *O* site. Under MAS conditions the lines narrow and only the central transition is seen. In a nutation experiment the three lines exhibit the same Rabi frequency, which leads to the conclusion that no spectral information is lost in the whole temperature range for all three lines. Therefore the electric field gradient at the Rb sites must be rather small requiring highly symmetric sites which is indeed expected from the crystal structure. The origin of the *T'* line seems to be still unsettled, however.

Temperature dependence of the line shifts

We find that the ^{87}Rb spectra show about the same temperature dependence as the electronic susceptibility and the ^{13}C line shift [3, 4] (Fig. 2). This linear scaling leads us to conclude that the observed ^{87}Rb line shift is caused by the hyperfine interaction of the ^{87}Rb nuclei with the conduction electrons due to polarization effects. The scaling of the shift with the electronic susceptibility [3] leads to the following hyperfine coupling constants: $a_{iso}(O) = -2\pi \cdot 25\,\text{MHz}$, $a_{iso}(T) = -2\pi \cdot 30\,\text{MHz}$, $a_{iso}(T') = -2\pi \cdot 16\,\text{MHz}$. They are quite small compared to the hyperfine coupling constant of a Rb atom (6.8 GHz). For the sites *O* and *T* the different T_1^{-1} relaxation rates correspond to the determined hyperfine coupling constants indicating that the main relaxation path at low temperatures is caused by hyperfine interaction.

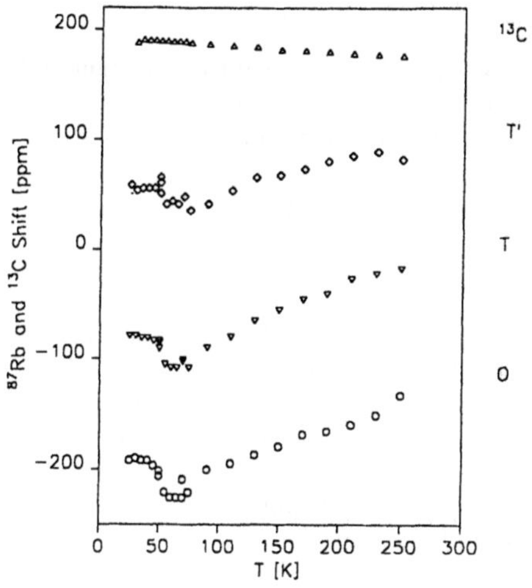

Figure 2: ^{13}C [4] and ^{87}Rb line shifts as a function of temperature.

Low temperature range ($T < 80$K)

In this temperature range, a rather complex behaviour is found: The line shift as well as the line width depend strongly on cooling and heating rates $\Delta T/\Delta t$ and the sample history as can be seen in fig. 3. This behaviour is a fingerprint of a first order phase transition occurring at about 60 K. The observed broadenings and shifts can be interpreted in terms of a change in the hyperfine interaction due to a rearrangement of the C_{60} molecules. Details will be published in a forthcoming publication.

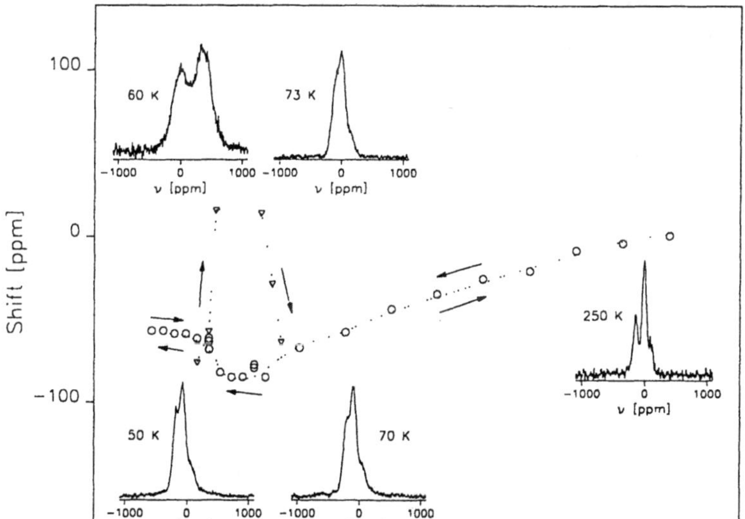

Figure 3: Temperature dependence of the line shift of line *b*. Typical spectra at different temperatures and after different thermal sequences are inserted. Arrows indicate the direction of temperature variation.

^{13}C and ^{87}Rb NMR for $T > 290\,K$

At room temperature the ^{13}C spectral width is given by chemical shift and Knight shift anisotropy (see also ref. [4]). The maximal hyperfine coupling tensor can be estimated by subtracting the chemical shift tensor [5] from the observed total shift ($\chi = 2.8 \cdot 10^{-4}$emu/mole C_{60}):

$$\underline{\underline{A}} = a_{iso} + \underline{\underline{A}}_{dip} = 2\pi \cdot 2\,\text{MHz} + 2\pi \begin{pmatrix} 9.2 & 0 & 0 \\ 0 & -3.9 & 0 \\ 0 & 0 & -5.3 \end{pmatrix} \text{MHz}.$$

These values are similar to the ones observed in other organic conductors. Note that the dipolar part is larger than the isotropic part. At about 380 K the ^{13}C line shows clearly motional narrowing. From an Arrhenius law (using $\tau_0 = 10^{-12}$s) an activation energy of $\Delta E \approx 600$ meV is estimated for the rotation barrier of the C_{60} molecules in Rb_3C_{60} which is larger than in K_3C_{60} [7]. The MAS spectra (Fig. 1) taken at different temperatures show an additional narrow line due to the Rb_4C_{60} phase at 182 ppm. This demonstrates the use of MAS-NMR to determine microphases in these materials.

3 Preliminary results on K_4C_{60}

We find extremely rapid ^{13}C NMR relaxation in this phase (T_1 of about 100 ms) (see also [1, 6]) and isotropic Knight shift of about 30 ppm. This implies that in insulating K_4C_{60} strong hyperfine interactions are present due to delocalized elctron spins, i.e. electron-electron correlations must play an important role as proposed in [8]. At about 150 K, we observe a broadening (powder pattern), indicating the slowing down of C_{60} motion. On cooling down further, a very broad line appears at about 120 K, which shifts to higher frequencies with increased broadening down to 40 K (Fig. 4). This observation hints possibly to the occurence of a spin density wave at about 120 K, which would be a further indication of the importance of electron-electron correlations. However, we do not know currently which role phase purity plays for the occurrence of this phenomenon.

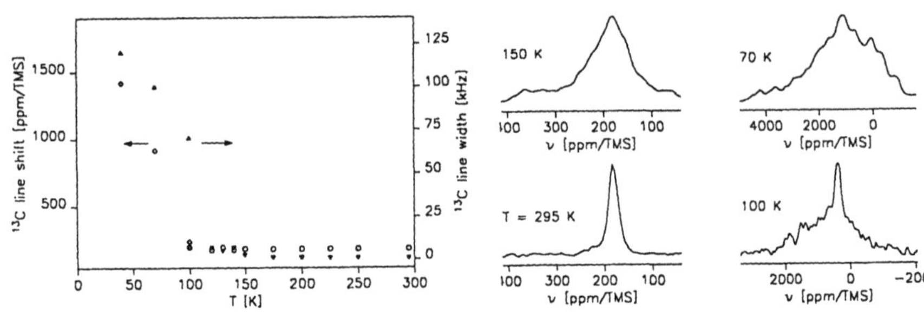

Figure 4: Temperature dependence of the ^{13}C line shift and line width for K_4C_{60}(left) and the corresponding spectra (right). Note the different scales at different temperatures.

Acknowledgments We would like to acknowledge C. Kessler, B. Nebendahl, J. Groß, and N. Winzek for discussions and assistance with the experiments.

References

[1] R. F. Kiefl, L. T. Duty, J. W. Schneider, A. MacFarlane, K. Chow, J. W. Elzey, P. Mendels, G. D. Morris, J. H. Brewer, E. J. Ansaldo, C. Niedermayer, D. R. Noakes, C. E. Stronach, B. Hitti, and J. E. Fischer, Phys. Rev. Lett. **69**, 2005 (1992)

[2] R. E. Walstedt, D. W. Murphy, and M. Rosseinsky, submitted to Nature

[3] A. P. Ramirez, M. J. Rosseinsky, D. W. Murphy, and R. C. Haddon, Phys. Rev. Lett. **69**, 1687 (1992)

[4] R. Tycko, G. Dabbagh, M. J. Rosseinsky, D. W. Murphy, A. P. Ramirez, and R. M. Fleming, Phys. Rev. Lett. **68**, 1912 (1992)

[5] C. S. Yannoni, R. D. Johnson, G. Meijer, D. S. Bethune, and J. R. Salem, J. Phys. Chem. **95**, 9 (1991)

[6] D. W. Murphy, M. J. Rosseinsky, R. M. Fleming, R. Tycko, A. P. Ramirez, R. C. Haddon, T. Siegrist, G. Dabbagh, J. C. Tully, and R. E. Walstedt, J. Phys. Chem. Solids, **53**, 1321 (1992)

[7] S. E. Barrett and R. Tycko, Phys. Rev. Lett. **69**, 3754 (1992)

[8] S. Chakravarty, and S. Kivelson, Europhys. Lett. **16**, 751 (1991)
S. Chakravarty, M. P. Gelfand, and S. Kivelson, Science **254**, 970 (1991)

EPR and Overhauser Shift on $[P\Phi_4^+]C_{60}^-$ and K_xC_{60}

H. Klos[1], *U. Becker*[1], *V. Dyakonov*[1], *G. Rösler*[1], *B. Gotschy*[1], *G. Denninger*[2], *and A. Hirsch*[3]

[1]Phys. Inst. und BIMF, Universität Bayreuth, D-95447 Bayreuth, Germany
[2]Phys. Inst. 2, Universität Stuttgart, D-70569 Stuttgart, Germany
[3]Institut für Organische Chemie, Universität Tübingen, D-72076 Tübingen, Germany

Abstract. Radical anion salts of C_{60} exhibit striking electronic and magnetic properties ranging from superconductivity to ferromagnetism. In this contribution we report on EPR and ^1H and ^{13}C-Overhauser shift measurements in the stable ^{13}C-radical anion salt $[P\Phi_4]^+C_{60}^-$ and ^1H-Overhauser shift measurements in K_xC_{60} ($x<3$). We performed Overhauser shift measurements to determine the spin density distribution of the unpaired electrons using ^{13}C and ^1H nuclei as local probes for the electronic wavefunction. In $[P\Phi_4]^+C_{60}^-$ the ^1H-Overhauser shift indicates spin density on the $P\Phi_4^+$-counterion. In K_xC_{60} three distinct ^1H-Overhauser shift signals are observed. Two signals are attributed to solvent molecules included during the extraction of C_{60} from the soot, and one signal is attributed to protons attached to C_{60}.

1. Introduction

Magnetic resonance techniques [1,2,3] have been successfully applied both to pure C_{60} crystals [4,5] and to the class of C_{60}-compounds like K_xC_{60} or $[TDAE]^+C_{60}^-$. A new compound Tetraphenylphosphonium$^+$-C_{60}^- (denoted as $[P\Phi_4]^+C_{60}^-$) excited our interest, because at low temperatures it exhibits an exchange narrowed [1,6] EPR-line of 300 mG linewidth. In radical ion salts and doped polymeres with comparable narrow EPR-lines, we could experimentally determine with Overhauser shift experiments more information about the electronic spin density distribution, the nuclear relaxation paths and coupling mechanisms between nuclear and electronic spins. For a review see [6]. The advantage of the Overhauser shift experiments compared to conventional NMR-investigations is, that only small samples with dimensions down to 0.1mm are required.

2. Experimental results

$[P\Phi_4]^+C_{60}^-$ was prepared by electrochemical crystallisation [1], yielding an air stable polycrystalline black shining material. The results of the elemental analysis can be best reconciled with $[P\Phi_4]^+C_{60}^-\cdot 2(P\Phi_4^+Cl^-)$. From polycrystalline samples small pieces can be separated with single-crystal shape. $[P\Phi_4]^+C_{60}^-$exhibits a single 300 mG exchange narrowed EPR-line [1,7] at 4.2 K (ΔH_{pp} = 45 G at R.T.).

Springer Series in Solid-State Sciences, Volume 117
Electronic Properties of Fullerenes Eds.: H. Kuzmany, J. Fink, M. Mehring, and S. Roth

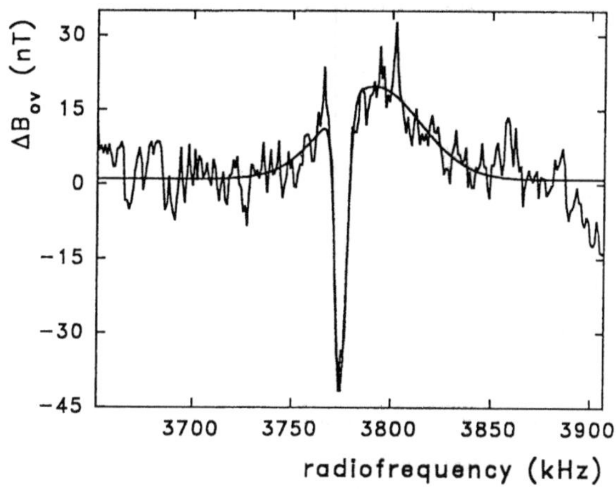

Figure 1: ^{13}C-Overhauser shift at high EPR saturation ($s \approx 0.8$) in $[P\Phi_4]^+C_{60}^-$.

The Overhauser shift [5] is the electronic analogue of the Knight shift which are both detectable in metals, organic metals, semiconductors and polymeres. One observes a shift of the EPR-line position (denoted as ΔB_{ov}) which is proportional to the nuclear spin polarisation P and to the hyperfine coupling (HFI). Figure 1 shows a typical ^{13}C-Overhauser shift spectrum of $[P\Phi_4]^+C_{60}^-$ which is composed of two lines. A wide one with 30 kHz FWHM and a narrow one with 4 kHz FWHM. A.W. Overhauser first predicted, that the nuclear ensemble can be polarised by saturation of the EPR: $\Delta B_{ov} = \Delta B_{ov}^{th} \cdot (1+V\cdot s)$. In our Overhauser shift experiments [6] we reduce the nuclear spin polarisation of e.g. ^{13}C or ^1H by a saturating radio-frequency field at the nuclear Larmor frequency. In order to determine the Overhauser shift in the thermal equilibrium ΔB_{ov}^{th} we performed Overhauser shift measurements at various incident microwave powers. $s = (S_o - \langle S_z \rangle)/S_0$ is the EPR saturation parameter ($0<s<1$) and V is the enhancement factor of dynamic nuclear polarisation (DNP). From the linear dependence between saturation parameter s and the Overhauser shift ΔB_{ov} we obtained for the narrow line in Fig. 1: $\Delta B_{ov}^{th} = -0.21 \pm 0.12$ nT , $V = 375 \pm 200$ (Figure 2) and for the broad line: $\Delta B_{ov}^{th} = 0.23 \pm 0.1$ nT , $V = 93 \pm 30$.

It was also possible to measure ^1H-Overhauser shifts. The spectra exhibit two lines with different sign. It was possible to determine the enhancement factor V and the Overhauser shift in thermal equilibrium: $\Delta B_{ov}^{th} = 3.8 \pm 1.2$ nT, $V = 72 \pm 23$ (positive shift); $\Delta B_{ov}^{th} = -2 \pm 1$ nT, $V = 181 \pm 90$ (negative shift).

After a saturating rf-pulse at the Larmor frequency of the investigated nuclei the temporal evolution of the Overhauser shift directly probes the nuclear spin-lattice relaxation. Figure 3 shows the ^1H-nuclear relaxation rate vs temperature of $[P\Phi_4]^+C_{60}^-$. It is possible to describe the relaxation rate as a sum of two contributions. One contribution from proton spin diffusion to localised paramagnetic centers (temp. independent, $C_{spin} = 0.059 \pm 0.002\ ms^{-1}$) and a second thermal activated process ($\Delta E = 44 \pm 4$ meV).

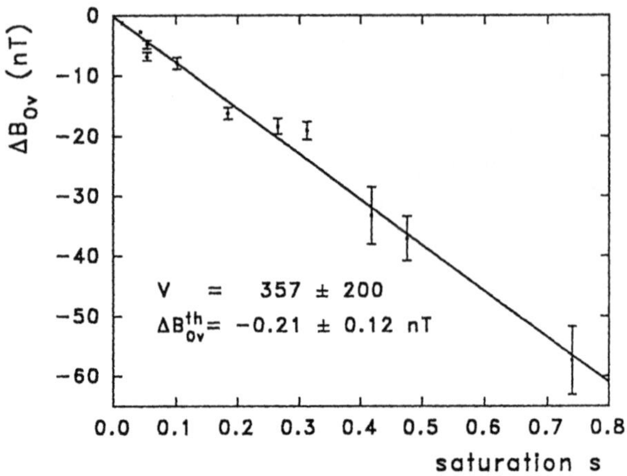

Figure 2: ^{13}C-Overhauser shift of the negative, 4 kHz wide signal vs EPR saturation.

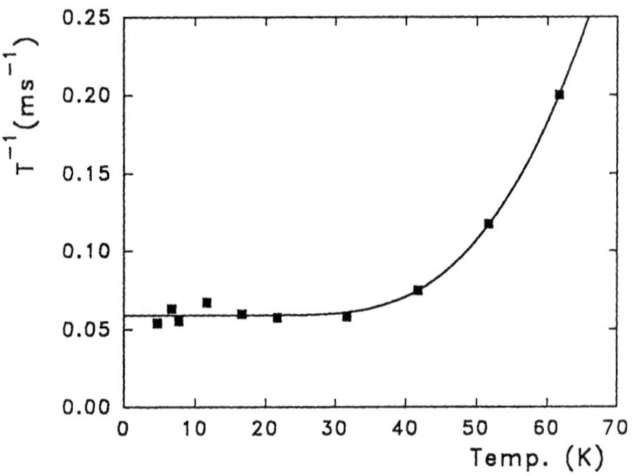

Figure 3: Temperature dependence of the ^{1}H-nuclear spin-lattice relaxation measured via the temporal evolution of the Overhauser shift after a saturating rf-pulse.

Unexpected ^{1}H-shift signals were found in K_xC_{60}. The H-signals detected in this substance were dominated by two positive Overhauser shift signals. A third negative could be possible, but it is hardly visible. For the two signals with positive signs we could determine the Overhauser shift in the thermal equilibrium and also the enhancement factors by using the DNP. $\Delta B_{ov}{}^{th}$ = 0.47+0.21 nT,V = 322±145 (1. pos.component); $\Delta B_{ov}{}^{th}$ = 1.17±0.67 nT, V = 261±93 (2. pos. component)

The spin-lattice relaxation time T_1 of K_xC_{60} is in the order of milliseconds. In comparison the $[P\Phi_4]^+C_{60}{}^-$ -proton T_1-time is about one order of magnitude shorter.

3. Discussion and Conclusion

The ^{13}C-Overhauser shift spectrum of $[P\Phi_4]^+C_{60}^-$ is composed of two lines. A wide one with 30 kHz and a narrow one with 4 kHz. 30 kHz linewidth is typical for C-H bonds. Therefore we assume, that this negative shift belongs to carbons at the counterion $[P\Phi_4]^+$. The 4 kHz wide Overhauser shift is assigned to the carbons on the C_{60}-ball (no dipolar broadening). Via analysis of the paramagnetic shifts of the ^{13}C-resonance frequencies we could confirm our Overhauser shift results.

The relaxation of the nuclear spins is dominated by electronic processes. Thus DNP is very effective. If we assume, that the electronic wavefunction is concentrated on the bucky ball, we expect an isotropic HFI on the bucky balls and a dipolar HFI outside (e.g. solvent molecules). Both types of HFI produce different signs of V. The sign of V gives thus information about the type of coupling. In the $[P\Phi_4]^+C_{60}^-$ we found an isotropic HFI of protons indicating partial spin density on the $P\Phi_4^+$, and a dipolar HFI which is attributed to solvents. In K_xC_{60} we found two dipolar and one isotropic HFI. The latter one is attributed to a proton attached to a ball. Both others are attributed to solvents.

4. References

[1] P. Allemand, G. Srdanov, A. Koch, K. Khemani, F. Wudl: J. Am. Soc. **113**, 2780 (1991)

[2] P. Allemand, K. Khemani, A. Koch, F. Wudl, K. Holczer, S.Donovan, G. Grüner, J. Thompson, Science: **253**, 301 (1991)

[3] R. Tycko, R. Haddon, G. Dabbagh, S. Glarum, D. Douglass, A. Mujsce: J. Phys. Chem., **95**, 518 (1991)

[4] H.W. Kroto, J. Heath, S. O'Brien, R. Curl, R. Smalley: Nature, **318**, 354 (1985)

[5] W. Krätschmer, L. Lamb, K. Fostiropoulos, D.R. Huffmann: Nature, **347**, 354 (1990)

[6] G. Denninger: Mol. Cryst. Liq. Cryst., **171**, 315 (1989)

[7] U. Becker, G. Denninger, V. Dyakonov, B. Gotschy, A. Hirsch, H. Klos, G. Rösler, H. Winter: Europhys. Lett., **21**(3), 267 (1993)

[8] A.W. Overhauser: Phys. Rev., **92**, 411 (1953)

NMR Studies of Fullerene C_{60}-Based Compounds

P. Bernier, F. Rachdi, M. Ribet, J. Reichenbach, L. Firlej, J.M. Lambert, A. Zahab, Z. Belahmer, and R. Aznar

Groupe de Dynamique des Phases Condensées
(Laboratorie Associé au CNRS no. 233), Université de Montpellier II,
F-34095 Montpellier Cédex 05, France

Abstract. We present NMR studies performed on A_nC_{60} compounds, where A can be K, Rb, I_2, O_2 and n can be 1, 2, 3 and 6 depending on the system. Using high resolution NMR we have obtained information on the molecular dynamics and the electronic properties of these compounds. The shape of the ^{13}C spectra shows that intercalation of species like K and Rb, with n=6, totally block the molecular reorientation of the C_{60} at room temperature. On the contrary the effect of O_2 is only partial while I_2 does not hinder these reorientations. The ^{13}C resonance positions are discussed in the light of existing theories concerning the NMR shifts in insulating and conducting media.

1. Introduction

NMR has been shown to be a valuable technique for studying the molecular dynamics in solid C_{60} [1, 2]. At room temperature the rapid molecular rotation leads to a ^{13}C NMR spectrum composed of a single narrow (3 ppm width) resonance at 143.6 ppm from TMS. If these movements are blocked (at the NMR time scale), the spectrum presents a broad shape (200 ppm wide) reflecting the chemical shift anisotropy, as it is the case in frozen benzene [3]. This happens at temperatures well below the orientational order transition temperature (259 K).

Intercalation of atomic or molecular species in the structure of solid C_{60} is also a way to alter the free reorientation of the C_{60} molecules. For instance, T_1 measurements show that solvent molecules, such as toluene, present as impurities in the solid, significantly increase the ^{13}C relaxation rate [4]. In the present paper we study the influence of atomic or molecular species on the molecular dynamics in solid C_{60} at room temperature.

2. Experimental

C_{60} powder has been prepared following standard procedures [5]. The purity, checked by HPLC, mass spectrometry and ^{13}C NMR gives more than 99.8% of C_{60}. Residual organic molecules, such as toluene, have been removed by heating the powder at 240°C during 6 hours under dynamic vacuum. ^{13}C enriched C_{60}

Springer Series in Solid-State Sciences, Volume 117
Electronic Properties of Fullerenes Eds.: H. Kuzmany, J. Fink, M. Mehring, and S. Roth

samples have been produced using cored carbon rods packed with 99% ^{13}C enriched amorphous carbon.

Intercalation of K or Rb has been performed by contact of the vapor phase of the alkali metal with the C_{60} powder at high temperature. Intercalation of iodine was obtained with the I_2 vapor at high temperature [6]. Intercalation of oxygen occured spontaneously by exposing the C_{60} powder to air at room temperature [7].

The NMR experiments have been performed using a Bruker CXP200 spectrometer working at 50.3 MHz for the ^{13}C. A standard $\pi/2$ pulse has been used to generate the NMR signal, while magic-angle spinning (MAS) at various rates (from 0.3 to 4 kHz) was needed to achieve high resolution. All the spectra are referenced to the TMS ^{13}C resonance.

All air sensitive samples had to be stored in sealed glass tubes. Consequently a special homebuilt NMR probehead had to be used in order to spin these samples.

3. Results and discussion

The ^{13}C NMR spectrum observed on K_6C_{60} without MAS presents a broad resonance (total width of more than 200 ppm) whose extent is equivalent to the one reported by Tycko et al. [8]. The observed width is due to the chemical shift anisotropy and reflects the random orientation of the K_6C_{60} crystallites. We also note that this width corresponds roughly to the one observed on pure frozen C_{60} [3], which indicates that in K_6C_{60} the molecules are reorientating at a rate smaller than 10 kHz at room temperature.

With such a slow motion, one should expect to resolve unequivalent carbon sites on the C_{60} molecule. According to the structural model of Zhou et al. [9] for K_6C_{60} only three of these sites are non-equivalent. Then we should observe three distinct ^{13}C resonances with intensities in the ratio 1:1:0.5. In order to check this possibility we have recorded the ^{13}C resonance with MAS, allowing us to :
(i) identify the position of the isotropic part of the spectrum and analyse its structure, and (ii) observe the whole spectrum through the envelope of the spinning side-band distribution. Fig 1 presents the high resolution spectrum obtained by spinning the sample at 2.8 kHz. Following the procedure outlined by Herzfeld and Berger [10], the following tensor values were determined: σ_{11} = 224 ppm, σ_{22} = 222 ppm and σ_{33}=18 ppm. A closer look at the isotropic part of the spectrum (see top trace of Fig.1), whose center is close to 155 ppm, shows that it is in fact composed of three lines at 156, 155 and 153 ppm. Their intensities are in the ratio 1:1:0.5, as expected from the above model. We then conclude that *in the case of K_6C_{60} the molecules are almost totally immobile at room temperature*. A very similar behaviour has been observed in the case of Rb_6C_{60} [11].

A somewhat different situation occurs in the case of C_{60} intercalated with Iodine. Zhu et al. [6] have evidenced from structural studies the compound $C_{60}I_4$ with no charge transfer between the intercalated species and C_{60}. Nevertheless, from the X-rays study, it is not clear if there is orientational disorder at room temperature or not. Our ^{13}C NMR spectrum on $C_{60}I_4$, obtained with MAS at 2.5 kHz, shows a single

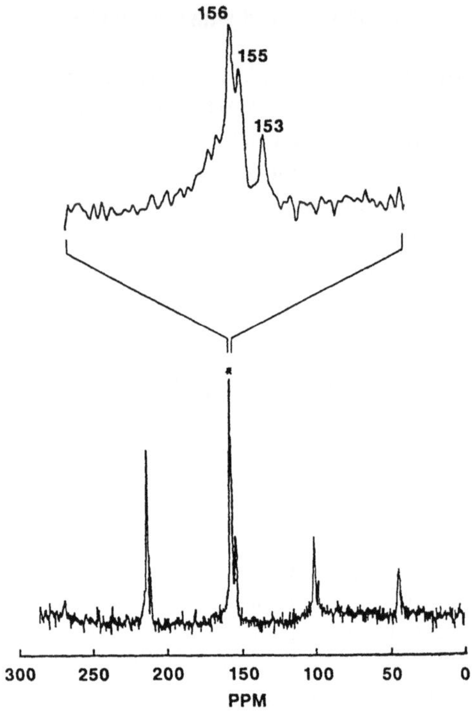

Figure 1. ^{13}C high resolution MAS-NMR spectrum of K_6C_{60} at room temperature with a spinning of the sample at 2.8 kHz. The star identify the isotropic part of the spectrum, which is expanded in the top trace.

sharp and structureless resonance (0.2 ppm width) at 143.6 ppm. These characteristics are almost exactly those observed on pure C_{60} at room temperature. Furthermore the spectrum recorded with no MAS presents a single resonance whose width is not significantly different from that of pure C_{60} at room temperature. The above results indicate that *in $C_{60}I_4$ the fullerene molecules are reorientating very fast,* probably as fast as in pure C_{60} at room temperature. The reason why such a large steric disturbance like I_2 does not affect the C_{60} rotation is not totally clear at the moment. Nevertheless, following Zhu et al [6], we can note that while the structure goes from fcc (pure C_{60}) to simple hexagonal ($C_{60}I_4$), the available volume per C_{60} changes from 711 Å^3 to 857 Å^3.

The case of intercalated O_2 represents an intermediate situation. Assink et al. [7] and Samara et al. [12] have identified and explained the influence of molecular oxygen on the ^{13}C resonance of solid C_{60}. We have also shown [13] that oxygen spontaneously intercalate C_{60}, filling the octahedral sites of the fcc structure. Fig 2 presents the ^{13}C MAS-NMR spectrum observed on a 9 % ^{13}C enriched sample which has been kept in air at room temperature. In addition to the expected resonance at 143.6 ppm, five weak lines appear at lower field (paramagnetic shift), at positions

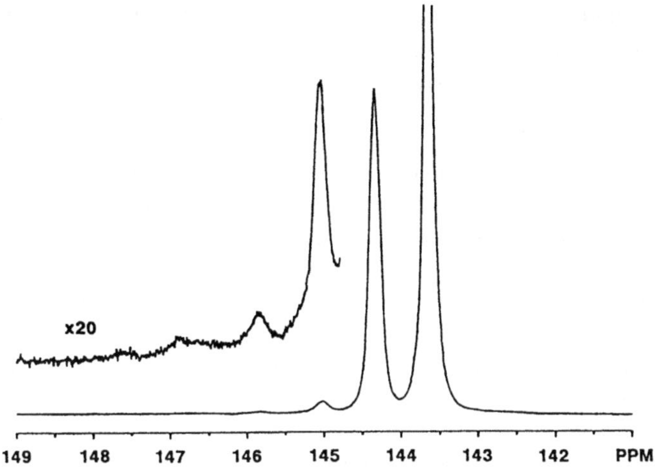

Figure 2. ^{13}C high resolution MAS-NMR spectrum of pure C_{60} at room temperature with a spinning of the sample at 2 kHz, showing the intense resonance of pure C_{60} and the additional 5 lines due to the presence of O_2.

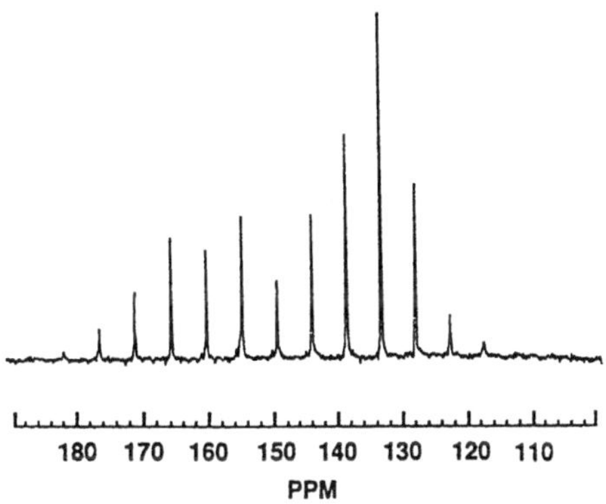

Figure 3. ^{13}C high resolution MAS-NMR spectrum of pure C_{60} at room temperature with a spinning of the sample at 265 Hz. The spectrum shows only the resonance, with its spinning side-bands, of the C_{60} molecules surrounded by one O_2 molecule.

which are 0.7 ppm from each other. These resonances correspond to C_{60} molecules having up to five O_2 molecules as nearest neighbours. Unfortunately these lines are too weak to be observed without MAS and consequently the anisotropic nature of the spectrum can be checked only through the observation of a low frequency MAS spectrum. Figure 3 presents such a spectrum for the first additional line (144.3 ppm corresponding to one O_2 per C_{60}) the sample spinning at 265 Hz. Following the same procedure as above [10] we can estimate the tensor components of the chemical shift anisotropy: σ_{11}=113 ppm, σ_{22}=156 ppm and σ_3=163 ppm. The extent of anisotropy, of the order of 50 ppm, is at least 10 times larger than the motionally narrowed resonance of pure C_{60}, but 4 times smaller than the value measured for immobile molecules. We then conclude that the *reorientational movement of C_{60} molecules having one O_2 in one of the surrounding octahedral sites is considerably hindered.*

A detailed study of the other additional resonances is difficult due to their very small intensity. Nevertheless, an attempt to estimate the anisotropy of the resonance associated with the presence of two O_2 (isotropic resonance at 145.0 ppm), from the observation of the spinning side-bands, gives a *value of the order of 30 ppm which is significantly smaller* than the anisotropy extent of the resonance due to one O_2. From this result we can conclude that C_{60} molecules surrounded by only one oxygen reorient at a smaller rate than C_{60} surrounded by two O_2 molecules. The origin of such a difference is not clearly understood at the moment but we can suggest an increasing environmental symmetry as one goes from one to two oxygen molecules beside C_{60}.

The origin of the shift of the isotropic resonance observed on the above compounds (155 ppm for K_6C_{60} and Rb_6C_{60}, 143.6 ppm for I_4C_{60}, 144.3 ppm for O_2C_{60}, 145.0 ppm for O_4C_{60}) is a very fundamental question. The same position observed for pure C_{60} and I_4C_{60} is consistent with the fact that no charge transfer is observed in the compound. As the chemical shift measured on C_{60} and its compounds is essentially an intramolecular property, no influence on the position is expected from the presence of iodine. In the case of oxygen in C_{60}, Assink et al. have interpreted the observed shift as due to the weak paramagnetism of O_2 molecules [7]. Nevertheless in this scope it is difficult to understand the molecular reorientation hindrance as described above. Let us suggest that there could be some partial charge transfer between the O_2 and C_{60} species giving rise to a corresponding paramagnetic shift and to the electrostatic interaction necessary to understand the molecular reorientation hindrance. Finally, the origin of the shift observed in K_6C_{60} is not so easy to identify. One should stress that the additional density corresponding to the six electrons transfered, shared by the 60 carbons of the molecule, induces such a shift. Nevertheless, using the simple rule established for aromatic molecules (one electron transfered by carbon induces a shift of -160 ppm) [14], yields a shift of the right order of magnitude (-16 ppm) *but of the wrong sign.* There is in fact no reason to apply directly simple rules relevant for aromatic systems to this molecule where the π electron system is not symmetric across the molecular surface [15]. Finally let us mention the possible importance of ring currents which strongly influence the amplitude and sign of the chemical shift, as suggested by Pasquarello et al. [16].

Acknowledgments. We thank K.Kniaz and J.E.Fischer who supplied the iodine intercalated sample. One of us (JR) thanks the MPI-CNRS agreement for financial support.

REFERENCES

1. R.D.Johnson, D.Bethune,C.S.Yannoni, Acc.Chem.Res., 25, 169 (1992).
2. R.Tycko, R.C.Haddon, G.Dabbagh, S.H.Glarum, D.C.Douglass, A.M.Musjce, J.Phys.Chem., 95, 518 (1991).
3. - C.S.Yannoni, R.D.Johnson, G.Meijer, D.S.Bethune, J.R.Salem, J.Phys.Chem.,95, 9 (1991).
   - R.Tycko, G.Dabbagh, R.M.Fleming, R.C.Haddon, A.V.Makhija, S.M.Zahurak, Phys.Rev.Lett., 67, 1886 (1991).
4. A.Zahab, L.Firlej, P.Bernier, R.Aznar, Solid St.Comm., 84, 429 (1992).
5. A.Zahab, J.L.Sauvajol, L.Firlej, R.Aznar, P.Bernier, J.Phys. I France, 2, 7 (1992)
6. Q.Zhu, D.E.Cox, J.E.Fischer, K.Kniaz, A.R.McGhie, O.Zhou, Nature, 355, 712 (1992).
7. R.A.Assink, D.A.Loy, J.E.Schirber, B.Morosin, Novel Forms of Carbon, MRS Symposium Proceedings, 270, 255 (1992).
8. R.Tycko, G.Dabbagh, M.J.Rosseinsky, D.W.Murphy, R.M.Fleming, A.P.Ramirez, J.C.Tully, Science, 253, 884 (1991).
9. O.Zhou, J.E.Fischer, N.Coustel, S.Kycia, Q.Zhu, A.R.McGhie, W.J.Romanow, J.P.McCauley, A.B.Smith III Jr., D.E.Cox, Nature, 351, 462 (1991).
10. J.Herzfeld, A.E.Berger, J.Chem.Phys, 73, 6021 (1980).
11. F.Rachdi, J.Reichenbach, L.Firlej, P.Bernier, M.Ribet, R.Aznar, G.Zimmer, M.Helmle, M.Mehring, Solid St.Comm., submitted.
12. G.A.Samara, L.V.Hansen, R.A.Assink, B.Morosin, J.E.Schirber, D.Loy, to be published.
13. Z.Belahmer, P.Bernier, L.Firlej, J.M.Lambert, M.Ribet, Phys.Rev.B, in press.
14. H.Spiesecke, W.G.Schneider, Tetrahedron Lett.,468 (1961).
15. P.Fowler, Nature, 350, 20 (1991).
16. - A.Pasquarello, M.Schluter, R.C.Haddon, Science, 257, 1660 (1992)
    - A.Pasquarello, M.Schluter, R.C.Haddon, to be published

QENS and NMR Investigation of Reorientational Dynamics in $C_{61}H_2$

L. Cristofolini[1], M. Riccò[1], G. Viola[1], and E. Dalcanale[2]

[1]Dipartimento di Fisica, Università di Parma, Parma, Italy
[2]Dipartimento di Chimica Organica ed Industriale, Università di Parma, Parma, Italy

Abstract. The molecular dynamics of the dihydrofulleroid $C_{61}H_2$ [1] have been studied. Protonated and deuterated samples have been synthesized and characterized by ^{13}C, ^{1}H, ^{2}H high resolution NMR, HPLC, TGA, DSC and mass spectrometry. The molecular dynamic has been studied by powder NMR, and Quasielastic Neutron Scattering. The unitary interpretation of the results obtained from these two techniques gives a picture of the molecular reorientation in this system which is completely different from that observed in the rotor phase of C_{60}, where the molecules are described to reorient isotropically, and should be compared to $C_{60}O$ [2] and C_{70}. In $C_{61}H_2$ the molecular units rotate almost uniaxially around the line joining the centre of the molecule and the methylic carbon. Much slower reorientation of the axis itself among different lattice directions also takes place.

Sample Preparation and Characterization. $C_{61}H_2$ was prepared by reacting our home synthesized C_{60} with an equimolar amount of diazomethane in dry toluene at room temperature, followed by thermal decomposition (80C, 30 min) of the intermediate $(CH_2N_2)C_{60}$ [1]. The result was a black powder composed of about half C_{60} and half $C_{61}H_2$, analyzed by HPLC. The deuterated sample $C_{61}D_2$ was produced in the same way, using deuterated diazomethane. The structure of $C_{61}H_2$ is shown in figure 1.

Residual crystallization solvent (toluene) was exchanged with CS_2 by dissolving the solid in CS_2, then the samples were dried and heated under dynamic

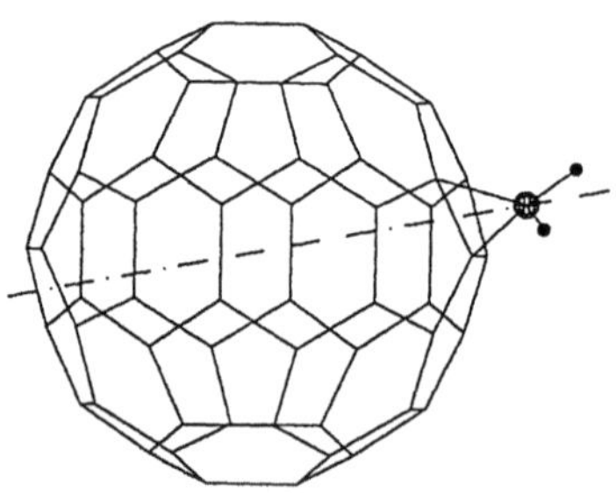

Fig. 1 *The structure of $C_{61}H_2$ molecule. The rotation axis passing across the molecular centre and the methylic carbon is shown*

Springer Series in Solid-State Sciences, Volume 117
Electronic Properties of Fullerenes Eds.: H. Kuzmany, J. Fink, M. Mehring, and S. Roth

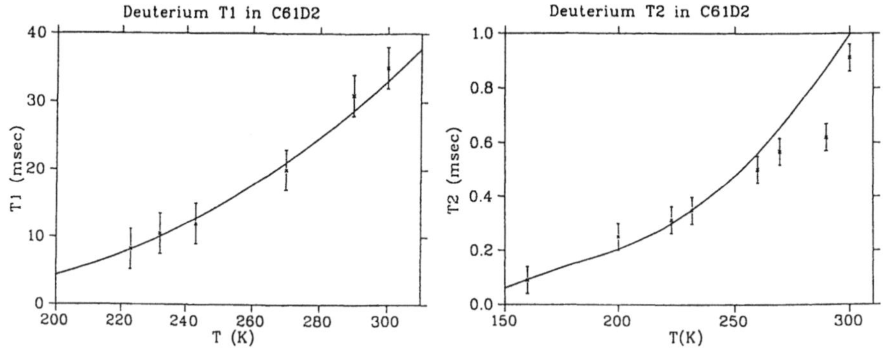

Fig. 2 *Deuterium relaxation times in $C_{61}D_2$ as a function of temperature. Left: longitudinal relaxation time T_1. Right: transverse relaxation time T_2. The solid lines are deduced from the result of the fit of an Arrhenius law for the correlation times τ_r and τ_c respectively.*

vacuum at 80C for one night. In this way we prepared 1.7 g of $C_{61}H_2$ and 0.9 g of $C_{61}D_2$. The samples have been characterized by ^{13}C and ^{1}H high resolution NMR, TGA, HPLC, DSC and mass spectrometry.

From the ratio of the integrated intensity of the 31 ^{13}C NMR lines of $C_{61}H_2$ (in agreement with literature [1,5]) versus the one of C_{60} we can state that our sample was composed for 54% of $C_{61}H_2$ and for 46% of C_{60} . In addition ^{2}H NMR showed that in the deuterated sample 1/3 of $C_{61}D_2$ molecules are subject to chemical exchange of a deuterium with a proton, resulting in a fraction of $C_{61}HD$. The exchange was selective, taking place only at the site with higher chemical shift. Desorption Chemical Ionization mass spectrometry revealed the molecular ion of $C_{61}H_2$ with mass MH^+=735 Da.

No phase transition was detected by DSC in the range from 170K to 340K. This is consistent with the fact that no change nor anomaly was observed in the T dependence of NMR relaxation times , as will be explained later.

NMR relaxation measurements. Deuterium longitudinal (T_1) and transverse (T_2) relaxation times have been measured on a home made spectrometer, operating at 45.77 MHz. The covered temperature range was from 170 K to RT. The results are shown in figure 2. Both relaxation times follow an Arrhenius law without appreciable anomalies, corroborating the DSC indication that no phase transition occurs in this temperature range.

The most striking feature of these data is the impossibility of explaining both relaxation times with just one dynamic, characterized by a single correlation time τ. For this reason we adopt a picture in which two kinds of motion are possible:

a) the molecule rotates almost uniaxially around the axis joining the centre of the molecule and the methylic carbon, the correlation time for this motion is τ_r ;

b) much slower reorientation of the axis itself takes place with correlation time τ_c . The two motions are assumed to be uncorrelated and can be treated independently.

It is well known [6] that for quadrupolar relaxation of spin 1, in the case of isotropic rotational diffusion:

$$\frac{1}{T_1} = \frac{3}{80}\, \nu_Q \, \{J(\omega_0) + 4J(2\omega_0)\}$$
$$\frac{1}{T_2} = \frac{1}{160}\, \nu_Q \, \{9J(0) + 15J(\omega_0) + 6J(2\omega_0)\},$$

where $J(\omega) = \frac{2\tau}{1+\omega^2\tau^2}$ is the spectral density function for rotational diffusion motion with correlation time τ, and ν_Q is the quadrupole frequency, which takes the value ν_Q= 180 KHz for deuterons in methylic groups (the principal axis of the quadrupole coupling tensor is directed parallel to the C-D bonding, and the asymmetry parameter is η=0).

We deduced the correlation time τ_r from the longitudinal relaxation time. To obtain qualitative results we made use of the formula above, even if it was deduced in the case of isotropic rotation. Each value of T_1 can be fitted by two values of τ_r , but the ambiguity is lifted considering that T_1 decreases on cooling, when τ_r is supposed to increase. The values of τ_r obtained range from 67ps at RT to 285ps at 220K. We fitted an Arrhenius behaviour and we obtained: τ_r =1.21 $exp(\frac{1219}{T(K)})$ [ps].

On the other hand the correlation time τ_c for the slower diffusion of the molecular axis was deduced from T_2, making use of the formula above. We obtained values of τ_c ranging from 2.6ns at RT to 76ns at 160K. The fitted Arrhenius law was: τ_c =46.6 $exp(\frac{1180}{T(K)})$[ps],

The values of τ_r in $C_{61}D_2$ are higher than the correlation times in C_{60} as measured by NMR [7,8] and μSR [9], by about a factor of 2 in the same temperature range (ratchet phase). However a more complex, self-consistent model, in which both the motions are considered simultaneously, should be taken into account [10]. For the application of this model it is crucial to determine *exactly* the angles between the C-D bonds and the molecular rotation axis. These angles are close to the magic angle, but a distortion of the structure is expected, because one proton is over a pentagon and the other is over an hexagon (their chemical shifts are quite different too: 6.35 vs 2.87 ppm [1,5]).

Quasielastic Neutron Scattering. The experiment has been performed on IRIS spectrometer operating at ISIS neutron source at Rutherford Appleton Laboratory (UK). Pyrolitic Graphite PG004, PG002 and Mica 002 reflections were exploited.The Q range was limited to 3.7, 1.8, 0.6 $Å^{-1}$ respectively. Both protonated and deuterated samples have been measured in order to discriminate between coherent and incoherent scattering.

We observed diffuse coherent scattering at 200K in the whole Q range, with a reduction of the intensity in the high Q region. This is a signature of the uniaxial rotation of the molecule, in analogy with C_{70} at 300K [11] and differently from the isotropic phase of C_{60} [12]. On the other hand we did no detect any quasielastic incoherent broadening at any temperature in the range from 170K to 365K. This may be related to the small value of the gyration radius of the protons that puts the inelastic broadening out of the experimentally reachable Q range.

Conclusions. In conclusion we have a picture of the molecular dynamics of $C_{61}H_2$ which is fairly different from C_{60} and is more similar to C_{70}. We have observed two kinds of molecular motion in $C_{61}H_2$: fast uniaxial rotations and slower diffusion of the rotation axis. The correlation time for the fast rotation is slower than in C_{60} and the Arrhenius activation temperature is reduced to about 1/2. Much slower reorientation of the molecular rotation axis itself among different lattice directions also takes place. Moreover no phase transition in the range 170-340K was detected.

References.

1 T. Suzuki, Q. ("Chan") Li, K. C. Khemani, F. Wudl, J. Am Chem. Soc. 114 (1992) 7301.
2 G. B. M. Vaughan, P. A. Heiney, D. E. Cox, A. R. McGhie, D. R. Jones, R. M. Strongin, M. A. Cichy, A. B. Smith III, Chem Phys. 168 (1992) 185.
3 D. H. Parker, P. Wurz, K. Chatterjee, K. R. Lykke, J. E. Hunt, M. J. Pellin, J. C. Hemminger, D. M. Gruen, L. M. Stock, J. Am. Chem. Soc. 113 (1991) 7499.
4 W.A. Scrivens, P.V. Bedworth, J.M. Tour, J. Am. Chem. Soc. 114 (1992) 7917
5 F. Wudl, private communication.
6 A. Abragam, "The Principles of Nuclear Magnetism" (Clarendon Press, Oxford 1961)
7 R.D. Johnson, C. S. Yannoni, H. C. dorn, J. R. Salem, D. S. Bethune, Science 255 (1992) 1235
8 R. Tycko, G. Dabbagh, R. M. Fleming, R.C. Haddon, A. V. Makhija, S.M. Zahurak, Phys. Rev. Lett. 67 (1991) 1886
9 R. f. Kiefl, J. W. Schneider, A .MacFarlane, K. Chow, T. L. Duty, T. L. Estle, B. Hitti, R. Lichti, E. J. Ansaldo, C. Schwab, P. Percival, G. Wei, S. Wlodek, K. Kojima, W. J. Romanow, J. P. McCauley, N. Coustel, J. E. Fischer, B. Smith, Phys. Rev. Lett. 68 (1992) 1347 (erratum, reprinted in Phys. Rev. Lett. 68 (1992) 2708)
10 D.E.Woessner, Jou. Chem. Phys, 36 (1962) 1
11 C. Christides, T.J. Dennis, K. Prassides, R.L. Capelletti, D.A. Neumann, J.R.D. Copley, submitted to Phys. Rev. B
12 J.R.D. Copley, D.A. Neumann, R.L. Cappelletti, and W.A. Kamitakahara, J. Phys. Chem. Solids 53 (1992) 1353

NMR Study of Orientational Ordering in C_{60} and the Magnetic Transition in TDAE-C_{60}

R. Blinc, J. Dolinšek, D. Arčon, D. Mihailovič, and P. Venturini

J. Stefan Institute, University of Ljubljana, 61111 Ljubljana, Slovenia

Abstract. The nature of the orientational ordering transition at T_c= 256 K in C_{60} as well as the microscopic nature of the magnetic ordering transition at T_M= 16.1 K in tetrakis (dimethylamino) ethylene-C_{60} (TDAE-C_{60}) have been studied by ^{13}C and proton NMR.

Whereas the currently accepted model of the structural phase transition in C_{60} assumes that we deal with isotropic rotational diffusion of the C_{60} molecules above T_c and with uniaxial jump reorientations between two nearly degenerate orientations - separated by 60° - below T_c, our data clearly show that this is not true: The reorientational motion below T_c in C_{60} is quasi-isotropic rather than uniaxial and C_{60} is significantly more disordered than assumed so far.

The temperature dependence of the paramagnetic respectively ferromagnetic shifts of the two proton NMR lines in TDAE-C_{60} demonstrate the existence of a magnetic ordering transition in this system around 15 K.

1. Introduction

Here we wish to report on:

i) a study of the orientational ordering transition in C_{60} by ^{13}C NMR, and

ii) a study of the microscopic nature of the magnetic transition in TDAE-C_{60} with proton NMR.

2. ^{13}C NMR and Orientational Ordering in C_{60}

The currently established model of the orientational ordering transition in solid C_{60} at T_c = 260 K involves [1, 2] isotropic molecular rotational diffusion of the nearly spherical C_{60} molecules above T_c and uniaxial jump reorientations around the < 111 > axis between two nearly degenerate orientations, "a" and "b" - separated by 60° - below T_c. Because of the $\bar{3}$ - fold rotational axis pointing along the < 111 > direction this type of the motion

Springer Series in Solid-State Sciences, Volume 117
Electronic Properties of Fullerenes Eds.: H. Kuzmany, J. Fink, M. Mehring, and S. Roth

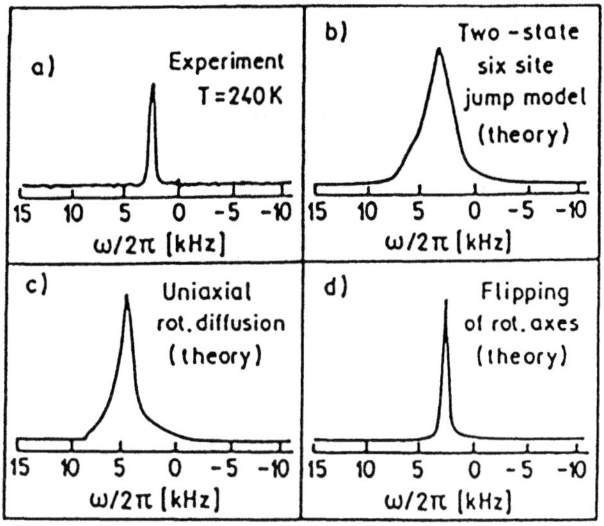

Figure 1: Comparison between experimental and theoretical 1D ^{13}C NMR spectra of powdered C_{60}: a) T=240 K, b) simulated powder spectra for the two state model, i.e. for uniaxial 60° reorientations between three pairs of nearly degenerate states, c) simulated powder spectra for uniaxial small amplitude rotational diffusion, d) simulated powder spectra for a random sequence of successive uniaxial 36°-step reorientations around three different rotational axes [6] separated by 60°. All simulations were performed in the fast motion limit.

takes the C_{60} molecule through six symmetry related sites: three "a" sites and three "b" sites, separated by 60°. The phase transition at T_c is accompanied by an abrupt decrease in the reorientational correlation time from 9.2×10^{-12} s to 2×10^{-9} s [3]. The two nearly degenerate orientations below T_c should differ in energy by $\Delta E = 11.4$ meV [1] and should be separated by an energy barrier of E= 290 meV [1]. This two state reorientational model is based on high resolution powder neutron diffraction data [1] and has been successfully used to describe a large number of macroscopic properties of solid C_{60} at lower temperatures such as the time - temperature dependence of the thermal conductivity [4] and thermal expansivity [5].

Here we wish to report that the above two state jump model is incompatible with ^{13}C NMR lineshapes observed by one - dimensional (1D) and two - dimensional (2D) NMR. A comparison between experimental and theoretical ^{13}C NMR spectra of powdered C_{60} (Fig.1) definitely rules out the uniaxial jump reorientational model [1, 2] of molecular motion below T_c. This is true for the two - state, six site 60° jump reorientational model [2] (Fig. 1b) as well as for any model involving reorientations or small amplitude rotational diffusion around a single axis (Fig. 1c). Our data show that even far below T_c the molecular reorientations are quasi - isotropic (Fig. 1d) rather than uniaxial and that the low temperature phase is more highly orientationally disordered than previously assumed [1, 2].

Our data support the recent observation by David et al. [6] that purely uniaxial reorientations below T_c would be inconsistent with the muon spin rotation (μSR) linewidth data [7] recorded for the labeled adduct Mu - C_{60}. They proposed an alternative model [6] involving a pseudo - random sequence of uniaxial reorientations leading to quasi - isotropic behavior below T_c, which is compatible with our data (Fig. 1.d).

3. ^{1}H NMR of the Magnetic Ordering Transition in TDAE-C_{60}

The unusual magnetic properties of TDAE-C_{60} below T_c= 16.1 K have recently attracted a great deal of attention [8,9]. It has been suggested that this system forms a kind of a soft organic molecular ferromagnet with a Curie temperature higher than any other reported for non-polymeric organic ferromagnets [8]. Here TDAE stands for a strong organic reducing agent tetrakis(dimethylamino)ethylene, $C_2N_4(CH_3)_8$, which is supposed to transfer one electron from its highest occupied molecular orbital to the lowest unoccupied molecular orbital of C_{60}. In contrast to the superconducting A_3C_{60} structures (A=K,Rb,...), which are all cubic [10], TDAE-C_{60} has a c-centered monoclinic unit cell [9]. The distance of nearest C_{60}-TDAE approach is 2.58 Å, whereas the inter-fullerene distances are 10.25 Å within the $a-b$ plane and 9.98 Å along the c axis [9]. The fact that the same basic molecule C_{60} should support superconductivity as well as ferromagnetism - depending on the doping - is highly unusual and at present poorly understood.

In order to throw some additional light on the nature of the unusual magnetic phase transition in TDAE-C_{60} we decided to perform a ^{1}H NMR lineshift study.

Our aim was to see on a microscopic level whether there is a spontaneous magnetization in TDAE-C_{60} or not. The shift of the NMR line in the local field H_L below T_c in a ferromagnet or an antiferromagnet directly measures the magnitude and temperature dependence of the spontaneous magnetization or the sublattice magnetization respectively and is an unambiguous sign of the onset of magnetic ordering.

Two inhomogeneously broadened proton NMR lines - designated as A and B lines - are observed in TDAE-C_{60} at room temperature. The position of the B-line is temperature independent whereas the position of the A-line shifts with decreasing temperature (Fig.2) according to a Curie law

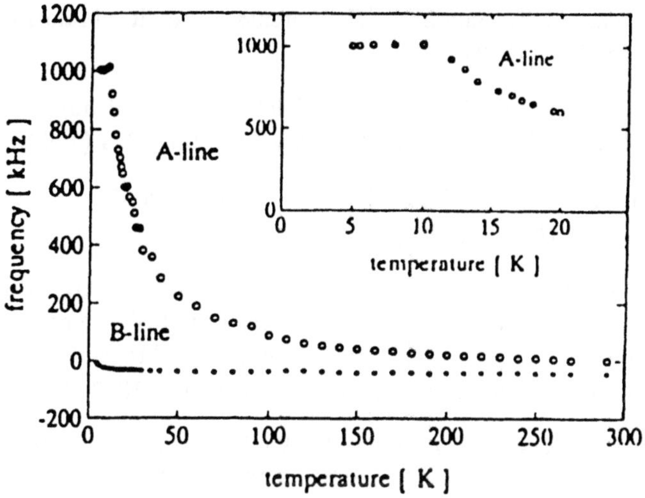

Figure 2: Temperature dependence of the frequencies of the A and B proton NMR lines in TDAE-C_{60}.

$$\delta\nu \propto \chi \propto \frac{C}{T - T_c} \qquad (1)$$

where $T_c \approx 0$ for $T > 20$ K. $(\delta\nu \cdot T)$ starts to deviate from a straight line below 20 K demonstrating the occurrence of a magnetic ordering transition.

The shift saturates below 10 K. The temperature dependence of the shift of the A-line below 15 K is similar to the one expected for a weak ferromagnet below T_c in an external magnetic field.

References

[1] W.I.F. David, R.M. Ibberson. T.J.S. Dennis, J.P. Hare, and K. Prassides, Europhys.Lett. **18**, 219 (1992).

[2] P.A. Heiney, J.E. Fischer, A.R. McGhie, W.J. Romanow, A.M. Denenstein, J.. McCauley, A.B. Smith, and D.E. Cox, Phys.Rev.Lett. **66**,2911 (1991). See also R.Moret, S.Ravy and J.M.Godard, J.Phys.I (France) **2**, 1699 (1992).

[3] R. Tycko, G. Dabbagh, R.M Fleming, R.C Haddon, A.V. Makhija, and S.M. Zahurak, Phys.Rev.Lett. **67**, 1886 (1991); C.S. Yannoni, R.D. Johnson, G. Meijer, D.S. Bethune, and J.R. Salem, J.Phys.Chem. **95**, 9 (1991).

[4] R.C. Yu, N. Tea, M.B. Salamon, D. Lorents, and R. Malotra, Phys.Rev.Lett. **68**, 511 (1992).

[5] F.Gugenberger, R Heid, C. Meingast, P. Adelmann, M. Braun, H. Wuhl, M. Haluska, and H. Kuzmany, Phys.Rev.Lett. **69**, 3774 (1992).

[6] W.I.F. David, R.M. Ibberson, T.J.S. Dennis, J.P. Hare, and K. Prassides, Europhys.Lett. **18**, 735 (1992).

[7] R.F. Kiefl, W. Schneider, A. MacFerlane, K. Chow, T.L. Duty, T.L. Estle, B. Hitti, R.L. Lichti, E.J. Ansaldo, C. Schwab, P.W. Percival, G. Wei, S. Wladek, K. Kojima, W.J. Romanow, J.P.Jr. McCauley, N. Coustel, J.E. Fischer, and A.B. Smith, Phys.Rev.Lett. **68**, 1347 (1992).

[8] P.M. Allemand, K.C. Khemani, A. Koch, F. Wudl, K. Holczer, S. Donavan, G. Gruner, and J.D. Thompson, Science **253**, 1154 (1991).

[9] P.W. Stephens, D. Cox, J.W. Lauher, J.B. Wiley, P.M. Allemand, A. Hirsch, K. Holczer, Q. Li, J.D. Thompson, and F. Wudl, Nature **355**, 331 (1992).

[10] P.W. Stephens, L. Mihaly, P. Lee, R.L. Whetten, S.M Huang, R.B. Kaner, F. Diedrich, and H. Holczer Nature **351**, 632 (1991); R.M Fleming, A.P. Ramirez, M.J. Rosseininsky, D.W. Murphy, R.C. Haddon, S.M. Zahurak, and A.V. Makhija, Nature **352**, 787 (1991).

The Quest for New Magnetic Fullerene Material

B. Gotschy[1,*], *H. Klos*[1], *W. Schütz*[1], *G. Denninger*[2], *A. Hirsch*[3], *and H. Winter*[4]

[1]Physikalisches Institut Universität Bayreuth, LS EP II, D-95447 Bayreuth, Germany
[2]2. Physikalisches Institut Universität Stuttgart, D-70569 Stuttgart, Germany
[3]Institut für organische Chemie, Universität Tübingen, D-72076 Tübingen, Germany
[4]Physikalisches Institut, Universität Karlsruhe, D-72076 Karlsruhe, Germany
*Author to whom correspondence should be addressed

Abstract. Two different routes to new magnetic fullerene materials where investigated: Preparation of organic charge transfer complexes of C_{60} via chemical reduction of C_{60} with strong organic donors and electro-crystallisation of C_{60} in the presence of appropriate counterions. In all systems a remarkable spin-spin exchange coupling was found, which depends on the size of the counterions. Short range magnetic order was seen up to about 100K, which is nearly two orders of magnitude higger than what is usually observed in organic materials.

1. Introduction

Fullerites exhibit a variety of interesting phases. They span from superconductivity to ferromagnetism. In this work we will concentrate on the magnetic properties. Two different routes for the preparation of new fullerites were investigated.

1.1 Chemical reduction of C_{60} with stong organic donors like tetrakis-(dimethylamino)-ethylene ([TDAE]C_{60}, [1]), (1,8)diazabicyclo(5,4,0)undec-7-en ([DBU]C_{60}, [2]) and (1,5)diazabicyclo(4,3,0)non-5-en ([DBN]C_{60}, [2]). These [X]C_{60} complexes where prepared by the treatment of C_{60} either directly or out of benzene solutions with an excess of X. The insoluble [X]C_{60} complex falls out. The samples are washed, decanted and finally dried under vacuum at room temperature. Polycrystalline material or long black needles are formed depending on the conditions of preparation. All systems are very senstive towards moisture and oxygen. The chemical reduction of C_{60} is reversible: $C_{60}+X \Leftrightarrow [X]^{\delta+}C_{60}^{\delta-}$; $0<\delta<1$. The redox potential of X determines whether the equilibrium of the reaction is on the left or right side.

1.2 Elektro-crystallisation of C_{60} in the presence of appropriate counterions like tetraphenylphosphonium ([$P\Phi_4$]C_{60}, [3]) yields black shining, polycrystalline radical anion salts. The results of the elemental analysis are best reconciled with a formula unit of $[P\Phi_4]C_{60}\times([P\Phi_4]Cl)_2$. Preliminary hyperfine interaction studies indicate an incomplete charge transfer from [$P\Phi_4$] to C_{60} [4].

Springer Series in Solid-State Sciences, Volume 117
Electronic Properties of Fullerenes Eds.: H. Kuzmany, J. Fink, M. Mehring, and S. Roth

2. Experimental results

2.1 [TDAE]C_{60}

[TDAE] is the strongest and biggest donor investigated. It was the only one which gave homogeneously doped samples. Charge transfer was one electron per formula unit. The EPR susceptibility χ_{EPR} shows a strong increase below about 17K. This was supposed to be the indication of a ferromagnetic ordering [1,5]. However, the lack of a spontaneous magnetisation and a shift of the NMR resonance frequency in the ordered phase are pointing rather towards a spin glass state or ferromagnetic grains. Several samples under various conditions where synthesised. Some samples showed a drastic reduction of the spin concentration and exhibited only a paramagnetic phase. It is not clear if this was due to the reversibility of the reaction or according to unsuitable preparation conditions.

It is well known in radical anion and cation salts, that the size of the counterion has an important influence on the size of the unit cell and thus on the spin-spin exchange coupling. This induced us to use smaller donors to diminish the lattice constant. Thus a higger spin-spin exchange and higger Curie-temperatures should be observable.

2.2 [DBU]C_{60}

DBU is an organic donor like TDAE but weaker and smaller. Pristine polycrystalline [DBU]C_{60} exhibits two EPR lines. A broad one with a line width of about 35G at 300K (labeled ∇) and a small one with a line width of 2G (labeled •). After several days a third one emerges with a line width of about 2G (labeled o). The EPR spectra where numerically separated. The basis of the simulation was a sum of three Lorentzian lines. Fit parameters are amplitude, line width and g-factor of each line. Within the noise of the experiment simulated and measured spectra coincide. The temperature dependence and the width of the broad line are typical for fullerites (figure 1). Thus it is attributed to coupled C_{60}^{-}. $\chi_{EPR}\times T$ vs. temperature T in [DBU]C_{60} is shown in figure 2. Data are in agreement with measurements of the static susceptibility. Obviously line ∇ gives the dominant contribution to χ. Within the error bare line • and o exhibit a Curie-behaviour in the whole temperature range. Above 170K the susceptibility of the C_{60}^{-} can be fitted by a Pauli plus a Curie like contribution. The Pauli part is suggested by the charge transfer of one electron. According to the band structure [6] C_{60}^{-} is metallic (unless it is a Mott-Hubbard isolator). The Curie-part takes account of the DBU^{+} counterions which have an unpaired spin. However, they are not visible in the EPR similar to the situation in [TDAE]C_{60}. Either the EPR of C_{60}^{-} and DBU^{+} overlap or a strong exchange between both leads to a common EPR resonance. Another possible description of the high temperature χ is a Curie-Weiss susceptibility with a Curie-temperature of Θ=-290K (!). This corresponds to an antiferromagnetic coupling. At about 90K a cusp in $\chi_{EPR}\times T$ is the sign of antiferromagnetic fluctuations. The insert of figure 1 shows the difference between the g-factors of line ∇ and •. While the position of line • is nearly unaffected, line ∇ is stronly shifted apart. This indicates a strong local field. A calibration of χ_{EPR} against a

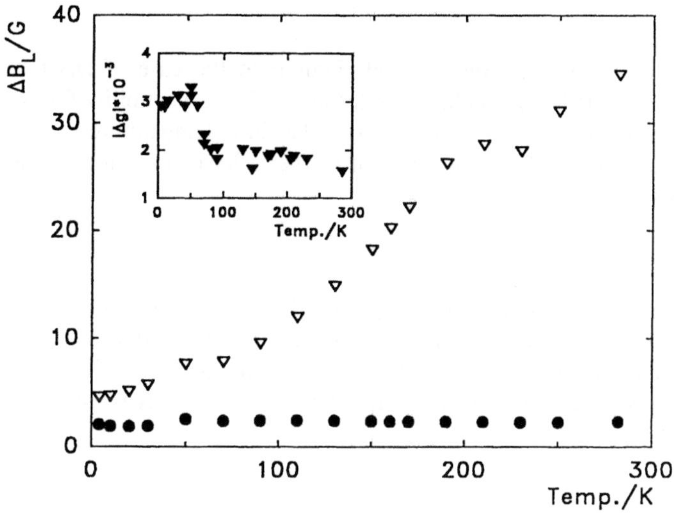

Figure 1: EPR line width vs. temperature in $[DBU]C_{60}$. Line ∇ exhibits an appreciable narrowing and is attributed to a C_{60}^-. The insert shows the difference of the g-factors of line ∇ and •. Below 100K line ∇ is shifted by an internal field.

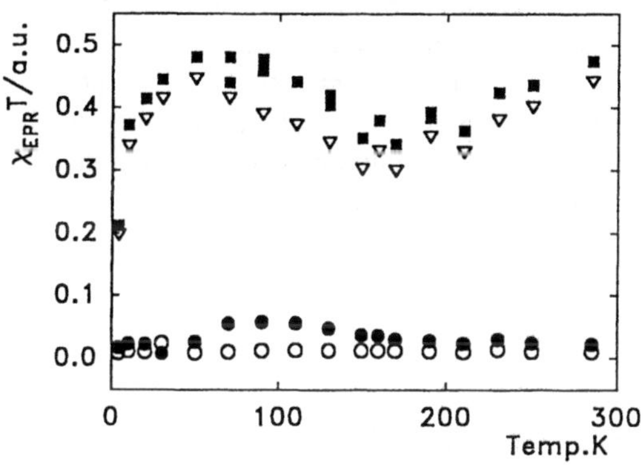

Figure 2: $\chi_{EPR} \times T$ vs. T in $[DBU]C_{60}$ in agreement with static susceptibility data. The C_{60}^- give the dominant contribution to χ_{EPR}. • and o are Curie-like contributions within the error bare.

ruby standard gave only 0.02 Curie-equivalent spins per formula unit. This seems not enough to explain the high Θ and the antiferromagnetic fluctuations around 100K. Thus we propose the picture of a very inhomogeneous sample with electron rich domains which exhibit a local order. Line • is attributed to isolated $[DBU]C_{60}$ within a diamagnetic surrounding. The line width of about 2G is typical for isolated C_{60}^-. The third line is a defect line of so far unknown origin and apears only in older samples.

2.3 [DBN]C_{60}

DBN was the smallest counterion investigated. Similar to the case of DBU no homogeneous sample of [DBN]C_{60} could be synthesised. The susceptibility follows a Curie-Weiss law in the whole temperature range. The Curie temperature is +7K. Here, the spin-spin coupling is ferromagnetic but neither a local nor a long range magnetic order was seen.

2.4 [PΦ_4]C_{60}

Extensive work has already been published [3]. [PΦ_4]C_{60} was the only stable system. The susceptibility is composed of a Curie-part and a contribution of a spin singulet with a low lying activated triplet. The activation energy of only 10K indicates, that the system is close to a magnetic instability or a spin density wave state.

3. Conclusion

All systems investigated show, that magnetic interactions in fullerites are important. We found, that spin-spin exchange depends on the fullerene-fullerene distance. Magnetic fluctuations could be seen up to 100K. Most of the samples are so far only poor characterised and strongly inhomogeneous. Further success in chemical preparations of fullerites can result in new organic materials which exhibit long range magnetic ordering at ambient temperatures. This competition between magnetic and superconducting ground states reveals a similarity of the fullerenes to the high T_c's or the Bechgaard salts [7] and magnetic correlations will have to be included in each theory of superconductivity in C_{60} compounds.

4. References

[1] P. Allemand, K. Khemani, A. Koch, F. Wudl, K. Holczer, S. Donovan, G. Grüner, J. Thompson: Science **253**, 301 (1991)

[2] H. Klos et.al., to be published

[3] P. Allemand, G. Srdanov, K. Koch, K. Khemani, F. Wudl: J. Am. Chem. Soc. **113**, 2780 (1991);
U. Becker, G. Denninger, V. Dyakonov, B. Gotschy, H. Klos, G. Rösler, A. Hirsch, H. Winter: Europhys. Letts. **21**, 267 (1993)

[5] U. Becker et.al., to be published

[6] K. Tanaka, A. Zakhidov, K. Yoshizawa, K. Okahara, T. Yamabe, K. Yakushi K. Kikuchi, S. Suzuki, I. Ikemoto, Y. Achiba: Solid State Com. **85**, 69 (1993)

[7] R. Haddon, L. Brus, K. Raghavachari: Chem. Phys. Letts. **125**, 459 (1986)

[8] D. Jérome: Springer Series in Solid State Science **Vol. 90**, 113 (1990)

Pulsed EPR on the Photoexcited Triplet State of C_{60}

A. Grupp, M. Bennati, and M. Mehring

2. Physikalisches Institut, Universität Stuttgart,
Pfaffenwaldring 57, D-70569 Stuttgart, Germany

Abstract. We report on the electronic structure of photoexcited triplet states of C_{60} molecules in dilute solutions. The analysis of the temperature dependence and kinetics of EPR spectra clearly indicates the existence of energetically close lying triplet states with different symmetry. Zero field splitting parameters |D| between 50 G and 123.5 G are observed. The local spin density distribution on the molecule is estimated from ^{13}C hyperfine coupling constants.

Introduction

Photoexcitation of isolated C_{60} molecules results in the population of the triplet state via inter-system-crossing (ISC) with a quantum yield Φ_T close to unity [1]. As the electron affinity of C_{60} in its triplet state is rather high, this state plays the dominant role in photoreactions of C_{60}, e.g. photoexcited electron transfer reactions [2]. Analysis of the fine structure and hyperfine structure provides direct information about the electronic spin density distribution.

Triplet EPR spectra of C_{60} in glassy matrices have been observed by several authors [3-8]. The fine structure parameters D and E are reported to be 'similar' to those determined the first time by Wasielewski et al.[3], but all spectra - recorded with various techniques and after optical excitation with different wavelengths and intensities - are obviously different from each other. The question is whether these differences result from experimental conditions, or due to intrinsic properties of C_{60}. From theoretical calculations a series of triplet states with different symmetry and small gap between the states is expected [9], i.e. experimental spectra could display the superposition of different triplet states.

Experimental

Solutions of C_{60} in toluene ($c=5*10^{-4}$ mol/l) and benzonitrile ($c=1*10^{-4}$ mol/l) were degassed by several freeze-thaw cycles and sealed off on a high vacuum line. The samples were irradiated at 337 nm by a nitrogen laser (0.5 mJ/pulse) in the temperature range between 5 K and 320 K. Two-pulse echo and free induction decay (FID) experiments were performed using a home- built X-band ESE spectrometer with a slotted tube resonator. Typical micro- wave excitation pulses were 16 ns and 32 ns at a power of 1 kW. EPR spectra were

Springer Series in Solid-State Sciences, Volume 117
Electronic Properties of Fullerenes Eds.: H. Kuzmany, J. Fink, M. Mehring, and S. Roth

recorded either by integrating the echo intensity with a boxcar integrator while sweeping the magnetic field, or by Fourier transformation of the FID signal.

Results

We start with the low temperature regime. Besides minor details, which will be discussed later, spectra of $^3C_{60}$ in frozen toluene and benzonitrile are identical. Spectra were recorded with respect to variation of the temperature and of the delay time τ between the laser pulse and the first microwave pulse. At 7 K a triplet powder spectrum with absorptive low field part and emissive high field part is observed (fig. 1 top).

This means that the triplet eigenstates are populated very selectively, resulting in a high spin polarization. With increasing delay time τ the shape of the spectrum changes significantly. At long delays ($\tau \geq 1$ ms) the spectra can be theoretically calculated with a single set of finestructure parameters D = 123.5 G (0.0115 cm^{-1}) and E = 0.5 G ($4.67*10^{-5}$ cm^{-1}), but at zero delay the spectrum consists of a superposition with another, different triplet state with D = 93.5 G and E = 1 G. Obviously the two components have a different triplet lifetime. Indeed, careful analysis of the triplet decay kinetic recorded at

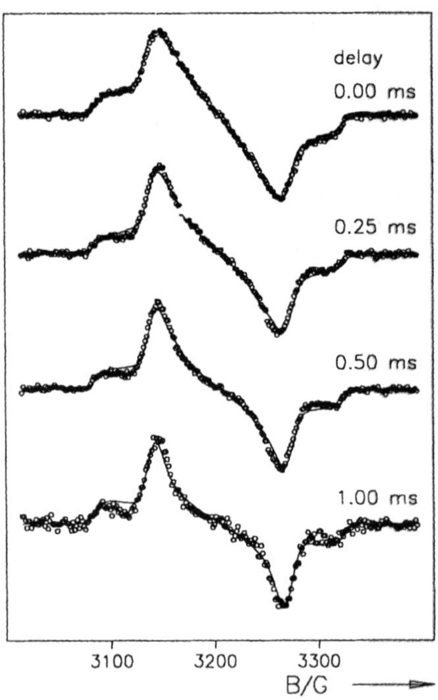

Fig. 1. EPR powder spectra at 7 K (circles) and theoretical spectra (solid lines) of $^3C_{60}$ at various delays after the laser pulse.

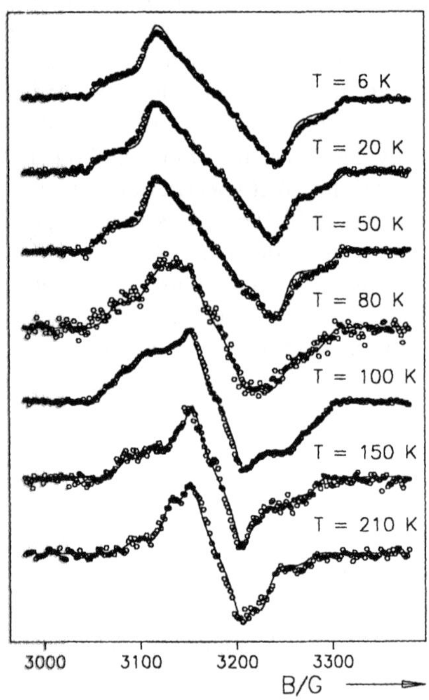

Fig. 2. Temperature dependence of the EPR spectra at zero delay (circles) and theoretical spectra (solid lines).

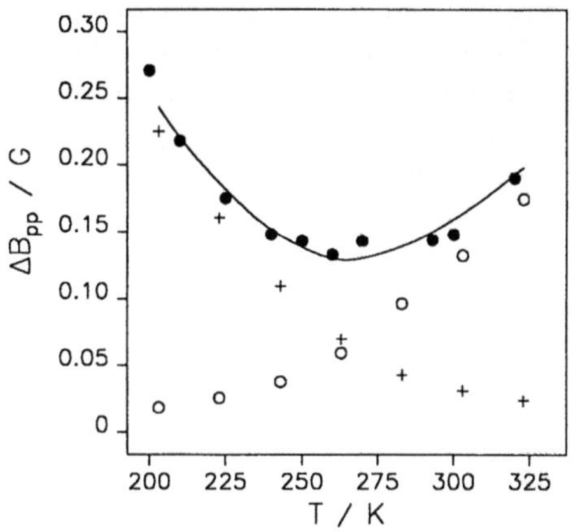

Fig. 3. EPR peak-to-peak linewidth (full cirles), motional narrowing contribution (+), spin-rotation contribution (o) and the sum of both (solid line)

various magnetic field positions within the powder spectrum yields a biexponential decay rate, except at the edges of the spectrum, where only one triplet contributes to the total spectrum. More details will be discussed elsewhere [10].

With increasing temperature the contribution of the second triplet decreases but above 50 K a third component (D $\simeq$ 50 G, E $\simeq$ 0 G) arises thermally activated with an activation energy of about 10 meV. Moreover a general narrowing of the spectra due to librational motion of the C_{60} molecules is observed (fig. 2). All these measurements were performed in benzonitrile because of its high melting point of 260 K.

So far we reported on $^3C_{60}$ in frozen solutions. In the following we present some results of measurements in liquid solution of C_{60} in toluene. Above 200 K a FID signal is observed. The Fourier transform of it results in a single narrow line with a g-factor of 2.0014 in agreement with other authors [5,7,11]. From our measurements of the kinetic of this signal and electron transfer reactions we conclude, that this signal originates from C_{60} in the triplet state. Details are presented in another contribution to this volume [12]. We now focus our interest on the temperature dependence of the linewidth (fig. 3).

As expected the fast rotation of the molecules leads to a Lorentzian line with decreasing linewidth with increasing temperature. In the case of fast rotation of triplet spins with $E^2 \ll D^2$ the longitudinal and transversal relaxation rate is given by

$$1/T_1 = 1/T_2 = 2/3\ D^2\ \tau_c$$

with the rotational correlation time

$$\tau_c = (4\pi\eta r^3)/(3k_bT)$$

where η is the viscosity and r is the molecular radius. However, at higher temperatures the linewidth is observed to increase again. This can be explained by spin- rotation interaction which contributes to the relaxation rate as

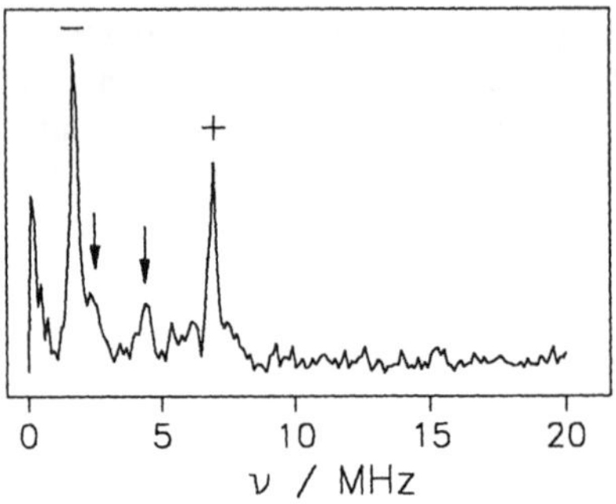

Fig. 4. Fourier transform of the echo modulation at 7 K. Sum (+) and difference peak (-) of one hyperfine interaction component (arrows) are marked.

$$1/T_1 = (C^2 I^2 k_b T)/(4\pi r^3 \hbar^2 \eta)$$

where C is the spin-rotation coupling constant and I is the moment of inertia of the C_{60} molecule. The total relaxation rate is given by the sum of these two contributions. A least squares fit procedure yields the best fit to the temperature dependence of the linewidth ΔB_{pp} with D $\simeq$ 62 G and C $\simeq$ 12.5 MHz. This result is in good agreement with the observation of D $\simeq$ 50 G in frozen benzonitrile.

Fine structure data can be used to evaluate the spin density distribution by a dipol-dipol interaction calculation. Much more direct access to spin densities however is given by the hyperfine interaction of ^{13}C nuclei with the electron spin. Indeed, at very low temperatures we were able to detect a two-pulse echo modulation pattern due to ^{13}C hyperfine interaction. Generally the modulation function of a two pulse experiment contains all hyperfine transition frequencies as well as the sum and difference of them. Fourier transformation of the modulation pattern results in the hyperfine spectrum (fig. 4), which has two prominent peaks at the sum and difference frequencies and two corresponding smaller peaks with a hyperfine splitting of 1.7 MHz. This hyperfine interaction is attributed to a spin density at one carbon position of $\rho \simeq 1/20$ to $1/30$, i.e. the total spin density is distributed over 20 to 30 carbon atoms of the sphere. More details of the analysis will be presented in a forthcoming paper [10].

Discussion

The appearance of different triplet states seems to be an intrinsic property of the C_{60} molecule, as the same states are observed in toluene and in benzonitrile matrix. The narrowing of the spectra due to molecular motion is different in the two solvents, because toluene has a glass transition temperature at about 117 K, whereas benzonitrile has a melting point at 260 K. Theoretical calculations [9] predict a series of Jahn-Teller distorted triplet states with different symmetries (D_{5d}, D_{2h}, D_{3d}) and slightly different energy levels relative to each other (0 meV, 49 meV, 175 meV). This is at least in qualitative agreement with our observation, however does not explain the temperature dependence of the D = 93.5 G component. Furthermore, our high temperature linewidth analysis supports the model of additional triplets with smaller D values. The individual population of the different triplet states after singlet excitation may be a complicated process depending on the symmetry of the excited singlet state, the intersystemcrossing rates and the internal conversion rates within the singlet and triplet level manifold. Thus it becomes plausible, that measurements with different excitation wavelength and different time resolution detect different superpositions of triplet states. From our investigations we conclude the D = 123.5 G state to be the lowest triplet state. For this state a simple dipol-diplol interaction model results in two triplet electron spins 6.1 Å apart from each other. This distance is close to the molecular diameter of 7.1 Å, i.e. the fine structure parameter is in accordance with a symmetric spin density distribution over opposite pole caps of the molecule. Due to the ^{13}C hyperfine interaction the average extension on one cap is on the order of 10 to 15 carbon atoms. From our limited spectral resolution of the hyperfine spectrum it is not possible to determine an accurate spin density distribution and the symmetry of the state. More detailed investigations with ^{13}C enriched material are in progress.

Acknowledgements: Financial support by the Deutsche Forschungsgemeinschaft (SFB 329) and the Fonds der Chemischen Industrie is gratefully acknowledged.

References

[1] J.W. Arbogast, A.P. Darmanyan, C.S.Foote, Y.Rubin, F.N. Diederich, M.M. Alvarez, S.J. Anz, R.L. Whetten, J. Phys. Chem. (1991), 95, 11

[2] J.W. Arbogast, C.S. Foote, M. Kao, J. Am. Chem. Soc. (1992), 114, 2277

[3] M.R. Wasielewski, M.P. O'Neil, K.R. Lykke, M.J. Pellin, D.M. Gruen, J. Am. Chem. Soc., (1991), 113, 2774

[4] P.A. Lane, L.S. Swanson, Q.-X. Ni, J. Shinar, L.P. Engel, T.J. Barton, L. Jones, Phys. Rev. Lett. (1992), 68, 887

[5] G.L. Closs, P. Gautam, D. Zhang, P.J. Krusic, S.A. Hill, E. Wasserman, J. Phys. Chem. (1992), 96, 5228

[6] M. Terazima, N. Hirota, H. Shinohara, Y. Saito, Chem. Phys. Lett. (1992), 195, 333

[7] H. Levanon, V. Meiklyar, A. Michaeli, S. Michaeli, A. Regev, J. Phys. Chem. (1992), 96, 6128

[8] M. Bennati, A. Grupp, M. Mehring, K.P. Dinse, J. Fink, Chem. Phys. Lett. (1992), 200, 440

[9] P.R. Surjan, K. Nemeth, L. Udvardi, this volume

[10] A. Grupp, M. Bennati, M. Mehring, to be published

[11] M. Rübsam, K.-P. Dinse, M. Plüschau, J. Fink, W. Krätschmer, K. Fostiropoulos, C. Taliani, J. Am. Chem. Soc. (1992), 114, 10059

[12] M. Bennati, A. Grupp, P. Bäuerle, M. Mehring, this volume

Electron Transfer from End-Capped Quaterthiophene to Photoexcited C_{60}: A Pulsed EPR Investigation

M. Bennati[1], *A. Grupp*[1], *P. Bäuerle*[2], *and M. Mehring*[1]

[1]2. Physikalisches Institut, Universität Stuttgart, Pfaffenwaldring 57, D-70569 Stuttgart, Germany

[2]Institut für Organische Chemie, Universität Stuttgart, D-70569 Stuttgart, Germany

Abstract. We report on a pulsed EPR investigation of the photoinduced electron transfer from end-capped quaterthiophene to the photoexcited triplet state of C_{60} in liquid solutions. Fourier transform of the free induction decays results in spectra of the transient triplet state and of the transient C_{60} monoradical anion. Electron transfer rates were deduced from the observed triplet quenching rates at different temperatures. The results give evidence for the formation of an intermediate donor-acceptor exciplex in the charge transfer mechanism.

Introduction

The electron acceptor properties of C_{60} have received a lot of attention because of the high electron affinity of C_{60} due to its particular molecular structure. Cyclic voltammetry studies report on reversible electrochemical reduction of C_{60} with up to six electrons [1]. First optical investigations on photoexcited charge transfer processes of C_{60} with electron donor molecules show that the electron affinity of C_{60} increases strongly in its photoexcited triplet state which is generated with a quantum yield of nearly 1 [2]. Photoinduced electron transfer from strong electron donors to $^3C_{60}$ in polar solvents proceeds with rates which are in the diffusion controlled limit [3,4].

Although several EPR investigations have characterized the photoinduced triplet state [5-8] less is known about the paramagnetic properties of transient photoproducts generated via electron transfer from electron donors to $^3C_{60}$. In this work we report on pulsed EPR studies on photoexcited C_{60} in solutions with end-capped quaterthiophene (EC4T) added as strong electron donor. With a time resolution of nanoseconds after pulsed laser excitation we are able to identify the transient triplet and radical anion EPR signals of C_{60}. From an analysis of their build-up and decay kinetics at different temperatures we are led to a charge transfer mechanism model, which is discussed in the following.

Experimental

Solutions of C_{60} in toluene ($c=1*10^{-3}$mol/l) with different concentrations of the electron donor EC4T were degassed carefully by several freeze-thaw cycles and sealed off on a high vacuum line. The samples were irradiated at 337 nm by

Springer Series in Solid-State Sciences, Volume 117
Electronic Properties of Fullerenes Eds.: H. Kuzmany, J. Fink, M. Mehring, and S. Roth

a pulsed N_2-laser ($\simeq$ 0.5 mJ/pulse) in a temperature range between 200 - 300 K. Free induction decay (FID) experiments were performed using a home-built X-band ESE spectrometer with a slotted tube resonator. FIDs were recorded after one $\pi/2$ microwave excitation pulse of 16 ns duration at a power of 1kW. We have determined the build-up and decay kinetic of the signals by varying the time delay τ between laser and microwave excitation. EPR spectra were obtained after complex Fourier transform of the FID using quadrature detection.

Results

Photoexcitation of pure C_{60} in toluene at room temperature leads to a single FID component. The Fourier transform of the FID gives a spectrum with a narrow EPR line (ΔB_{pp}=0.144 G) as obtained previously [6,7,9]. We determined a g-factor of 2.0014 by comparison with DPPH radicals measured simultaneously.

We have measured the build-up and decay kinetic of this transient by evaluating the FID height after the microwave pulse in dependence on the delay time τ. The resulting curve at room temperature is shown in fig.1a. The assignment of the transient EPR signal as due to a triplet state or to a photoproduct has led to controversies in the literature. Anticipating the results of the following charge transfer experiments we are able to identify this transient as $^3C_{60}$. The build-up and decay kinetic can be explained considering spin-lattice relaxation and additional triplet decay and is given by a multiexponential relaxation with three characteristic time constants τ_1, τ_2, τ_3. The fit parameters and their temperature dependence are listed in table 1.

When adding the electron donor EC4T to the C_{60} solutions we observe two components in the FID after laser excitation. The Fourier transform spectrum (fig.2) shows two lines at g=2.0014 and g=1.9999. We attribute the line at

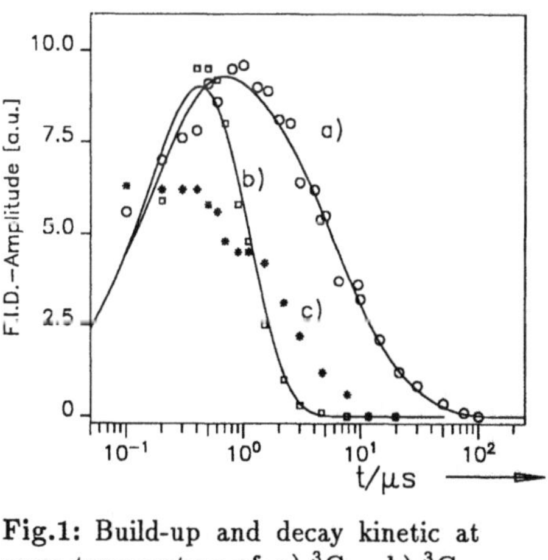

Fig.1: Build-up and decay kinetic at room temperature of: a) $^3C_{60}$, b) $^3C_{60}$ in C_{60}/EC4T, c) C_{60}-radical anion

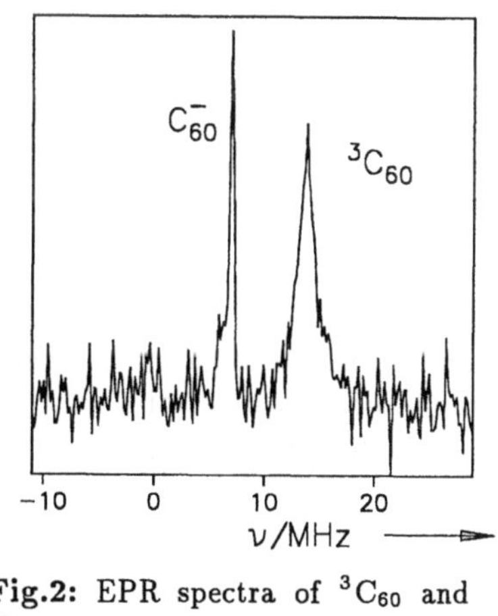

Fig.2: EPR spectra of $^3C_{60}$ and C_{60}^-

Table 1: Simulation parameters for $^3C_{60}$ decay kinetic

T [K]	τ_1 [μs]	τ_2 [μs]	τ_3 [μs]
293	0.2	5.3	20
270	0.22	2.8	27
250	0.27	4.5	39
225	0.3	6.8	44
200	0.3	10.5	56

Table 2: Quenching rates k_q and diffusion rates k_D in dependence on T

T [K]	k_q 10^{-9} $[M^{-1}s^{-1}]$	k_D 10^{-9} $[M^{-1}s^{-1}]$
293	0.31	18
270	0.47	13
250	0.62	7.6
225	0.82	4.3
200	0.27	2.2

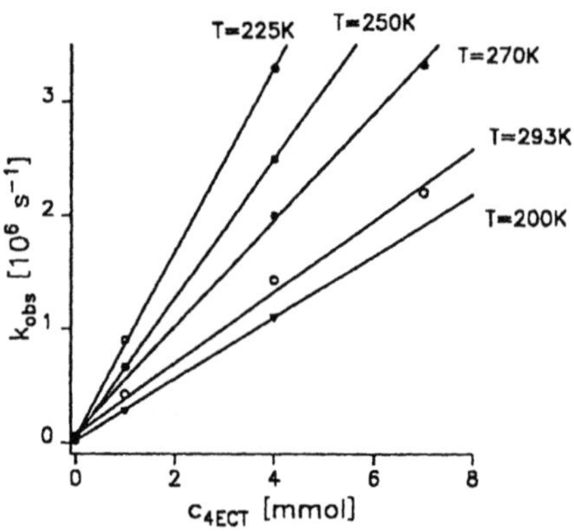

Fig.3: Concentration dependence of the observed triplet decay rates k_{obs} in C_{60}/EC4T at different temperatures

g=2.0014 to the triplet state. The second narrow line (ΔB_{pp}=0.103 G) has a g-factor which corresponds to that found in electrochemically prepared C_{60} monoradical anions [10]. The build-up and decay kinetic of the two different transients ($^3C_{60}$ and C_{60}^-) is shown in figure 1b and 1c and is compared with the kinetic of pure $^3C_{60}$. The addition of EC4T leads to a quenching of the triplet lifetime. When plotting the observed triplet decay rates k_{obs} versus the concentration of added EC4T we obtained for every temperature a linear correlation (fig.3). The quenching follows the relation (Stern-Volmer):

$$k_{obs} = k_0 + k_q[EC4T]$$

where k_0 is the triplet decay rate without quencher. From the slopes of the straight lines we determined the quenching rate k_q. The quenching and electron transfer rates k_q and k_{et} are correlated by the following relation:

$$1/k_q = 1/k_D + 1/(K_{dif} * k_{et})$$

We estimated the bimolecular diffusion rate k_D for the C_{60}/EC4T system in toluene at the relevant temperatures using the expression derived by Smoluchowski. k_D and k_q values are listed in table 2.

Discussion

Comparing the k_D and k_q values one observes that k_D is orders of magnitude larger than k_q leading to $1/k_q \simeq 1/(K_{dif} * k_{et})$. By plotting $\ln k_q$ versus 1000/T we determined a negative activation quenching energy of $E_a \simeq$ -10 kJ/mol. As the electron transfer rate k_{et} is expected to be thermally activated and the diffusion equilibrium constant K_{dif} is nearly temperature independent the correct interpretation of the results requires the consideration of a second quenching mechanism. Many fluorescence quenching studies show that in nonpolar solvents like toluene ($\epsilon = 2.7$) the formation of an exciplex intermediate plays an important role for the charge separation mechanism [11]. In this path the equili-

$$D + A^* \underset{k_{-1}}{\overset{k_1}{\rightleftharpoons}} {}^*[{}^+D{-}A^-] \xrightarrow{k_{ip}} D^+ + A^-$$

brium constant $K_{exc} = k_1/k_{-1}$ involves exciplex dissociation which for a bound complex is expected to be strongly temperature dependent. The experimental activation energy for k_q can therefore be separated into two contributions:

$$E_a = \Delta H_{exc} + \Delta H_{ip} + RT$$

For a large negative exciplex formation enthalpy ΔH_{exc} and a small ion solvation enthalpy ($\Delta H_{ip} < -\Delta H_{exp}$) the activation energy becomes negative as we observed here.

Acknowledgements: Financial support by the Deutsche Forschungsgemeinschaft (SFB 329) and the Fonds der Chemischen Industrie is gratefully acknowledged.

References

[1] Q. Xie, E. Perez-Cordero, L. Echegoyen, J. Am. Chem. Soc. (1992), 114, 3978
[2] J.W. Arbogast, A.P. Darmanyan, C.S.Foote, Y.Rubin, F.N. Diederich, M.M. Alvarez, S.J. Anz, R.L. Whetten, J. Phys. Chem. (1991), 95, 11
[3] J.W. Arbogast, C.S. Foote, M. Kao, J. Am. Chem. Soc. (1992), 114, 2277
[4] L. Biczok, H. Linschitz, Chem. Phys. Lett. (1992), 195, 339
[5] M.R. Wasielewski, M.P. O'Neil, K.R. Lykke, M.J. Pellin, D.M. Gruen, J. Am. Chem. Soc., (1991), 113, 2774
[6] G.L. Closs, P. Gautam, D. Zhang, P.J. Krusic, S.A. Hill, E. Wasserman, J. Phys. Chem. (1992), 96, 5228
[7] H. Levanon, V. Meiklyar, A. Michaeli, S. Michaeli, A. Regev, J. Phys. Chem. (1992), 96, 6128
[8] M. Bennati, A. Grupp, M. Mehring, K.P. Dinse, J. Fink, Chem. Phys. Lett. (1992), 200, 440
[9] M. Ruebsam, K.-P. Dinse, M. Plueschau, J. Fink, W. Kraetschmer, K. Fostiropoulos, C. Taliani, J. Am. Chem. Soc. (1992), 114, 10059
[10] D. Dubois, M.J. Jones, K.M. Kadish, J. Am. Chem. Soc. (1992), 114, 6446
[11] A. Weller in M. Gordon and W.R. Ware, The Exciplex, Academic Press New York, 1975

Part VI

Electrical Transport and Superconductivity

Transport Measurements of the Normal State and Superconducting Properties of Fulleride Superconductors

X.-D. Xiang, J.G. Hou, and A. Zettl

Department of Physics, University of California at Berkeley, and Materials Sciences Division, Lawrence Berkeley Laboratory, Berkeley, CA 94720, USA

Abstract. The intrinsic, reproducible temperature dependent functional forms of the resistivity for high quality K and Rb intercalated single crystals of C_{60} are obtained. Important parameters of normal and superconducting states have been obtained through transport and magnetotransport measurements. Extremely sharp transition width ~0.2K and pure 3-D superconducting fluctuation conductivity observed in these crystals have further guaranteed that the measured normal and superconducting properties are intrinsic. These results have put stringent experimental constrains on various models of the superconductivity. The effects of intercalation induced imperfection, such as 0-D fluctuation effect and upper critical field enhancement, are discussed.

1. Introduction

The discovery of fullerene-based superconductors has generated much excitement in the scientific community.[1] In contrast to the 2-dimensional (2-D) features of copper oxides, fulleride superconductors have isotropic 3-D crystal structures. The T_c's of new fulleride superconductors, although still significantly lower than the record of copper oxides, have set a new record (33K) for 3-D superconductors. The fascinating structure of the C_{60} molecule and the unique features of this molecular solid have made these materials very attractive and puzzling to theorists. Both unconventional and conventional mechanisms have been suggested to account for the superconductivity, but these remain controversial. Although initial studies on powder specimens have provided important information, the intrinsic normal and superconducting properties obtained on high quality crystals are crucial for the discussion of the superconductivity mechanisms.

2. Synthesis and Characterization

We grow single crystals of pure C_{60} with a thermal sublimation method.[2] X-ray diffraction studies performed on our crystals are consistent with previous report on the crystal structure.[3] Specific heat measurement on these crystals have revealed a very sharp (~3K-wide) hysteretic (~0.6K) lambda-shaped anomaly at 260K due to the well-known rotational ordering phase transition.[4]

Electrical contacts to the sample with either a Van der Pauw or in-line four-probe geometry were made prior to intercalation. The mounted samples were then sealed together with fresh K or Rb in a high vacuum glass-metal apparatus with electrical

Springer Series in Solid-State Sciences, Volume 117
Electronic Properties of Fullerenes Eds.: H. Kuzmany, J. Fink, M. Mehring, and S. Roth

feed-through wires. Samples were mounted on sapphire substrates and thermally sunk to a copper post to which the thermometer was attached. A repetitive high temperature (~200 °C) vapor-intercalation /anneal cycle process has been used to achieve uniform intercalation.[2]

3. Normal State Transport

The electrical resistivity ρ was measured versus temperature for crystals of K_3C_{60} and Rb_3C_{60}. Crystals from different preparation batches yielded similar results.

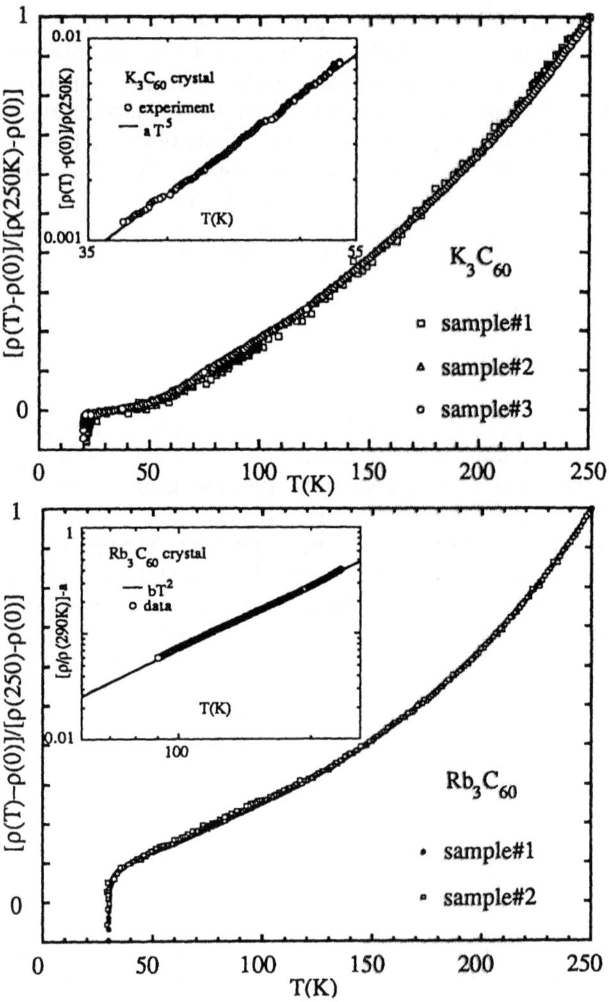

Fig.1. Normalized temperature dependent part of resistivity versus T for K_3C_{60} (a) and Rb_3C_{60} (b). Insets: Log-log plots of normalized resistivities (with extrapolated residual resistivity subtracted) versus temperature. Solid line corresponds to aT^5 for K_3C_{60} (a) and bT^2 for Rb_3C_{60} (b).

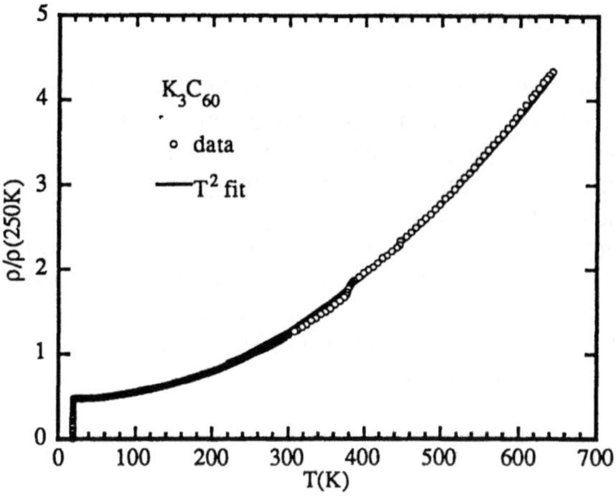

Fig.2. Normalized resistivity versus temperature for KC_{60} measured up to 650K.

Resistivities estimated from direct measurements vary from sample to sample even though their normalized resistivities follow the same functional form with temperature. We believe the values of the high temperature resistivity [$\rho(250)$-$\rho(T\rightarrow 0)$] ~130mΩ/cm and 240 mΩ/cm, for K_3C_{60} and Rb_3C_{60} respectively, obtained from the indirect measurements discussed below, are more intrinsic parameters independent of sample geometry. As shown in Fig.1 (a) and (b), $\rho(T)=\rho-\rho(0)$ varies with temperature in a metal-like fashion for both K_3C_{60} and Rb_3C_{60}. To demonstrate that the measured behaviors are indeed true intrinsic properties, we plotted in Fig. 1 normalized $\rho(T)$ for multiple samples of both compounds. The reproducibility of $\rho(T)$ also suggest that the superconducting phase is a line phase rather than a solid solution with variable dopant level. Roughly speaking, $\rho(T)$ varies with temperature in an AT^2 form for both compounds above 80K as shown in the inset of Fig.1 (b) for Rb_3C_{60}. This temperature dependence extends up to 650K for K_3C_{60} as shown in Fig. 2 with a coefficient A of 2.0 x 10^{-9} $\Omega cm/K^2$. In contrast to the saturation behavior of $\rho(T)$ in A-15 compounds, which has been attributed to the approximation of mean free path to the interatomic spacing [5], we did not observe any sign of saturation in $\rho(T)$ for K_3C_{60} up to 650K. At about 400K, we observed a small drop in the resistivity (see Fig. 2). We attribute this discontinuity to the effect of a phase separation from a small percentage of KC_{60} existing at high temperatures (if sample is under global stoichiometry of K_3C_{60}) into α–C_{60} and K_3C_{60} .[6] At low temperatures, a remarkable T^5 term is observed for K_3C_{60} as shown in the inset of Fig,1 (a). T^5 dependence is expected from the Bloch-Gruneisen expression at low temperatures for an electron-phonon scattering mechanism.[7] This observation strongly suggest that the normal state scattering is dominated by electron-phonon scattering at low temperatures. T^2 dependence at high temperatures may be speculated as a signature of electron-electron scattering. We note that the coefficient A is about one order of magnitude smaller than that of TiS_2 in which the electron-electron scattering has been shown to dominate the normal state transport.[8] However, a careful examination of the effect of thermal expansion and a realistic estimation of electron-electron scattering contribution are needed to draw a definitive conclusion.

4. Granularity and Fluctuation Conductivity

Just above T_c, thermodynamic fluctuations produce small transient regions of the superconducting state, giving rise to an anomalous increase in the normal-state conductivity known as paraconductivity. The paraconductivity effect is a critical phenomenon and is strongly dependent on the dimensionality of the system. Measuring this effect allows us to determine the dimensionality of the superconductivity and the extent of granularity of the samples. The latter is especially important since the intrinsic properties are often severely distorted by granularity. In addition, these measurements allow an indirect determination of the normal state resistivity.

As first derived 25 years ago by Aslamazov and Larkin (AL),[9] the excess conductivity σ' is directly related to the dimensionality [10]

$$\sigma'_{AL} = \sigma_{exc}\, t^{-(4-D)/2}, \qquad (1)$$

where D is the dimensionality of the specimen and t is the reduced temperature defined by $t = (T-T_c)/T_c$. Dimensionality, therefore, can be easily determined through measurement of paraconductivity.

For a 3-dimensional superconductor with grain size much lager than superconducting coherence length ξ, we expect to observe 3-D paraconductivity. Indeed, our high quality samples that have extremely sharp superconducting transition of ~0.2K, show 3-D paraconductivity.[11] For a granular superconductor with a grain size of L, the functional form of the excess conductivity will crosses over from 3-D behavior to 0-D behavior when the Ginzburg-Landau coherence length $\xi(T)$ exceeds L/3 as T approaches T_c.[12] The absence of a 0-D crossover down to $t \sim 0.0005$ in our in our best samples gives a lower limit for a domain size of about 0.6 μm. This dimension is at least 100 times larger than the measured $\xi(0)$'s [13,14] and, therefore, guarantees that physical parameters measured on these samples reflect intrinsic properties. However, imperfect intercalation process do introduce granularity in some samples. The granularity manifests itself in those samples as a broader transition width (> 0.3K) and most clearly 0-D paraconductivity, even though the overall temperature dependence of normal resistivity has not been affected. The observation of 0-D paraconductivity in those samples indicates the existence of grain size in the order of ~ 100 Å (this may not be observable in an x-ray study), which could severely affect various measured parameters (see below).

The measured normalized fluctuation conductivity (σ'/σ_n) is independent of the effective sample dimensions and is reproducible in different samples. We can use this quantity to determine the normal state conductivity σ_n since the excess conductivity prefactor σ_{exc} in 3-D is related to the independently determined coherence length $\xi(0)$ by

$$\sigma_{exc} = \frac{\pi e^2}{16 h \xi(0)}. \qquad (2)$$

Using Eq. (2) together with the experimentally determined values $(\sigma_{exc}/\sigma_n)_K \sim 2 \times 10^3$, $(\sigma_{exc}/\sigma_n)_{Rb} \sim 7.3 \times 10^{-3}$, $\xi(0)_K \sim 45$ Å ,[13] and $\xi(0)_{Rb} \sim 23$ Å,[14] we obtain the zero-temperature residual normal state resistivities $\rho(0)_K \sim 0.12$ mΩ-cm and $\rho(0)_{Rb} \sim 0.22$ mΩ-cm where the subscripts K and Rb represent K_3C_{60} and

Rb_3C_{60}, respectively. The estimated uncertainties in ρ(0) mainly resulting from different extrapolations of the normal state resistivity are about ± 20%.

5. Magnetotransport measurements [13,14]

Fig. 3 (a) and (b) shows the temperature dependence of the resistivity of K_3C_{60} and Rb_3C_{60} crystals, respectively, near T_c for different applied magnetic fields. Samples used in these measurements have extremely sharp transition width and show 3-D paraconductivity. For K_3C_{60}, the transition broadens slightly under high magnetic

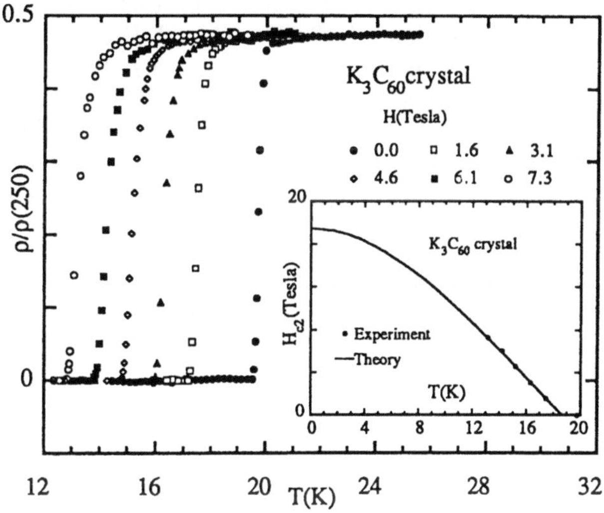

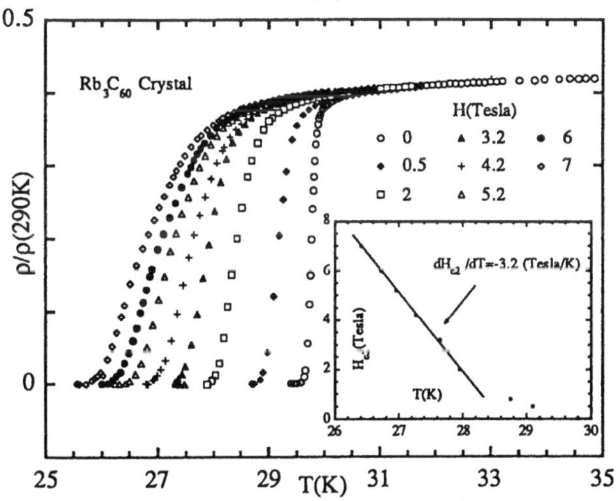

Fig.3. Normalized magneto-resistivities as a function of T for K_3C_{60} (a) and Rb_3C_{60} (b). Insets: H_{c2} versus T. Solid lines are a theoretical fit for K_3C_{60} (a) and a linear fit for Rb_3C_{60} (b).

field and the sharpness of the transition allows a clear identification of H_{c2} (T). The inset of Fig.2 (a) shows the temperature dependence of H_{c2} where we have used the transition midpoints to define the T-H critical line. Other methods of defining the T-H line produce similar results. A linear fit to H_{c2} (T) yields dH_{c2}/dT =-1.34 T/K. Using the WHH formula [15], we obtain $H_{c2}(0)$ =17.2 Tesla.

For Rb_3C_{60}, application of the field causes an apparent broadening of the resistive transition in addition to a depression of T_c. The field-induced broadening of the resistive transition near the onset of transition can be understand as field-enhanced fluctuation conductivity.[14] A fit to the theoretically predicted expression for this effect determines the T_c (H). The inset of Fig. 2 (b) shows the temperature dependence of H_{c2} for Rb_3C_{60} and a linear fit yields dH_{c2}/dT =-3.2 T/K.

There are considerable differences among existing results of H_{c2} measurements, especially for K_3C_{60} with dH_{c2}/dT ranging from -1.34 T/K to -3.7 T/K.[16] We believe that the granularity in various samples is the cause of this diversity. It is well-known that granularity may seriously alter the measured properties of superconductors, in particular, the measurement of H_{c2}. Usually ξ is determined through the measurement of H_{c2} and the standard relation $H_{c2}(0) = \Phi_0/2\pi\xi^2$. However, in a granular superconductor, if the grain size is comparable to or smaller than coherence length, measured upper critical field will depend on grain size L.[17] As a control study, we synthesized K_3C_{60} crystals with different degrees of imperfection measured by the transition width and dimensionality of fluctuation conductivity. We found a systematic trend in that the samples having 0-D fluctuation conductivity behavior (which indicates that L is comparable to or smaller than ξ) and a wider transition width (~0.33 K to ~1K) have higher dH_{c2}/dT ranging from -1.7 T/K to 3.4 T/K.

6. Discussions

It is instructive to calculate various important normal and superconducting parameters with a simple model. First, we use the relation $\sigma = 2e^2 k_f^2 \ell/3\pi h$, where ℓ is electron mean free path. If we determine the Fermi wave vector k_f and electron density n within a free electron model $k_f = (3\pi^2 n)^{1/3}$, $n = 4.2 \times 10^{21}$ / cm^3, we obtain ℓ = 41Å and 22Å for K_3C_{60} and Rb_3C_{60}, respectively. Using the value from the band calculation [18] $v_f = 1.8 \times 10^7$cm/s for K_3C_{60} and 1.5×10^7cm/s for Rb_3C_{60}, we estimate $\tau = \ell/v_f$ of 2.3×10^{-14} s and 1.5×10^{-14} s, for K_3C_{60} and Rb_3C_{60}, respectively. We note that the product $k_f \ell$ is about 20 for K_3C_{60} and about 10 for Rb_3C_6. Therefore, the Ioffe-Regel condition is well satisfied in these compounds. The plasma frequency ω_p, effective mass m^* , Fermi energy ε_f, and London penetration depth λ_L can be estimated through the relations : $\rho = 4\pi/\tau\omega_p^2$, $\omega_p^2 = ne^2/m^*$, $\varepsilon_f = v_f^2/2m^*$, and $\lambda_L = c/\omega_p$. To summarize the analysis, we list the superconducting and normal state parameters of K_3C_{60} and Rb_3C_{60} from our experimental measurements and simple calculations in Table 1.

Table 1. Superconducting state and normal state parameters of Rb_3C_{60} and K_3C_{60} single crystals

Parameters	Rb_3C_{60}	K_3C_{60}
T_c	30 K	19.7 K
dH_{c2}/dT	-3.2 Tesla/K	-1.34 Tesla/K
$H_{c2}(0)$	62 Tesla	17.5 Tesla
$\xi(0)$	23 Å	45 Å
$\rho(T\rightarrow 0)$	0.22 mΩcm	0.12 mΩcm
ℓ	22 Å	41 Å
τ	$1.5\text{x}10^{-14}$ sec.	$2.3\text{x}10^{-14}$sec.
$\hbar\omega_p$	1.23 eV	1.33 eV
λ_L	1600 Å	1480 Å
m^*	$3.7m_e$	$3.2m_e$
ε_f	0.29 eV	0.23 eV

Acknowledgements

We would like to acknowledge W. A. Vareka, and G. Briceño for technical assistance, and Vincent H. Crespi and Marvin L. Cohen for theoretical contributions to this work. This research was supported by the Office of Energy Research, Office of Basic Energy Sciences, Materials Sciences Division of U.S. Department of Energy under contract DE-AC03-76SF00098.

References

1. A. F. Hebard, M. J. Rosseinsky, R. C. Haddon, D. W. Murphy, S. H. Glarum, T. T. M. Palstra, A. P. Ramirez and A. R. Kortan, Nature **350**, 600 (1991).
2. X.-D. Xiang, J. G. Hou, G. Briceño, W. A. Vareka, R. Mostovoy, A. Zettl, V. H. Crespi, and M. L. Cohen, Science **256**, 1190 (1992).
3. P. A. Heiney,J. E. Fischer, A. R. McGhie, W. J. Romanow, A. M. Denenstein, J. P. McCauley, A. B. Smith, D. E. Cox, Phys. Rev. Lett. **66**, 2911 (1991).
4. M. Chung, Yiqin Wang, J. W. Brill, X.-D. Xiang, R. Mostovoy, J. G. Hou, and A. Zettl, Phys. Rev. **B23**, 13831 (1992).
5. P. B. Allen, W. E. Pickett, K.-M. Ho, and M. L. Cohen, Phys. Rev. Lett. **40**, 1532 (1978).
6. D. M. Poirier and J. H. Weaver, (unpublished).
7. J. M. Ziman, Principles of the theory of solids, 2nd Ed. Cambridge: University Press, P225 (1972).
8. Thompson R. S., Phys. Rev. **B1**, 327(1970); Physica **55**, 296 (1971).
9. Aslamasov L. G. and Larkin A. I. , Phys. Letters **26A**, 238(1968).
10. Skocpal W. J. and Tinkham M. , Rep. Prog. Phys. **38**, 1049 (1975)
11. X.-D. Xiang, J. H. Hou, A. Zettl, V. H. Crespi and M. L. Cohen, Nature **361**, 54 (1993).

12. Schmidt H., Z. Phys. **216**, 336 (1968).
13. J. G. Hou, V. H. Crespi, X.-D. Xiang, W. A. Vareka, G. Briceño, A. Zettl, and M. L. Cohen, (unpublished).
14 J. G. Hou, X.-D. Xiang, A. Zettl, V. H. Crespi and and M.L. Cohen, (unpublished).
15. N. R. Werthamer, E. Helfand, and P. C. Hohenberg, Phys. Rev. **147**, 295 (1966).
16. .S. Foner, E. J. McNiff, Jr., D. Heiman, S.-M. Huang and R. B. Kaner,Phys. Rev. **B46**, 14936 (1992).
17. G. Deutcher, O. Entin-Wohlman, and Y. Shapira, Phys. Rev. **B22**, 4264 (1980).
18. Steven C. Erwin and Warren E. Pickett, Science **254**, 842 (1991).

Magnetic Investigations of Superconducting Fullerenes

M. Baenitz[1], *E. Straube*[1], *M. Kraus*[1], *M. Kanowski*[1], *H. Werner*[2], *R. Schlögl*[2], *and K. Lüders*[1]

[1]Institut für Experimentalphysik, Freie Universität Berlin, Arnimallee 14, D-14195 Berlin, Germany
[2]Institut für Anorganische Chemie, Universität Frankfurt/Main, Niederurseler Hang, D-60439 Frankfurt, Germany

Abstract. The superconducting properties of several doped fullerenes were investigated using dc and ac susceptibility measurements. The superconducting volume fractions, the lower and upper critical magnetic fields were determined. Low field analyses of the imaginary part of the ac susceptibility show several dissipation peaks allowing the interpretation in terms of inter- and intragrain flux motion and, applying Beans's critical state model, an estimation of j_c .

1. Introduction

The discovery that doped fullerenes show superconductivity with surprisingly high transition temperatures [1] has stimulated the research activities considerably within this new class of materials. The main results concerning superconducting properties are colleted in a series of review articles [2-5] . Doping of C_{60} with alkali metals leads to superconducting compounds with transition temperatures of $T_c = 17.5 - 20$ K for K_3C_{60}, 23 - 30 K for Rb_3C_{60} and up to 33 K for $RbCs_2C_{60}$. Alkali earth doped C_{60} showed lower T_c values: 7 K and 8.3 K for Ba_6C_{60} [6, 7] and Ca_5C_{60} [8], respectively.

Many of these investigations were performed by means of dc susceptibility measurements yielding some fundamental characteristics of the superconducting specimen. They allow the determination of the Meissner effect and in the case of granular samples it is possible to obtain the temperature dependence of the magnetic penetration depth $\lambda(T)$ by using the zero field cooling ZFC curve [9, 10]. However, more important from a technological point of view are ac susceptibility measurements. This technique has been used frequently for the investigation of high-T_c superconductors [11] especially to deduce critical parameters of the intrinsic and coupling components of these granular superconductors.

In this contribution the application of both types of measurements to several superconducting C_{60} compounds is reported.

2. Sample Preparation

Using the method of Krätschmer [12], C_{60} molecules were synthesized from the soot of evaporated graphite. The samples used in this investigation are listed in

Springer Series in Solid-State Sciences, Volume 117
Electronic Properties of Fullerenes Eds.: H. Kuzmany, J. Fink, M. Mehring, and S. Roth

Table 1. Critical and characterizing quantities of superconducting fullerenes. The index j indicates the intergrain (Josephson coupled) values.

Compound	1. Rb_3C_{60}/C_{70} + Tl	2. Rb_3C_{60}	3. K_3C_{60}
T_c (K)	27.5	29.5	17.8
Particle size (μm)	0.1-1	10 ± 5	10 ± 5
Volume fraction (%)	7 (19)	77 (51)	109 (73)
$B_{c1}^{j}(0)\,(G)$	—	60 ± 10	6 ± 1
$B_{c1}(0)\,(G)$	—	200 ± 50	—
$j_c^{j}\left(A/cm^2\right)$	—	10^4 (17 K)	10^4 (4 K)
$j_c\left(A/cm^2\right)$	—	10^5 - 10^6 (28 K)	—
$\left(\partial B_{c2}/\partial T\right)_{T_c}(T/K)$	-2.1 ± 0.3	-2.3 ± 0.3	-1.4 ± 0.3
$B_{c2}(0)\,(T)$	37 ± 6	44 ± 3	17 ± 4
ξ_{GL} (Å)	30 ± 3	27 ± 1	44 ± 1

Table 1. The first Rb_3C_{60} sample was prepared using C_{60} powder containing approximately 10 % C_{70} by annealing it in close contact with the respective amount of the alloy $RbTl_{1.5}$. As deduced by low temperature ac measurements, X-ray and ^{13}C NMR investigations the alloy decomposes during the preparation procedure and the resulting sample consists of a mixture of two major phases: Rb_3C_{60} and metallic Tl [13].

The other samples were prepared from pure C_{60} as starting material. Solid C_{60} was carefully degassed in a high vacuum chamber for 6 hours, applying a temperature ramp from room temperature to 300 °C. After this procedure C_{60} was exclusively handled under pure argon in a glove box with an oxygen and water content < 1 ppm. Rb_3C_{60} was prepared according to the method of McCauley et al. [14], however, with slight modifications. One equivalent of C_{60} was sealed with excess Rb in a quartz ampoule ($\cong 10^{-6}$ mbar) and reacted via the two-zone method known from graphite intercalation reactions at 300 °C for 5 days. Then the resulting phase-pure Rb_6C_{60} was ground with another equivalent of C_{60} and afterwards pelletized by applying a pressure of 20 MPa (200 bar) to provide a fast adjustment between rubidium-rich and rubidium-poor grains. The crushed pellet was sealed again in a small quartz ampoule and annealed at 350 °C for 5 days to result in phase-pure Rb_3C_{60}. The average particle size of the material could be determined to be (10 ± 5) μm by applying scanning electron microscopy.

3. Results and Discussion

3.1 Dc Measurements

The temperature dependence of the dc susceptibility for all samples was measured with a Squid magnetometer. First, the sample was cooled down in zero field (ZFC) from room temperature to 6 K (5 K/min). The ZFC curve was obtained after applying a low magnetic field (8 G) to the sample and warming up slowly (2 K/h) to a temperature above T_C. By cooling again slowly in the same magnetic field the FC curve was obtained. The irreversible behaviour of these two curves clearly indicates type-II superconductivity with flux pinning.

Analysis of the low temperature branch of the ZFC curve allows a rough estimate of the superconducting volume fraction. A difficulty with it is that the curvature depends on the geometry and the grain structure of the samples due to the finite magnetic penetration depth λ, its ratio to the sample dimensions, its temperature dependence and finally the demagnetization factor. λ reduces the effective diamagnetic volume fraction of the superconducting volume. In the case of spheres of a radius a, the reduction factor is [15]

$$\mathrm{P(x)} = 1 - \frac{3}{x}\coth x + \frac{3}{x^2};\ x = \frac{a}{\lambda} \tag{1}$$

For the first sample the ratio x could be determined experimentally [10]: $x = 3.34$ leading to a volume correction factor of 2.7 (Table 1).

The influence of the demagnetization factor N,

$$\chi = \frac{\chi_{\mathrm{exp}}}{1 - N\chi_{\mathrm{exp}}} \tag{2}$$

results in a correction just in the opposite direction. However, the effective demagnetization factor decreases with decreasing particle size because of the inhomogeneous field distribution inside the particles which results from the finite penetration depth. Therefore, this correction was only applied for samples 2 and 3.

3.2 Ac Measurements

In magnetic measurements using alternating magnetic fields the magnettization of the sample changes periodically in response to an applied ac field $B(t) = B_0 \cos \omega t$. Under the assumption that the material response is linear the magnetization can be described by

$$\mu_o M(t) = \chi' B_o \cos \omega t + \chi'' B_o \sin \omega t \tag{3}$$

where χ' and χ'' are the in-phase and the out-of-phase components of the magnetic susceptibility, respectively. Combining this to the complex susceptibility, $\chi = \chi' - i\chi''$, the real part χ' describes the dispersive response which corresponds to the ZFC curve of dc measurements under certain conditions, whereas the imaginary part χ'' describes the energy dissipation.

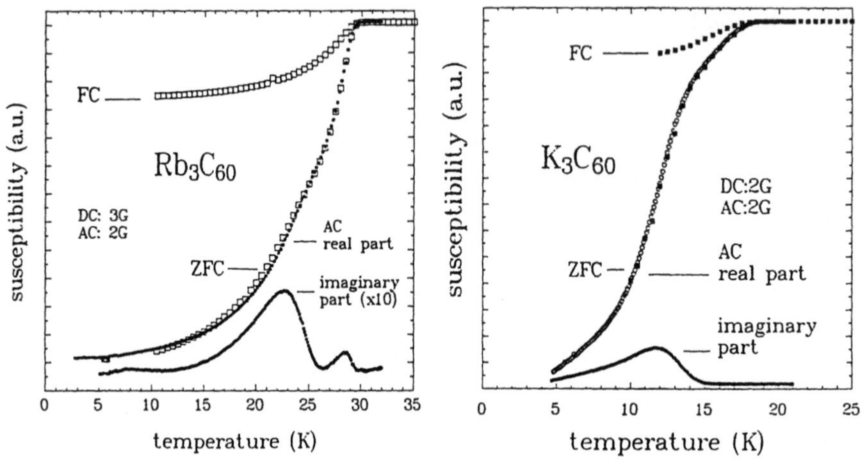

Fig. 1. Temperature dependence of the dc and ac susceptibility for samples 2 and 3.

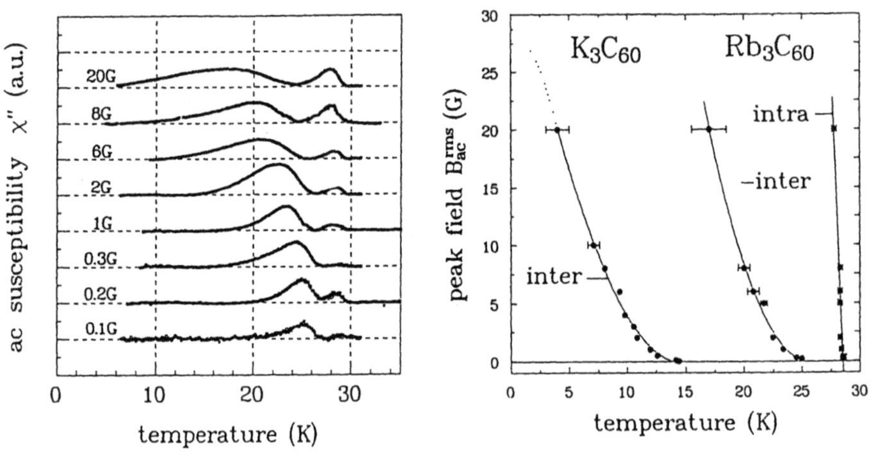

Fig. 2. Temperature dependence of the imaginary part of the ac susceptibility measured with different ac field amplitudes for Rb_3C_{60} (left). Right: Temperature dependence of the peak field values for K_3C_{60} and Rb_3C_{60}.

Fig. 1 shows the comparison of the ZFC curves with the real parts of the ac susceptibility for two samples. In the case of Rb_3C_{60} two significant peaks occur in the imaginary part. Such a behaviour indicates a two-phase structure of the sample. Varying ac amplitude (Fig. 2) does not affect the first peak at higher temperature whereas the second is shifted to lower temperatures with increasing amplitude. This is typical for samples with a microscopic structure consisting of an array of superconducting grains embedded in a matrix with different superconducting properties and coupled by a network of Josephson junctions. Similar effects are frequently observed in high-T_c ceramics.

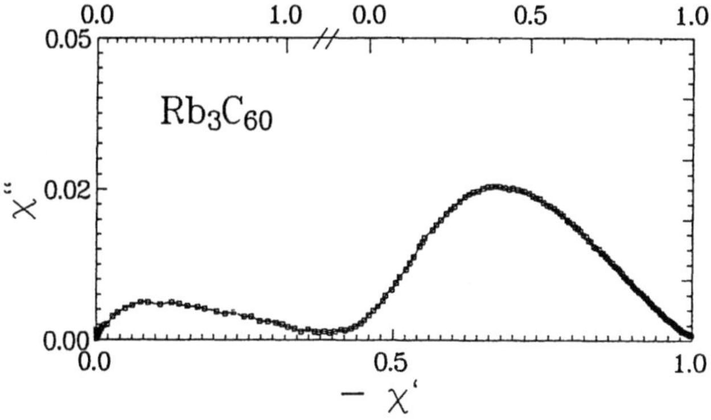

Fig. 3. Plot of χ'' vs. χ' for Rb_3C_{60} showing a Bean model behaviour with the maxima at about 0.4 (upper scale).

According to this picture the first peak is due to intragrain dissipation in the Shubnikov phase of the grains between the upper and the lower critical magnetic field. B_{c1} values can be estimated approximately from the minimum between the two peaks (Table 1). The second peak represents the dissipation in the Shubnikov phase of the matrix material where the diamagnetic signal is superimposed by that of the grains. Below B_{c1} of the matrix the complete macroscopic sample is in the Meissner state.

These results clearly demonstrate the granular character of this sample. The reason for this behaviour is the application of pressure during the preparation of this sample (section 2). Although this procedure yields higher T_c values in some cases, the sample structure is more complicated.

In order to get more information about the dissipation mechanism the χ'' results were compared with different theoretical models. Best description is achieved by application of Bean's critical state model [16] (Fig. 3). This model describes the magnetic response of type-II superconductors in the Shubnikov phase. As soon as pinning mechanisms are present, a finite critical current density j_c exists. j_c depends on the magnetic field profile in the interior of the sample. B decreases linearly with the distance from the edge to the inside. This leads to

$$j_c = \frac{1}{\mu_o} \frac{B_p}{a} \tag{4}$$

where B_p is the magnetic field causing just full flux penetration and a is the half extension of the sample. Under an applied ac magnetic field the dissipation due to flux rearrangements reaches its maximum if B_{ac} is equal to B_p. Therefore, the maximum values of the χ'' peaks (Fig. 2) can be used to get information about j_c using equation 4 (Table 1).

The influence of external dc fields up to 14 T is shown in Fig. 4. At all fields a sharp onset of the superconducting transition exists allowing a precise determination of the temperature dependence of the upper critical magnetic field B_{c2} (T) (Fig. 5).

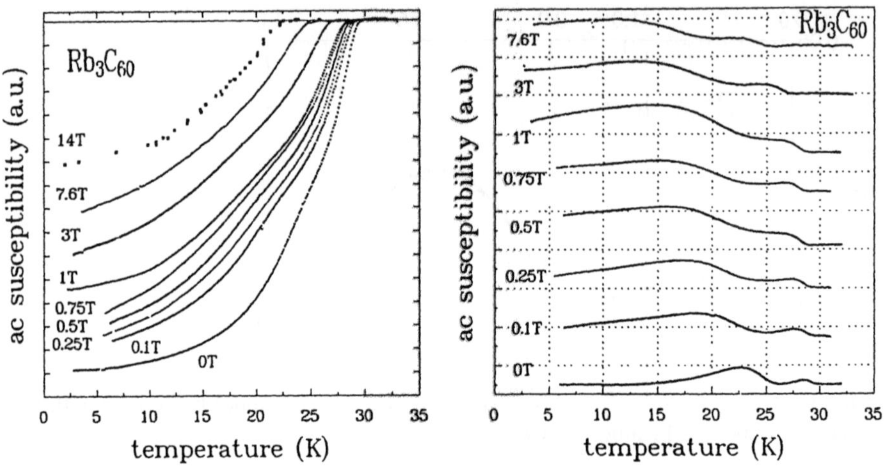

Fig. 4. Influence of external dc fields on χ' (left) and χ'' (right) for Rb_3C_{60}.

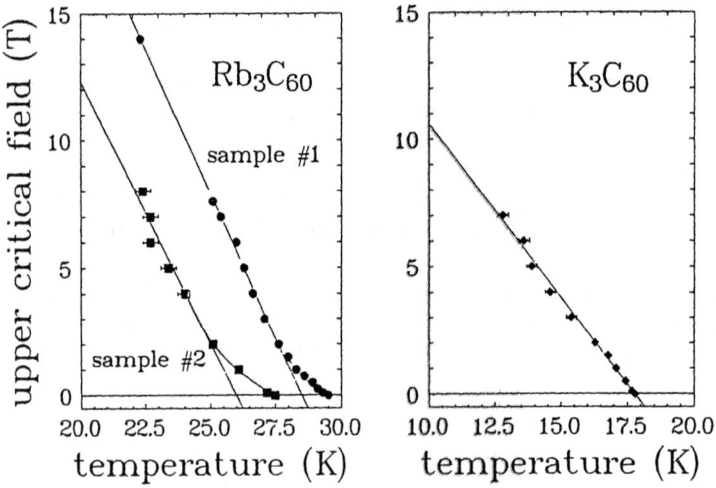

Fig. 5. Temperature dependence of the upper critical magnetic field.

All curves exhibit a linear behaviour (for the Rb_3C_{60} samples above approximately 1.5 - 2 T) (Table 1). The resulting values of B_{c2} at zero temperature and the Ginzburg-Landau coherence length ξ_{GL} are determined with the most common method, that is employing the relations [17]

$$B_{c2}(0) = 0.69\, T_c \left(\partial B_{c2} / \partial T\right)_{T=T_c} \tag{5}$$

$$B_{c2}(0) = \Phi_o / \left(2\pi\xi_{GL}^2\right). \tag{6}$$

Here $\Phi_o = h / 2e_o$ is the flux quantum. The values are also listed in the Table. They agree rather well with those reported by Xiang [18].

The influence of an external dc field to χ'' is shown in Fig. 4b. The peak structure of the curves changes with increasing field to a more step-like structure which is due to passing above the lower critical field B_{c1}.

Acknowledgement. This work was supported by the Bundesministerium für Forschung und Technologie (F + E-Vorhaben 13 N 60 74 1).

References

[1] A.F. Hebard, M.J. Rosseinsky, R.C. Haddon, D.W. Murphy, S.H. Glarum, T.T.M. Palstra, A.P. Ramirez, A.R. Kortan, Nature **350**, 600 (1991).

[2] S. Gärtner, Festlörperprobleme/Advances in Solid State Physics **32**, 295 (1992).

[3] A.F. Hebard, Physics Today **45**, No. 11, 26 (1992).

[4] D.W. Murphy, M.J. Rosseinsky, R.M. Fleming, R. Tycko, A.P. Ramirez, R.C. Haddon, T. Siegrist, G. Dabbagh, J.C. Tully, R.E. Walstedt, J. Phys. Chem. Solids **53**, 1321 (1992).

[5] K. Lüders, Proc. NATO-ASI "Intercalation Compounds II", Bonas/France, July 1992, in print.

[6] A.R. Kortan, N. Kopylov, S. Glarum, E.M. Gyorgy, A.P. Ramirez, R.M. Fleming, O. Zhou, F.A. Thiel, P.L. Trevor, R.C. Haddon, Nature **360**, 566 (1992).

[7] M. Kraus, M. Kanowski, M. Baenitz, H. Werner, R. Schlögl, K. Lüders, to be published.

[8] A.R. Kortan, N. Kopylov, S. Glarum, E.M. Gyorgy, A.P. Ramirez, R.M. Fleming, F.A. Thiel, R.C. Haddon, Nature **355**, 529 (1992).

[9] E.-W. Scheidt, C. Hucho, K. Lüders, V. Müller, Solid State Commun. **71**, 505 (1989).

[10] M. Kraus, M. Baenitz, M. Kanowski, E. Straube, E.-W. Scheidt, S. Gärtner, H.M. Vieth, H. Werner, R. Schlögl, W. Krätschmer, K. Lüders, Applied Superconductivity **1**, 901 (1993).

[11] See for instance: R.B. Goldfarb, M. Lelental, C.A. Thompson , in "Magnetic Susceptibility of Superdonductors and Other Spin Systems" (R.A. Hein, T.L. Francavilla, D.H. Liebenberg, eds.), Plenum Press, New York and London 1991, p. 49.

[12] W. Krätschmer, D.L. Lamb, K. Fostiropoulos, D.R. Hufman, Nature **347**, 345 (1990).

[13] M. Baenitz, M. Kraus, S. Gärtner, H.M. Vieth, H. Werner, R. Schlögl, W. Krätschmer, M. Kanowski, K. Lüders, Springer Series in Solid-State Sciences **113**, 475 (1993).

[14] J.P. McCauley Jr., Q. Zhu, N. Coustel, O. Zhou, G. Vaughan, S.H.I. Idziak, J. E. Fischer, S.W. Tozer, D. Groski, N. Bykovetz, C.L. Lin, A.R. McGhie, B.H. Allen, W.R. Romanow, A.M. Denenstein, A.B. Smith III, J. Am. Chem. Soc. **113**, 8537 (1991).
[15] D. Schoenberg, Superconductivity, Cambridge 1960, p. 234.
[16] C.P. Bean, Rev. Mod. Phys. **36**, 31 (1964).
[17] M. Tinkham, Introduction to Superconductivity, McGraw-Hill, New York 1975.
[18] X.-D. Xiang, this issue.

The Electrodynamic Response in Superconducting Fullerenes

L. Degiorgi

Laboratorium für Festkörperphysik, ETH-Zürich, CH-8093 Zürich, Switzerland

Abstract. We have measured the optical reflectivity of single phase K_3C_{60} and have evaluated the optical conductivity both below and above the superconducting transition temperature.

1. Introduction

The recent discovery [1] of superconductivity, found at relatively high temperatures in the alkali metal-C_{60} compounds has created considerable interest. Despite a variety of experiments, several issues remain controversial either for the normal or for the superconducting state properties. Optical investigation is a powerful experimental tool in order to evaluate various important parameters, like the plasma frequency ν_p and the relaxation time τ for the normal state, and the single particle gap Δ, if any, and the penetration depth λ for the superconducting state [2,3]. From optical investigations, one can also single out, if the investigated system is in the so-called clean or dirty limit and furthermore whether the weak or strong coupling limit of the BCS theory is more appropriate to describe the superconducting state.

In this paper, we review our optical investigations on the complete excitation spectrum of K_3C_{60}, both above and below T_c. Particularly, we will evaluate several experimental parameters, characterizing the normal and superconducting state properties, and compared them with similar quantities arrived at by other investigations.

2. Experiment and results

The samples used in the present work were prepared from a starting composition of K_4C_{60} heated in vacuum at 200 ^{0}C for one week, and subsequently annealed under He atmosphere over several days at the same temperature. X-ray diffraction experiments demonstrated the single crystalline phase nature of our powder samples. The samples have a superconducting phase transition temperature T_c=19 K. The powder was then pressed in a He-chamber in order to obtain pellet of about 3 mm diameter and 1 mm thick.

Reflectivity measurements ($R(\omega)$) were performed between 14 and 50000 cm^{-1} using three different spectrometers. In the far infrared (FIR) spectral range we measured $R(\omega)$ also as function of the temperature. A gold mirror was used as reference in the FIR and mid-IR. After all measurements have been performed, the pellets were also covered with a 400 Å layer of gold in order to take into account the surface roughness (particularly important here, since the pellet can not be polished). Since the materials are highly air sensitive, the measurements in various

Springer Series in Solid-State Sciences, Volume 117
Electronic Properties of Fullerenes Eds.: H. Kuzmany, J. Fink, M. Mehring, and S. Roth

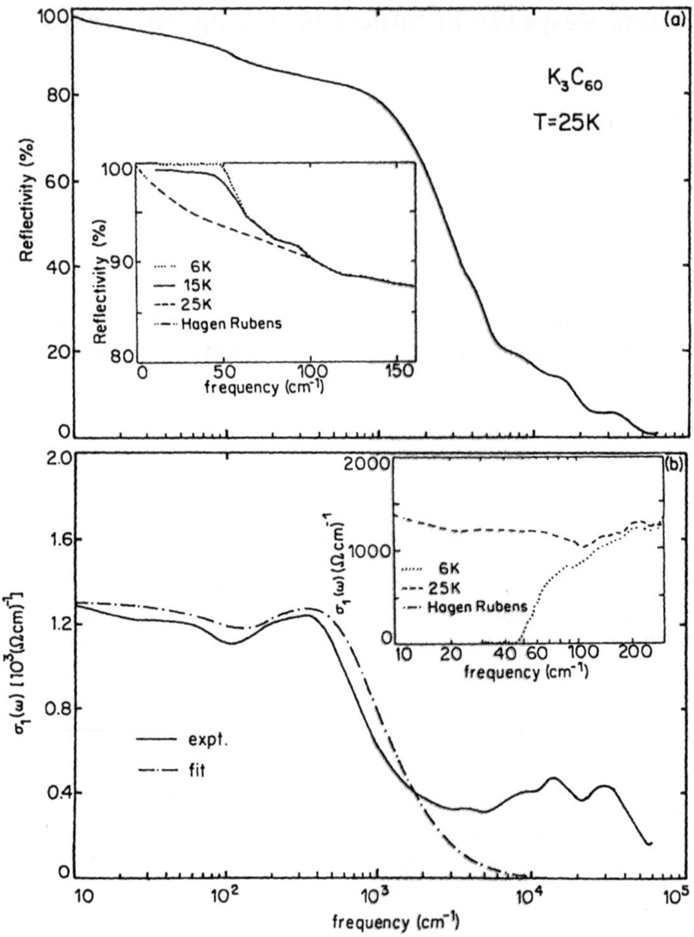

Fig. 1: (a) Reflectivity spectrum of R(ω) of K_3C_{60} at 25 K. The inset displays R(ω) below and above T_c in the far infrared (FIR). (b) Optical conductivity $\sigma_1(\omega)$ at 25 K obtained from Kramers-Kronig transformation of R(ω) in (a). The dot-dashed line represents the phenomenological fit (see text). Inset: $\sigma_1(\omega)$ in the normal and superconducting state in FIR.

spectrometers were performed in vacuum or He gas atmosphere conditions, so that the samples were never exposed to air contamination [2,3].

Fig. 1a shows the R(ω) spectra on K_3C_{60} at 25 K on the entire energy spectral range investigated. This R(ω) spectrum extending over four decades is subsequently used for Kramers-Kronig (KK) transformation. The frequency range of the measured reflectivity spectrum has been extrapolated towards zero frequency by means of the Hagen-Rubens relation using the ρ_{dc} data at 25 K published by Klein et al. on K_3C_{60} [4], and for energies larger than 50000 cm^{-1} (where R(ω) is already approaching zero) the reflectivity has been assumed to drop off as ω^{-2}. Fig. 1b displays the optical conductivity $\sigma_1(\omega)$ at 25 K obtained from our KK analysis.

In the inset of Fig. 1a, we display the optical reflectivity measured both above and below the superconducting transition temperature, in the spectral range which covers the (expected) superconducting gap values. The frequency dependent conductivity $\sigma_1(\omega)$ measured both above and well below T_c is displayed in the inset of Fig. 1b. The low frequency extrapolation of $\sigma_1(\omega)$ (dashed-dot-dot line) is that of the Hagen-Rubens form in the normal state, and this leads to a dc conductivity $\sigma=ne^2\tau/m$ of 1.3×10^3 $(\Omega cm)^{-1}$, in full agreement with the dc resistivity values [4]. Recently, we have also reported on our optical investigations of the Rb-doped C_{60} [2,3]. In view of the qualitative similar results, the present discussion will be then limited to the potassium compound only.

3. Discussion

We first note, that $\sigma_1(\omega)$ as shown in Fig. 1b is remarkably different from that of a simple metal where the Drude model is appropriate. Instead, we observe a Drude like response together, however, with an overlapped mid-infrared excitation, and several absorptions at higher frequencies. At high frequencies (i.e. from the visible to the UV) our excitation spectrum is characterized by broad transitions peaked at 18000 cm^{-1} (with a shoulder at approximately 10000 cm^{-1}) and at 35000 cm^{-1}. We ascribe these excitations to the electronic interband transitions involving s and p carbon orbitals and s, p and d alkali orbitals. Our measured conductivity or correspondingly the imaginary part of the dielectric function ($\varepsilon_2(\omega)$) of K_3C_{60} is in good accord with the theoretical predictions extracted from a band structure calculation [5], which lead to broad peaks centered at 1 eV, 2.5 eV and 4 eV (see Fig. 3b of Ref. 5).

The low frequency part of $\sigma_1(\omega)$ (i.e. from FIR up to mid-IR) is thus characterized by two main features: the broad band centered at approximately 500-800 cm^{-1} and the Drude like contribution to $\sigma_1(\omega)$ at FIR frequencies. We model the dielectric function with the following expression:

$$\varepsilon(\nu) = \varepsilon_\infty + \nu_p^2\left(\frac{f_G}{\nu_G^2 - \nu^2 - i\nu\gamma_G} - \frac{1-f_G}{\nu(\nu + i\gamma_D)}\right) \tag{1}$$

where ε_∞ describes the high frequency contribution to the dielectric function, while the second and third term in the brackets describe the mid-IR broad excitation and the Drude like response at low frequencies, respectively. The best fit is obtained with the following parameters: $\varepsilon_\infty=5$, $\nu_p=9468$ cm^{-1}, $\nu_G=500$ cm^{-1}, $\gamma_G=1210$ cm^{-1}, $\gamma_D=147$ cm^{-1} and $f_G=0.87$ (see dashed-dot line in Fig. 1b).

Starting with formula (1) there are two main directions that can be followed to interpret our spectroscopic data. The first one, which we will call the one-component picture, is based on the assumption that the low frequency Drude like behaviour and the mid-IR harmonic oscillator are due to the response of the conduction band, and crudely speaking the charge carriers (which from simple counting arguments have a density $n= 4.1\times10^{21}$ cm^{-3}) behave as free carriers at low frequencies and as bound ones at high frequencies. The total spectral weight associated to the plasma frequency $\nu_p=13471$ cm^{-1} would correspond to an effective band mass $m_b= 2m_e$ with $n= 4.1\times10^{21}$ cm^{-3}. This value is somewhat smaller than the value arrived at from spin susceptibility measurements, $m_b=6.5m_e$ [6].

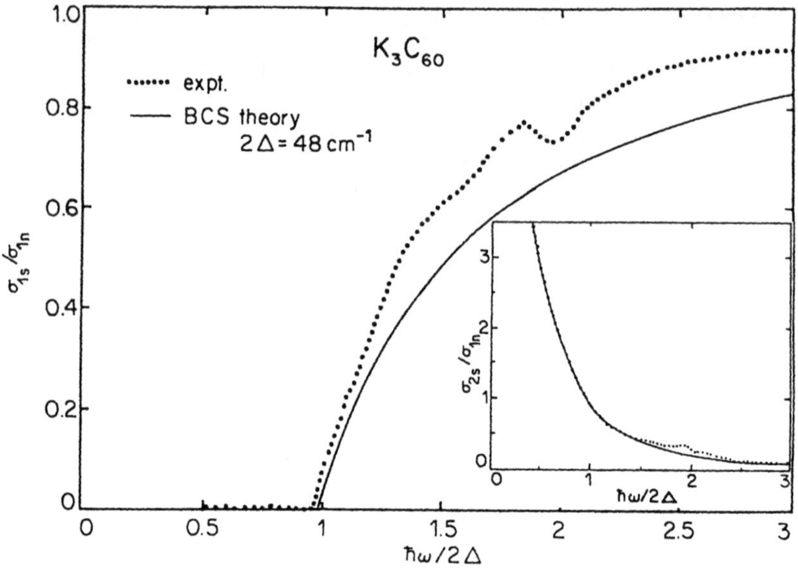

Fig. 2: The measured ratio $\sigma_{1s}(\omega)/\sigma_{1n}(\omega)$ together with the calculated one using the Mattis -Bardeen theory (BCS). Inset: the imaginary part $\sigma_{2s}/\sigma_{1n}(\omega)$.

The second way to interpret equation (1) is to consider the two contributions independently, using a two component picture. Then, the Drude part is characterized by a plasma frequency of $\tilde{\nu}_p$=1205 cm^{-1} (i.e. $(1-f_G)^{1/2}\nu_p$), and the harmonic oscillator is viewed as a separate mid-IR excitation. First, we note that the reduced Drude spectral weight ($\tilde{\nu}_p$) implies $m^*=33m_e$ and that the scattering rate (γ_D) of the Drude component is slightly larger than the superconducting gap 2Δ as evaluated from tunneling spectroscopy measurements (100 cm^{-1}) [7] and from nuclear magnetic resonance experiments [8]. This suggests that the K_3C_{60} compound is in the dirty or nearly dirty limit. We can use our estimation of $1/\gamma_D$ to evaluate the mean free path (ℓ). With the Fermi velocity v_F=5x10^6 cm/sec, obtained from magnetic measurement [6] (or 1.8x10^7 cm/sec, from band structure calculation [9]) and the expression $\ell=v_F\tau$ we obtain ℓ=18 Å (65 Å), which is comparable with the coherence length ζ [6]. The fit parameters ν_p, τ and consequently the mean free path, obtained within the two components picture, are also in agreement with the evaluation of these quantities arrived at by the analysis of the dc electrical resistivity data [4]. As far as the broad mid-IR excitation is concerned, we suggest to assign it to an electronic interband transition, in accord also with the band structure calculation results [5]. A remarkable phonon activity is moreover expected in this mid-IR frequency range. Although this can be an important source of absorptions, at present we are not able to resolve features clearly associated to phonon excitations.

The optical reflectivity below T_c shows a well defined gap feature, which becomes progressively sharper when the temperature is decreased. In the superconducting state at T=6 K, our reflectivity data are, up to a well defined threshold frequency of 48 cm^{-1}, within the experimental error of 0.5% equal to 100%, and we have used this value to evaluate the optical conductivity. This procedure gives zero conductivity up to a threshold frequency which we identify as the superconducting gap and we obtain $\Delta = 24\ cm^{-1}$. Together with the superconducting transition temperatures T_c=19 K this leads to the ratio $2\Delta/k_BT_c$= 3.6, in good agreement with the weak coupling BCS result of $2\Delta/k_BT_c$= 3.52. Our conductivity result can be compared furthermore with the measured [10] penetration depth values using the missing spectral weight argument [11]. We obtain λ= 8000±500 Å, in satisfactory agreement with other evaluation of the penetration depth of 6000 Å [8] or 4800 Å [10].

The functional form of $\sigma(\omega)$ within the BCS framework below T_c has been evaluated first by Mattis and Bardeen [12].The measured ratio of the optical conductivity in the superconducting state over the normal state one is displayed in Fig. 2 with the Mattis-Bardeen calculation. Given the fact, that the calculation neglects the role played by mean free path effects, the agreement between our results and the theory based on a BCS ground state is excellent. We also note, that the functional form also reflects the so called case II coherence factors [11] which depend sensitively on the symmetry of the superconducting wave function, and therefore the agreement between theory and experiment gives clear evidence for a singlet ground state.

4. Conclusion

We have examined the electrodynamic response of the normal and superconducting ground state in K_3C_{60}. In the normal state we have observed a Drude contribution and a fairly broad excitation in the mid-IR. We have discussed these features with a harmonic oscillator phenomenological approach and have found that the parameters evaluated within the two components picture are in broadest agreement with those arrived at by other experiments [4]. The measured optical conductivity below T_c is in full agreement with that of a BCS singlet superconductor, and the magnitude of the single particle gap also agrees with the weak coupling limit.

However, the gap feature does not decrease remarkably by raising the temperature from 6 K up to T_c, as it would be expected according to a simple BCS picture. While this anomalous temperature dependence of the reflectivity below T_c is similar to what has been found by other investigations [13,14], it still remains to be seen whether it is an indication of measurement uncertainty or the consequence of an intrinsic physical mechanism (like e.g. a gap distribution [14]). Moreover, the good agreement with the weak coupling limit of the BCS theory in the superconducting state contrasts with the strongly renormalized optical properties of the normal state. Although this is still a puzzling issue, it might be suggestive of the important role played by correlation effects.

Acknowledgment

This work was performed in collaboration with G. Grüner and R.B. Kaner at UCLA and supported by the Defense Advanced Research Projects Agency through a grant monitored by the Army Research Office. The author also thanks O. Klein, S. Cho and P. Wachter.

References

[1] A. F. Hebbard, M.J. Rosseinsky, R.C. Haddon, D.W. Murphy, S.H. Glarum, T.T.M. Palstra, A.P. Ramirez and A.R. Kortan, Nature **350**, 320 (1991) and
K. Holczer, O. Klein, S.-M. Huang, R.B. Kaner, K.-J. Fu, R.L. Whetten and F.N. Diederich, Science **252**, 600 (1991)

[2] L. Degiorgi, G. Grüner, P. Wachter, S.-M. Huang, J. Wiley, R.L. Whetten, R.B. Kaner, K. Holczer, F. Diederich, Phys. Rev. **B46**, 11250 (1992)

[3] L. Degiorgi, P. Wachter, G. Grüner, S.-M. Huang, J. Wiley and R.B. Kaner, Phys. Rev. Lett. **69**, 2987 (1992)

[4] O. Klein, G. Grüner, S.-M. Huang, J.B. Wiley and R.B. Kaner, Phys. Rev. B46, 11247 (1992)

[5] Y.-N. Xu, M.-Z. Huang and W.Y.Ching, Phys. Rev. **B44**, 13171 (1991)

[6] W.H. Wong, M. Hanson, W.G. Clark, K. Holczer, G. Grüner, J.D. Thompson, R.L. Whetten, S.-M. Huang, R.B. Kaner and F. Diederich, Europhys. Lett. **18**, 79 (1992)

[7] Z. Zhang, C.-C. Chen, S.P. Kelty, H. Dai and C.M. Lieber, Nature **353**, 333 (1991)

[8] R. Tycko, G. Dabbagh, M.J. Rosseinsky, D.W. Murphy, A.P. Ramirez and R.M. Fleming, Phys. Rev. Lett. **68**, 1912 (1992)

[9] S.C. Erwin and W.E. Pickett, Science **254**, 842 (1991)

[10] Y. J. Uemura, A. Keren, L.P. Le, G.M. Luke, B.J. Sternlieb, W.D. Wu, J.H. Brewer, R.L. Whetten, S.-M. Huang, S. Liu, R.B. Kaner, F. Diederich, S. Donovan, G. Grüner and K. Holczer, Nature **352**, 605 (1991)

[11] see for example: M. Tinkham, "Introduction to superconductivity", R.E. Krieger (Ed.)

[12] D.C. Mattis and J. Bardeen, Phys. Rev. **111**, 412 (1958)

[13] L.D. Rotter, Z. Schlesinger, J.P. Cauley, N. Courtel, J.E. Fischer and A.B. Smith, Nature **355**, 532 (1992)

[14] S.A. Fitzgerald, S.C. Kaplan, A. Rosenberg, A.J. Sievers and R.A.S. McMordie, Phys. Rev. **B45**, 10165 (1992)

Pairing on Small Clusters in the Peierls-Hubbard Model: Implications for C_{60}

J.T. Gammel[1], *F. Guo*[2], *D. Guo*[2], *K.C. Ung*[2], *and S. Mazumdar*[2]

[1]Materials Research Branch, NRaD, San Diego, CA 92152-5000, USA
[2]Department of Physics, University of Arizona, Tucson, AZ 85721, USA

Abstract. We study the pairing within the Peierls-Hubbard Model for electron- and hole-doped analogs of C_{60} accessible to exact diagonalization techniques (cube, truncated tetrahedron, *etc.*). We discuss how inclusion of electron-phonon interactions can substantially modify the conclusions about pairing obtained when this coupling is neglected. We also discuss potential pitfalls in the extrapolation from these small system calculations to C_{60}, and stress the necessity of having the correct intuitive picture.

1. Introduction

A great deal of numerical effort has gone into investigating whether pairing of the Cooper type occurs in the weakly doped two-dimensional Mott-Hubbard insulator. More recently, such a pairing mechanism has been proposed [1] for doped C_{60}. Specifically, these calculations involve determination of the ground state energies of the half-filled band system, and the electronic energies needed to add one and two electrons (holes). Pairing is supposed to occur if the quantity

$$\Delta E = 2E(N \pm 1) - E(N) - E(N \pm 2) \tag{1}$$

is positive. As approximate analytic techniques have been known to lead to qualitatively incorrect predictions for solid state systems with strong electron correlation, exact numerical techniques – exact calculation for a (small) finite clusters followed by extrapolation to the system size of interest – have been employed [2,3]. Within the Hubbard model,

$$H_{Hub} = \sum_{\langle ij \rangle, \sigma} t_{ij} c^{+}_{i,\sigma} c_{j,\sigma} + U \sum_{i} n_{i,\uparrow} n_{i,\downarrow} \ , \tag{2}$$

where $\langle ij \rangle$ are nearest neighbors and $t_{ij}{=}{-}t_0$, pairing is found for the cube and the truncated tetrahedron [2,3,4].

2. Geometry, Distortion, and Pairing

For the geometries shown in Fig. 1, we have calculated the pair binding energy given in Eq. 1. The result is shown in Fig. 2. We see that there appear to be

Springer Series in Solid-State Sciences, Volume 117
Electronic Properties of Fullerenes Eds.: H. Kuzmany, J. Fink, M. Mehring, and S. Roth

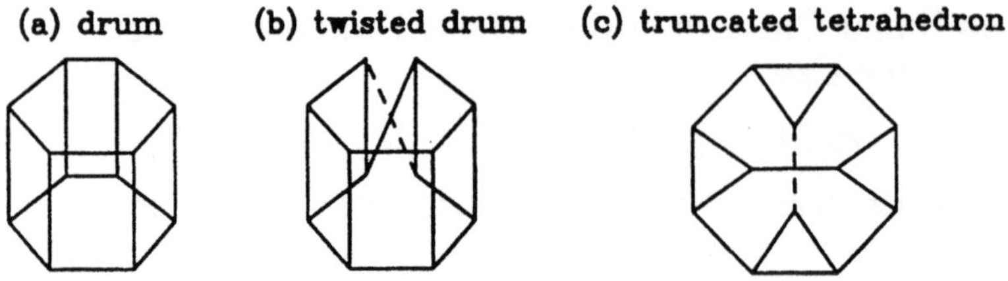

Fig. 1. Geometries for which the pair binding energy has been calculated: (a) drum, N=8,10,12; (b) twisted drum, N=8,10,12; (c) truncated tetrahedron.

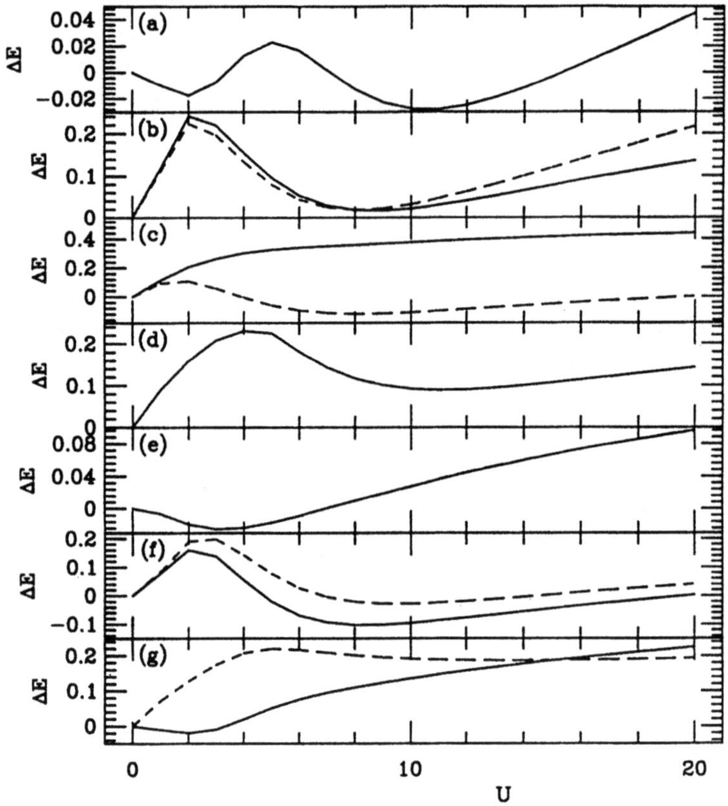

Fig. 2. Pairing in undistorted clusters for electrons (solid) and holes (dashed): (a,c,e) drums, (b,d,f) twisted drums, (g) truncated tetrahedron, (a,b) N=8, (c,d) N=10, (e,f,g) N=12. Note there are two regions, small and large U, where pairing can occur.

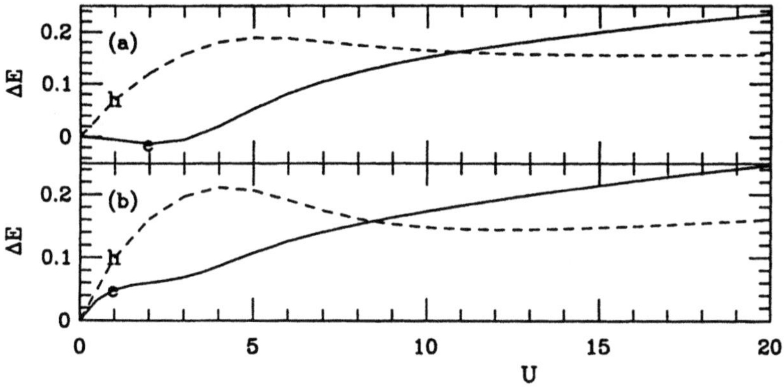

Fig. 3. Pair binding energies for the uniformly distorted truncated tetrahedron: (a) Peierls-distorted; (b) Jahn-Teller distorted. Notice the absence of pairing at small U in (b).

two regions where pairing can occur: one at small U and one near U/t_0=10. Our investigation indicates this oscillation may be related to (avoided) energy level crossings. Interestingly, large U pairing is found in the pentagonal prism, indicating that it is not limited to bipartite lattices. The pairing at small U appears to be related to the presence of energy level degeneracies at U=0.

In Ref. 4 we discussed how the variation in U-dependence of the Jahn-Teller distortion with filling masks this Jahn-Teller driven pairing (present on finite chains only!) as an electron-electron driven pairing at small U. Basically, by not letting the system distort, we are overestimating the energies in each case, but this overestimation is much stronger for the odd electron case than for the even electron case. The calculated energy required to add one hole (or electron) to the undistorted system is therefore too high, giving "pairing".

To investigate the effects of distortion numerically, we first add a fixed external distortion δ_{ij} from uniform bond length d_0 and use $t_{ij}=-t_0+\alpha\delta_{ij}$ in the Hamiltonian Eq. 2. In Fig. 3 we show the results of our calculation for the truncated tetrahedron. For the undistorted case (Fig. 3), we merely reproduced the results of White *et al.* [2] — pairing was seen to occur at small U. With short bonds connecting the triangles – the distortion preferred at 1/2-filling – the pairing still exists. This is however not the Jahn-Teller mode, the noninteracting limit still having degeneracies. The pairing at small U disappears for the Jahn-Teller distorted truncated tetrahedron (calculated as the self-consistent distortion for $N_{el}=N\pm1$, see below). This clearly indicates the relationship between pairing at small U and the degeneracies in the single particle limit: in the absence of distortion, the energy to add one hole or electron is being overestimated. In Ref. 4 we showed for the cube that the pairing at large U survives the U=0 Jahn-Teller distortion. However, since there appears to be an energy-level crossing it may be that a different geometry than the U=0 one should be considered the "large-U Jahn-Teller" mode, and we are investigating this further.

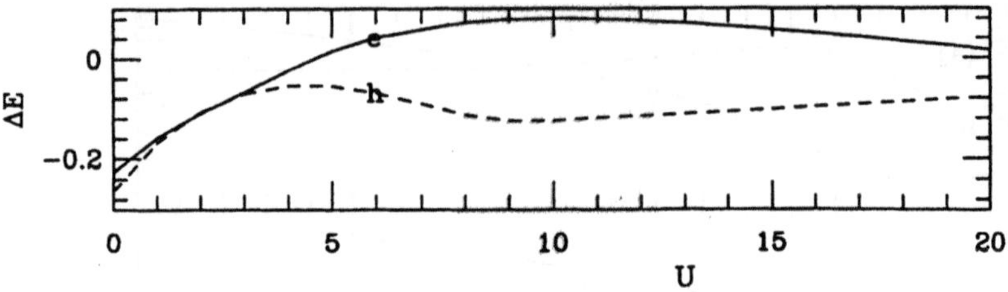

Fig. 4. Pair binding energies in the self-consistently distorted truncated tetrahedron. Note that the pairing is strongest at U=0.

To calculate the effects of variable distortion we use the Peierls-extended Hubbard Hamiltonian,

$$H = \sum_{\langle ij\rangle,\sigma} [-t_0 + \alpha\delta_{ij}](c^{+}_{i,\sigma}c_{j,\sigma} + c^{+}_{j,\sigma}c_{i,\sigma}) + U\sum_{i} n_{i,\uparrow}n_{i,\downarrow} + \frac{1}{2}\kappa \sum_{\langle ij\rangle}(\delta_{ij} - d_0)^2 \quad (2')$$

where the δ_{ij} are now self-consistently chosen to minimize the total energy, and thus depend on filling. In Fig. 4 we show the results of our calculation for the truncated tetrahedron. d_0 and κ were chosen to depend on U such that the minimum energy distortion at $N_{el}=N$ was independent of U: δ_{ij}=0.1 on the triangular faces and δ_{ij}=-0.2 on the bonds connecting triangular faces. Using these values the self-consistent distortion for each filling was then found and the pairing as given in Eq. 1 calculated. Here, we find a greatly enhanced pairing, though maximum at U=0 indicating again that it is driven by the Jahn-Teller distortion and not the electron-electron interaction ("bipolaronic" pairing).

3. Conclusions

We have calculated the pair-binding energy for a number of three dimensional molecules, and find the following. For an N site system, pairing at small U occurs only if the $(N-1)$ (or $(N+1)$) electron system has a strong tendency to have a Jahn-Teller distortion. This pairing is destroyed when a fixed Jahn-Teller distortion is introduced. Thus the mechanism of pairing in the undistorted cluster at small U in all these cases is related to the suppression of Jahn-Teller distortion. Since the Jahn-Teller distortion results from the discrete level degeneracies in finite molecules, we conclude that the observed pairing in the small U region is a finite size effect.

Pairing at large U is unrelated to the tendency to have Jahn-Teller distortion [4], and is not necessarily accompanied by pairing at small U. The occurrence of such large U pairing in nonbipartite systems (*e.g.*, in a pentagonal prism) indicates that it is also apparently unrelated to antiferromagnetism. It may be related to energy level crossings. Currently we are investigating whether the large U pairing is also a finite size effect, and if so, the nature of the finite size effect.

The arguments here and in Ref. 4 indicate the absence of pairing in the two dimensional Hubbard model, as well as in weakly coupled layers. However, as C_{60} is a finite molecule, this "on-ball" pairing mechanism cannot be ruled out. Sawatzky [5] has argued, though, that since the long-range Coulomb interaction in a C_{60} molecule is essentially constant – for this finite system screening moves charge to the other side of the ball, not infinity, enhancing the long range interaction while suppressing the short range interaction – a (multi-orbital) Hubbard model where the "site" is a C_{60} in the molecular crystal rather than a C on an individual ball should be used. In this case the above conclusion that this pairing mechanism does not operate in infinite systems holds.

Finally, we note that the definition of pairing in Eq. 1 clearly is not synonymous with superconducting pairing; it may indicate *e.g.* phase segregation [4,6] instead. We also have not discussed the consequences of longer range Coulomb interactions. We note that V has been found to destroy pairing [2], though potentially it can be restored by other nearest neighbor Coulomb terms [7]. Clearly the effect of Jahn-Teller interactions must also be included before such conclusions can be considered final.

Acknowledgements. JTG was supported by a NRC-NRaD Research Associateship through grants from the ONR and DOE.

REFERENCES

1. S. Chakravarty, M.P. Gelfand, and S. Kivelson, *Science* **254**, 970 (1991).
2. S.R. White, S. Chakravarty, M.P. Gelfand, and S.A. Kivelson, *Phys. Rev. B* **45**, 5062 (1992).
3. R.M. Fye, M.J. Martins, and R.T. Scalettar, *Phys. Rev. B* **42**, 6809 (1990).
4. F. Guo, D. Guo, K.C. Ung, S. Mazumdar, and J.T. Gammel, to appear in the Proceedings of the Discussion Meeting on Strongly Correlated Systems in Chemistry, Bangalore, India, 4-8 January 1993 (Springer) (cond-mat/9303003).
5. G.A. Sawatzky, these proceedings.
6. V.J. Emery, S.A. Kivelson, and H.Q. Lin, *Phys. Rev. Lett.* **64**, 475 (1990).
7. D.K. Campbell, private communication.

Broadening and Anomalous Infrared Activity of the H_g-Derived Phonons in the Metallic Fullerides

M.J. Rice[1], *H.-Y. Choi*[2], *E.J. Mele*[3], *and M. Deshpande*[3]

[1]Xerox Webster Research Center, 0114-39D, 800 Phillips Road, Webster, NY 14580, USA

[2]Department of Physics, Sung Kynn Kwan University, 440-7446 Suwon, Korea

[3]Department of Physics, University of Pennsylvania, PA 19104, USA

Abstract. Broadening of the H_g phonons in the fulleride superconductors is due to interband electron-phonon coupling. Orientational disorder can render these phonons infrared active.

The metallic fullerides have formula A_3C_{60}, where A denotes an alkali atom [1]. Each of the latter has donated an electron to a partially filled complex of three narrow overlapping conduction bands derived from intermolecular interaction of the threefold degenerate t_{1u} molecular orbitals (MO) ϕ_υ (r) (υ = 1, 2, 3) of the C_{60} molecule [2]. In these states the conduction electrons couple to phonons derived from the Raman-active H_g and A_g vibrations of the C_{60} molecular ion. This electron-phonon coupling is relevant [3] to the remarkable occurrence of superconductivity [4] in these compounds at relatively high temperatures. Experimental information on the magnitude of this coupling is available from Raman [5] and inelastic neutron [6] scattering spectroscopies.

In this paper we make the following points with regard to the observability of the electron-phonon interaction. First, the *longwavelength* ($q \to 0$) H_g-derived phonons do not couple to *intra*-t_{1u}-band excitations but do couple to *inter*-t_{1u}-band excitations. Thus, Raman spectroscopy provides information only on the *inter*band electron-phonon interaction and not on the intraband electron-phonon interaction. The main consequence of the former is to dampen the H_g phonons whose frequencies overlap those of the inter-t_{1u}-band excitation continuum. Second, the longwavelength A_g phonons couple neither to intra- or inter-t_{1u}-band excitations and are, therefore, left unrenormalized by the electron-phonon interaction. Third, we point out that the H_g phonons can become weakly infrared (IR) active due to the breaking of the usual selectron rules by orientational disorder.[7]

In a tight-binding band description [3] of A_3C_{60} the frequency-dependent conductivity $\sigma(\omega)$ can be derived to be of the following form

$$\sigma(\omega) = \frac{e^2}{3} \sum_n N_F(n) v^2_{F,n} \tau_n / (1 - i\omega\tau_n) - i\,\omega\Omega_0^{-1} \chi(\omega) \tag{1}$$

where the first term is the Drude contribution from the intraband excitations and $\chi(\omega)$ is the total electronic polarizability of the crystal due to the *coupled* electron-phonon system. In the Drude term, $N_F(n)$ is the density of states at the Fermi level contributed by the band n, $v^2{}_{F,n}$ is an average over the Fermi surface n of the square of the Fermi velocity, and τ_n are phenomenological transport lifetimes. Ω_0 denotes the volume of the solid. $\chi(\omega)$ consists of two parts,

$$\chi(\omega) = \chi_e(\omega) + \sum_{\alpha\beta} V_\alpha(\omega + i\delta)\, D_{\alpha\beta}(\omega)\, V^*_\beta(\omega - i\delta) \quad . \tag{2}$$

Springer Series in Solid-State Sciences, Volume 117
Electronic Properties of Fullerenes Eds.: H. Kuzmany, J. Fink, M. Mehring, and S. Roth

$\chi_e(\omega)$ is the direct inter-t_{1u}-band contribution,

$$\chi_e(\omega) = \sum_k \sum_{n'n}{}' \frac{|\mu_{n'n}(\mathbf{k})|^2 \, \nu_{n'n}(\mathbf{k})}{\omega_{n'n}(\mathbf{k}) - (\omega + i\delta)} \tag{3}$$

which has already been investigated in [8]. In the second part of Eq. (2) the function

$$V_\alpha(\omega) = \sqrt{\frac{1}{N}} \sum_k \sum_{n'n}{}' \frac{\mu_{n'n}(\mathbf{k}) \, g^{\alpha}_{n'n}(\mathbf{k})^* \, \nu_{n'n}(\mathbf{k})}{\omega_{n'n}(\mathbf{k}) - \omega} \tag{4}$$

determines the effective charge [9] acquired by the phonon mode α. In Eqs. (3) and (4), $\mu_{n'n}(\mathbf{k})$ is the transition dipole moment between Bloch states. It may be expressed as $\mu_{n'n}(\mathbf{k}) = i\hbar e \, v_{n'n}(\mathbf{k})/\omega_{n'n}(\mathbf{k})$, where $v_{n'n}(\mathbf{k})$ is the x-component of the interband matrix element of the one-electron velocity, and $\hbar\omega_{n'n}(\mathbf{k})$ is the energy of an interband transition. The quantity $n_{n'n}(\mathbf{k}) = f_{\mathbf{k},n} - f_{\mathbf{k},n'}$, where $f_{\mathbf{k},n}$ denotes the Fermi-Dirac occupation function for the Bloch state of energy $\varepsilon_{\mathbf{k},n}$. In Eq. (2) $D_{\alpha\beta}(\omega)$ is the phonon-propagator given by the Dyson equation

$$(1/D)_{\alpha\beta} = (1/D^0_{\alpha\alpha}) \, \delta_{\alpha\beta} - \Pi_{\alpha\beta} \tag{5}$$

in which

$$\Pi_{\alpha\beta}(\omega) = \frac{1}{N} \sum_k \sum_{n'n}{}' \frac{g^{\alpha}_{n'n}(\mathbf{k}) g^{\beta}_{n'n}(\mathbf{k})^* \, \nu_{n'n}(\mathbf{k})}{\omega_{n'n}(\mathbf{k}) - (\omega + i\delta)} \tag{6}$$

is the phonon self-energy and $D^0_{\alpha\alpha} = (2\omega_\alpha) / (\omega^2_\alpha - (\omega + i\delta)^2)$ defines the unperturbed phonon propagator. Finally, $g^{\alpha}_{n'n}(\mathbf{k})$ is the longwavelength limit ($\mathbf{k}' = \mathbf{k}$) of the interband electron-phonon coupling constant $g^{\alpha}_{n'n}(\mathbf{k}',\mathbf{k})$.

It is important to note that in Eqs. (3), (4), and (6), the prime over the summation symbol indicates that terms corresponding to $n' = n$ are to be excluded from the sums over the band indices. This is because at the frequencies (ω_α, $\omega_{n'n}(\mathbf{k})$) and wavevectors ($q \to 0$) of interest, the intraband contributions have the form

$$\lim_{q \to 0} \sum_{k'n} \frac{(\partial f_{k,n} / \partial \varepsilon_{k,n}) \, (q \cdot v_{F,n})^{2-s}}{(q \cdot v_{F,n})^2 - (\omega + i\delta)^2} \tag{7}$$

where, on using the relation between $\mu_{n'n}(\mathbf{k})$ and $v_{n'n}(\mathbf{k})$, $s = 2$, 1 and 0 for the right-hand-sides of Eqs. (3), (4) and (6), respectively. Thus, the *intra*band contributions to the effective phonon charge $V_\alpha(\omega)$ and phonon self-energy $\Pi_{\alpha\beta}(\omega)$ both vanish, [10] while the intraband contribution to $\chi_e(\omega)$ survives, and has, in fact, been factored out as the Drude contribution in Eq. (5) (in which we carried out the replacements $\delta \to (1/2\ \tau_n)$. Consequently, unlike the corresponding situation in a conventional metal ($qv_F/\omega \sim v_F/s >> 1$, where s is the sound velocity) [11], the renormalization of the H_g phonons is due entirely to their coupling to *inter*-t_{1u}-band excitations. Equations (5) and (6) show that the dampings of the H_g phonons are determined by the strength of the interband electron-phonon coupling constants and the density of the inter-t_{1u}-band *transitions*,

$$N(\omega) = \frac{1}{N} \sum_k \sum_{n'>n}{}' \nu_{n'n}(\mathbf{k}) \, \delta(\omega_{n'n}(\mathbf{k}) - \omega) \tag{8}$$

rather than the density of states at the Fermi surface, N_F. Significant damping arising from a high value of $N(\omega)$ is probably the reason why the highest frequency H_g modes are apparently absent from the observed Raman spectra of the A_3C_{60} compounds. We note that the A_g modes remain entirely unrenormalized since $g^{\alpha}{}_{n'n}(\mathbf{k}) = 0$ for A_g modes. The A_g phonons should therefore appear as sharp Lorentzian features in the Raman spectrum..

For the H_g modes to become IR active the effective charge $V_\alpha(\omega)$ defined in Eq. (8) must be non-vanishing. Since for the crystal we have the symmetry properties $g^{\alpha}{}_{n'n}(-\mathbf{k}) = g^{\alpha}{}_{n'n}(\mathbf{k})^*$ and $v_{n'n}(-\mathbf{k}) = -v_{n'n}(\mathbf{k})^*$, $V_\alpha(\omega)$ is non-vanishing only if the product $v_{n'n}(\mathbf{k})\ g^{\alpha}{}_{n'n}(\mathbf{k})^*$ has an *imaginary* part. However, both $v_{n'n}(\mathbf{k})$ and $g^{\alpha}{}_{n'n}(\mathbf{k})$ are real if the conduction bands are derived purely from t_{1u} orbitals, in which case $V_\alpha(\omega) = 0$. This selection-rule, however, is broken by orientational disorder which can be regarded to introduce imaginary contributions to $v_{n'n}(\mathbf{k})$ and $g^{\alpha}{}_{n'n}(\mathbf{k})$. A presumably weak IR activity of the H_g phonons can therefore be expected in orientationally disordered A_3C_{60}.

The above conclusions regarding the phonon renormalization and the IR activity can be explicitly illustrated by the introduction into Eqs. (6) and (4) of the simplifications $g^{\alpha}{}_{n'n}(\mathbf{k}) = g_\alpha$, Re $\mu_{n'n}(\mathbf{k}) = \delta\mu$, where g_α and $\delta\mu$ are real constants independent of the labelling of the band states. The Dyson Eq. (5) may then be solved, resulting in the phonon contribution to $\sigma(\omega)$

$$\sigma_{ph}(\omega) = -(i\omega/\Omega_0) N\, \delta\mu^2\, (F(\omega)^2/F(0))\, \lambda\, D_{ph}(\omega) \tag{9}$$

Here, $D_{ph}(\omega)$ is the composite phonon propagator satisfying $D_{ph}(\omega)^{-1} = D_0(\omega)^{-1} - \lambda\, \tilde{F}(\omega)$, where

$$D_0(\omega) = \sum_\alpha (\lambda_\alpha \omega_\alpha^2/\lambda)\,(\omega_\alpha^2/(\omega_\alpha^2 - (\omega + i\delta)^2)) \tag{10}$$

and $F(\omega)$ is the interband density-density response function

$$F(\omega) = N^{-1} \sum_{k,n'n}{}' v_{n'n}(\mathbf{k}) / (\omega_{n'n}(\mathbf{k}) - (\omega + i\delta)) \tag{11}$$

The imaginary part of $F(\omega)$ is just $N(\omega)$ defined in Eq. (8). $\tilde{F}(\omega) = F(\omega)/F(0)$. The quantities $\lambda_a = (2\, g^2{}_\alpha/\omega_\alpha)\, F(0)$ denote the dimensionless interband electron-phonon coupling constants while $\lambda = \Sigma_\alpha \lambda_\alpha$. The frequencies Ω_m of the renormalized H_g phonons are given by the solution of Re $D_{ph}(\omega)^{-1} = 0$. If the exciting frequency virtually promotes a conduction electron to the lowest lying t_{1g}-band, the H_g Raman spectrum is proportional to λ Im $D_{ph}(\omega)$ which is suppressed as $1/N(\Omega_m)$ in the region of Ω_m. On the other hand, Fano-like indentations in $\sigma_1(\omega)$ arise from dominant (negative) contributions in $\sigma_{ph}(\omega)$ which are proportional to $-N(\omega)^2\ \lambda$ Im $D_{ph}(\omega)$. Thus, the renormalized H_g modes may be observed in the infrared spectrum.

Illustrative calculations of λ Im $D_{ph}(\omega)$ and $\sigma_1(\omega)$ (excluding the Drude contribution) as a function of ω are shown in Figs. 1 and 2, respectively. We have taken representative values [13] for the frequencies ω_α of the six highest-frequency C_{60} H_g vibrations and assumed $\lambda_\alpha = 0.1$ for all α. $\chi_e(\omega)$ is expressed as $\mu^2 F(\omega)$ and $F(\omega)$ taken to be the sum of two broadened Lorentz oscillators centred at 780 cm^{-1} and 3000 cm^{-1}, respectively. We have assumed the value $(\delta\mu/\mu)^2 = 0.1$.

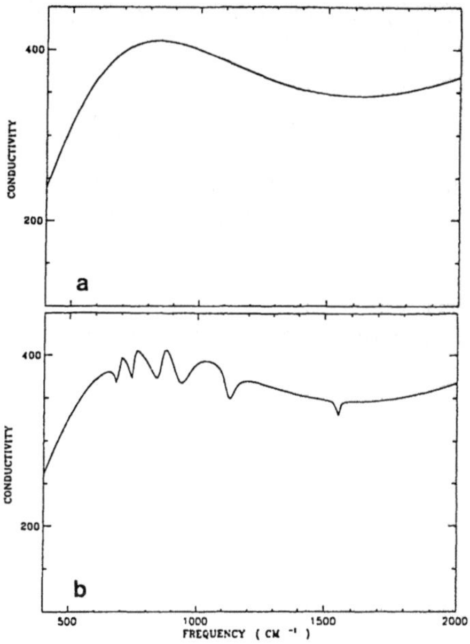

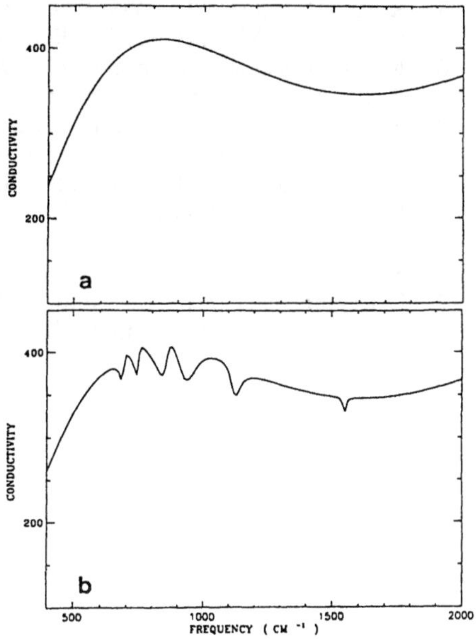

Fig. 1. H_g Raman spectrum (arbitrary units) calculated for (a) no coupling and (b) coupling to interband excitations.

Fig. 2. $\sigma_1(\omega)$ vs. ω (arbitrary units) calculated for (a) $\delta\mu = 0$ and (b) $\delta\mu \neq 0$.

A preliminary measurement of $\sigma_1(\omega)$ for Rb_3C_{60}, which reveals a broad absorption band centered at 0.1 eV with no vibrational features resolved, has been recently reported.[12] During the preparation of this MS, however, we have been informed of measurements [13] of $\sigma_1(\omega)$ for K_3C_{60}, which show that the Raman active H_g derived phonons are indeed present in the IR spectrum in accordance with the discussion of this paper.

We thank T. Pichler, M. Matus, and H. Kuzmany for much stimulating discussion. We also thank M.P. Gomes Da Costa for discussions at the Xerox Webster Research Center. One of us (H.-Y.C.) was supported by the Korean Science and Engineering Foundation (KOSEF) through the Center for Theoretical Physics (CTP), Seoul National University and through grant no. 931-0200-003-2.

References

1. R. C. Haddon et al., *Nature 350*, 320 (1991).
2. S. Saito and A. Oshiyama, *Phys. Rev. Lett. 66*, 2647 (1991).
3. C.M. Varma, J. Zaanen and K. Raghavachari, *Science 254*, 898 (1991); See also, M. Schluter et al., *Phys. Rev. Lett. 68*, 526 (1992).
4. A.F. Hebbard et al., *Nature 350*, 320 (1991); K. Holczer et al., *Science 252*, 1159 (1991).

5. K.-A. Wang et al., *Phys. Rev. B44*, 1955 (1992); S. Duclos et al., *Science 254*, 1625 (1991); M.G. Mitch et al., *Phys. Rev. Lett. 68*, 883 (1992).
6. K. Prassides et al., *Nature 354*, 462 (1991).
7. P. Stevens et al., *Nature 351*, 632 (1991).
8. Y.-N. Xu, M.-Z. Huang, and W.Y. Ching., *Phys. Rev. B44*, 13171 (1992).
9. M.J. Rice, *Phys. Rev. Lett. 37*, 36 (1976).
10. S. Chakravarty, S. Khlebnikov, and S. Kivelson, *Phys. Rev. Lett. 69*, 212 (1992), have also noted that there is no phonon broadening due to coupling to intra t_{1u}-band transitions.
11. See, e.g., P.B. Allen, *Phys. Rev. B6*, 2577 (1972).
12. L. Degiorgi et al., Phys. Rev. B46, 11250 (1992).
13. T. Pichler, M. Matus, and H. Kuzmany, private communication, and see this volume.

Langmuir-Blodgett Films of C_{60}: Structure of Pure Films and Their Superconductivity After Exposure to Potassium Vapor

R.M. Metzger and P. Wang

Department of Chemistry, University of Alabama,
Tuscaloosa, AL 35487-0336, USA

Abstract. C_{60} forms very rigid insoluble films ("monolayers") at the air-water interface (collapse pressure Π_c=62.5 mN/m). The "area per molecule" values A_0= 21.6 Å^2 (at Π=0) and A_c=11.5 Å^2 (at Π_c) suggest that more than 4 C_{60} molecules are piled atop each other, per "monolayer". These films were transferred to glass and HOPG as Langmuir-Blodgett (LB) films, and to a Cu grid as Langmuir- Schaefer films. These films have 2-D and 3-D crystalline regions, seen by scanning tunneling microscopy (STM) and high resolution elecron microscopy (HRTEM). The STM image is a {013} face-centered cubic (FCC) projection , a= 14.2 Å. The HRTEM micrographs and a selected-area diffraction pattern show an FCC lattice, a = 14.20 ± 0.05 Å. When optimally "doped" with potassium, an LB film of 50 "monolayers" of C_{60} on polyethylene terephthalate shows a low-field susceptibility signal (LFS) at or below 8.1 K: this is the first report of superconductivity in an LB film.

1. Introduction

The first report of the observation of C_{60} by laser desorption mass spectrometry [1] was followed, 5 years later, by its first practical synthesis [2]: C_{60} and its related higher homologs are the largest molecules ever synthesized by an astrophysicist. This has stimulated many studies of the structural, electronic, and electrical properties of these novel forms of carbon [1-6].

We summarize here our own studies of the structure of Langmuir-Blodgett and Langmuir-Schaefer films of C_{60} [7], and of the finding of superconductivity in LB films of C_{60} exposed to potassium vapor [8].

Studies of C_{60} by X-ray diffraction (XRD), STM, atomic force microscopy (AFM), and HRTEM have proved that the C_{60} molecule is a truncated icosahedron with 20 six-membered rings and 12 five-membered rings, with the carbon atoms roughly distributed on the surface of a sphere. In a crystal, the C_{60} molecules form a close-packed array, which, with equal density, could achieve either a face-centered cubic (FCC) structure, or a hexagonal close-packed (HCP) structure.

Most work on C_{60} involves crystals or vapor-deposited films, but C_{60} can also form [9-14] an insoluble film at the air-water interface, because the molecule is

Springer Series in Solid-State Sciences, Volume 117
Electronic Properties of Fullerenes Eds.: H. Kuzmany, J. Fink, M. Mehring, and S. Roth

hydrophobic; this film, or "monolayer", can be transferred onto solid substrates by the vertical Langmuir-Blodgett (LB) method, or by the horizontal Langmuir-Schaefer (LS) method.

While the early work attributed an HCP structure to bulk C_{60} [2], later work confirmed an FCC structure with lattice spacing **a**=14.11 Å [4] or 14.04(1) Å [5]; below 260 K, the structure transforms to simple cubic, with **a**=14.04 (1) at 11 K [5]. Similarly, an early study attributed to the LB films an HCP structure [10]. XRD and AFM studies of thick films of C_{60} on CaF_2(111) show FCC {311} planes [15].

2. Monolayers and Multilayers of Pure C_{60}

C_{60} (Texas Fullerenes, stated purity 99.9%) in benzene (1.297 x 10^{-4} mol/L) was spread on a purified water subphase (conductivity 16 MΩ) at room temperature in a Lauda film balance, and compressed after waiting for 20 minutes. The resulting pressure(Π)-area (A) isotherm [7] for the resulting Pockels-Langmuir (PL) film at the air-water interface is shown in Fig. 1. The collapse pressure Π_c = 62.5 mN/m indicates a very rigid film. Two estimates of the "area per molecule" are: A_0= 21.6 $Å^2$ / molecule (at Π=0) and A_c=11.5 $Å^2$ / molecule (at Π_c) [7]. These estimates, and the van der Waals volume of 78.5 $Å^3$ for C_{60} suggest that molecular aggregation

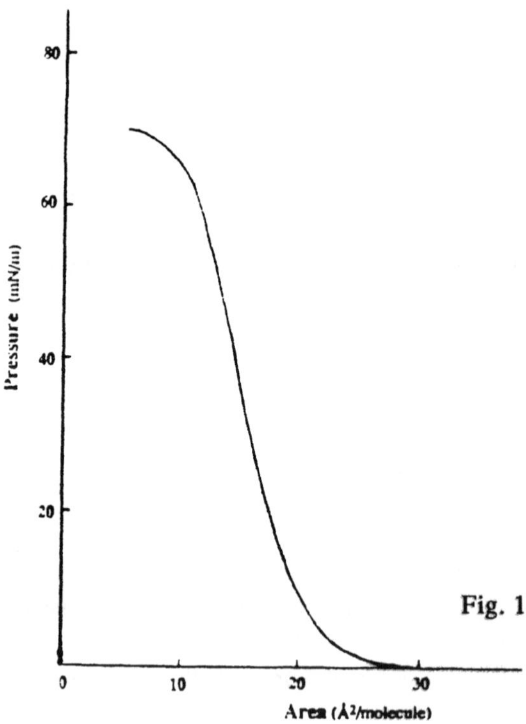

Fig. 1. A typical surface pressure-area isotherm of C_{60} on pure water at 20°C (Pockels-Langmuir film) [7].

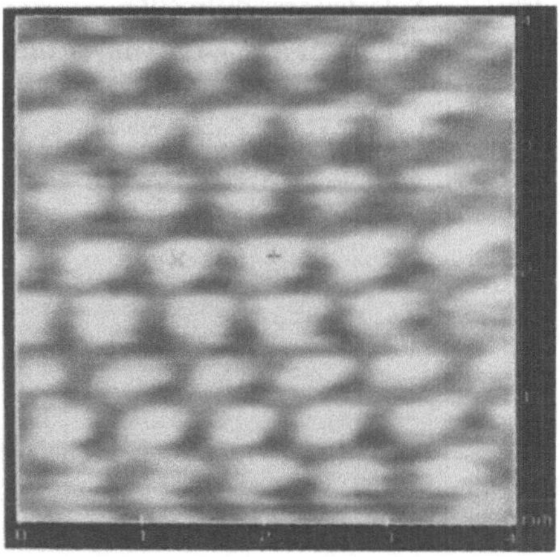

Fig. 2.
STM image of a C_{60} LB film on HOPG: a {013} projection of an FCC lattice, with distances 7.1 Å ((200]) × 5.0 Å ([131]) (4 x 4 nm^2 area) [7].

occurs before compression ("soap scum" formation): there are 4 to 7.5 C_{60} molecules per PL "monolayer" [7]. Larger areas are obtained only when very dilute spreading solutions [13], or a different subphase [14] are used.

The PL films were transferred vertically (LB method) onto glass and highly oriented pyrolytic graphite (HOPG), and horizontally (LS method) onto a Cu grid, with Z-type transfer assumed [11,12]. The thickness of two yellowish C_{60} LB films on glass substrates (n_s=1.424, k_s=0.001) was measured by ellipsometry (Rudolph Auto-EL-III, λ= 632.8 nm); for C_{60}, the complex part of the index of refraction was taken to be zero. The index of refraction n_f and the thickness t_f for the film varied in different regions of the films; the averages of 52 readings were n_f = 1.80(6), t_f = 19.0±1.2 Å/layer.

The C_{60} LB film on HOPG was studied by STM in air at room temperature (Digital Instruments Nanoscope II, A head,Pt/Ir tip, V= 0.2 to 1 V, i = 2 to 5 nA, tilt correction). The STM images showed amorphous, polycrystalline and crystalline regions. Fig 2 shows a crystalline part: the parallelogram pattern of roundish peaks 7.1(1) × 5.0(1) Å with angle 72±2° is consistent with a slightly distorted {013} face-centered pattern (7.1 Å = [200]; 5.0 Å =[131]), whence **a**=14.20 Å (if successive peaks along [200] are not at the same height) [7].

The C_{60} LS film growth on a Cu grid was examined in a 200 KV Hitachi H-800 transmission electron microscope near optimum defocussing, at a magnification of 2 × 10^5 diameters. Regions of faulty crystallinity, {111} and {200} lattice fringes, twin boundaries along the {111} plane, and other evidence of local formation of an FCC lattice by recrystallization were seen [7]. A selected-area diffraction pattern (Fig. 3) shows an FCC {211} plane with **a** = 14.20±0.05 Å [7].

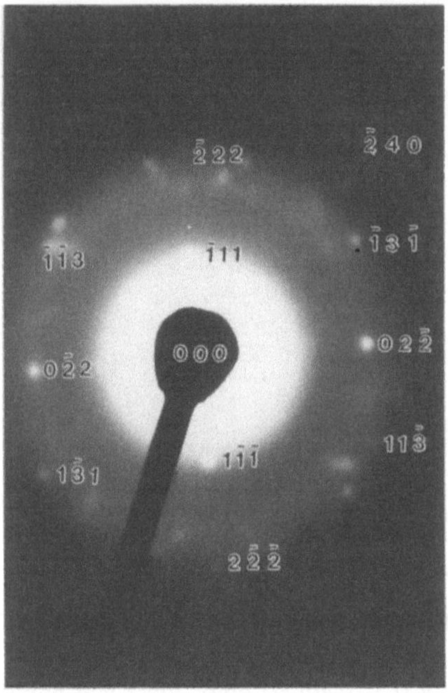

Fig. 3. Selected area diffraction pattern of a C_{60} LS 1-layer film, which is characteristic for the [211] plane of an FCC crystal, with a=14.20±0.05 Å [7].

3. Superconductivity in LB Films Doped with Potassium

The obvious next issue [8] was whether LB multilayers of C_{60} can become superconducting, in line with the "doping" of bulk C_{60} with K, Rb or Cs [6]. Although conducting LB films of partially ionic salts of amphiphilic molecules are known [16,17], no superconductivity had been reported heretofore in LB films. C_{60} is not amphiphilic, and its packing as an LB multilayer is similar to that of bulk C_{60}; thus the difference between LB multilayers and bulk films of C_{60} is academic.

Superconductivity was indeed found at or below 8.1 K, as evidenced by a low magnetic field microwave absorption signal, or low-field signal (LFS), in a K-doped Langmuir-Blodgett (LB) 50-layer film of C_{60} [8]. The LFS technique was used in an earlier study of bulk K_3C_{60} [18].

C_{60} (stated purity 99.9%, dissolved in benzene, 2×10^{-5} mol L^{-1}) was spread on a purified water subphase at 20℃, allowed to sit for 20 minutes, compressed to 25 mN/m pressure, and transferred onto strips of polyethylene terephthalate by the LB method (50 Z-type transfers). Five such strips were placed in one side of a quartz electron spin resonance (ESR) tube; an excess amount of potassium metal was rapidly introduced in the other side of the tube, then the tube was evacuated, the K was distilled, and the tube was sealed during pumping. The tube, with K in one end and the C_{60} LB films in the other, was heated to 150℃ (K side) and 200℃ (C_{60} LB films side) for "doping" times t_d = 0 - 700 min [8]. ESR and LFS spectra were

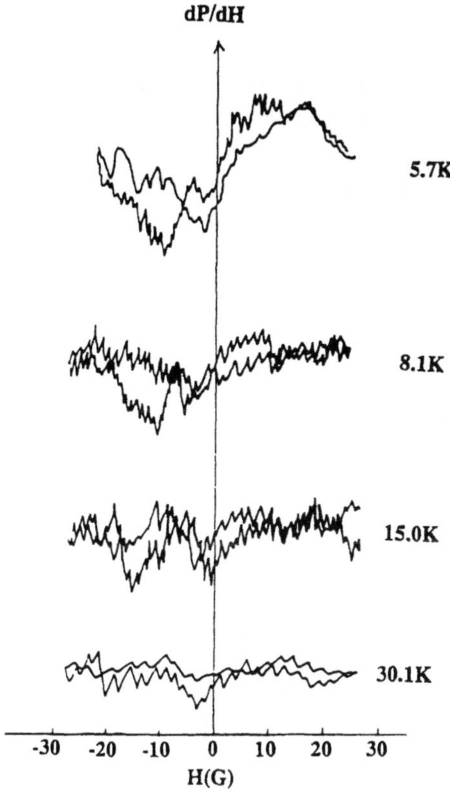

Fig. 4. Low-field signal (LFS) as a function of temperature T for a 50-layer Langmuir-Blodgett film film of C_{60} doped with K for t_d =455 min [8].

measured with a Bruker ESP300E spectrometer operated at X band, equipped with an Oxford Instruments ESR 900 liquid He flow cryostat (T = 2.7 - 300 K).

No LFS was seen for t_d < 300 min. At t_d = 300 min, a weak LSF signal appears at T = 2.7 K. With increasing t_d, the LSF signal intensity increases. Figure 4 shows the T-dependence of the LFS signal for t_d = 455 min: an LFS signal with hysteresis (and phase opposite to the ESR high-field signal) can be seen for T ≤ 8.1 K [8]. This indicates that the K-doped LB film superconducts at or below 8.1 K. Compared with a bulk K_3C_{60} sample, the LFS intensity is very weak, probably because of the very thin LB film. This superconducting transition occurs below the critical temperature T_c = 18.6 K of bulk K_3C_{60} [6,18]. We have no estimate of the superconducting fraction. The thickness of our superconducting film is presumably at least 500 Å. When the sample is doped for t_d > 500 min, the LFS becomes gradually weaker, presumably as K_4C_{60} and/or K_6C_{60} form [8].

The four-probe resistivity of conventional (non-LB) 1000 Å films of K-doped C_{60}, after two days of annealing, shows T_c = 8 K (midpoint), ΔT_c= 10 K [19].

4. Acknowledgement

We thank our excellent collaborators M. Shamsuzzoha, X.-L. Wu, W.-J. Lee (Univ. of Alabama) S. Bandow, and Y. Maruyama (Institute for Molecular Science, Okazaki, Japan), and are grateful for the financial support of JVC America and the United States Air Force Office of Scientific Research (F49620-92-J-0529DEF).

5. References

[1] H. W. Kroto, J. R. Heath, S. C. O'Brien, R. F. Curl, and R. E. Smalley, *Nature* **318:** 162 (1985).

[2] W. Krätschmer, L. D. Lamb, K. Fostiropoulos, and D. R. Huffman, *Nature* **347**: 354 (1990).

[3] P. J. Benning, J. L. Martins, J. H. Weaver, L. P. E. Chibante, and R. E. Smalley, *Science* **252:** 1417 (1991).

[4] P. W. Stephens, L. Mihály, P. L. Lee, R. L. Whetten, S.-M. Huang, R. Kaner, F. Deiderich, and K. Holczer, *Nature* **351:** 632 (1991).

[5] P. Heiney, J. E. Fischer, A. R. McGhie, W. Romanow, A. M. Denenstein, J. P. McCauley, Jr., and A. B. Smith, III, *Phys. Rev. Lett.* **66**: 2911 (1991).

[6] A. F. Hebard, M. J. Rosseinsky, R. C. Haddon, D. W. Murphy, S. H. Glarum, T. T. M. Palstra, A. P. Ramirez, and A. R. Kortan, *Nature* **350**: 600 (1991).

[7] P. Wang, M. Shamsuzzoha, X.-L. Wu, W.-J. Lee and R. M. Metzger, *J. Phys. Chem.* **96**: 9025 (1992).

[8] P. Wang, R. M. Metzger, S. Bandow, and Y. Maruyama, *J. Phys. Chem.*, accepted and in press.

[9] Y. S. Obeng and A. J. Bard, *J. Am. Chem. Soc.* **113:** 6279 (1991).

[10] C. Long, Y. Xu, F. Gou, Y. Li, D. Xu, Y. Yao, and D. Zhu, *Solid State Commun.* **82:** 381 (1992).

[11] T. Nakamura, H. Tachibana, M. Yumura, M. Matsumoto, R. Azumi, M. Tanaka, and Y. Kawabata, *Langmuir* **8**: 4 (1992).

[12] G. Williams, C. Pearson, M. R. Bryce, and M. C. Petty, *Thin Solid Films* **209**: 150 (1992).

[13] L. O. S. Bulhoes, Y. S. Obeng, and A. J. Bard, submitted to *Adv. Mater.*

[14] Y. Tomioka, M. Ishibashi, H. Kajiyama, and Y. Taniguchi, *Langmuir* **9**: 32 (1993).

[15] J. L. Wragg, J. E. Chamberlin, H.W. White, W. Krätschmer, and K. Huffman, *Nature* **384:** 623 (1991).

[16] A. Ruaudel-Teixier, M. Vandevyver, and A. Barraud, *Mol. Cryst. Liq.Cryst.* **120**: 319 (1985).

[17] T. Nakamura, H. Tachibana, M. Matsumoto, M. Tanaka, and Y. Kawabata, *NATO ASI* **B248**: 519 (1991).
[18] A. A. Zakhidov, A.Ugawa, K. Imaeda, K. Yakushi, H. Inokuchi, K. Kikuchi, I. Ikemoto, S. Suzuki, and Y.Achiba, *Solid State Commun.* **79**: 939 (1991).
[19] Y. Maruyama, private communication.

Steady-State Photoconductivity of Fullerene Films

M. Kaiser, W.K. Maser, H.J. Byrne, J. Reichenbach, J. Anders, A. Mittelbach, and S. Roth

Max-Planck-Institut für Festkörperforschung,
Heisenbergstr. 1, D-70569 Stuttgart, Germany

Abstract. The spectral and temperature dependence of the photoconductive response of polycrystalline fullerene films is reported. The temperature and excitation energy dependence of the photoconductivity reveals the transport to be thermally activated, limited by transport through the crystalline grain boundaries. An additional anomalous behaviour in the temperature dependence of the photoconductivity is observed and discussed.

1. Introduction

In the past, some work on transport in solid films of C_{60} has been reported [1-5], but since Pichler et al. [6] demonstrated the differing transport behaviour of purified C_{60} and C_{60}/C_{70} mixtures, there have been no further reports of transport in highly purified material. Yonehara and Pac [7] have reported measurements in a polycrystalline C_{60}-film containing 2.7% C_{70}, concluding that transport is controlled by impurity sites. In polycrystalline samples, measurement of transport properties can be strongly sample dependent and so caution is necessary. Thus, it is surprising that investigation of well defined films of high crystalline order has been so badly neglected. In this letter, the photoconductive response of polycrystalline films of highly purified fullerene is reported.

2. Experimental

C_{60} was prepared following the procedure described by Mittelbach et al. [8]. The fullerenes were separated by MPLC and the eluted C_{60} was dried under dynamic vacuum and characterised by time of flight mass-spectroscopy. The purity was found to be better than 99.9%. Films of a few hundred nm were evaporated onto thin sapphire plates with gold contacts in a meander structure of 200μm gap width. Electron microscopy shows the films to be polycrystalline with domain size of the order of 200 nm. The samples were exposed to air only during the transfer to a cryostat. To minimise the reported influence of oxygen [9], the samples were heated at 430 K for several hours under vacuum. The sample was mounted in a cryostat which allowed controlled variation of temperature between 450 and 77 K under a vacuum of ~ 10^{-5} mbar. The photoconductivity action spectrum (PAS) was

Springer Series in Solid-State Sciences, Volume 117
Electronic Properties of Fullerenes Eds.: H. Kuzmany, J. Fink, M. Mehring, and S. Roth

taken with a 500 W Xenon lamp, chopped at 10 Hz, using lock-in detection. To enable variable excitation energy, a 1/4 m monochromator combined with an edge filter system was used. The spectral resolution was 5 nm. The photocurrent was corrected for the system spectral response. The applied electric field was of the order of 5000 V/cm and the light intensity was always below 2 mW/cm^2.

3. Results and Discussion

Fig. 1 shows the normalised PAS compared to the absorption spectrum obtained by a PDS (photothermal deflection spectroscopy) measurement of a C_{60} film in silicon oil. The PAS follows the absorption with roughly the same features in the high energy regime, which may be assigned to the higher optical transitions and the onset for both photocurrent and absorption is in the region of 1.6 eV. The absorption edge is roughly exponential, a behaviour which is characteristic of an amorphous semiconductor. The absorption profile is given by $\alpha = \alpha_0 \exp(h\nu/E_U)$ with an Urbach energy E_U representing the degree of disorder. The observed value of 90meV for fullerenes is comparable to that of amorphous silicon [10] although a direct comparison of the transport mechanism to that of amorphous silicon, dominated by dangling bond defects, seems inappropriate. The sub-gap regime is dominated by defect states. The influence of a midgap absorption can clearly be seen in the PAS. The photocurrent has an onset at 0.6 eV and goes through a maximum at 0.8 eV. The Fermi level appears to be pinned by defect states in the midgap and the subgap photocurrent is a result of excitation from and into midgap states. Fig. 2 shows the effective quantum efficiency of the photocurrent, i.e. the number of charge carriers per absorbed photon. The onset of intrinsic photocurrent remains at the same excitation energy of 1.6 eV. The ascent over one order of magnitude is much weaker than in other well defined molecular crystals [11], but this behaviour can be a result of the comparatively high degree of disorder. Nevertheless, there is an intrinsic photoconductivity, i.e. the probability of

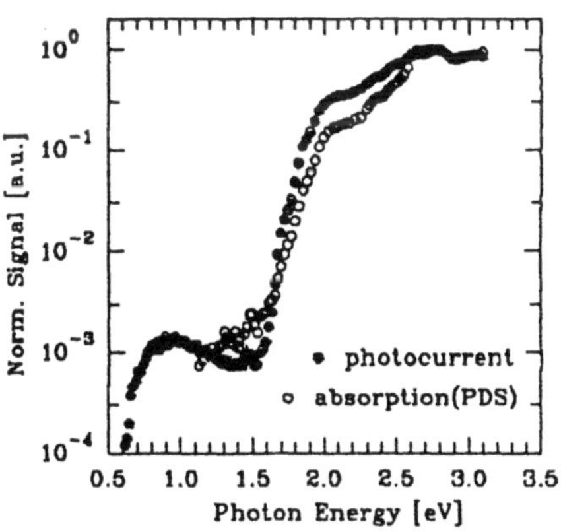

Fig. 1; Photoconductivity action spectrum

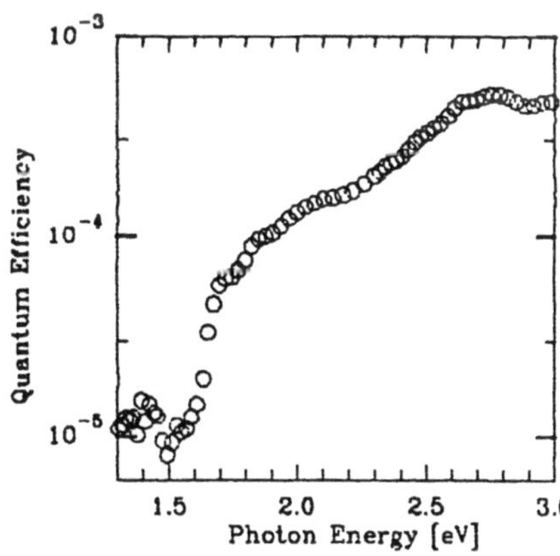

Fig. 2; Effective quantum efficiency.

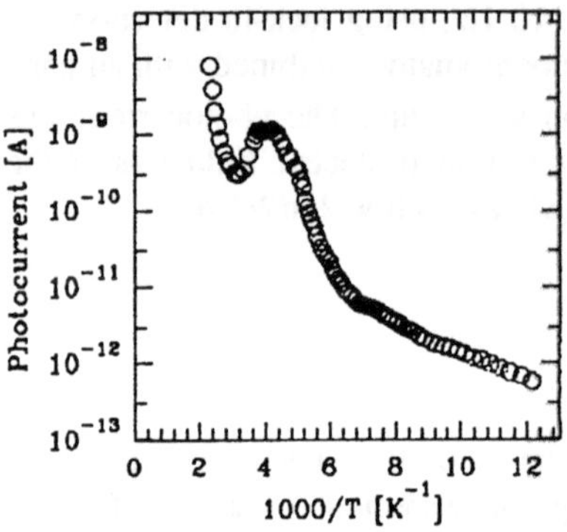

Fig. 3; Dependence of the photoconductive response on inverse temperature.

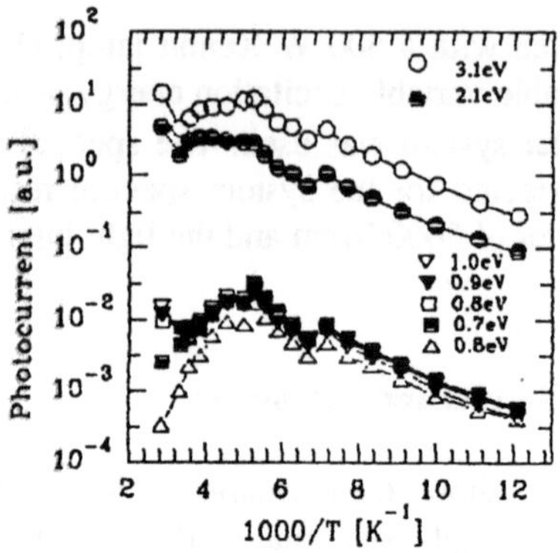

Fig. 4; Temperature dependence of the photoconductivity for different excitation energies

dissociation for the electron-hole pair created by the absorbed photon depends on the excitation energy. Such a behaviour has been explained by the Onsager theory of dissociation [11], but in contrast to the behaviour observed by Mort et al [4], the C_{60} films showed no clear agreement with Onsager's theory. The influence of the electric field on photoconductivity was found to be linear in the regime of 2000 up to 20000 V/cm. The temperature dependence of photoconductivity on the other hand looks much more exciting.

Fig. 3 shows an Arrhenius plot of the photocurrent at a photon energy of 2.1 eV. In the low temperature regime, the photoconductivity increases slowly with temperature, which may be due to a hopping process previously described [12]. In the intermediate temperature regime the photocurrent increases with temperature up to a local maximum at about 250 K. The high temperature regime is dominated by an activated process with an energy of 0.5 eV. Recently, a similar feature was observed but dismissed as a systematic error by Mort et al. [3]. To examine this feature in more detail, the temperature dependence of the photoconductive response was monitored at varying excitation energies and is shown in Fig. 4. Apparent differences to fig. 3 are the result of differing experimental conditions. The important features of fig. 3, however, remain. In the regime of medium and low temperatures, the temperature dependence does not vary with excitation energy between 0.6 and 3.1 eV, implying that the local maximum is independent of light. Indeed, it appears also in the dark conductivity [2]. In the high temperature regime a change in the sign of the slope at about 0.8 eV is observed. Thermal activation occurs only from states excited with 0.8 eV or more, an energy which may be associated with midgap states (Fig 1). Absorption above 0.8 eV creates carriers in the conduction band and so the activated process at higher temperatures may be connected to these charge carriers.

In polycrystalline films the current, or photocurrent, is always limited by the weakest link, the boundary between the C_{60} grains. In Fig. 1, it was demonstrated that the onset and spectral dependence of the photocurrent is characteristic of an amorphous material although the films are polycrystalline. Spectroscopic studies indicate that the first, symmetry forbidden, transition and the second transition at 2.7 eV are somewhat independent in films and largely influenced by crystallite surfaces [13]. It may be inferred, therefore, that the onset of the photoconductivity at 1.6 eV is associated with molecules on the grain surfaces, the random perturbation of which gives rise to a dependence on excitation energy which is characteristic of amorphous materials. The intrinsic dissociation probability of electron hole pairs may be a combination of surface and bulk dissociation, the relative contributions of which cannot be separated conclusively, but the onset at 1.6 eV will be influenced mainly by the C_{60} molecules in the grain surface and the higher energy region will be dominated by the bulk material.

A further feature of the spectral dependence is the transport maximum at excitation energies of about 0.8 eV. Skumanich [14] reported the presence of a midgap absorption and discussed it in terms of dangling bonds. In a π-conjugated molecular material, however, such a phenomenon is difficult to conceptualise. In polymeric materials, subgap states of solitonic or polaronic nature are well known [15], but in this case, an explanation in terms of damaged molecules may be more plausible.

The dependence of the photocurrent on temperature should be discussed in terms of the different temperature regions. The high temperature regime in Fig. 3, described by an activated process, is connected to charge carriers in the conduction band that may thermally overcome the potential barrier between the grains. At an excitation energy of about 0.8 eV, the onset of generation of charge carriers in the conduction band out of mid gap states, a thermally activated transport across grain boundaries resulting in an exponential increase of photocurrent is observable. If the excitation energy is not high enough to create the charge carriers in the conduction band and if the thermal energy is too small to overcome the grain boundary, the transport is extrinsic, dominated by hopping, or tunnelling processes This is strikingly demonstrated in Fig. 4, where the thermally activated process at high temperatures disappears at excitation energies below 0.8 eV while the low temperature regime retains its characteristics independent of excitation energy.

The local maximum in the temperature dependence of photocurrent at about 250 K has been proposed to be connected to the phase transition at 249 K [16]. However, the local maximum in the photoconductive response is observed to be surprisingly dependent on sample preparation. A more satisfactory explanation may result from comparison to the photoconductivity of amorphous silicon. There an increase of temperature effects the mobility of hole defects, rendering a new recombination process with electrons more probable [17]. An onset of such a recombination path in the region of the grain boundaries with increasing temperature would compete with the transport through the barrier until the thermal energy is high enough to overcome it.

4. Conclusions

In summary, this study elucidates the importance of both extrinsic and intrinsic transport mechanisms in the photoconductive response of C_{60}. At excitation energies above 1.6 eV, an intrinsic photoconductivity is observed which is activated via the forbidden optical transitions and so may be ascribed to a crystallite surface effect. A strong feature in the spectral dependence of the photocurrent is seen at midgap energies and is discussed in terms of defect trapping levels, possibly caused by chemically modified molecules. Although bulk conductivity is observed, the limiting factor at all energies is transport through the grain boundaries. This process is thermally activated at photon energies above 0.8 eV, below these energies, and at low temperatures, the transport is reduced to a tunnelling and/or hopping process. At intermediate temperatures, both the photocurrent and the dark current go through a maximum, which may be interpreted as the turn on of a new temperature dependent recombination process, possibly at the grain boundaries. The most notable observation is the strong dependence of the transport properties of the films on the polycrystalline nature of the samples. The features of the photoconductive response are strongly determined by the influence of grain boundaries and defect states. Implied, thus, is that the assignment of measured transport phenomena to bulk, intrinsic material properties should be performed cautiously and the need for improved quality films and studies on single crystalline materials is emphasised.

5. Acknowledgement. The authors would like to acknowledge M. Brandt and M. Stutzmann for help with the PDS and L. Akselrod for film preparation.

6. References

[1] J. Mort et al., Chem. Phys. Lett. **186**, 281 (1991)
[2] J. Mort et al., Chem. Phys. Lett. **186**, 284 (1991)
[3] J. Mort et al., Appl. Phys. Lett. **60**, 1735 (1992)
[4] J. Mort et al., Appl. Phys. Lett. **62**, 1 (1992)
[5] N. Minami, Mol. Cryst. Liq. Cryst. Sci. Technol. A **217**, 231 (1992)
[6] K. Pichler et al., J. Phys.: Condens. Matter **3**, 9259 (1991)
[7] H. Yonehara, C. Pac, Appl. Phys. Lett. **61**, 575 (1992)
[8] A. Mittelbach et al., Angew. Chem. in press
[9] H. Werner et al., Angew. Chem, **104**, 909 (1992)
[10] G. D. Cody in *Semiconductors And Semimetals*, edited by J. I. Pancove (Academic Press, New York, 1984), Vol 21B, pp 11-79
[11] E. A. Silinsh, *Organic Molecular Crystals - Their Electronic States*, Springer Ser. Solid-State Sci. Vol 16 (Springer, Berlin, 1980).
[12] M. Kaiser et al., Solid State Commun. **81**, 261 (1992).
[13] L. Akselrod et al., these proceedings
[14] A. Skumanich, Chem. Phys. Lett. **182**, 486 (1991)
[15] W. P. Su, J. R. Schrieffer and A. J. Heeger, Phys. Rev. B, **22**, 2099 (1980)
[16] P. A. Heiney et al., Phys. Rev. Lett. **66**, 2911 (1991).
[17] H. Dersch, L. Schweitzer and J. Stuke, Phys Rev. B **28**, 4678 (1983)

Charge Transient Spectroscopy of Fullerites

D. Barančok[1], *M. Haluška*[2], *H. Kuzmany*[2], *and V. Nádaždy*[3]

[1]Department of Physics, EF STU, Ilkovičova 3, 812 00 Bratislava, Slovakia
[2]University of Wien, Institute for Solid-State Physics,
Strudlhofgasse 4, A-1090 Wien, Austria
[3]Institute of Physics, SAS, Dúbravská cesta 9, 842 28 Bratislava, Slovakia

Abstract Suitability of the charge transient spectroscopy - QTS [1] for the investigation of fullerites is demonstrated. The principle of this method consists in correlating the charge transient induced by periodic rectangular voltage pulses, while scanning temperature. QTS can yield an information on the electrical properties of both the bulk and the interface of fullerites interfaced with metals (insulators, semiconductors) as well as on its bulk phase transition in the former.

1. Experimental technique

The charge sensitive amplifier (Fig. 1.) is the heart of the QTS apparatus. It converts both emitted and displacement charges $Q(t)$ of the sample under test into an output voltage $U_{out}(t)$,

$$Q(t) = C_s(t)\ \Delta U = C_f\ U_{out}(t)\ ,$$

where $C_s(t)$ is capacitance of the sample and ΔU is an input voltage step. The feedback capacitor C_f determines the amplifier sensitivity $S = 1/C_f$ which is defined as the ratio of ΔU_{out} and the related change of the detected charge in response to ΔU. If the sample was an ideal capacitor, the shape of the ΔU_{out} would be identical to that of the rectangular voltage source. In case of real system, the response to the bias voltage step would not settle immediately. Instead, only a portion of the charge that corresponds to geometric capacitance would be transfered immediately and any slow processes, such as trap limited charge emission or dielectric polarization, would gradually transfer the remainig charge to the output. A suitable sampling of the transient allows implementation of at least two regimes. First of them represents the classical boxcar rate window concept [2] for assessing parameters of the thermally activated relaxation. The second regime uses sampling of the output voltage before and after the small voltage step ΔU to measure the sample capacitance $C_s(t)$, as it was introduced by Mego [3]. As was shown earlier [4], there is a relationship between the time domain capacitance $C_s(t)$ and its frequency domain counterpart $C_s(\omega)$ measured by the standard ac technique.

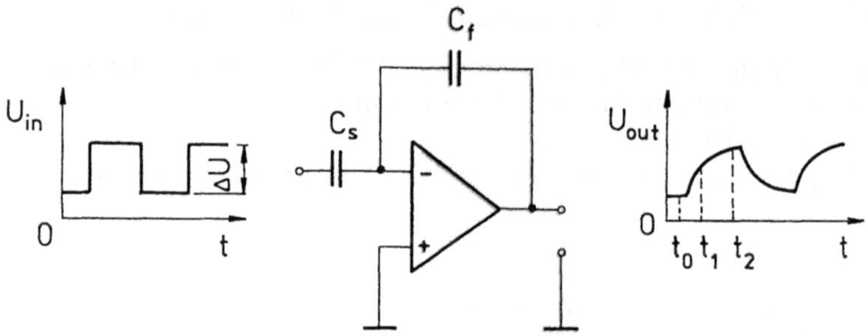

Fig. 1. The charge sensitive amplifier (CSA), ΔU - excitation pulse amplitude, C_s - capacitance of the sample, C_f - feedback capacitance, $\tau = \tau_0 \exp(\Delta E/kT)$ describes the thermally activated relaxation.

2. Experiment and Results

Our measurements were carried out on three C_{60} single crystals. After the growth the crystals were kept in air at room temperature. Their approximate dimensions were 1 x 1 x 0.3 mm^3. Gold contacts were evaporated onto the larger parallel planes of the crystals in order to create capacitors. One of the contacts was formed through a mask with the area of 0.34 mm^2. The geometrical capacitance of such a system, supposing a relative static permittivity of 4.4 [5], is about 50 fF.

At room temperature the resistivity of all crystals was on the order of 10^8 Ωcm. The expected phase transition is visible at $T_C = 255$ K as a step in the signal for either of the crystals. Otherwise, there were two different responses. One crystal showed peak A (above the phase transition) which is not detected on the remaining crystals. However, the low temperature peak B was measured only on the later ones. Peak A is characterized by an activation energy 0.63 eV determined from Arrhenius plot and a preexponetial factor $\tau_0 = 10^{-9}s$. Further we were interested particullarly in peak B. Analogically determined parameters of the latter are $\Delta E = 0.23$ eV and $\tau_0 = 2$ x $10^{-9}s$. The activation energy calculated from the full width at half maximum is only 0.11 eV. It points to an energy distribution of the relaxation process. The range of available rate windows allowed us to monitor this peak also around the phase transition temperature. Its evolution is depicted in Fig. 2. The peak amplitude decreases but the relaxation survives even after the phase transition.

In addition, we followed the effect, the sample annealing in both vacuum and air has on the peak behaviour. We began with the annealing in vacuum (10 Pa) for 100 minutes. It caused an increase in the activation energy of the relaxation B up to 0.45 eV with $\tau_0 = 1$ x $10^{-14}s$. Then the sample was exposed to air at 300 K for 12 hours. The peak B moved to lower temperatures and its amplitude decreased slightly (Fig. 3). It corresponds to $\Delta E = 0.27$ eV and $\tau_0 = 3$ x $10^{-11}s$. Next two separate annealings in air at 400 K for 100 min. gave cause for a further decrease in the activation energy ($\Delta E = 0.21$, 0.14 eV) and increase in the frequency factor ($\tau_0 = 7$ x 10^{-10}, 2 x $10^{-8}s$). The

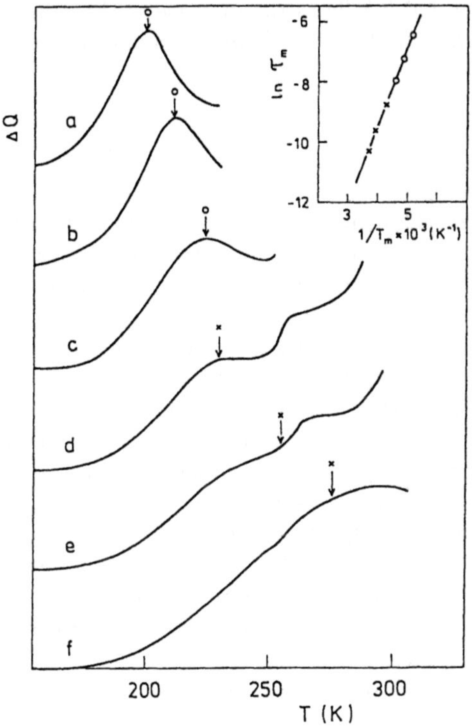

Fig. 2. The evolution of the peak B at various values of the ratio t_1/t_2: a) 1/2, b) 0.5/1, c) 0.25/.5, d) 0.1/0.2, e) 0.05/0.1 f) 0.025/0.05 *ms*. Note the peak survival after the phase transition. The inset depicts Arrhenius plot (ln τ_m vs $1/T_m$) for a set of peaks B. $\tau_m = t_1/\ln(t_2/t_1)$ is the relaxation time for the peak maximum at T_m. Daggers correspond to extrapolated values of T_m.

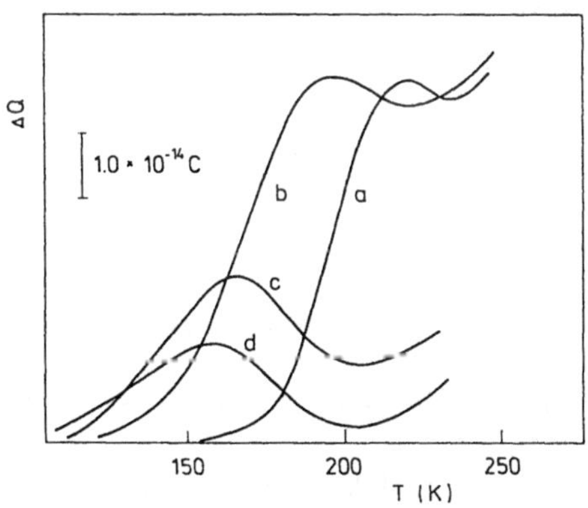

Fig. 3. Behaviour of the peak B after thermal treatments: a) annealing in vacuum at 400 K for 100 minutes, b) following storage in air for 12 hours, c) the first annealing in air at 400 K for 100 minutes, d) the second annealing under the same conditions as for c).

decrease of the ramp of the spectra in fig. 3 reflects the increase of the sample resistivity at 250 K from 3.2 x 10^8 Ωcm to 9.7 x 10^8 Ωcm. It should be noted that this annealing impact is reversible.

3. Discussion and Conclusion

In principle, there are three possible mechanisms for the origin of the thermally activated relaxation. First, dipolar polarization seems to be the least probable one. The reason consits in the fact that the measured capacitance dispersion ($\Delta C_{Tmax} = \Delta Q_{Tmax}/\Delta U$) is on the order of the static capacitance while usually value around $\sim 10^{-2}$ is found. Other possibilities are connected with trap limited charge emission, either from bulk traps or those located in the contact regions. From the presented data it is difficult to decide unambiguosly. However, the experiment with annealing in air enables to make some preliminary conclusions. A few works [6 - 8] provided namely convincing evidence of a strong diffusion of oxygen into C_{60}. Incorporation of O_2 molecules in the crystal lattice is attended with increasing resistivity of C_{60}. Therefore the peak B probably reflects the charge emission from bulk traps. Since at the temperatures employed the crystal has the high resistivity, a more profound analysis should be based on models like that of the semi - insulating GaAs. As for the peak B, we found three remarkable facts:

- after the phase transition at 255 K the corresponding thermally activated relaxation survives,
- a continuos decrease of its activation energy (from 0.45 eV to 0.17 eV) and the simultaneous increase of its preexponential factor τ_0 (from 1 x $10^{-14}s$ to 2 x $10^{-8}s$) at annealing in air,
- storation of its primary parameter values at annealing in vacuum.

References

[1] K.I. Kirov, K.B. Radev, phys. stat. sol. (a), **63**, 711 (1981).
[2] D.V. Lang, J. Appl. Phys. **45**, 3014 (1974).
[3] T.J. Mego, Rev. Sci. Instrum. **52**, 2798 (1986).
[4] I. Thurzo, M. Grendel, J. Meas. Sci. Technol. **3**, 726 (1992).
[5] A.F. Hebard, R.C. Haddon, R.M. Fleming, A.R. Kortan, Appl. Phys. Lett. **59** , 2109 (1991).
[6] J.E. Schirber, R.A. Assink, D. Loy, B. Morosin, Bull. Am. Phys. Soc. **37**, 400 (1992).
[7] S.J. Duclos, R.C. Haddon, S.H. Glarum, A.F. Hebard, K.B. Lyons, Solid State Commun. **80**, 481 (1991).
[8] T. Arai, Y. Murakami, H. Suematsu, K. Kikuchi, Y. Achiba, I. Ikemoto, Solid State Commun. **84**, 827 (1992).

Characterisation of Fullerene Schottky Diodes

S. Curran[1], *J. Callaghan*[1], *D. Weldon*[2], *E. Bourdin*[1], *K. Cazini*[1], *W.J. Blau*[1], *E. Waldron*[3], *D. McGoveran*[1], *M. Delamesiere*[2], *Y. Sarazin*[1], *and C. Hogrel*[1]

[1]Department of Pure and Applied Physics, Trinity College Dublin, Dublin 2, Ireland
[2]Chemistry Department, Trinity College Dublin, Dublin 2, Ireland
[3]Microelectronics and Electrical Engineering Department, Trinity College Dublin, Dublin 2, Ireland

Abstract.

The fabrication and characterisation of the fullerenes C_{60} and C_{70} as semiconductors in Schottky diodes are reported. The preparation of these devices involved forming a Schottky barrier between heavily doped n-Si and Ca with C_{70}, and Mg-In with C_{60}. Electrical diode characteristics are determined and from these measurements both the device quality and the electron affinity of C_{60} and C_{70} were calculated. The value obtained for the electron affinity of C_{60} is in good agreement with other work reported in this area. In the case of n-Si/C_{70} Schottky diode the efficiency as a photovoltaic device was also investigated.

1 Introduction

Since the discovery of this new class of materials the fullerenes have been investigated in great detail towards practical technological application. Studies have ranged from electronic devices such as MIS[1] and FET[2] to the area of superconductors[3]. The following is a report for their application as Schottky diodes.
A Schottky diode is a device which is formed when a metal and a semiconductor come into contact. These diodes are normally majority carriers except on rare occasions when a drift current is introduced. The materials used for these devices are:

1) Au/C_{70}/n-Si,
2) ITO/C_{70}/Ca/Al,
3) ITO/C_{60}/Mg-In.

The Schottky diode formed is between the fullerene and the low work function materials: n-Si, Ca, Mg-In, where the work functions vary between 2.5 - 3.9 eV. From these measurements it is possible to obtain the diode quality factor and the electron affinity value.

Springer Series in Solid-State Sciences, Volume 117
Electronic Properties of Fullerenes Eds.: H. Kuzmany, J. Fink, M. Mehring, and S. Roth

2 Schottky Diode Theory

When a metal makes contact with a semiconductor the fermi levels of the two materials must be coincident at thermal equilibrium. The equilibrium charge distribution is in the region of the metal/semiconductor interface. For n-type diode characteristics the Schottky diode is similar to that of a one sided abrupt p-n junction. Unlike their counterparts the p-n junction diodes of the Schottky diode is dependent on majority carriers for current transport. There are a number of methods of analysing the current characteristics of the diodes.

1) Thermionic Emission: This theory is based on the assumption that the barrier height $q\Phi_{Bn}$ is much larger than kT, and thus is highly temperature dependent.
2) Isothermal Diffusion: This theory neglects the effect of temperature and is dependent on electron collisions within the depletion region as well as the fact that the carrier concentration within the depletion region is unaffected by the current flow (i.e. they are in equilibrium).
3) Drift Carrier Transport: In this case the minority carriers dominate the transport mechanism in the device.

The barrier height between metal and semiconductor is given by:

$$q\Phi Bn = q(\Phi_m - \chi),$$

where Φ_m is the metal work function and χ is the electron affinity value. The general expression which incorporates all these theories can be described as follows:

$$Jn = Jsat \exp\frac{qV}{kT} - 1 \quad .$$

V is the applied voltage and J_{sat}, the saturation current, can be described by:

$$Jsat = A * T^2 \exp\frac{-q\Phi Bn}{kT}$$

The above expression is a generalised form of the saturation current for the three theories, where A^* is the Richardson's constant for thermionic emission into a vacuum ($A^* = 120$ mA/cm^2/K) and Φ_{Bn} is the barrier height. The barrier height, and hence the electron affinity can be obtained as follows:

$$\Phi Bn = \frac{kT}{q} Ln(\frac{A * T^2}{Jsat})$$

The effect of the minority carriers must be analysed. At large forward bias the minority carrier injection ratio γ (the ratio of minority carrier current to the total current) increases due to the drift field component which becomes larger than the diffusion current. The concentration of the number of carriers available for the conduction process plays a large part in dictating whether or not which process will dominate.

3 Device Preparation

Evaporation of C_{60} and C_{70} was performed at a pressure of $2x10^{-6}$ mbar in the temperature range of 690K to 750K for two hours. This procedure resulted in a film thickness of 200 nm. The metals Ca, Mg-In and Al were all evaporated at a pressure of $2x10^{-6}$ mbar. This procedure took five minutes and resulted in a metal thickness of 500 nm. To obtain an Au layer a sputtering device was used. The sputtering lasted 30s and gave a film thickness of 200 $\mathring{A}$. The following are the areas of the diodes:

Area of Mg-In/C_{60}/ITO: 4mm,
Area of n-Si/C_{70}/Au: 3.8cm,
Area of Al/Ca/C_{70}/ITO: 4.2cm.

4 Analysis

It can be seen clearly that all three diodes show quite good characteristics not dissimilar to most other organic diodes[4]. The exponential expansion is in the region of 2 to 2.75 volts, depending on the size of the device. The overall area difference is not that great so that the response of all three would be expected to be somewhat similar. When considering electrode materials, the most obvious choice are those incorporating very high and very low work functions. The diode characteristics shown in Fig.1 is the Mg-In/C_{60} device. By using Mg-In alloy it was possible to produce a stable electrode contact with a low work function. From the diode characteristics it was thus possible to derive the C_{60} electron affinity value. However, this value is larger than expected probably due to the alloy used. The higher the work function used the weaker the Schottky effect and thus the exactness of the electron affinity value. The most common diode fabricated to date using organic materials[5] is one which has metal contacts consisting of ITO (Indium Tin Oxide) and calcium (diode characteristics are shown in Fig.2). The result of using these is that the value of the electron affinity is quite good. However, this device is not as stable as others that may be used.

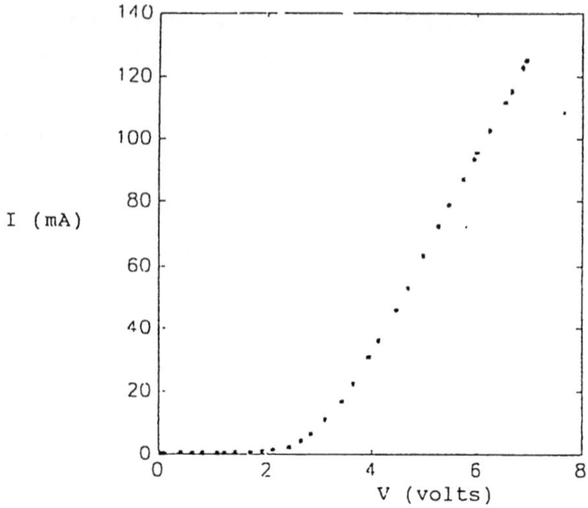

Fig. 1: MgIn/C_{60} Schottky Diode

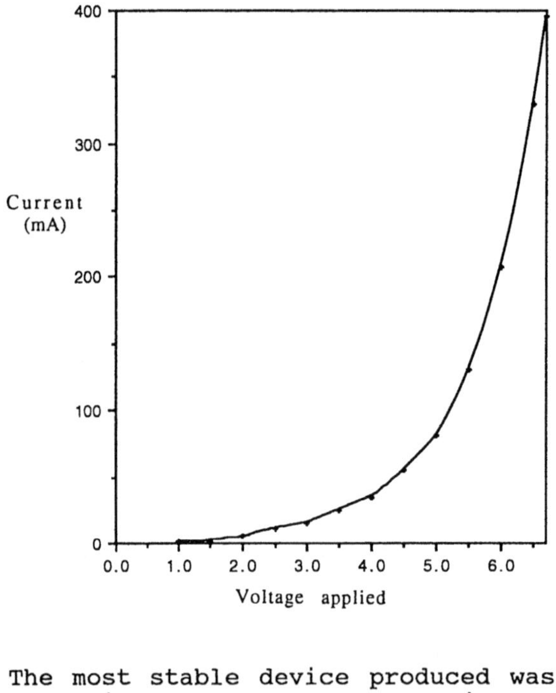

Fig.2: Ca/C_{70} Schottky Diode

The most stable device produced was the n-Si/C_{70} device. The main advantage of the n-Si/C_{70} Schottky diode is its ease of fabrication. After full examination of its characteristics the device still worked well. The peak current of 108 mA was quite stable and reproducible over a period of weeks. A reference diode of Au/n-Si was prepared and showed a typical ohmic response, which would

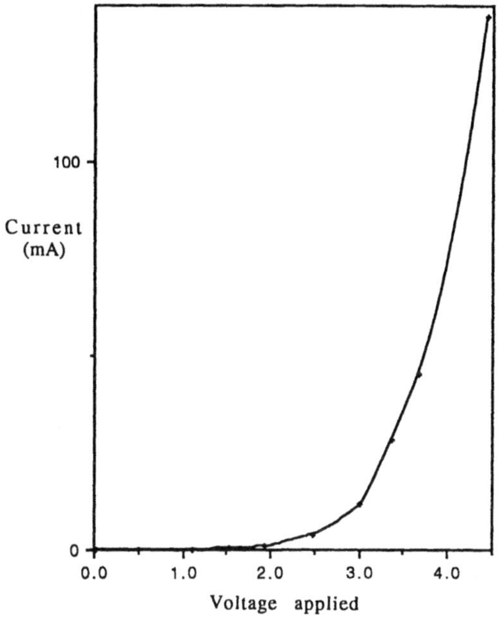

Fig. 3: Si/C_{70} Schottky Diode

be expected from two materials which have low resistivity. To ensure that the fullerenes formed a Schottky diode with the low work function metals and not the high ones, a sandwich layer was prepared consisting of $Au/C_{60}/Au$ and $Au/C_{70}/Au$. The result of this was a bulk conductivity analysis. However, it did not form a Schottky diode thus proving that the device characteristics observed were indeed due to the Schottky diode formed between low work function materials and the fullerenes. Normally a p-type semiconductor makes a Schottky contact with a low work function metal[6]. However if the Schottky contact is made utilising an n-type semiconductor and a low work function metal this is then due to a drift current induced by the minority carriers.

Another important point to note is that the n-Si/C_{70} diode shows a photovoltaic response shown in Fig.4. A short circuit current of 1.5 A and an open circuit voltage of 106 mV is produced when exposed to 1 mW broadband light intensity. The efficiency of this diode was in the region of 0.1 but it is an indication of where fullerenes may be made useful practically.

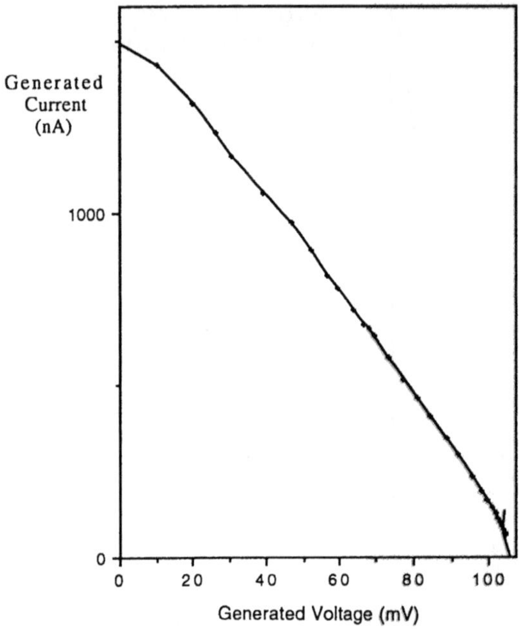

Fig. 4: Generated Current Voltage Characteristic of Si/C_{70}

5 Conclusion

The J vs V characteristics of fullerene devices all show Schottky diode behaviour which are quite stable and reproducible over a number of weeks. The diode behaviour is dominated by a drift current induced by the presence of the minority carriers and as a result the Schottky diode is formed at the low work function metal/fullerene interface. By calculating the barrier height at the junction between the metal and the semiconductor the electron affinity of both fullerenes is evaluated as 3.25 eV for C_{60}/Mg-In, 3.14 eV for C_{70}/Si and a more realistic value of 2.36 eV for C_{70}/Ca probably due to the better balancing of the metal contact work functions. As the more stable device produced was the n-Si/C_{70} it was studied for its photovoltaic ability. It produced a short circuit current of 1.5 A and an open circuit voltage of 106 mA which was stable for a number of weeks.

References

1. K.Pichler, R.H.Friend: Synth. Met. **51**(1992)**257-265**
2. J.Kastner, J.Paloheimo, H.Kuzmany: In Electronic Properties of High-T_c Superconductors, Springer Ser, Solid-State Sci., Vol.113 (Springer, Berlin, Heidelberg 1993) p.512
3. J.Mort, R.Ziolo, M.Machonkin, D.R.Huffman, M.I.Ferguson: Chem.Phys.Lett.**186**(1991)**284**
4. S.Anthoe, L.Tugula: phys.stat.sol. (a) **128**,253(1991)
5. D.Braun, A.J.Heeger: Thin Solid Films **216**(1992)**96-98**
6. J.B.Whitlock, P.Panayotatos: Thin Solid Films **205**, (1991)**69-75**

References

Index of Contributors

Springer Series in Solid-State Sciences

Editors: M. Cardona P. Fulde K. von Klitzing H.-J. Queisser

1 **Principles of Magnetic Resonance** 3rd Edition By C. P. Slichter
2 **Introduction to Solid-State Theory** 2nd Printing By O. Madelung
3 **Dynamical Scattering of X-Rays in Crystals** By Z. G. Pinsker
4 **Inelastic Electron Tunneling Spectroscopy** Editor: T. Wolfram
5 **Fundamentals of Crystal Growth I** Macroscopic Equilibrium and Transport Concepts. 2nd Printing By F. Rosenberger
6 **Magnetic Flux Structures in Superconductors** By R. P. Huebener
7 **Green's Functions in Quantum Physics** 2nd Edition By E. N. Economou
8 **Solitons and Condensed Matter Physics** 2nd Printing Editors: A. R. Bishop and T. Schneider
9 **Photoferroelectrics** By V. M. Fridkin
10 **Phonon Dispersion Relations in Insulators** By H. Bilz and W. Kress
11 **Electron Transport in Compound Semiconductors** By B. R. Nag
12 **The Physics of Elementary Excitations** By S. Nakajima, Y. Toyozawa, and R. Abe
13 **The Physics of Selenium and Tellurium** Editors: E. Gerlach and P. Grosse
14 **Magnetic Bubble Technology** 2nd Edition By A. H. Eschenfelder
15 **Modern Crystallography I** Symmetry of Crystals. Methods of Structural Crystallography By B. K. Vainshtein
16 **Organic Molecular Crystals** Their Electronic States. By E. A. Silinsh
17 **The Theory of Magnetism I** Statics and Dynamics. 2nd Printing By D. C. Mattis
18 **Relaxation of Elementary Excitations** Editors: R. Kubo and E. Hanamura
19 **Solitons**. Mathematical Methods for Physicists. 2nd Printing By G. Eilenberger
20 **Theory of Nonlinear Lattices** 2nd Edition By M. Toda
21 **Modern Crystallography II** Structure of Crystals By B. K. Vainshtein, V. M. Fridkin, and V. L. Indenbom
22 **Point Defects in Semiconductors I** Theoretical Aspects By M. Lannoo and J. Bourgoin
23 **Physics in One Dimension** Editors: J. Bernasconi, T. Schneider
24 **Physics in High Magnetic Fields** Editors: S. Chikazumi and N. Miura
25 **Fundamental Physics of Amorphous Semiconductors** Editor: F. Yonezawa
26 **Elastic Media with Microstructure I** One-Dimensional Models. By I. A. Kunin
27 **Superconductivity of Transition Metals** Their Alloys and Compounds By S. V. Vonsovsky, Yu. A. Izyumov, and E. Z. Kurmaev
28 **The Structure and Properties of Matter** Editor: T. Matsubara
29 **Electron Correlation and Magnetism in Narrow-Band Systems** Editor: T. Moriya
30 **Statistical Physics I** Equilibrium Statistical Mechanics 2nd Edition By M. Toda, R. Kubo, N. Saito
31 **Statistical Physics II** Nonequilibrium Statistical Mechanics 2nd Edition By R. Kubo, M. Toda, N. Hashitsume
32 **Quantum Theory of Magnetism** 2nd Edition By R. M. White
33 **Mixed Crystals** By A. I. Kitaigorodsky
34 **Phonons: Theory and Experiments I** Lattice Dynamics and Models of Interatomic Forces. By P. Brüesch
35 **Point Defects in Semiconductors II** Experimental Aspects By J. Bourgoin and M. Lannoo
36 **Modern Crystallography III** Crystal Growth 2nd Edition By A. A. Chernov
37 **Modern Crystallography IV** Physical Properties of Crystals Editor: L. A. Shuvalov
38 **Physics of Intercalation Compounds** Editors: L. Pietronero and E. Tosatti
39 **Anderson Localization** Editors: Y. Nagaoka and H. Fukuyama
40 **Semiconductor Physics** An Introduction 5th Edition By K. Seeger
41 **The LMTO Method** Muffin-Tin Orbitals and Electronic Structure By H. L. Skriver
42 **Crystal Optics with Spatial Dispersion, and Excitons** 2nd Edition By V. M. Agranovich and V. L. Ginzburg
43 **Structural Analysis of Point Defects in Solids** An Introduction to Multiple Magnetic Resonance Spectroscopy By J.-M. Spaeth, J. R. Niklas, and R. H. Bartram
44 **Elastic Media with Microstructure II** Three-Dimensional Models By I. A. Kunin
45 **Electronic Properties of Doped Semiconductors** By B. I. Shklovskii and A. L. Efros
46 **Topological Disorder in Condensed Matter** Editors: F. Yonezawa and T. Ninomiya
47 **Statics and Dynamics of Nonlinear Systems** Editors: G. Benedek, H. Bilz, and R. Zeyher
48 **Magnetic Phase Transitions** Editors: M. Ausloos and R. J. Elliott
49 **Organic Molecular Aggregates** Electronic Excitation and Interaction Processes Editors: P. Reineker, H. Haken, and H. C. Wolf

Springer Series in Solid-State Sciences

Editors: M. Cardona P. Fulde K. von Klitzing H.-J. Queisser

Springer Series in Solid-State Sciences

Editors: M. Cardona P. Fulde K. von Klitzing H.-J. Queisser

Managing Editor: H. K.V. Lotsch

Volumes 1–89 are listed at the end of the book

90 **Earlier and Recent Aspects of Superconductivity**
Editors: J. G. Bednorz and K. A. Müller

91 **Electronic Properties of Conjugated Polymers III** Basic Models and Applications
Editors: H. Kuzmany, M. Mehring, and S. Roth

92 **Physics and Engineering Applications of Magnetism** Editors: Y. Ishikawa and N. Miura

93 **Quasicrystals** Editors: T. Fujiwara and T. Ogawa

94 **Electronic Conduction in Oxides**
By N. Tsuda, K. Nasu, A. Yanase, and K. Siratori

95 **Electronic Materials**
A New Era in Materials Science
Editors: J. R. Chelikowsky and A. Franciosi

96 **Electron Liquids** By A. Isihara

97 **Localization and Confinement of Electrons in Semiconductors**
Editors: F. Kuchar, H. Heinrich, and G. Bauer

98 **Magnetism and the Electronic Structure of Crystals** By V. A. Gubanov, A. I. Liechtenstein, and A.V. Postnikov

99 **Electronic Properties of High-T_c Superconductors and Related Compounds**
Editors: H. Kuzmany, M. Mehring, and J. Fink

100 **Electron Correlations in Molecules and Solids** 2nd. edn., ed. by P. Fulde

101 **High Magnetic Fields in Semiconductor Physics III** Quantum Hall Effect, Transport and Optics By G. Landwehr

102 **Conjugated Conducting Polymers**
Editor: H. Kiess

103 **Molecular Dynamics Simulations**
Editor: F. Yonezawa

104 **Products of Random Matrices** in Statistical Physics By A. Crisanti, G. Paladin, and A. Vulpiani

105 **Self-Trapped Excitons** By K. S. Song and R. T. Williams

106 **Physics of High-Temperature Superconductors** Editors: S. Maekawa and M. Sato

107 **Electronic Properties of Polymers**
Orientation and Dimensionality of Conjugated Systems Editors: H. Kuzmany, M. Mehring, and S. Roth

108 **Site Symmetry in Crystals** Theory and Applications By R. A. Evarestov and V. P. Smirnov

109 **Transport Phenomena in Mesoscopic Systems** Editors: H. Fukuyama and T. Ando

110 **Symmetry and Optical Phenomena in Superlattices and Other Heterostructures**
By E. L. Ivchenko and G. E. Pikus

111 **Low-Dimensional Electronic Systems**
New Concepts
Editors: G. Bauer, F. Kuchar, and H. Heinrich

112 **Phonon Scattering in Condensed Matter VII**
Editors: M. Meissner and R. O. Pohl

113 **Electronic Properties of High-T_c Superconductors**
Editors: H. Kuzmany, M. Mehring, and J. Fink

114 **Interatomic Potential and Structural Stability**
Editors: K. Terakura and H. Akai

115 **Ultrafast Spectroscopy of Semiconductors and Semiconductor Nanostructures** By J. Shah

116 **Electron Spectrum of Gapless Semiconductors**
By J. M. Tsidilovski

117 **Electronic Properties of Fullerenes**
Editors: H. Kuzmany, J. Fink, M. Mehring, and S. Roth

MIX
Papier aus verantwortungsvollen Quellen
Paper from responsible sources
FSC® C105338

If you have any concerns about our products,
you can contact us on
ProductSafety@springernature.com

In case Publisher is established outside the EU,
the EU authorized representative is:
Springer Nature Customer Service Center GmbH
Europaplatz 3, 69115 Heidelberg, Germany

Printed by Libri Plureos GmbH
in Hamburg, Germany